# Nonparametric Methods
# for Quantitative Analysis

# INTERNATIONAL SERIES IN DECISION PROCESSES

Ingram Olkin, Consulting Editor

# Nonparametric Methods for Quantitative Analysis

Jean Dickinson Gibbons

*The University of Alabama*

HOLT, RINEHART AND WINSTON
*New York*     *Chicago*     *San Francisco*     *Atlanta*
*Dallas*     *Montreal*     *Toronto*     *London*     *Sydney*

**Library of Congress Cataloging in Publication Data**
Gibbons, Jean Dickinson, 1938–
Nonparametric methods for quantitative analysis.
(International series in decision processes ; 23)
Bibliography: p. 445
Includes index.
1. Nonparametric statistics.   I.   Title.
QA278.8.G498     1976     519.5'3     74–28910
                        **ISBN** 0–03–007811–3
   7 8 9     0 3 8     9 8 7 6 5 4 3 2


This book was set in Times Roman by Trade Composition

Editor:   Holly Massey
Designer:   Katrine Stevens
Printer and Binder:   Maple Press
Drawings:   Eric G. Hieber Associates Inc.

To my husband Jack

# Preface

This book is designed to introduce the reader to an area of statistical analysis known as nonparametric methods. The reader should be assured that little mathematical sophistication is needed to understand and to use the techniques presented here. He or she needs only an ability to deal with simple formulas and algebraic substitutions. The reader should have an acquaintance with the basic terms and techniques of elementary classical statistics, such as those covered in books at the level of *Modern Elementary Statistics* by John E. Freund, *Introduction to Statistical Analysis* by W. J. Dixon and F. J. Massey, *Elementary Statistical Methods* by H. M. Walker and J. Lev, or *Basic Statistics: A Modern Approach* by Morris Hamburg.

The reader should also be assured that a formal course in statistics is not a strict prerequisite to benefiting from this book, since many persons engaged in quantitative analysis may acquire sufficient background knowledge through independent reading and experience. The ultimate prerequisite therefore is a desire to learn how to use nonparametric techniques and apply them properly in an empirical investigation.

This book should be useful for any of the following three purposes:

1. As a textbook for students, both graduate and undergraduate, in education and in those fields of the behavioral, social, and physical sciences where quantitative techniques of analysis are considered relevant to the curriculum. This includes courses offered for programs in Psychology, Educational Psychology, Sociology, Education, Anthropology, Political Science, Marketing Research, Management Science, Operations Research, Industrial and Operations Management, Industrial Engineering, Computer Science, Agricultural Economics, Econometrics, Home Economics, Social Work, Geology, Biology, Biostatistics, and Business Statistics, and undoubtedly others as well.

2.  As a textbook for beginning students in applied or experimental statistics.
3.  As a handbook for researchers in any field where the subject matter under investigation requires statistics as a tool for quantitative analysis.

For formal classroom use, this book contains ample materials for a one-semester or one-quarter course. Included are a large number of examples and problems that have been carefully chosen to represent or simulate actual situations of empirical investigation. The examples not only illustrate the methods of analysis; they also introduce and discuss some of the practical problems of experimental design, sampling, validity of inference, and the like, which are important for any investigator who wants to maximize the information derived from an empirical study. Hopefully, these examples will motivate the reader to master statistical techniques, and help him or her to recognize the power and limitations of statistics.

This book introduces each technique by an example. Then the general technique is explained in accordance with the sequential steps a trained investigator would follow in attacking such problems in research and statistical analysis. Specifically, we first state the problem or type of questions to be answered, determine what kind of data and measurement are needed, and then explain how to perform the statistical analysis. The methodology of each inference procedure is developed by a logical rationale that is neither mathematical nor theoretical.

In addition to its elementary level and practical approach to statistical inference, this book has a number of significant features that differentiate it from other books on nonparametric statistics. For example, in hypothesis testing we do not usually conclude with a decision as to whether to accept or reject the null hypothesis. Rather, the associated probability, or $P$-value, is found. This practice is followed since most important decisions should take into account other factors in addition to the purely statistical analysis of the data. As a result, the statistical tables here give the complete probability distribution whenever practical, rather than only critical values for hypothesis tests at selected levels. Methods of confidence interval estimation of parameters and simultaneous multiple comparisons are also presented whenever such procedures are possible and relevant. Moreover, selection of appropriate and reliable statistical procedures is discussed.

A complete compendium of nonparametric techniques would not be appropriate to the purpose of this book. Instead, the book focuses on those inference situations that are very basic, are commonly faced by researchers, and lend themselves to the application of simple techniques. Those procedures not covered, therefore, either are primarily of academic interest, have relatively poor performance properties, apply to very special situations, or are not practical for common use because of their difficulty or the relative inaccessibility of tables necessary for their application. On the other hand, because the definition adopted here for nonparametric techniques is quite broad, this book does include many procedures that are seldom covered in other nonparametrics books (such as goodness-of-fit and other tests for categorical data).

The author would like to express her sincere appreciation to Ingram Olkin for his many helpful comments and suggestions for improving the manuscript; to Professors James K. Brewer, Janet Elashoff, Alan Lasater, Joan Welkowitz, Frederick C. Weston, Jr., and Robert L. Winkler for their careful reviews of the manuscript; to Gottfried Noether for assistance with Chapter 5; to William H. Etheredge for assistance in finding examples and working problems; to Kitty Jo Elledge for typing the entire manuscript; to the graduate students who helped debug the manuscript and identify potential problems for the readers; and to The University of Alabama for cooperating with the author in carrying this project to completion.

Jean Dickinson Gibbons

*Tuscaloosa, Alabama*
*August 1975*

# Contents

# A Note to the Reader

This book consists of eight chapters with problems, an Appendix, a Bibliography and an Index. A two-digit internal cross-reference system is used within each chapter. The first digit always designates the section number. Within each section, the subsections, examples, equations, figures, and tables inserted in the text are numbered sequentially by a second digit, but the first digit is always repeated to eliminate confusion. To refer to part of another chapter, the chapter number is specified. For example, Section 3.4 of Chapter 2 is interpreted as Subsection 4 of Section 3 in Chapter 2. If a section has no subsection, or the entire section is relevant, this reference would be Section 3 of Chapter 2. The numerical reference is always preceded by the appropriate word label, such as Section, Table, and so on, except for equations. Equation numbers always appear in parentheses and are referred to in this way. Accordingly, (3.4) of Chapter 2 means Equation (3.4), which is the fourth numbered equation in Section 3 of Chapter 2. References made within the same chapter omit the chapter number. Problems are given only at the end of each chapter and are numbered sequentially by a digit following the chapter number. Answers to even-numbered problems are given at the back of the book. The tables in the Appendix at the back of the book are designated by Latin letters, as A, B, and so on.

Numerous references are given throughout the text and problems. These are all referred to by surname of author and date. The full citations for these and other pertinent references are all given in the Bibliography at the end of the book. The Bibliography is divided into two parts, a List of References and a List of Experiments. The latter group includes actual investigations reported in the literature and referred to in this book in a problem or example.

# 1

# Introduction to Nonparametric Statistics

## 1 Introduction

The word *statistics* is an all-inclusive term, whose meaning now extends far beyond its original definition. For, oddly enough, the term statistics used to mean simply a mass of numbers. A statistician now calls such numbers *data*.

The science of statistics is defined as an area of applied mathematics devoted to a particular means of problem solving or scientific investigation. Statistical analysis is useful whenever a problem can be defined in such a way that data of some kind, when properly analyzed, lead to its solution. These analytical techniques have been applied to empirical investigations in almost every field of man's intellectual activities.

Today the world is experiencing a quantitative revolution. Old intuitive or guesswork approaches to problem solving are rapidly becoming out-moded. There is a continual quest for objective bases for all kinds of deci-sions. The resources devoted to data collection are staggering. This is partly a result of the computer revolution, for the use of the computer has elimi-nated many of the practical problems and limitations of data collection and processing. Moreover, as computer techniques become increasingly sophisti-cated and efficient, more and more fields of study, formerly thought of as philosophical, or subjective, have come to rely on quantitative methods of analysis.

But the computer itself cannot think; it can only process according to the program a human mind has stored in it. An investigator cannot depend on the computer to provide more than the arithmetic manipulations needed for the solution to a problem. As a result, the widespread use of computers has increased, rather than decreased, society's need, not only for statisticians, but also for doctors, psychologists, business managers, sociologists, lawyers, and the like, who understand and can apply modern statistical tools for analysis.

What choices face such subject-matter investigators when they approach problems requiring statistical analysis?

1. They can become full-fledged statisticians in their own right.
2. They can use the services of a statistical consultant.
3. They can acquire enough knowledge of statistical methods to be able to perform the analysis themselves (or tell the computer what to do) and to interpret the results.

The first alternative is usually not practical, nor is it appealing to specialists in most subject-matter fields of investigation. Why? Because it is difficult enough these days to keep abreast of developments in one's own field, without trying to become an expert in another field as well.

The second alternative may not be practical either, not only because of the cost, or the lack of availability, of an appropriate consultant, but also because of problems of communication. In order to make effective use of a consultant, an investigator must be able to communicate the problem to the statistician in comprehensible terms. And that is not always so simple, because the statistician may be no more expert in the investigator's field than the investigator is in statistics. This communication gap can be bridged much more easily if the investigator has some acquaintance with statistical methods, for it is unreasonable to expect the statistical expert to be sophisticated in all the possible fields of intellectual inquiry that rely on statistics.

There is another reason why investigators should be acquainted with statistical methods. Many problems require decisions or judgments that are ultimately the responsibility of the investigator, even though statistical conclusions provide valuable input. (The statistician, for example, may be able to show the business executive that one machine clearly outperforms another, but he is hardly in a position to judge whether the company can afford the cost of the new machine, the worker retraining required, or the temporary disturbance of the production process.) An investigator who fully understands the statistical results is in a much better position to make a sound judgment on these other matters.

There are other caveats to consider when one contemplates complete reliance on an expert in statistics. Today, with knowledge expanding exponentially, subject-matter researchers and practitioners must keep up with the burgeoning literature in their field. A mere cursory glance at recent issues of professional and learned journals, or even newspapers and trade journals, proves conclusively that there is an increasing use of statistical analysis to substantiate research conclusions and to provide quantitative bases for previously held, but subjective, opinions. Such journals, crowded for space, cannot afford to provide lengthy explanations of statistical methods used. Thus, they leave it to the reader to judge the appropriateness and validity of the techniques used by the author. The subject-matter expert will soon not be "expert" at all if he is not able to read and judge the value of information derived from statistical analyses.

All that is left among the three foregoing choices is the third. The subject-matter investigator *must* learn enough statistical techniques relevant to his particular field for him to be comfortable and at ease. It is not recommended that such investigators resort to their own devices in analyzing problems that require the use of nonroutine methods of analysis. However, many problems can, and should, be analyzed by methods that are routine in the sense that they have been well developed and their properties have been thoroughly investigated by professional statisticians. While born of frequently complex and elegant methods of statistical theory, routine methods can be presented to quantitative analysts by statisticians in simple, usable forms, and nonexpert users can follow instructions and get satisfactory results in much the same way as we can produce an excellent meal by following the instructions of an expert chef.

Professional statisticians complain that cookbooks may contribute to the "misuses" of statistics by encouraging the reader to follow a list of step-by-step instructions for manipulation. The reader, unaware of the theoretical bases for these instructions, is prone to disregard assumptions, use inappropriate methods, or apply appropriate methods improperly. Therefore, expert statisticians frequently refuse to write a cookbook on the grounds that the effect will be to encourage more nonexperts to dabble in statistics without really understanding what they are doing.

There is, however, a strong argument in favor of such cookbook efforts. Investigators are finding it necessary to try their hand at statistics anyway, and will continue to do so. Since the resulting product can be no better than the quality of the recipes contained in the cookbook, the best solution is to provide good cookbooks, which state specifically the conditions under which a procedure is valid, emphasize the importance of checking assumptions, and encourage those methods of analysis that are least likely to be misused.

This book takes just such a stance. Frankly, its purpose is to provide an easy guide to the proper use of methods in a special area of statistical analysis called nonparametric statistics. It is hoped that this book will motivate and enliven the study of statistical methods in connection with other disciplines, so that statistical tools will be used properly in actual research studies. Numerous examples of such applications in all fields are presented throughout the book.

# 2   Analysis of Data by Statistical Inference

Even though this book assumes that the reader has had some exposure at an elementary level to traditional statistical methods, we precede the introduction to nonparametric methods by briefly reviewing those terms, concepts, and techniques that are basic to any kind of statistical inference. In this section we also introduce the important concept of *P*-values, which may be new to some readers.

### 2.1   Preliminaries

Whenever a statistical investigation is performed, the investigator must go through a regular sequence of steps. Although the order may vary somewhat, depending on the problem, the following outline is a general guide to accepted procedure.

1. *State the problem, the questions to be answered by the analysis, or the hypotheses of interest.*
2. *Decide what kind of data are relevant to the problem, and determine what data, if any, are available.* (The data may represent the results of a survey or an experiment. The group of all elements or outcomes under consideration in a particular investigation is called the *population*, and this group must be clearly identified. The data are called a *census* if they represent all elements of the population. If the data represent only a subgroup of the population, they are called a *sample*.)
3. *Collect those data that are needed but not already available.* (The method of collection, the type of measurement to be used, and the method of recording the data must also be decided. The sampling method or design of experiment must be chosen and carried out according to plan.)
4. *Analyze the data.* (This step may consist of a simple tabulation of results; a presentation of results in the form of a chart, graph, or table; the calculation of summary descriptive figures, such as a mean or median; and the like.)
5. *Interpret the analysis, draw conclusions, or perform inferences to answer the questions defined in step 1 as well as possible, given the limitations imposed by the type of investigation and the data available.*

The importance of steps 2 and 3 cannot be overemphasized. The results of a statistical study can be of little value if the data entering the analysis are not relevant to the problem or are not representative of the population about which questions are to be answered. Further, the validity of any inference procedure in statistics is dependent on the assumption that a *random experiment* is performed or a *random sampling procedure* is followed. This means that the data are the result of some random phenomenon, or that the elements in the sample are selected by some probabilistic method or chance mechanism. There are many acceptable types of experimental designs and random sampling techniques, but a complete discussion is beyond the scope

of this book. (Some books that give elementary discussions of sample surveys, sampling techniques, and experimental design are listed in Section 8.)

The examples cited in this book represent a legitimate use of statistical
inference techniques, but only if the stated sampling situation requirements
are satisfied. Otherwise, the results of the analysis may be highly questionable, or at least undependable. Because sampling in particular, and experimental design in general, are so crucial to the validity of any inference procedure, we occasionally criticize the method of design or data collection employed in particular examples in this book. Hopefully, this will encourage
the reader to adopt a similar critical posture before he accepts conclusions
based on questionable sampling procedures, or inadequate research designs.

The operations performed in step 4 are frequently called *descriptive
statistics*. This term refers to the calculation or presentation of figures,
whether visual or conceptual, that summarize or characterize a process or a
set of data. If a summary figure is calculated from data that represent an
entire population under study (or from a mathematical function that represents a theoretical distribution), the quantity is called a *parameter*. On the
other hand, a figure calculated from sample data is called a *statistic*. A mean,
for example, may be either a parameter or a statistic, depending on the numbers entering into its calculation and the definition of the population in a
particular investigation, but it is always a descriptive measure.

Important types of descriptive statistics are those that serve as *measures
of location* or *central tendency* (mean, median, and so on), and those that are
*measures of dispersion, spread*, or *variation* (standard deviation, range, and
so on). (Measures of *kurtosis* describe the property of peakedness.) *Quantile*
(or *percentile*) *points* are also useful as descriptive statistics. These are points
that divide the distribution or data into two parts of fixed percentages. For
example, if 95 percent of the values in a distribution are smaller than some
number, that number is the .95 quantile point; the median is the .50 quantile
point.

Although a statistic in any particular study is a single number (or
possibly a set of numbers), when the study represents the outcome of a random sampling process or a random experiment the statistic can be thought
of in the abstract as a *random* (or *chance*) *variable*. The particular value of
the variable is due entirely to chance because the sample is the result of a
chance mechanism. Any random variable, whether in a population or a
sample, has a *probability distribution* (or *function*). If the values of a variable
can be enumerated using the positive integers (that is, the variable can take
on only a finite or countably infinite number of values), then it is called a
*discrete random variable*, and its probability distribution is also called discrete. In this case the possible values of the variable, along with their respective probabilities (or theoretical relative frequency ratios) of occurrence

(which must be nonnegative and sum to one), can be listed to specify the probability distribution. For example, the probability function of a discrete random variable might be specified as

$$f(x) = \begin{cases} .49 & \text{if } x = -1 \\ .02 & \text{if } x = \phantom{-}0 \\ .49 & \text{if } x = \phantom{-}2 \\ 0 & \text{otherwise.} \end{cases}$$

If the variable takes on values on a continuum, the number of possible values is infinite and it is not possible to make a listing of values to specify the probability distribution. The random variable and its probability distribution are then called *continuous*, and the probability of any particular value in this continuum must be zero. The occurrence of a specified value is obviously not impossible, but it is highly improbable because it is just one of an infinite number of possibilities. In this case the probability distribution may be specified by a mathematical function called a *density*.

The probability of occurrence of an interval of values for a continuous random variable is the area under the curve (which is the locus of the density) and between the endpoints that define this interval. Of course, the probability distribution of a discrete random variable may also be given as a mathematical function instead of a listing. The values of the function are themselves the probabilities in the discrete case, whereas they are only ordinates of the curve whose areas are probabilities in the continuous case. (Although it will not concern us here, some random variables are continuous for certain values and discrete for others. Such variables, and their probability distributions, are called *mixed*.)

The probability function of a random variable can also be given as a *cumulative distribution function* (or *c.d.f.*), whose values are probabilities in both the discrete and continuous cases. This function, always denoted by a capital letter, such as $F(x)$ or $G(x)$, is defined for all real numbers $x$. Its value at $x$ is the probability that the value of the random variable is less than or equal to $x$. That is, $F(x)$ is the probability $P(X \leq x)$, where $X$ is the random variable.[1] It is found by cumulating probabilities for all values of the random variable between minus infinity and the number $x$ — that is, a cumulative sum (found by addition) for discrete random variables and a cumulative area (found by integration) for continuous random variables. Many of the especially useful cumulative distributions have been tabled, so these operations are not needed in practice. A cumulative distribution $F(x)$

---

[1] The uppercase letter $X$ is used to differentiate the random variable $X$ from the actual number $x$, which is the largest value assumed by the random variable in the event that $X \leq x$.

for a random variable $X$, whether discrete or continuous, is easily used to find probabilities because of the following relationships:

$$P(X \le a) = F(a)$$
$$P(a < X \le b) = F(b) - F(a)$$
$$P(X > b) = 1 - P(X \le b) = 1 - F(b).$$

In order to eliminate confusion, when the cumulative distribution is denoted by a capital letter, the corresponding probability function that has been or is to be cumulated is denoted by the same letter in lowercase, such as $F(x)$ and $f(x)$, or $G(x)$ and $g(x)$.

Since the concept of a c.d.f. is frequently difficult to understand, we include here a simple example of how it is found for a discrete random variable, and how it can then be used to find probabilities. Consider the discrete random variable that has nonzero probability for only three values of $x$, namely $x = -1$, 0, and 2, and corresponding $f(x)$ values .49, .02, and .49, respectively. Then the cumulative distribution values $F(x)$ for $x = -1$, 0, and 2 are found as follows:

| $x$ | $f(x)$ | $F(x)$ | |
|---|---|---|---|
| $-1$ | $f(-1) = .49$ | $F(-1) = f(-1)$ | $= .49$ |
| 0 | $f(0) = .02$ | $F(0) = f(-1) + f(0) = .49 + .02$ | $= .51$ |
| 2 | $f(2) = .49$ | $F(2) = f(-1) + f(0) + f(2) = .51 + .49$ | $= 1.00.$ |

The function $F(x)$ is now defined for $x = -1$, 0, and 2, but it must be defined for all $x$. What is the probability that the random variable is less than or equal to $-1.5$, or $-5$? The probability is 0, and similarly for any other number smaller than $-1$ since $-1$ is the smallest value of this random variable that has nonzero probability. We summarize this result by writing $F(x) = 0$ for $x < -1$. Now consider numbers $x$ such that $x \ge -1$. It is clear that $F(-.5) = .49$, but $F(.5) = .49 + .02 = .51$; these same results hold for all numbers $x$ in the appropriate intervals. That is, as long as $x$ is at least $-1$ but smaller than 0, we have $F(x) = .49$, and when $x$ is at least 0 but smaller than 2, $F(x) = .49 + .02$. The complete cumulative distribution function is then written as

$$F(x) = \begin{cases} 0 & \text{for } x < -1 \\ .49 & \text{for } -1 \le x < 0 \\ .51 & \text{for } 0 \le x < 2 \\ 1.00 & \text{for } x \ge 2. \end{cases}$$

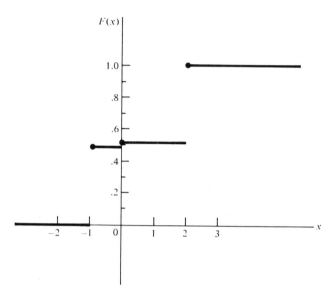

**Figure 2.1**

This function is graphed in Figure 2.1. It is called a step function. Note that the height of the jump at any number $x$ is equal to $f(x)$. If $f(x) = 0$ for the number $x$, the height of the jump at $x$ equals 0.

We now illustrate the use of this c.d.f. to find probabilities. The following result from simple substitution in the formulas stated earlier:

$$P(X \leq .60) = F(.60) = .51$$
$$P(-.45 < X \leq .75) = F(.75) - F(-.45) = .51 - .49 = .02$$
$$P(X > .21) = 1 - F(.21) = 1 - .51 = .49.$$

Now suppose that we want to find $P(X < 0)$. The cumulative distribution gives probabilities directly only for a "less than or equals" type of statement, but $(X < 0)$ means strictly "less than." We can still simply sum the relevant probabilities. The only value of $x$ that is less than 0 and has nonzero probability is $x = -1$. Hence, $P(X < 0) = f(-1) = .49$. Alternatively, we can express such probabilities by introducing the $\varepsilon$-terminology. We consider less than 0 as meaning less than or equal to some number slightly smaller than 0, that is, "$< 0$" means "$\leq 0 - \varepsilon$" for some sufficiently small amount $\varepsilon$, and then $P(X < 0) = P(X \leq 0 - \varepsilon) = F(0^-) = .49$, where $0^-$ indicates a number just to the left of 0. Some other applications of this type are as follows:

$$P(X < -.01) = F(-.01^-) = .49$$
$$P(X < .01) = F(.01^-) = .51$$
$$P(X \geq 2) = 1 - P(X < 2) = 1 - F(2^-) = 1 - .51 = .49$$
$$P(X \geq 2.01) = 1 - P(X < 2.01) = 1 - F(2.01^-) = 1 - 1 = 0.$$

Whether the distribution is discrete or continuous, we also frequently need to talk about *left-tail* and *right-tail probabilities*. A left-tail probability at some point is the probability cumulated from the smallest possible value or minus infinity to that point; for the point $a$, it is $P(X \leq a) = F(a)$. A right-tail probability at some point is cumulated from that point to the largest possible value or plus infinity; for the point $b$, it is $P(X \geq b) = 1 - F(b^-)$. These pieces are commonly called the *left tail* and *right tail* of the distribution, respectively.

Two other important concepts relating to both discrete and continuous random variables are independence and identical distributions. Although statistical independence has a formal mathematical definition, it is sufficient for our purposes to give the following more intuitive definition. Two random variables, or experimental or sample observations, are *independent* if the value of one observation does not have any effect, whether spurious or causal, on the value of the other observation. For example, successive tosses of a coin produce independent experimental results, since a coin has neither a memory nor a conscience. However, turning over cards in a standard deck results in dependent observations no matter how well shuffled the deck, since once a card is turned over it cannot appear again (unless it is replaced in the deck). If items produced on an assembly line are statistically independent, then the probability that one item is defective is not affected by whether some other item is defective. Two random variables are said to be *identically distributed* if their probability functions are exactly the same with respect to every aspect, such as the form, the shape, the values of all parameters, and the like.

The term *sampling distribution of a statistic* refers to the probability distribution of that statistic relative to all possible random samples of fixed size that could be drawn from a fixed population. The possible values of the statistic as a chance variable are all the values that would result when all possible samples are taken and the statistic is calculated from each sample. A sampling distribution may be discrete or continuous. If discrete, the probability of any value of the statistic is its relative frequency of occurrence, and this defines the probability distribution. If continuous, the probability of any value is zero, and the probability distribution must be determined by mathematical methods.

When sample descriptions are used to infer some information about the population, the operation is called *inductive statistics* or *statistical inference*.

The inference always refers to one or more characteristics of the population under study. The justification for any statistical inference procedure lies in probability theory, and it is this factor that differentiates statistical inference from mere guessing or ordinary judgment. The probability theory rests on the assumption that a random experiment, random sampling procedure, or some kind of probabilistic method is used to obtain the data on which the inference is based.

A *simple random sample* is the easiest type of probability sample to understand. It satisfies the property that, for a finite population and a fixed sample size, the sample is chosen in such a way that every element in the population has the same chance of being included in the sample and every combination of that number of elements has an equal chance of being the sample selected. For an infinite population, the observations in a random sample are independent and identically distributed. Any conclusion about a population that is based on a sample cannot be reliable if the sample is not representative of the population, because the conclusion is based on one or more descriptive statistics calculated from the sample data. By using random sampling, the investigator has a good, logical reason for believing that the sample does represent the population in miniature. In fact, he can attach a probability to measure the "accuracy" of an inference statement; the probability represents the likelihood that the inference procedure will lead to a correct statement about this population if an arbitrary random sample is used.

Statistical inference includes two types of procedures, called estimation and hypothesis testing. These techniques are explained in the next two subsections. We first discuss hypothesis testing, so that confidence interval estimation can be defined as the analogue of hypothesis testing.

## 2.2   Hypothesis Testing

When data collected by an experiment or sampling procedure are to be used for hypothesis testing, the steps to be taken are as in the general guide to accepted procedure given at the beginning of Section 2.1. Steps 1, 4, and 5 need further elaboration so that the explicit procedure will be more clear.

A statistical *hypothesis* is a statement about a population; for example, its form or shape, or some aspect thereof; the numerical value of one or more parameters; and so forth. The *hypothesis set* consists of two statements, called *the hypothesis, H,* and *the alternative, A. H* is frequently called the *null hypothesis* and *A* the *alternative hypothesis*. These statements must be mutually exclusive; that is, they must not overlap in any way. Except in tests for goodness of fit (and sometimes in tests for randomness), the alternative is frequently formulated to correspond to the research hypothesis or the motivated hypothesis — that is, the statement that the investigator hopes to conclude on the basis of the empirical evidence, or the statement he wishes to investigate the merits of. For statistical purposes, the converse statement, or some subset thereof, is then given as the null hypothesis. For example, if the alternative is "A motivated group exhibits more persistence for a task than a control group," then the hypothesis might be "The groups do not differ in persistence." As a mathematical statement, if $\mu_e$ and $\mu_c$ are parameters representing average persistence for the experimental and control groups respectively, then the alternative is $\mu_e > \mu_c$ and the hypothesis is $\mu_e = \mu_c$. The hypothesis could also be $\mu_e \leq \mu_c$. The alternative here is called one-sided (or one-tailed), since it states a particular direction of inequality, namely the upper tail or right tail of the distribution of experimental minus control differences. The alternative $\mu_e < \mu_c$ is also one-sided, but it is left-tailed; the alternative $\mu_e \neq \mu_c$ is two-sided.

The decision as to whether to accept or reject any hypothesis is made on the basis of the sample evidence according to some *statistical test procedure* and *test statistic*. The test procedure may also be called *one-sided* or *two-sided*, according to whether the alternative is one-sided or two-sided. One-sided tests are either left-tailed or right-tailed, depending on the direction of the alternative. The test statistic should be consistent with, and appropriate for, the type of alternative, the type of data available for analysis, and the assumptions the investigator is willing to make about the population. The sampling distribution of the test statistic is determined under the assumption that the null hypothesis is true. Hence it is sometimes called the *null distribution of the test statistic*. The sampling distributions of most test statistics that are useful in inference have been tabled. Such tables are given in the Appendix of this book for small sample sizes, as either cumulative distribution functions or as tail probabilities.

Once the test statistic is chosen, its value is calculated for the data obtained. Then the investigator can use this number and the sampling distribution of this test statistic to determine a quantity called the *P-value*, or the *associated probability*. The *P*-value is the probability, when the hypothesis $H$ is true, of obtaining a value of the test statistic which is equal to or "more extreme" (in the appropriate direction) than its observed value. The *P*-value is either a right-tail probability, a left-tail probability, or some combination thereof, in the sampling distribution. The investigator may then simply report the *P*-value and conclude the analysis there, with an explanation that the *P*-value is the probability of a sample result as extreme as that observed when $H$ is true.

In some cases the investigator may wish to make a statistical decision as to whether to reject or accept $H$. This decision can be based on the magnitude of the *P*-value in the following way. Since the *P*-value is found from the probability distribution of the test statistic under the null hypothesis, a *P*-value that is very small implies that a sample result this extreme when $H$ is true occurs only very rarely by chance alone. However, the random sample *results* are a fact, not subject to uncontrolled error except with regard to unknown aspects of the population sampled. Hence, the statement given in *H* about the population must be the factor that is inconsistent with the sample results. When the *P*-value is so small that a sample result this extreme occurs only very rarely by chance phenomena, the investigator can state that the data do not support the null hypothesis, or that the result is "significant." The statistical decision is then to *reject the null hypothesis.* On the other hand, suppose that the *P*-value is large. Then the sample result offers no convincing evidence that the statement made in *H* about the population is false; in other words, the data do not deny the null hypothesis. The statistical decision here might be to *accept the null hypothesis.* Some statisticians claim that the decision should be stated as "fail to reject *H*," "do not reject *H*," or "reserve judgment," as opposed to "accept *H*," since the analysis showed only insufficient evidence for rejection. (Similarly, when the research hypothesis is the alternative, the investigator might prefer to state that on the basis of the data obtained he was unable to prove beyond a reasonable doubt that the research hypothesis is correct.) This method of stating the conclusion is indeed justifiable, since although only one contradiction is needed to conclude the falsity of any absolute (not probabilistic) statement, accepting the truth of a statement requires more support than a single example that is consistent with the statement. Indeed, there may be many hypotheses that are consistent with the data obtained, so why should any particular one be accepted? This author would rather "reserve judgment" in such a semantic differentiation, since the practical effect of the decision is likely to be the same no matter how the conclusion is stated.

In discussing this decision process, we have so far avoided the important question as to "How rare is rare?" The probability value taken as the cutoff between a rare event and a likely occurrence is frequently called the *level* of the test. Its value is largely a matter of personal choice, but it should reflect the investigator's feeling about the cost and the consequences of error, and not be completely arbitrary. Two types of errors may occur: rejecting a null hypothesis that is true, or accepting a null hypothesis that is false (failing to reject a false hypothesis). The former is called a *Type I error*, and the latter a *Type II error*. These definitions are summarized below.

**Actual Situation**

|  |  | *H* true | *H* false |
|---|---|---|---|
| **Investigator's Decision** | Accept *H* | Correct Acceptance | Error (Type II) |
| | Reject *H* | Error (Type I) | Correct Rejection |

If rejecting a true hypothesis would be considered a very serious kind of error, a small cutoff point would be appropriate. However, a choice that is too small may not be good either, because the Type I and Type II errors interact in the following way. For a fixed sample size, when the probability of one type of error decreases, the probability of the other increases (but usually not by the same amount). (An increase in the sample size decreases both probabilities.) Thus if a Type II error has serious consequences, a larger cutoff point might be advisable. If a number $\alpha$ is chosen as the cutoff point or level and the test result gives a $P$-value of $P$, the null hypothesis would be rejected if $P \leq \alpha$ and not otherwise. The value of a test statistic for which the $P$-value equals a particular chosen level is called the *critical value* of the test statistic.

The numbers commonly used in practice as the maximum $P$-value that should lead to rejection (cutoff between a rare occurrence and a likely occurrence) represent a range of stringent to mild as .001, .01, .05, and .10; there is nothing sacred about these figures, except that some tables are prepared for use only with these probabilities in one or both tails. With such tables the precise $P$-value cannot be found unless the calculated value of the test statistic happens to equal one of the critical values for those levels covered in the table. However, an interval, or an upper or lower bound, on the $P$-value can usually be determined from such tables, and this range can be reported or used to reach a decision. As long as the sampling distribution is exact, the test, and the corresponding $P$-value, are called *exact* even if the precise $P$-value cannot be determined from the table available.

Thus we have two ways of treating a problem in hypothesis testing. We can either simply report the $P$-value, or we can use the $P$-value to form an opinion about $H$ and possibly also make a definitive decision. In conventional analysis of data from experiments or samples, persons with a decision-making orientation tend to regard the goal of analysis as rejecting the null hypothesis, finding a "significant" result or attaching an asterisk to the result to indicate significance. However, statistical analysis is more appropriately viewed as the reduction of data to the form of a brief condensation that contains the gist of the results. A "bare bones" statement like "reject $H$" is frequently not very informative unless the information about the $P$-value is also reported (then the $P$-value is sometimes called the *critical level*).

If a purely statistical decision is to be reached about the null hypothesis formulated, it can be based solely on the $P$-value if the cutoff point for a "rare event" is selected before the data are analyzed. That is, the $P$-value then provides an objective and complete basis for this kind of inference. However, in most practical problems the final decision should also be influenced by other factors, which may be strongly subjective. For example, frequently the population sampled differs from the target population — that is, the population about which information is desired. It is not really possible to take a sample of all persons who will vote in a certain election, or all parts produced by a machine, or all persons with a certain disease. Then the inference is a two-stage process, first from the data to the population actually sampled (a reasonably objective decision), and then to the target population (a highly subjective decision). Even in situations where the population sampled is the population of interest, the final decision may be influenced by many factors in addition to the $P$-value. Some of these factors, such as reliability of sampling procedure or validity of test procedure, are statistical, whereas others, such as economic or practical implications of the decision, are purely nonstatistical. Any or all of these factors may be more important than the $P$-value in reaching a practical decision. In essence, then, the $P$-value by itself generally should not be considered the sole basis for practical decision-making, but rather one of the many factors which are important and useful inputs. Thus, the $P$-value is an objective aid to the ultimate decision-maker, whether it is the investigator, his employer, the reader of a report or study, or anyone else.

When the purpose of an empirical study is to investigate a research hypothesis, decision-making is not usually a simple process either. First of all, the research questions posed must be translated into statements about parameter values or other aspects of a distribution that can be investigated by statistical techniques. In many cases a certain statistical model must be developed before statistical methods can be applied. The experiment must be designed in such a way that the data collected and analyzed are relevant to the question being investigated. For example, suppose that an investigator needs to decide which of a number of predictors (the Graduate Record Examination score, undergraduate grade record, recommendations, and so on) is most relevant for evaluating candidates for graduate study. What criteria should be used to compare the predictors? Success in graduate school might be measured by graduate grade record, faculty rating, whether or not the graduate program is completed, other factors, or some combination thereof. This is a complicated research problem, and will not be discussed further. The important point is that statistical analysis in general, and $P$-values in particular, can be an aid to making a research conclusion, but only in concert with the judgment and intelligence of the investigator.

Whether the decision is a practical one or a research conclusion, reporting the $P$-value in addition to the words "accept" or "reject" the hypothesis provides considerably more information and latitude. Reporting the $P$-value alone provides an objective measure of the extent of disagreement between the observed value and a value that could have resulted by chance alone, from which the reader can make his own decision as to whether to accept or reject the null hypothesis. The escape into $P$-values to avoid a decision cannot be recommended as a general practice, since the investigator is far more familiar with the data and the other factors than is a casual reader. However, in any textbook example, it is almost impossible to appreciate all aspects of a problem in decision-making. Thus, many problems in this book will be considered as solved statistically once the $P$-value is found. When a decision is actually reached, the reader should realize that it is to be interpreted as a purely statistical decision, which might be reversed or at least modified if information about other relevant factors were available and used as input. Hence the language used when the $P$-value is large is that the data support $H$, or, when the $P$-value is small that the data do not support $H$. When the research hypothesis is the statement in the alternative, a small $P$-value then allows the investigator to conclude that the empirical evidence appears to support his research hypothesis.

As mentioned above, the sampling distribution of the test statistic must be known before the $P$-value can be found. For either practical or theoretical reasons, it may not be possible to use the exact null distribution of the test statistic to obtain an exact test. Then the procedure is called an *approximate test* because the sampling distribution is only approximate; the corresponding $P$-value is also approximate. In many cases it is known that $P$-values based on that approximation are overestimated. Then one will be *less* likely to conclude that a rare event has occurred when it has not, or to reject the null hypothesis if it is true, and thereby commit a Type I error. When a $P$-value is reported for this kind of approximate test, it is called here a *conservative P-value*, whether given as a precise number or within an interval.

For large samples, many sampling distributions converge to another known distribution form. This is called an *asymptotic distribution*. For finite sample sizes, any test procedure based on an asymptotic distribution is an approximate test; its degree of accuracy depends on the sample size used and the speed at which the true distribution approaches the asymptotic distribution. Many asymptotic distributions are accurate enough for practical purposes with surprisingly small sample sizes. When a $P$-value, as reported, has been based on an asymptotic distribution, it is called here an *asymptotic approximate P-value*, whether given as a precise number or a number within an interval. When the sample size is large, the investigator frequently has a broader choice of test statistics since the assumptions required for the validity of many asymptotic sampling distributions are much less stringent than for exact sampling distributions.

## 2.3   Estimation

The objective of an inference procedure may be simply to obtain some information about the value of a population parameter rather than to test a specific null hypothesis concerning its value. In this case, the usual statistical inference procedure employed is *confidence interval estimation* or, more simply, interval estimation. A confidence interval for a parameter $\theta$ is an interval, with endpoints that can be calculated from sample data, such that the probability is a preselected number that the sampling and estimation procedures will lead to an interval which includes the true value of $\theta$. In symbols, we have an interval $(U, V)$, where $U$ and $V$ are random variables that satisfy the probability statement

$$P(U < \theta < V) = \gamma.$$

The number $\gamma$ is a probability for the random interval, but once the interval endpoints $(u, v)$ are found for the particular sample obtained,[2] the interval is no longer a chance variable. The number $\gamma$ is therefore called a *confidence coefficient*, rather than a probability, in order to emphasize that it represents an evaluation of the investigator's confidence in the numerical interval that results from the particular sample data and confidence procedure.

Let us interpret the endpoint numbers $u$ and $v$ as dividing the range of all possible values of the parameter into two nonoverlapping groups. The set $u < \theta < v$ is supported by the sample evidence, whereas values of $\theta$ that are outside this interval are not. The confidence interval then consists of all values of the parameter that, when formulated as a null hypothesis, would *not* be rejected by a corresponding test of this hypothesis. For the confidence coefficient to be equal to $\gamma$, the cutoff $P$-value between acceptance and rejection must be equal to $1 - \gamma$, that is, the complement of the probability of accepting the null hypothesis when it is true. If the confidence interval procedure corresponds to a test procedure for which the $P$-value is conservative, the confidence procedure is also called *conservative*. Then the confidence coefficient is known to be at least the stated value $\gamma$.

In general, trial-and-error methods can always be used to find the boundaries $u$ and $v$ of hypothesized parameter values that would lead to acceptance. However, in many cases a systematic procedure, either algebraic or geometric, can be followed instead.

Either of the endpoints, $u$ or $v$, may be infinite so that the confidence interval is open-ended on one side. Such intervals correspond to hypothesis sets with a one-sided alternative, and hence to one-sided tests, but all of the statements above still apply. Since we use only two-sided confidence intervals here (that is, both $u$ and $v$ are finite), this special case is not discussed further.

---

[2]As before, the lowercase letters $u$ and $v$ are used to differentiate the actual numbers referred to here from the random variables denoted by the corresponding capital letters $U$ and $V$.

With this definition of confidence interval estimation, it should be obvious that any hypothesis testing problem that is a statement about a parameter value could be handled just as well by the confidence interval approach. If the hypothesized value is included in the interval with confidence coefficient $\gamma$, the null hypothesis should not be rejected at level $1 - \gamma$. However, when the ultimate goal of the inference problem is hypothesis testing, a direct test procedure may be easier to carry out in practice.

# 3   Classification of Data

Although statistics is concerned with the collection and analysis of data, only certain analytical procedures are appropriately applicable to certain types of data. Therefore, preliminary to giving a formal definition of nonparametric methods, we need to discuss some classifications of data or the underlying variable.

We first distinguish between count data and other kinds of data.

*Count data* result from noting, for each unit of observation, whether it does or does not have some particular property or attribute that is a member of a set of properties or attributes. For example, we might divide hair color into four mutually exclusive and exhaustive groups, such as blond, brown, black, and red. If a number of individuals are classified according to hair color, the natural way to summarize the results would be to report how many individuals have each hair color. The same would be true in a study of incomes if we were to divide income level into three well-defined groups, such as low, medium, and high. We could also cross-classify individuals according to two different sets of properties, like hair color and eye color. Such data are all *categorical,* irrespective of what labels are placed on the categories. Although income is a quantitative property, and hair color is qualitative, that distinction is inconsequential so long as the focus is on the *frequency* of elements in each category or cross-category. In noting whether an element does or does not have a certain attribute, we are essentially making a measurement on that element. However, this measurement is simply an indicator, and it is not retained when the data are summarized by numbers that represent the category frequencies' subtotal. Therefore, in analysis such data are called *count data* or *enumeration data.*

For other kinds of data, the individual measurements are of interest in themselves and are retained. Such data may be classified according to certain characteristics of the random variable underlying the measurement. The dimension of a random variable is frequently of concern — that is, whether the measurements occur singly, or in pairs, triplets, and so forth. If the random variable has only one dimension, its distribution is called *univariate.*

It is called *bivariate* in the case of pairs, and *multivariate* when there are more than two dimensions. For example, measurements on the heights of a group of people form a univariate distribution, whereas measurements on both heights and weights make a bivariate distribution. Sample data from such distributions are classified the same way; the term *one sample situation* is used in the univariate case, *paired sample situation* in the bivariate case, and *k-related sample situation* in the general multivariate case. This property of data is called *sampling situation* in this book.

A univariate distribution, moreover, can be classified as *discrete* or *continuous*, according to whether the variable is discrete or continuous. As we have already noted, the variable is continuous if theoretically it can assume any value between some two points (even though, in practice, the value of the observation is limited by the degree of precision of the measurement). For example, weight should be considered a continuous variable, in spite of the fact that it may be observed and recorded only to the nearest pound, or the next full pound. Bivariate and multivariate distributions can represent variables that are all discrete, all continuous, or a mixture of the two kinds.

In some applications, simply knowing whether a variable is discrete or continuous is sufficient to determine whether a statistical procedure is appropriate. In other cases, we may want to know what *level of measurement* the data represent, since this affects the interpretation of the results. For example, a student's grade on a test can be measured according to the proportion of questions answered correctly, the proportion correct less some fraction of the proportion incorrect, standing relative to the others taking the test, a letter grade, or the like.

Measurements can be classified as one of four levels depending on the precision represented by the measurement procedure. In increasing order of precision, these levels are called nominal, ordinal, interval, and ratio scales.

1. *Nominal scale.* As the word "nominal" implies, a name is given to classify each element observed with respect to the property of interest. Thus, a part may be classified as defective or good, the outcome of a coin toss as a head or a tail, the color of hair as blond, brown, black, or red. Numbers may be chosen arbitrarily to record a nominal measurement, as 0 for defective and 1 for good, but they could just as well be called 43 and 21. We are simply giving numerical names to the types of outcomes, and the principle of order in the real number system is not relevant. In many (but not all) situations in which data are measured on a nominal scale, we deal with count data because the quantities of interest are the observed number of elements of each type.

2. *Ordinal or ranking scale.* When measurements are made on an ordinal scale, the elements can be arranged in some meaningful kind of order, which corresponds to their relative position or "size." In a taste test

for three different coffee blends, the tasters may evaluate the coffees as most preferred, next preferred, and least preferred. Students may receive grades A, B, C, and so on, according to their performance on a test. In a horse race, the horse who finished first ran faster than did the one who finished second, and so on. The factor that differentiates ordinal data from nominal data is that the elements can be judged not only as to whether they are the same or different, but also according to whether one is greater than or less than the other, or better than or worse than the other. (If some two or more elements have the same relative position or size, they are said to be tied; the data as a group are still called ordinal.) Thus, the data are essentially classificatory, but the classes have labels that imply a specific relative order of magnitude. No difference is inferred among elements of one class, but any two elements that are in two different classes can be ordered. Numbers may be assigned to indicate the relative magnitudes of the elements in different classes. For example, the outcome of the horse race might be indicated by 1, 2, 3, . . . , to show the order in which the horses finished. Although consecutive positive integers are often the most convenient indicators, the numbers 8, 25, 125, and so on, could just as well be assigned, since the same order is implied by their relative magnitude. Such data are frequently called *rank data*. It should be remembered that the element with rank 2 is not to be interpreted as twice as "good," twice as "large," or twice as "preferred," as the element with rank 1, or as one unit "better" or "larger." The difference between ranks 1 and 2 is not necessarily the same as the difference between ranks 2 and 3, or 10 and 11.

   3. *Interval scale.* When the elements can be differentiated and ordered, *and* the arithmetic difference between elements is meaningful, the data are measured on an interval scale. Thus we can say not only that one element is larger than or smaller than another, but also by how much. This scale of measurement is much more informative than either of the scales above, since the fact that the distance between elements can be determined implies that there is a fixed unit of measurement and a zero point (origin), even though the latter is arbitrary. Thus, interval scale data are quantitative in the ordinary sense of the word, since the numbers have a true meaning. A classic example of the use of an interval scale is in the measurement of temperature. The common unit of measurement is the degree, but the origin differs depending on whether the temperature is Fahrenheit or Celsius (centigrade). On the Celsius temperature scale, the zero point $0°C$ is taken as the freezing point of water, but the freezing point of any other solution (such as ethyl alcohol) could just as well be used. The freezing point of water in Fahrenheit temperature is $32°F$. The mathematical relationship is $C = 5(F - 32)/9$. Since the C scale is 5/9 of the F scale, a one-degree change in the F scale is equivalent to a 5/9-degree change in the C scale. The difference between

50°F and 53°F is the same as the difference between 41°F and 44°F, but these differences are each 5/9 of the difference between 5°C and 8°C.

4. *Ratio scale.* The highest level of measurement is the ratio scale. Here, we have not only the order property, a unit of measurement and a meaningful arithmetic difference between elements, but we also have a fixed origin or zero point as opposed to an arbitrary origin. For example, height, weight, and length are ratio scale measurements, as long as the number zero remains a fixed origin. A person who weighs 160 pounds is twice as heavy as a person who weighs 80 pounds, unless some origin other than zero is used. That is, if 100 pounds is the reference point, these weights would be recorded as 60 pounds and − 20 pounds and the ratio of 2 is no longer apparent from these figures. The term "ratio scale" is used because the ratio of two measurements on this highest scale is meaningful. Index numbers provide another good example of ratio scale measurements, as do Kelvin temperatures.

In order to clarify the difference between these four scales of measurement, let us consider a specific example. Suppose that a rubber company manufactures and distributes automobile tires. Observations on tire output could be taken and recorded in any one of the following methods:

1. Classify the product by some dichotomy (into two categories), such as good or bad, acceptable or not acceptable, salable or scrap, premium or not, blemished or perfect.
2. Classify the product into three or more categories according to some nominal characteristic, such as radial, tubeless or tube; blackwall, whitewall or snow; truck, tractor, or automobile.
3. Classify the product into three or more categories according to some ordinal characteristic, such as Grades AA, A, B, C; Premium, Choice, Standard, and Economy.
4. Inspect each tire to determine the number of surface irregularities or flaws.
5. Count the number of tires produced per day for a given number of days to determine the average daily production, and measure its variation.
6. Measure some characteristic precisely, such as tread contact width, to determine its typical magnitude and variability.
7. Subject some tires to a road test or laboratory torture test to determine average tread wear-out rate, typical load-carrying capacity, ability to withstand road hazards, impact resistance, amount of traction, and so forth, and also the variation in that characteristic measured.

Each of these methods involves a measurement of one kind or another. As described, methods 1 and 2 represent nominal measurement; method 3 is ordinal; and methods 4 and 5 are ratio. The measurements in methods 6 and 7 are probably interval or ratio, but they may be ordinal; the scale depends on the method used to measure the characteristics mentioned.

The four scales of measurement have been listed in increasing order of refinement. The levels themselves form a cumulative scale, in that a measurement at any level has all the properties of a measurement at a lower level plus an additional property — respectively, ordinality, a unit of measurement, and an absolute zero. The important implication of this property is that any measurement in one scale can be easily transformed to a measurement in a lower scale, as for example interval data can be converted to ordinal (or even nominal) data. However, the reverse process is *not* possible because the information necessary to perform the transformation to a higher scale is not available.

We can make use of this fact in selecting a procedure for statistical analysis once we know what minimum level is used in the arithmetic of each procedure. That is, if we have ordinal scale data, we can use a method of analysis appropriate for either ordinal or nominal scale data. (Operationally, methods appropriate for interval or higher scale data can be used with ordinal data. However, interpretation of the result is then difficult, since the conclusions can have no more meaning than the numbers entering the analysis.)

Because transformation of measurement scale can go in only one direction, it is important that measurement be made at a level not lower than that needed or desired for analysis. On the other hand, recording data at a scale higher than that needed for analysis is a waste of time, energy, and resources. Thus, if the investigator knows in advance that he is going to apply a procedure that uses only ordinal data, it is more convenient and economical to design the experiment with ordinal measurement even though a higher level may be possible.

Since our focus is on the data that are to enter the analysis directly, specification of minimum level of measurement required for the data is not relevant when enumeration data are used. In methods 1 and 2 of the tire example, how are the nominal measurements to be used? In method 3, how are the ordinal measurements to be used? Once the day's output is classified into salable and scrap, or is graded, if we count the number in each category and base the analysis on these totals, we have enumeration or count data. If data of this kind are used, there is no need to distinguish whether the categories are nominal as in methods 1 and 2, or ordinal as in method 3, since the individual measurements are not retained. On the other hand, if we look at the quality of each tire as it comes off the production line and analyze a time-ordered sequence of observations like GGGBBG ... (G = good, B = bad), counts are not involved, and the data are simply nominal scale (and also time-ordered). Further, it is usually not necessary for the purpose of statistical analysis to distinguish between interval and ratio scale data.

As a result of these factors, the most convenient way to designate "level of measurement" of data entering a statistical analysis is to add a group called count data, and combine the scales called interval and ratio, that is:

I. Count or enumeration data.
II. Nominal data.
III. Ordinal data.
IV. At least interval data.

Throughout this book we specify for each statistical procedure covered that the data should retain the properties of one of these "levels" of measurement.

## 4   Nonparametric Statistical Methods

The basic aspects of hypothesis testing and estimation (as described in Sections 2.2 and 2.3) are relevant to any statistical inference technique. With parametric statistics, the derivation of a procedure and its theoretical justification rest on fairly specific assumptions about the form of the population distribution — usually that it is the symmetric, bell-shaped distribution known as the normal (or Gaussian) curve. The validity of such a procedure, and any probability calculations based on the resulting sampling distribution, are dependent upon the validity of the assumptions from which the procedure is developed. In practice, however, stringent assumptions can seldom be completely justified. It may be that the sample size is too small, or that previous experience with this type of data is too limited, or that relevant information is simply insufficient for the investigator to know just what sort of assumptions might be tenable in the situation at hand. Sometimes reasonable assumptions can be determined, but they are not the ones for which the parametric methods are appropriate. Most methods in parametric statistics are designed primarily for use with interval (or ratio) scale data, at least for typical interpretations of results, and yet in many practical situations the only data available are measured at a lower level of precision.

These are the types of problems for which nonparametric methods provide a technique of analysis. The procedures are appropriate and valid in a wide variety of practical situations in which methods of parametric statistics should not or cannot be used.

For the purpose of this book, certain statistical procedures are termed nonparametric either because of the type of analysis used on the data or the type of data used in the analysis (level of measurement). The latter was explained in Section 3. Before considering a formal definition of nonparametric methods, the former must be discussed.

We first review briefly the nature of the inference procedures that are ordinarily introduced in a traditional study of statistics. In most cases the inference concerns the value of one or more population parameters, usually the mean or variance. Further, in each case (except for large sample procedures based on the Central Limit Theorem), the mathematical justification for the inference procedure is provided by information about the functional form (probability model) of the population from which the samples are drawn.

Those statistical inferences that are not concerned with the value of one or more parameters would logically be termed *nonparametric*. Those inferences whose validity does not rest on a specific probability model in the population would logically be termed *distribution free*. While these terms are not synonymous, and in fact refer to different aspects of a statistical inference, procedures of either type have now become known as *nonparametric methods*. This class of methods is sometimes also interpreted as including procedures, whether for inference or description, which use count data or data measured on a scale lower than interval, that is, either ordinal or nominal level. Since such procedures are especially useful in many different areas of application, they are included in this book.

Therefore, for the purpose of this book a statistical technique is called nonparametric if it satisfies at least one of the following five types of criteria:

1. The data entering the analysis are enumerative — that is, count data representing the number of observations in each category or cross-category.
2. The data are measured and/or analyzed using a nominal scale of measurement.
3. The data are measured and/or analyzed using an ordinal scale of measurement.
4. The inference does not concern a parameter in the population distribution — as, for example, the hypothesis that a time-ordered set of observations exhibits a random pattern.
5. The probability distribution of the statistic on which the analysis is based is not dependent upon specific information or assumptions about the population(s) from which the sample(s) are drawn, but only on general assumptions, such as a continuous and/or symmetric population distribution.

By this definition, the distinction of a procedure as nonparametric is accorded either because of level of measurement used or required for the analysis, as in types 1 through 3; the type of inference, as in type 4; or the generality of the assumptions made about the population distribution, as in type 5.

It follows that classical parametric statistics can then be defined here as those techniques that require data measured on at least an interval scale, *and* require specific population distribution assumptions, *and* relate to inferences concerning parameters. According to this definition, methods of types 1 through 4 are nonparametric and have no direct competitor in parametric statistics. Further, certain well-known procedures treated in most elementary courses in classical statistics are in fact nonparametric; for example, the test for independence in a contingency table is of type 1, the binomial test is of type 2, the rank correlation coefficient is of type 3, and methods based on Chebyshev's inequality or the Central Limit Theorem are of type 5.

It should be noted that there is no common agreement as to exactly which statistical techniques should be termed nonparametric, or even agreement on the name *nonparametric*. The definition given above is comparatively exacting; it serves to identify the areas of interest in this book, even though it may not be satisfactory for other purposes.

## 5   Advantages of Nonparametric Methods

Hopefully, the investigator will come to appreciate nonparametric methods as study progresses through this book. Nevertheless, some preliminary motivation may be supplied by the following brief examination of some of the more obvious general advantages.

The attribute of nonparametric methods that may be most persuasive to the conscientious investigator who is not a professional statistician is that he is somewhat less likely to "get into trouble" or to misuse statistics when applying nonparametric techniques than when using those methods that are parametric according to our definition. The easiest way to abuse any statistical technique is to disregard and/or violate the assumptions necessary for the validity of the procedure. Although the product of a statistical analysis can always be presented in an elegant manner and in meticulous scientific terminology, any inferential conclusions based on the technique are not exactly valid unless every assumption is satisfied. If the assumptions cannot be substantiated, or are disregarded, or are not even known to the investigator, then the inferences may be less reliable than a judicious opinion, or even an arbitrary guess.

This follows because the mathematical derivation of all sampling distributions is based on the complete satisfaction of all assumptions made. The $P$-values in hypothesis testing and related confidence coefficients in interval estimation are exact for these sampling distributions only, and may not even be close to the corresponding probability resulting from some other sampling distribution. Conclusions based on such $P$-values are then highly questionable. A decision is of little or no value when the investigator should have little faith in its accuracy.

Assumptions needed to develop techniques of statistical analysis can be divided into two types: those relating to the sampling procedure and those concerning the nature of the population from which the sample is drawn. The assumption of a random sample is basic to almost any statistical procedure, whether parametric or nonparametric, and cannot generally be relaxed. However, a careful investigator may be able to adopt sampling techniques that justify this assumption.

Nonparametric methods are sometimes called "weak assumption" statistics because the assumptions made about the population are fewer and much less stringent than those made in parametric statistics. In some situations, nothing need be assumed. Usually the underlying population distribution or variable is assumed to be continuous. (Occasionally, symmetry is also assumed.) An assumption of continuity is mild compared with the very specific assumptions made in parametric statistics. For example, the assumption of a normal distribution implies a continuous, symmetric, bell-shaped distribution with infinite domain and a specific mathematical function.

The population assumptions required for the absolute validity of nonparametric methods are quite general, and are indeed satisfied in most applications. Besides, in many cases even these assumptions can be relaxed, since they are sufficient, but not necessary, for complete validity of the procedure (that is, the procedure may remain valid under some weaker form of the assumptions). As a result, conclusions or inferences based on nonparametric methods need not be tempered by stringent qualifiers such as, "If the populations are normal with equal variances, then. . . ." Such qualifiers are frequently relegated to footnotes, if noted at all, and are ignored by many readers of a statistical report.

Since many practical research situations afford no basis for assuming a given distribution in the population, the aforementioned factor provides substantial inducement for the conscientious investigator to want to add nonparametric techniques to his repertoire of methods of analysis. However, there are many other persuasive advantages. Several of these relate to cost and speed of experimentation or analysis. The process of collecting and compiling sample data may be less expensive and time-consuming with nonparametric methods, because the sample sizes need not be large and the level of measurement needed for analysis is frequently less refined. For example, if a procedure is to be based on the ranks of the observations, only the ranks need be observed. There is no minimum sample size required for most methods to be valid and reliable. With so-called "dirty data," many nonparametric techniques are still appropriate. The methods are generally easy to use and understand, and the logic and rationale behind them are usually also simple to comprehend and appreciate. In fact, a careful study of nonparametric statistics provides considerable insight into the general ideas of statistical inference because the perspective is not clouded by as complicated mathematics.

As a final point, the broad range of applicability and types of inference possible should be mentioned. The scope of application is wider than with parametric methods, since the techniques may legitimately be applied to phenomena for which it is impractical or impossible to obtain precise measurements on at least an interval scale. Further, there are many important inference problems that do not concern a population parameter, such as tests for randomness and tests for symmetry, which are all classified as nonparametric methods.

Of course, there are also many statistical problems for which nonparametric techniques have not as yet been developed, such as regression, analysis of variance or covariance, design, and so on. The primary other disadvantages cited for nonparametric methods are that explanations of the techniques are scattered in the statistical literature and are therefore not readily available or comprehensible to many investigators, and that the manipulations are sometimes laborious in large samples. It is hoped that this book will alleviate the former problem. As to the latter, many of the "packaged computer programs" for statistical analysis now include the numerical manipulations needed for most nonparametric methods.

The reader should not interpret this discussion as implying that nonparametric methods provide a panacea for all problems of quantitative analysis. They do not eliminate the necessity of proper sampling procedures and/or experimental design, nor do they solve any of the problems raised by unknown measurement level, sloppy data, outliers, data that are not particularly relevant to the questions of primary interest, or completely untenable general assumptions. Nevertheless, they comprise an important and useful body of statistical techniques.

Performance and reliability of an inference procedure are also matters of concern to the conscientious investigator. How "good" are these nonparametric methods? The more objective criteria for measuring good performance in hypothesis testing are generally referred to as power and robustness. *Power* is defined as the probability of rejecting the null hypothesis when it is false. A test is called *robust* if inferences based on it remain valid despite the violation of one or more of the basic assumptions necessary for the theoretical development of the procedure. A test is called *valid* at a particular level $\alpha$ if the probability of a Type I error does not exceed $\alpha$ for all null distributions.

Since a Type II error is accepting a false hypothesis, the power of a test is equal to $1 - \beta$, where $\beta$ is the probability of a Type II error. Power can thus be interpreted as the probability of reaching a certain type of correct decision. The value of $\beta$ and the corresponding power in a particular test cannot be found unless some true situation is specified. However, we can still give a general interpretation of the concept of power as the ability of a test procedure to use sample evidence to detect whether or not the true situation differs from the hypothesized situation.

As would be expected, the most powerful tests are those that are based on the most stringent assumptions. On the other hand, the most robust tests are by definition those with the weakest assumptions. The conditions assumed for the derivation of parametric tests are the factors that make them powerful. When these assumptions are not satisfied, the tests may not even be valid, because the particular null distribution (on which the decision or $P$-value is based) is not accurate. Nonparametric tests are inherently robust because their construction and validity require only very weak assumptions. Does it then follow that a test can be powerful, or robust, but not both? The answer is no. Many parametric tests are robust against certain violations of assumptions, and many nonparametric tests are powerful against certain alternatives.

Although there are notable exceptions, particularly in the multivariate case, several parametric tests have nonparametric analogues. Performance comparisons are then of interest, even though it is difficult to weigh the importance of power versus robustness when the criteria are incompatible. Most comparison studies are made on the basis that if the conditions of the parametric test are met but an analogous nonparametric test is used instead, how much power is lost? If such loss is small and a nonparametric test is used in a situation where the validity of the assumptions is not determinable, the investigator can have confidence that (a) if the parametric assumptions are satisfied, there will be little loss in power, whereas (b) if the assumptions are not satisfied, the nonparametric test is still absolutely valid and its power is unchanged. In the latter case, the actual power cannot be determined because there is little, if any, information about the true situation. However, the nonparametric test may well be the more powerful, especially if the sample size is small.

In case (a) above, the relative loss in power (called relative efficiency) is small for many nonparametric tests when the sample size is large. The decrease can frequently be compensated for by taking a slightly larger sample for application of a nonparametric technique of analysis. The *asymptotic relative efficiency* (called the *ARE*) provides a single measure of relative performance of two tests, both using large sample sizes. The ARE figures have been calculated under the classical distribution assumptions for most of the nonparametric tests that have parametric analogues.[3] Suppose the ARE of a

---

[3] Values of the ARE have also been calculated under other distribution assumptions. In many cases, the nonparametric tests are more efficient than their parametric counterparts. The ARE can also be calculated to compare two nonparametric tests. An explanation of techniques to calculate an ARE value is beyond the scope of this book. The interested reader is referred to Gibbons (1971, Chapter 14).

nonparametric test relative to a parametric test is reported to be .98. Loosely interpreted, this means that the nonparametric test based on, say, 100 observations is as efficient as the corresponding parametric test based on 98 observations if the assumptions required for the strict validity of the parametric test are met and if these sample sizes are large enough for the tests to reflect their asymptotic properties. The ARE figures for some nonparametric tests relative to their parametric counterparts are given in the next section. Of course, these numbers may not be indicative of relative power for small sample sizes.

# 6   Guide for Selection of Test Procedures

On the assumption that the investigator wishes to select a procedure from all available techniques of statistical analysis, his first decision might well be between a parametric and a nonparametric type of procedure. This judgment is easy if the level of measurement available or the type of inference is such that only a nonparametric technique can be deemed appropriate. These are the types of problems covered by criteria types 1 through 4 of the definition of nonparametric statistics given in Section 4.

If the data are measured on at least an interval scale, and the inference situation concerns the value of one or more parameters, either a parametric or a nonparametric procedure may be appropriate. In this case the decision might depend on whether it makes a difference what parameter the inference concerns — for example, the mean or the median, if they cannot be presumed equal. The decision should also depend on whether the investigator can justify, to himself and to others, the requisite parametric assumptions. If he can, the parametric test is the better choice in theory because of its power. (In practice, cost, speed, or other factors may result in a decision to use the nonparametric test anyway.) If certain assumptions are not tenable, the investigator should consider whether the parametric test is known to be robust against violations of these assumptions. A choice between a robust parametric test and an analogous nonparametric test is largely a matter of personal preference, since any nonparametric procedure is inherently more robust, whereas the parametric procedure may well be more powerful, depending on the true population distribution. On the other hand, if the parametric test is not robust against a suspect assumption, there is little justification for using it in the hope that it will be more powerful.

If this decision process leads to a parametric procedure, usually only one test is appropriate for any given inference situation and set of assumptions, and thus there is no problem of selection. However, if a nonparametric procedure is desired, there may be several different techniques appropriate for the specific experimental and inference situation. The procedure that uses the highest practicable level of measurement for manipulations is usually a good selection, since it employs more of the available information. The test with the highest ARE under parametric assumptions should not provide a definitive choice, since these ARE figures reflect relative performance only in large samples and when the fixed distribution assumptions of the analogous parametric test are satisfied. A better basis for decision is provided by comparing the power functions of the nonparametric tests under other distribution assumptions. Such studies are frequently reported in the statistical literature. In this book the nonparametric techniques covered are only those that are well known and have been shown to perform efficiently for a variety of distributions. Other techniques may be of theoretical or esoteric interest, but a limited repertoire of techniques is probably sufficient for the practical user of nonparametric methods.

If an investigator is going to make a decision between a parametric and a nonparametric technique when either is applicable, he needs to know which tests are analogous for various types of inference problems. For all of the nonparametric tests covered in this book which are for inferences about population parameters, Table 6.1 lists those parametric tests that are almost their direct counterparts. We must say "almost" since, for example, in location tests the parametric tests are concerned with the value of a mean and the nonparametric tests concern the value of a median, and these parameters are not necessarily equal, especially in asymmetric distributions. The tests are classified in the table according to the type of hypothesis to be tested. However, the analogy is equally useful if the purpose of the inference is to find a confidence interval, since confidence procedures are known by the names of the corresponding tests. The values of the ARE of each nonparametric test relative to its analogous parametric test under normal distribution assumptions are given in the last column of this table.

In summary, the decision to use a nonparametric technique as opposed to a parametric technique is in practice based on either (a) a belief that the data obtained are such that the assumptions required for the parametric test are unlikely to be satisfied in this instance; (b) a general feeling that inferences can be regarded as sound only if they are based on the minimum possible number of unverifiable assumptions; or (c) a conclusion that a particular nonparametric test is likely to perform better than some parametric test for this null hypothesis and experimental situation. A small sample size may dictate the position stated in (a) or (b), whereas justification through (c) requires some feeling about what the true distribution is and some information as to what the power would be if that feeling could be justified — that is, information about ARE if the sample size is large, or otherwise about power under specific assumptions.

**TABLE 6.1**
**ANALOGY BETWEEN NONPARAMETRIC AND PARAMETRIC METHODS**

| Type of Hypothesis | Name of Nonparametric Test(s) | Analogous Parametric Test(s) | Asymptomatic Relative Efficiency for Normal Distributions |
|---|---|---|---|
| Location parameter(s): | | | |
| One sample or paired sample | Sign test | Normal or Student's $t$ test | .637 |
| One sample or paired sample | Wilcoxon signed rank test | Normal or Student's $t$ test | .955 |
| Two independent samples | Mann-Whitney-Wilcoxon test | Normal or Student's $t$ test | .955 |
| $k$ independent samples | Kruskal-Wallis test | $F$ test (one-way analysis of variance) | .955 |
| $k$ related samples | Friedman test | $F$ test (randomized blocks analysis of variance) | $.955k/(k+1)$ |
| Scale parameter(s): | | | |
| Two independent samples | Siegel-Tukey test | $F$ test | .608 |
| Association analysis: | | | |
| Two related samples | Spearman rank correlation or Kendall Tau | Pearson product-moment correlation | .912 |
| $k$ related samples | Kendall test | $F$ test (randomized blocks analysis of variance) | $.955k/(k+1)$ |
| | Kendall test | $F$ test (balanced incomplete blocks analysis of variance) | $.955k/(k+1)$ |

# 7   Organization of this Book

A general presentation of nonparametric techniques can be organized according to several different criteria. The primary criteria that serve to delineate natural groupings of nonparametric methods are as follows:

1. The type of sampling distribution on which the theory of the inference procedure is based.
2. The level of measurement of the data to be analyzed.
3. The number and/or type of samples — for example, one sample, paired samples, two independent samples, and so forth.
4. The type of inference situation or hypothesis set to be tested.
5. Some combination of the above.

Organization according to item 1 would be useful primarily for a study of the *theory* of nonparametric statistics, and item 2 would be too broad a criterion on which to base a survey book. Most of the more recent applied books have been organized according to item 3 insofar as possible; admittedly, this makes a rather neat pattern.

However, the first step in a scientific investigation should be to ask a question, pose a problem, or state a research hypothesis. Only then can one decide how many samples are appropriate and consider what level of measurement is feasible for recording the data. The final step is to select an appropriate inference procedure and use the sampling distribution that applies. In other words, the investigator must think through each of items 1 through 4 in planning a study, but he would approach the problem by examining these items in reverse order.

Since the purpose of this book is to train investigators to solve their own problems in research analysis by a correct application of nonparametric methods and to provide a useful handbook for practitioners, the basic pattern of organization followed here will be that which most closely resembles the thought processes of the investigator as he approaches his problem. We will begin by considering the type of inference desired (item 4). Then, for a given type of inference, we will proceed according to the sampling situation (item 3) and the minimum level of measurement required (item 2). For any fixed combination of these three factors, only the most useful and accepted nonparametric techniques are presented. The procedures not covered in this inventory are, for the most part, either those that are primarily of academic interest, have relatively poor performance properties, apply to very special situations, or are not practical for common use because of their difficulty or the relative inaccessibility of tables necessary for their application.

This organization scheme and the procedures covered in this book are depicted visually in Figure 7.1. The types of inferences, as shown in the large

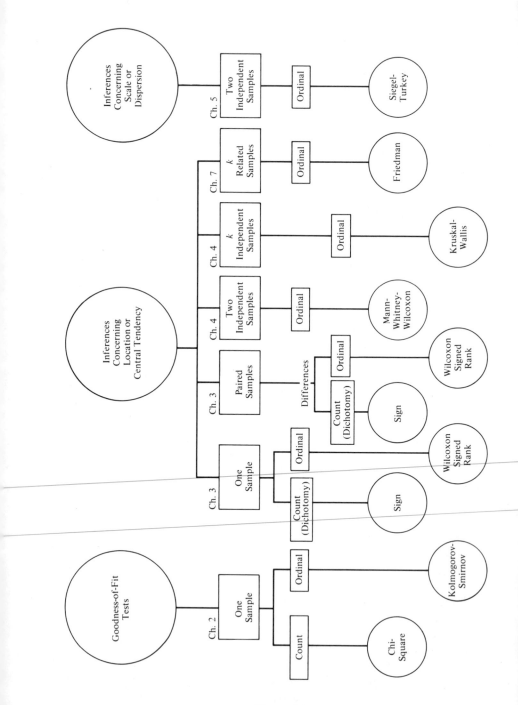

Figure 7.1

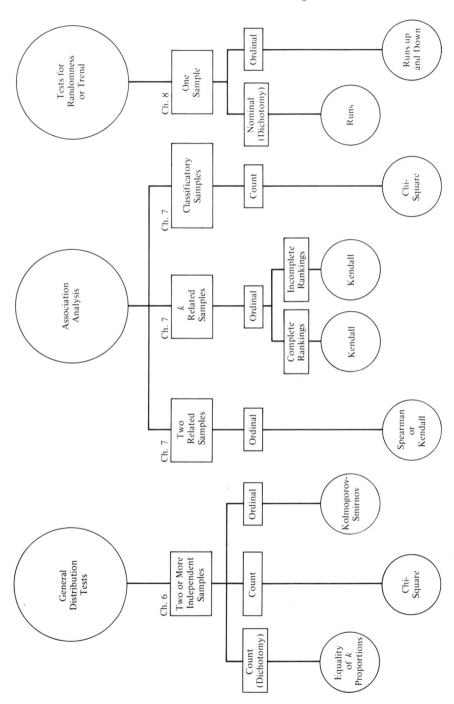

**Figure 7.1**

circles, provide the primary division into chapter groups. These groups are further subdivided according to number and/or type of samples on which the inference can be based, as shown in the first row of rectangles. For each combination of these two factors, the names of the appropriate nonparametric procedures (test or confidence interval) are shown in the smaller circles at the bottom of the figure. The minimum "level" of measurement required for the data entering the analysis is shown in the rectangle above that smaller circle for each procedure.

This systematic treatment of nonparametric methods begins with goodness-of-fit tests in Chapter 2. The inferences for these tests are concerned with the form or shape of the basic population; they are classed as nonparametric procedures because of the type of hypothesis. Chapters 3 and 4 discuss inferences about population parameters of location in the one-sample (or paired-sample) and two-sample situations, respectively, while Chapter 5 covers inferences regarding scale parameters. Chapter 6 treats general distribution tests for two or more independent samples, a sort of extension of the one-sample goodness-of-fit tests. Chapter 7 covers association analysis, for which descriptions of association are as important and useful as inferences concerning the existence of association in the population. In Chapter 8 a rather unusual type of inference problem is included, namely the question of randomness in a sequence of ordered observations.

For each procedure the logical rationale is discussed, but none of the underlying theory is presented. The methods of analysis are carefully described, both through formulas and numerical examples. These examples are not included solely for the purpose of illustrating the arithmetic manipulations. They have been selected with a view toward providing motivation for the investigator to learn statistical techniques in general and nonparametric techniques in particular, and also to recognize both the power and the shortcomings of statistical analysis whatever the subject matter of his field of specialization.

## 8   Suggestions for Further Reading

Some elementary books on general (mostly parametric) statistics are Anderson and Sclove (1974), Chou (1969), Dixon and Massey (1969), Freund (1974), Walker and Lev (1953), and Walker and Lev (1969). Hays (1973) and Hardyck and Petrinovich (1969) give special attention to applications in the behavioral sciences. Freund and Williams (1966) is a dictionary of statistical terms.

Some current books that cover only nonparametric statistical methods are Conover (1971) and Hollander and Wolfe (1973). Gibbons (1971) covers more of the theoretical aspects of nonparametric statistics.

Some books that give a detailed study of sample surveys are Cochran (1963); Deming (1960); Hansen, Hurwitz, and Madow (1953); Kish (1965); Mendenhall, Ott, and Scheaffer (1971); Raj (1972); and Yamane (1967). Very elementary and brief presentations of sampling techniques are given in Chapter 3 of Edgington (1969) and Chapter 11 of Chou (1969). Discussions of experimental design as applied to specific areas of research are given in Chapin (1960), Davies (1956), Edwards (1968), Kerlinger (1973), Kirk (1968) and Lindquist (1953).

The primary reference for a discussion of measurement scales is Stevens (1946). Some other pertinent papers are Stevens (1951), Gaito (1960), Anderson (1961), and Guilford (1954, pp. 8–18).

# 2

## Goodness-of-Fit Tests

## 1  Introduction

An important problem in statistical analysis relates to obtaining information about the form or shape of the basic population from which a random sample is obtained. If some particular probability distribution is postulated for the population, a goodness-of-fit test can be performed to compare this theoretical distribution with the distribution of the observed sample data.

In many kinds of quantitative studies a probability distribution is needed as a model to represent some phenomenon that can be investigated empirically. For example, queuing models are used to analyze waiting lines and thereby determine an appropriate service or output facility (for example, the number of tables in a restaurant or the number of telephone lines). Any queuing model depends on the assumption of a particular arrival or input process, frequently taken to be a *Poisson process*. In such a process the events occur independently, with a constant mean rate, and rarely occur more than once in a small interval of time or space. The adequacy of any input distribution model must be substantiated before a useful and valid queuing model can be developed. Similarly, the binomial distribution is frequently used as a model for the number of occurrences of one of two mutually exclusive types of events (dichotomous outcomes). The adequacy of this model depends on the assumption of a *Bernoulli process*, that is, a sequence of independent trials, each dichotomous, where the probability of each outcome is constant from trial to trial. In practical applications (for example, parts taken off a production line and classified as defective or satisfactory, or subjects given some drug or treatment and classified according to whether the treatment is effective or not), the process seldom meets exactly the stringent requirements of a Bernoulli process. The binomial distribution may be a satisfactory probability model anyway, but this property should be verified before conclusions based on this model are considered to be useful and valid. Another reason for needing to identify a probability distribution arises when one wants to use typical parametric test and confidence interval procedures. The absolute validity of such methods is dependent upon specific assumptions about the distribution of the population sampled, usually that it is the bell-shaped curve called the normal distribution. Before such an inference procedure is used, a preliminary study may be made to determine whether the population distribution is such that an exact (or almost exact) inference is possible with a parametric technique. Any results or conclusions based on a method that assumes a particular probability distribution may be more subject to error when that assumption is violated.

Thus, goodness-of-fit tests are useful in quantitative analysis for developing models that adequately describe empirical phenomena, and also for making preliminary investigations to justify the use of various statistical techniques of analysis. The null hypothesis in a goodness-of-fit test is a statement about the identity of a probability distribution that fits a set of data. Ideally, the hypothesized distribution is completely specified, including all parameter values. If the hypothesis states only some general family of probability distributions, the unknown parameters must be estimated from the data before these tests can be performed. In either case the alternative is necessarily quite broad. As a result, rejection of the null hypothesis gives no information about what the population form is, only what it is not. Goodness-of-fit tests then are primarily useful when the investigator wants to test a theory about the distribution, or when he has a strong suspicion about the form of the population and hopes to obtain statistical support for his suspicion by accepting the null hypothesis that a particular distribution fits the data at hand.

In this chapter we discuss the two best-known statistical tests for goodness of fit. These are called the chi-square test and the Kolmogorov-Smirnov test. Both tests are based on a comparison between the distribution of the observed data and the theoretical distribution that is hypothesized; however, the two tests use a different basis for comparison. In discussing each test, we consider first the situation in which the null hypothesis completely specifies the theoretical distribution, and then the case in which only a family of probability functions is specified.

# 2    The Chi-Square Goodness-of-Fit Test

We introduce the chi-square goodness-of-fit test by considering the following specific problem.

**example   2.1**

A bank frequently makes large installment loans to builders. At any point in time, outstanding loans are classified in the following four repayment categories:

> A:   Current
> B:   Up to 30 days delinquent
> C:   30 to 60 days delinquent
> D:   Over 60 days delinquent.

The bank has established the internal standard that these loans are "in control" as long as the percentage in each category is as follows:

> A:  80%    B:  12%    C:  7%    D:  1%.

They make frequent spot checks by drawing a random sample of loan files, noting their repayment status at that time, and comparing the observed distribution with the standard for control.

Suppose that a sample of 500 files produces the following data on number of loans in each repayment category:

> A:  380    B:  69    C:  43    D:  8.

Does it appear that installment loan operations are under control at this time?

**SOLUTION**

The hypothesis set is

> $H$:   Outstanding loans conform to the standard ratio 80:12:7:1
> for categories A:B:C:D, respectively
> $A$:   Outstanding loans are in some other ratio.

Thus the null hypothesis specifies that outstanding loans have the proportional distribution

> A:  .80    B:  .12    C:  .07    D:  .01,

and that among 500 loans the respective *numbers* in each category are expected to be

> A:   .80(500) = 400    B:   .12(500) = 60
> C:   .07(500) = 35    D:   .01(500) = 5.

These expected values, denoted by $e$, along with the corresponding observed numbers, denoted by $f$, are shown below.

| Category | Observed ($f$) | Expected ($e$) |
|----------|----------------|----------------|
| A | 380 | 400 |
| B | 69 | 60 |
| C | 43 | 35 |
| D | 8 | 5 |
|   | 500 | 500 |

Some observed frequencies are smaller than those expected under $H$, while some are larger. The question we wish to answer is whether the differences between corresponding frequencies, when considered as a group, are small enough to have occurred by chance. If so, the data support the conclusion that current loans outstanding are in control.

We compare the two sets of frequencies, $f$ and $e$, by calculating the differences and squaring them so that no differentiation is made according to whether too few or too many loans are in any category. These numbers are shown in the fourth and fifth columns of Table 2.1, the computation table. Notice that the sum of the differences in the fourth column is equal to zero, as it must be since the second and third columns each sum to 500. This provides an internal check on the arithmetic of computing the differences $f - e$. In the final column of the table, we obtain $(f - e)^2/e$; that is, we divide each squared difference by the respective expected number of loans. This computation reduces to original units the scale used to measure goodness of fit.

**TABLE 2.1**

| Category | $f$ | $e$ | $f - e$ | $(f - e)^2$ | $(f - e)^2/e$ |
|----------|-----|-----|---------|-------------|---------------|
| A | 380 | 400 | $-20$ | 400 | 1.00 |
| B | 69 | 60 | 9 | 81 | 1.35 |
| C | 43 | 35 | 8 | 64 | 1.83 |
| D | 8 | 5 | 3 | 9 | 1.80 |
|   |     |     |         |             | 5.98 |

The overall lack of fit, measured in original units, then is represented by the sum of the numbers in the last column, in this case 5.98. This is the value of the chi-square goodness-of-fit test statistic. We will use it to apply a formal test and reach a conclusion in this example after explaining the general procedure for statistical analysis based on the chi-square goodness-of-fit test statistic.

*Inference Situation*    The hypothesis set to be tested is

*H:*    The probability function of the observed random variable is $F_o(x)$
*A:*    The probability function of the observed random variable is
         different from $F_o(x)$.

Notice that the alternative does not specify the way in which the true distribution differs from the theoretical (hypothesized) distribution.

When the hypothesized distribution is discrete, with $r$ possible simple outcomes, these outcomes can be used to designate $r$ fixed categories, which are mutually exclusive and exhaustive. For any category, say the $i$th, there is a certain probability $p_i$ that an observation from this distribution is classified in category $i$. Since all observations must be in some category and the categories do not overlap, these $r$ probabilities sum to 1; that is,

$$p_1 + p_2 + \cdots + p_r = 1.$$

The hypothesis set above can then be written in terms of the theoretical values of these probabilities as follows:

*H:*   $p_i = p_{io}$     for all $i = 1, 2, \ldots, r$
*A:*   $p_i \neq p_{io}$     for some $i = 1, 2, \ldots, r$.

The values of $p_{io}$ (theoretical probabilities) are given numbers that sum to 1 and are each between 0 and 1.

*Sampling Situation*    The data consist of $N$ independent observations, which may be either (a) categorical data, classified into some $r$ categories, $r \geq 2$, that do not overlap and cover all classification possibilities, or (b) quantitative data of any sort that are (or can be) grouped into $r$ nonoverlapping numerical categories. (These categories may arise naturally due to the nature of the variable, or they may be arbitrarily chosen by the investigator.)

In either case, the data to be analyzed are in the form of count data, where the counts represent the numbers of observations classified in each of the $r$ categories. We denote these respective counts by $f_1, f_2, \ldots, f_r$; that is, $f_i$ is the number of sample observations classified in the category labeled $i$, or the observed frequency for the $i$th category.

*Rationale*    Denote by $e_i$ the *number* of sample observations that would be expected to be classified in category $i$ if the null hypothesis is true. These $e_i$ are usually called the expected frequencies. They are calculated for the data from the formula

$$e_i = Np_{io}, \tag{2.1}$$

where $p_{io}$ is the probability that a random observation in the population is classified in category $i$ if the null hypothesis is true. If the hypothesis states the values for these $p_{io}$, the calculation of $e_i$ is trivial. If the hypothesis does not specify the $p_{io}$, they must be calculated from the distribution $F_o(x)$ or read from a table of the distribution $F_o(x)$. In some cases, $F_o(x)$ may be specified as a family of distributions in such a way that all population parameters required to find the $p_{io}$ are not given. Then these parameters must be estimated from the sample data, and used to estimate the $p_{io}$. This situation will be discussed in more detail later in this section. Regardless of how the $p_{io}$ are found, the $e_i$ are obtained using Eq. (2.1).

The best evidence we have about the population distribution is the sample distribution, which is reflected by the numbers $f_1, f_2, \ldots, f_r$. If the sample data "fit" the theoretical distribution (that is, if the null hypothesis is true), there should be close agreement between the observed and expected frequencies, $f_i$ and $e_i$, for all categories. The decision regarding fit can be based on the magnitudes of the $r$ deviations, $f_i - e_i$. These differences sum to zero except for rounding, since the sum of each set of frequencies is $N$

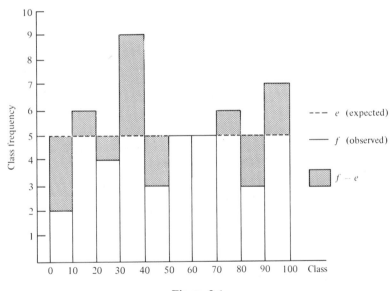

**Figure 2.1**

Figure 2.1 provides a visual interpretation of those numbers, which are compared for a particular problem (Example 2.3). The hypothesized distribution in this figure is the discrete uniform distribution for $r$ categories. This distribution is called uniform because the probabilities, and hence also the expected frequencies, are identical for each category. With $r$ categories, each probability is $1/r$ and each expected frequency is then $N/r$. The particular case represented is where $N = 50$ and $r = 10$, and the category names are the intervals between the numbers 00, 10, 20, . . . , 100. The frequencies observed in the sample of Example 2.3 are plotted on the same graph. The deviations $f_i - e_i$ are depicted as the shaded portions; that is, they are the differences between heights of corresponding rectangles in the two superimposed histograms. A deviation is negative when $e_i$ is above $f_i$.

For a measure of disagreement between the observed and hypothesized population distributions, equal weight should be given to positive and negative differences. This can be accomplished by squaring the differences, and the scale can be reduced to original units by dividing the respective squares by $e_i$. For the $i$th category then, $(f_i - e_i)^2/e_i$ is a measure of the "goodness of fit." Note that division by $e_i$ has the effect of weighting the squared deviation inversely according to the magnitude of the $e_i$. Thus, a departure from expectation has relatively more weight if the expected frequency is small. An overall measure, which considers the agreement for all categories simultaneously, is provided by the sum of these individual measures over all categories. The sum is called the chi-square goodness-of-fit test statistic. A small value supports $H$, while a large value reflects an incompatibility between the observed and expected frequencies.

*Test Statistic*   Using the notation introduced above, the chi-square goodness-of-fit test statistic is defined as

$$Q = \sum_{i=1}^{r} \frac{(f_i - e_i)^2}{e_i} = \frac{(f_1 - e_1)^2}{e_1} + \cdots + \frac{(f_r - e_r)^2}{e_r}. \qquad (2.2)$$

In order to perform the arithmetic needed to calculate $Q$ from this formula, the tabular presentation used in Table 2.1 is recommended. The second and third columns in Table 2.1 are labeled $f$ (observed) and $e$ (expected); the other column labels then indicate the arithmetic performed. The subscripts are dropped for convenience; that is, $f$ is the generic label for $f_1, f_2, f_3, f_4$.

Equation (2.2) can be simplified for calculation to

$$Q = \sum_{i=1}^{r} \frac{f_i^2}{e_i} - N = \frac{f_1^2}{e_1} + \frac{f_2^2}{e_2} + \cdots + \frac{f_r^2}{e_r} - N, \qquad (2.3)$$

but this form does not clearly indicate which categories make the major contributions to the lack of agreement, and hence is not generally recommended. When using (2.3), the columns needed in the computation table besides $f$ and $e$ are simply $f^2$ and $f^2/e$. (A brief review of the sigma notation $\Sigma$ is given in the Appendix.)

*Finding the P-value*   The exact sampling distribution of the test statistic $Q$ defined in (2.2) or (2.3) is quite complicated, but for large samples it may be approximated by the chi-square distribution with $r - 1$ degrees of freedom. We use the symbol "df" to denote degrees of freedom.[1] In Example 2.1 there

---

[1] The df number may be interpreted as the parameter of the chi-square distribution.

are four categories, so $r = 4$ and the correct value of df is df $= 4 - 1 = 3$. The chi-square distribution is given in Table B. Any table entry on a particular df row is the value of that chi-square variable such that the right-tail probability is the number given at the top of that column.

The calculated value of $Q$ is itself a measure of the goodness of fit, since it measures the agreement (or lack thereof) between the hypothesized distribution and the true distribution as reflected by the sample data. In particular, when the data support the hypothesized distribution, most of the $f_i$ will be close to the corresponding $e_i$ and hence $Q$ will be small. On the other hand, a large value of $Q$ indicates a lack of agreement between the $f_i$ and $e_i$; therefore the appropriate $P$-value for lack of agreement is the probability that a chi-square variable is greater than or equal to the observed value of $Q$, that is, a right-tail probability. However, since $Q$ is only approximately chi-square distributed, this $P$-value is always asymptotic approximate.

**example   2.1 (continued)**
Returning to Example 2.1, we can now find the $P$-value and reach a decision about the null hypothesis. The asymptotic approximate $P$-value is a right tail probability from Table B with 3 degrees of freedom. On the line for df $= 3$, we find that 5.98 lies between the two table entries 4.64 and 6.25, which correspond to right-tail probabilities of .20 and .10, respectively. Hence the $P$-value lies between these two probabilities, and can be given as the interval $.10 < P < .20$. This is interpreted as saying that the probability of observing a measure of overall lack of fit as large as 5.98 through sampling variation alone is between .10 and .20 when $F_o(x)$ is the true distribution. With a $P$-value this large, we can justify attributing the lack of fit to chance and conclude that these sample data do support the stated null hypothesis. Hence the installment loans appear to be under control at this time.

The reader should be warned that the statement $.10 < P < .20$ is not to be interpreted as a confidence interval on $P$. A precise value of $P$ could be found from more extensive tables of the chi-square distribution. In general, the precision in stating $P$ is a function only of the completeness of the table employed. A different table would give an interval that always overlaps but may differ in either or both endpoints. The decision will generally not differ, however.

*Additional Notes on Procedure*   As mentioned above, the sampling distribution of $Q$ is only approximately the chi-square distribution. In most applications $N$ is large, say at least 30, so this approximation can be considered satisfactory. If $N$ is not large, the approximation can usually be improved by certain procedural refinements, which we now discuss.

The chi-square approximation improves as the sizes of the expected frequencies increase and can certainly be used with confidence whenever each $e_i$ is at least equal to 5. A common rule of thumb is that the approximation is reliable as long as no more than 20 percent of the $e_i$ are less than 5 and no $e_i$ is less than 1. If either of these conditions is not satisfied, the customary procedure is to combine adjacent categories until this rule is met. The number of degrees of freedom then must be reduced so that $r$ represents the *actual* number of categories used in the analysis. (The null hypothesis may need a corresponding restatement since the categories are changed.) Although this "rule of five" is a useful guideline, it should not be considered inflexible. Some statisticians consider it too conservative, since the chi-square approximation is often reliable even for expected frequencies as small as 1.5, as long as all the $e_i$ are of similar size and $N$ is large in relation to $r$. The reader should be warned that pooling categories is an arbitrary practice, which may markedly affect the inference and its interpretation. A poor approximation to the true sampling distribution may be preferable to a questionable interpretation of results.

When the hypothesized distribution is discrete, the categories are ordinarily the designations for the types of outcomes recorded for the sample data, although they may be combinations thereof. For example, if the variable observed is the number of spots on a die, the categories might be the numbers 1, 2, . . . , 6, or just even numbers and odd numbers. However, when the data reflect quantitative measurements, the investigator may receive the data in an ungrouped form. These original data must be grouped into numerical interval categories *before* the chi-square goodness-of-fit test can be performed, and the investigator usually may choose these categories at will. The recommendation here is to choose the bounds on the $r$ intervals such that the expected frequencies are the same (or nearly so) for all categories; that is, $e_i = N/r$ for $i = 1, 2, \ldots, r$. In many cases this procedure improves the approximation as well as the power of the chi-square test. It also eliminates the arbitrariness of grouping. Further, the calculation of $Q$ is simplified when $e_i = N/r$ for each $i$, since (2.3) then reduces to

$$Q = \left(r\sum_{i=1}^{r} f_i^2/N\right) - N = \frac{r}{N}(f_1^2 + \cdots + f_r^2) - N. \qquad (2.4)$$

Equation (2.4) is particularly useful when the hypothesized distribution is uniform over $r$ specified categories, since then $p_{io} = 1/r$, and thus $e_i = N/r$, for each $i$. If the hypothesized distribution is uniform over a fixed interval of length $L$, choosing the categories as subintervals of length $L/r$ makes all the expected probabilities equal to $1/r$, and each expected frequency is again $e_i = N/r$. In the case of other hypothesized distributions, however, the simplification aspect is admittedly misleading, especially for interval scale data that are ungrouped. Then the values of the endpoints of each category, which is a numerical interval, may have to be calculated in order to ensure equal expected frequencies. When the investigator chooses the categories arbitrarily, the endpoints need no calculation, but then the $e_i$ may not be equal. The shortcut formula in (2.4) is not used in the examples here so that the reader will not be misled into applying it improperly. It will be illustrated in a footnote, however.

**example   2.2**
Consider again the situation described in Example 2.1, but suppose that a sample of only 200 is taken. Then the expected frequencies in categories A, B, C, and D, respectively, are 160, 24, 14, and 2. Since 25 percent of the $e_i$ are less than 5, category D might be combined with some other category. In this case, the logical combination is with C, and then the null hypothesis should be modified accordingly as follows:

*H:* Outstanding loans conform to the ratio 80:12:8 for categories A:B:(C and D), respectively.

For the observed data, A: 136, B: 38, C: 21, and D: 5, the steps in the calculation of $Q$ from (2.3) are shown in Table 2.2.

**TABLE 2.2**

| Category | $f$ | $p$ | $e = Np$ | $f^2$ | $f^2/e$ |
|----------|-----|-----|----------|-------|---------|
| A | 136 | .80 | 160 | 18496 | 115.6 |
| B | 38 | .12 | 24 | 1444 | 60.2 |
| C and D | 26 | .08 | 16 | 676 | 42.2 |
|  | 200 |  | 200 |  | 218.0 |

$$Q = 218.0 - 200 = 18.0 \qquad df = 3 - 1 = 2$$

The asymptotic approximate *P*-value from Table B is $P < .001$. Thus the probability of such a sample outcome when *H* is true is very small, and we must conclude that *H* is not true. An examination of the data reveals that too few loans are current; the bank should take steps to correct this problem.

In the next example the chi-square goodness-of-fit test is used to verify the randomness of data that are purportedly randomly generated. As we pointed out in Chapter 1, a random sample is crucial to the validity of almost all statistical inference techniques. How does one select a simple random sample from a population? For a finite population, one of the easiest procedures is based on the use of a table of random numbers or random digits. These tables consist of many pages with listings of integers from 0 through 9. A page of such a table is reproduced in the Appendix as Table O so that the reader will be familiar with its appearance.

Suppose that a population consists of $V$ elements, and a random sample of $N$ elements, $N < V$, is to be chosen. Then the procedure is to assign a number, from 0 to $V - 1$, to each element in the population. The investigator selects an arbitrary starting point in a table of random numbers, and reads to the right, or left, or up, or down, until he obtains $N$ different numbers or groups of numbers. If $V \leq 10$, he reads only the digits 0 through 9. If $11 \leq V \leq 100$, he reads pairs of digits, 00 through 99; if $101 \leq V \leq 1000$, he reads triples of digits, 000 through 999; and so on. The elements in the population whose numbers correspond to the numbers read from the table comprise those to be included in the sample.

Random number tables are also useful in designing experiments that involve randomization. For example, in an experiment designed to compare two or more different types of treatments, the treatment groups should be determined by chance in order to balance out effects of factors not related to the treatments. If we number the elements that comprise the total sample, we can assign the elements randomly to different treatment groups by using a random number table.

Random number tables are constructed so that each digit is a random selection from the set 0, 1, 2, ... , 9, each pair of digits from 00, 01, 02, ... , 99, and so on. That is, a set of any $k$ digits follows the discrete uniform distribution over the possible values 0, 1, ... , $10^k - 1$, and also over equal-width, nonoverlapping intervals of these values.

**example   2.3**

Suppose that 50 pairs of digits are read from a random number table; the following list shows these numbers after rearrangement in increasing order of size:

|    |    |    |    |    |    |    |    |    |    |
|----|----|----|----|----|----|----|----|----|----|
| 03 | 15 | 25 | 33 | 39 | 52 | 61 | 71 | 81 | 93 |
| 07 | 16 | 29 | 34 | 43 | 54 | 63 | 73 | 85 | 95 |
| 10 | 18 | 30 | 35 | 46 | 55 | 66 | 74 | 88 | 96 |
| 12 | 21 | 31 | 36 | 48 | 59 | 68 | 76 | 90 | 97 |
| 14 | 23 | 31 | 38 | 51 | 60 | 70 | 77 | 92 | 99. |

Verify that these two-digit numbers are truly random.

**SOLUTION**

If the pairs of digits are random numbers, the observed frequencies in 10 equal decade classes covering the range 00 to 99 should be compatible with a discrete uniform distribution over these classes. The hypothesis set then is

$H$:   The digit pairs are uniformly distributed among ten intervals evenly spaced between 00 and 99
$A$:   The digit pairs are not uniformly distributed.

Since there are 10 intervals of numerical values that form the categories, we have $r = 10$. Each probability under $H$ is $p_{io} = 1/10$, and each expected frequency is $e_i = 50(1/10) = 5$. The observed data and the necessary steps in the statistical analysis are shown in Table 2.3.[2]

$$\Sigma f^2 = 2^2 + 6^2 + \cdots + 7^2 = 290$$
$$Q = [10(290)/50] - 50 = 8.0.$$

**TABLE 2.3**

| Decade Class | $f$ | $e$ | $(f - e)^2$ | $(f - e)^2/e$ |
|---|---|---|---|---|
| 00–09 | 2 | 5 | 9 | 1.8 |
| 10–19 | 6 | 5 | 1 | .2 |
| 20–29 | 4 | 5 | 1 | .2 |
| 30–39 | 9 | 5 | 16 | 3.2 |
| 40–49 | 3 | 5 | 4 | .8 |
| 50–59 | 5 | 5 | 0 | 0.0 |
| 60–69 | 5 | 5 | 0 | 0.0 |
| 70–79 | 6 | 5 | 1 | .2 |
| 80–89 | 3 | 5 | 4 | .8 |
| 90–99 | 7 | 5 | 4 | .8 |
| | 50 | 50 | | 8.0 |

$$Q = 8.0 \qquad df = 10 - 1 = 9$$

The asymptotic approximate $P$-value from Table B is a little larger than .50. The null hypothesis that the digit pairs are uniform is supported by these data.

It should be noted that the selection of 10 decade classes for investigation was arbitrary. The general null hypothesis is randomness, and this could be adequately tested using some other categories, such as odd and even numbers for $r = 2$, or intervals 00–24, 25–49, 50–74, 75–99, with each expected frequency $50/4 = 12.5$ and $r = 4$, or even with intervals of unequal width. The particular null hypothesis statement must be changed to correspond to the categories chosen, but the choice is entirely the prerogative of the investigator. With other categories, the value of $Q$ could be different, even to the extent of changing the conclusion. Another point worth mentioning is that while it is true that random numbers *imply* the uniform distribution as stated in $H$, accepting $H$ does *not necessarily imply* that the numbers were generated by a random process.

[2]Since the expected frequencies are all equal here, the formula for $Q$ in (2.4) is applicable. The calculations are as follows:

In the next example the distribution hypothesized is the binomial. Hence we first review briefly this probability distribution. Suppose an experiment consists of a series of $n$ trials. On each trial there can be exactly two different kinds of outcomes (for example, success or failure of a treatment, a good or defective part). If the probabilities of these outcomes remain constant for all trials and the trials are independent (that is, we have a Bernoulli process), then the probability of exactly $x$ occurrences of one type of outcome is given by

$$f(x) = \binom{n}{x} \theta^x (1 - \theta)^x \qquad x = 0, 1, \ldots, n,$$

where $\theta$ is the probability of that type of outcome on any one trial and $\binom{n}{x}$ is the so-called binomial coefficient, calculated for any $x = 0, 1, \ldots, n$ as

$$\binom{n}{x} = \frac{n(n - 1)\cdots(n - x + 1)}{x(x - 1)\cdots(2)(1)}.$$

The probability function $f(x)$ is called the binomial distribution for $n$ trials with parameter $\theta$. Table E in the Appendix gives the complete cumulative distribution function of the binomial for selected $\theta$ and $n \le 20$, and Table F gives both left-tail and right-tail binomial probabilities for $\theta = .5$ and $n \le 20$.

**example 2.4**
  A manufacturer of stainless steel and leather chairs wants to make a new style, which is heavy and sturdy and yet gives the appearance of being light and graceful. A designer creates two different styles of approximately the same weight; one style has basically curved lines and the other has basically straight lines. The manufacturer likes both styles. He wants to make the one that appears to be lighter in weight. The designer feels that straight lines create the illusion of bulkiness more than do curved lines. An industrial psychologist designs the following simple experiment to investigate this research hypothesis. Four pairs of pieces of stainless steel are made. Each piece has exactly the same weight, but the shapes are different. Within each pair, one piece has basically straight lines (as a cube or rectangular box) and the other piece is basically curved (as a ball or egg). A subject is presented with these four pairs and told that the weights within each pair are different. He is asked to indicate which one he perceives as lighter in weight. The null hypothesis is that there is no difference in visual perception.
  We consider each pair a "trial," and define $x$ as the "score" of the curved shape — that is, the number of times the curved shape is designated as the lighter in weight. If there is no difference in visual perception, the subject is as likely to choose the curved piece as the straight piece, and the variable $x$ follows the binomial distribution for $n = 4$ with parameter $\theta = .5$. Then its cumulative distribution function $F(x)$ is given in Table E.

Let $p(x)$ denote the probability of exactly $x$ designations of the curved piece, that is, the "point" probability function. Since $x$ can be only integer-valued, for any $x$ we find $p(x)$ from $p(x) = F(x) - F(x - 1)$, where $F(x)$ is the table entry for $x$ when $n = 4$, $\theta = .5$. For example, $p(3) = F(3) - F(2) = .9375 - .6875 = .2500$. The complete "point" probability function is shown in the last column of Table 2.4 in both decimal and fraction form.

**TABLE 2.4**

| Score of Curved Piece $x$ | Cumulative Probability $F(x)$ | "Point" Probability $p(x)$ |
|---|---|---|
| 0 | .0625 | $.0625 = 1/16$ |
| 1 | .3125 | $.2500 = 4/16$ |
| 2 | .6875 | $.3750 = 6/16$ |
| 3 | .9375 | $.2500 = 4/16$ |
| 4 | 1.0000 | $.0625 = 1/16$ |

If this experiment were carried out with 16 randomly chosen subjects' under the null hypothesis we expect one score of 0, four scores of 1, six scores of 2, and so on. The probabilities $p(x)$ can be used in a similar way for any number of repetitions; that is, the expected number of times that the score is $x$ in $N$ repetitions is $Np(x)$. However, if the curved shape is perceived as lighter in weight, it will be chosen more frequently than expected and the binomial model for $n = 4$ trials with $\theta = .5$ will not fit the data for $N$ repetitions.

Suppose that the experiment is repeated 64 times and the observed frequency of scores of the curved piece is as shown in Table 2.5. Test the adequacy of the binomial model with $\theta = .5$, $n = 4$.

**TABLE 2.5**

| Score of Curved Piece $x$ | Observed Frequency $f$ |
|---|---|
| 0 | 1 |
| 1 | 12 |
| 2 | 24 |
| 3 | 18 |
| 4 | 9 |
| | 64 |

**SOLUTION**

The expected frequencies are easily calculated as $e = Np = 64p$. These numbers are compared with the corresponding values of $f$ by computing $(f - e)^2/e$, and $Q$ is calculated from (2.2) as usual; see Table 2.6.

**TABLE 2.6**

| $x$ | $f$ | $e$ | $(f - e)^2$ | $(f - e)^2/e$ |
|---|---|---|---|---|
| 0 | 1 | 4 | 9 | 2.25 |
| 1 | 12 | 16 | 16 | 1.00 |
| 2 | 24 | 24 | 0 | 0.00 |
| 3 | 18 | 16 | 4 | .25 |
| 4 | 9 | 4 | 25 | 6.25 |
| | | | | 9.75 |

$$Q = 9.75 \qquad \text{df} = 5 - 1 = 4$$

The asymptotic approximate $P$-value from Table B is slightly smaller than .05, which most people would consider small enough to conclude that $H$ should be rejected.

Now what can we say about the question posed by the manufacturer? In the $(f - e)^2/e$ column of Table 2.6, the first and last entries, particularly the last, account for the size of $Q$. Since there are so many more scores of 4 than expected, and correspondingly fewer scores of 0 than expected, it appears that the curved shape is perceived as lighter than the straight-line shape. (Note that this kind of analysis would not be possible if $Q$ had been calculated from (2.3).)

*Estimation of Parameters*    As mentioned earlier, it is sometimes necessary to estimate some parameter values before the $p_{io}$ or $e_i$ can be found. These parameters are usually estimated using the procedure called "maximum likelihood estimation," but discussion of this technique is beyond the scope of this book. For many of the distributions commonly hypothesized, this estimate is the sample analogue of the population parameter. That is, if the parameter is the mean of the population, the sample mean is used as its estimate; if the parameter is the population variance, the sample variance is used. However, since the chi-square goodness-of-fit test must be performed on the data *after* grouping into categories, and the grouping is admittedly arbitrary in many cases, any estimate should be calculated from these grouped data. Otherwise, the procedure used in estimation may bias the test results toward rejection of $H$ and therefore give an anticonservative $P$-value. If the data are sensitive to grouping, the estimate based on the ungrouped data may differ from the same type of estimate based on the data as grouped to such an extent that the expected frequencies computed using these estimates are not representative of the grouped data regardless of the population family. Hence, a rejection may reflect inaccuracy in estimating the parameters, as opposed to a nonacceptable population form. It is therefore recommended that all parameters be estimated from the data after they are grouped into categories.

Most parameters in the commonly hypothesized distributions are either means, functions of means, or variances. The formulas for calculating sample estimates from data grouped into $r$ categories are as follows:

$$\text{Mean:} \quad \bar{x} = \sum_{i=1}^{r} f_i x_i / N \tag{2.5}$$

$$\text{Variance:} \quad s^2 = \sum_{i=1}^{r} f_i (x_i - \bar{x})^2 / N = \left( \sum_{i=1}^{r} f_i x_i^2 - N\bar{x}^2 \right) / N \tag{2.6}$$

$$\text{Standard deviation:} \quad s = \sqrt{s^2}.$$

In these expressions $x_i$ is the midpoint value of that category whose observed frequency is $f_i$, and $r$ is the number of categories. If the parameter is some function of the population mean, the same function of the sample mean is usually the maximum likelihood estimate.

Once the parameters are estimated and used to estimate the $p_{io}$ and $e_i$, $Q$ is calculated as before. The distribution of $Q$ is still approximately the chi-square distribution, but the approximation is improved by decreasing the number of degrees of freedom. The correct value is df $= r - w - 1$, where $w$ is the number of independent parameters in $F_o(x)$ that were unspecified and had to be estimated from the data in order to find the $p_{io}$ or $e_i$.

Some typical hypothesized families of distributions, their parameters and appropriate estimates, the value of $w$, the correction in degrees of freedom, and the corresponding values of df $= r - w - 1$ when these parameters are estimated are listed in Table 2.7. The symbols commonly used to denote the relevant parameters and their maximum likelihood estimates are also given. The distributions named are used so frequently that tables of their probability functions for selected values of the parameters are readily available in most elementary books on classical statistical methods, as well as in most sets of mathematical tables.

The correct procedure to be followed when $H$ does not completely specify the distribution is illustrated for the binomial and Poisson distributions in the following two examples. Since the normal and exponential distributions are continuous, the chi-square test is, strictly speaking, not appropriate. (Application of the chi-square test is frequently justified by considering a continuous population distribution as grouped into a finite number of nonoverlapping class intervals.) The Kolmogorov-Smirnov test of Section 3 is specifically designed for continuous theoretical distributions. Therefore, examples of testing goodness of fit with these other distributions are deferred until that test is presented.

**TABLE 2.7**
**GUIDE FOR ESTIMATION OF PARAMETERS IN GOODNESS-OF-FIT TESTS**

| Hypothesized Distribution | Parameters | Estimates | $w$ | df |
|---|---|---|---|---|
| Binomial with $n$ trials | Probability of success on any trial $\theta$ | Sample proportion of successes $\Sigma xf/n\Sigma f$ | 1 | $r - 2$ |
| Poisson | Mean rate of occurrence $\lambda$ | Sample mean rate $\bar{x}$ | 1 | $r - 2$ |
| Normal | Mean, variance $\mu, \sigma^2$ | Sample mean, variance $\bar{x}, s^2$ | 2 | $r - 3$ |
| Exponential $F(x) = 1 - e^{-\lambda x}$ | Mean time to occurrence $1/\lambda$ | Reciprocal of sample mean $1/\bar{x}$ | 1 | $r - 2$ |

**example 2.5**

In order to determine preference between two different brands of tea, each member of a panel of 30 professional tasters is asked to state a preference in six different taste tests. The order in which the brands are presented is randomized on each trial for each taster in order to eliminate bias in the results. Using the letters A and B to designate the two brands, the stated preferences are given in Table 2.8.

Brand A could have been chosen 0, 1, 2, 3, 4, 5, or 6 times by any one taster. If we summarize these data according to the number of times A is chosen, the result is as shown in Table 2.9.

Test the null hypothesis that the choice between pairs of teas by any given taster is a random variable with probability of preference for A the same on each taste test.

**TABLE 2.8**

| Taster | Taste Test | | | | | | Numer of Times A is Preferred |
|--------|---|---|---|---|---|---|---|
|  | 1 | 2 | 3 | 4 | 5 | 6 |  |
| 1 | A | A | B | A | B | B | 3 |
| 2 | A | A | A | A | B | B | 4 |
| 3 | B | A | B | A | A | B | 3 |
| 4 | B | B | B | B | B | B | 0 |
| 5 | A | B | B | B | B | A | 2 |
| 6 | B | A | B | B | B | B | 1 |
| 7 | A | B | B | B | B | A | 2 |
| 8 | A | A | B | A | B | B | 3 |
| 9 | B | A | A | B | B | B | 2 |
| 10 | B | A | A | B | B | B | 2 |
| 11 | B | B | A | B | B | B | 1 |
| 12 | B | A | A | B | B | B | 2 |
| 13 | A | B | A | B | B | B | 2 |
| 14 | A | B | A | A | B | B | 3 |
| 15 | B | A | B | A | B | B | 2 |
| 16 | A | B | B | B | B | B | 1 |
| 17 | A | A | B | B | B | B | 2 |
| 18 | B | A | B | B | B | A | 2 |
| 19 | A | B | A | B | B | B | 2 |
| 20 | A | B | A | A | B | B | 3 |
| 21 | B | B | B | B | B | B | 0 |
| 22 | B | B | B | A | B | B | 1 |
| 23 | A | A | B | A | A | B | 4 |
| 24 | B | B | A | B | B | B | 1 |
| 25 | B | A | A | A | B | B | 3 |
| 26 | A | B | B | B | B | B | 1 |
| 27 | A | A | B | A | B | B | 3 |
| 28 | A | B | A | A | B | A | 4 |
| 29 | B | A | B | B | B | B | 1 |
| 30 | B | A | A | A | B | B | 3 |
|  |  |  |  |  |  |  | 63 |

**TABLE 2.9**

| Times Chosen | Frequency |
|:---:|:---:|
| 0 | 2 |
| 1 | 7 |
| 2 | 10 |
| 3 | 8 |
| 4 | 3 |
| 5 | 0 |
| 6 | 0 |
|  | 30 |

**SOLUTION**

If the probability of choosing A is the same for each taste test, then the probability distribution of $x$, the number of times A is chosen among the six varieties, is binomial. We formulate this as the null hypothesis. Since the parameter $\theta$ is not specified, it must be estimated from the data. Table 2.7 indicates that the interpretation of $\theta$ is the probability of success, which here means an individual preference for A; the maximum likelihood estimate is the sample proportion of successes, here the relative number of times A is preferred in all repetitions. This proportion is found by dividing the total number of times A is preferred by the total number of paired comparisons in the study, which is six taste tests for each of 30 tasters, or $6(30) = 180$. The estimate of $\theta$ is then $\Sigma xf/6\Sigma f = 63/180 = .35$. The format recommended for finding $\Sigma xf$ and $\Sigma f$ is shown in the first three columns of Table 2.10.

**TABLE 2.10**

| $x$ | $f$ | $xf$ | $p_o$ | $e = Np_o$ | $f - e$ | $(f - e)^2$ | $(f - e)^2/e$ |
|---|---|---|---|---|---|---|---|
| 0 | 2 | 0 | .0754 | 2.26 | −.26 | .0676 | .0299 |
| 1 | ·7 | 7 | .2437 | 7.31 | −.31 | .0961 | .0131 |
| 2 | 10 | 20 | .3280 | 9.84 | .16 | .0256 | .0026 |
| 3 | 8 | 24 | .2355 | 7.06 | .94 | .8836 | .1252 |
| 4 | 3⎫ | 12 | .0951 | 2.85⎤ | | | |
| 5 | 0⎬3 | 0 | .0205 | .62⎬3.52 | −.52 | .2704 | .0768 |
| 6 | 0⎭ | 0 | .0018 | .05⎦ | | | |
| | 30 | 63 | 1.0000 | 29.99 | | | .2476 |

We now need to find the values of $p$, the probabilities for each category — that is, for each of the $x$ number of times $A$ is chosen among the six tests by any taster. Under the null hypothesis, $x$ is binomial with $n = 6$ trials (not to be confused with $N = \Sigma f = 30$, the number of tasters). Thus the entries in the $p_o$ column are found from Table E with $n = 6$, $\theta = .35$. As in Example 2.4, successive cumulative probabilities are subtracted to find the point probabilities $p_o$. The expected frequencies $e$ are obtained by multiplying the entries in the $p_o$ column by $N = 30$. Since the last two entries in the $e$ column are each less than one, the last three categories are combined as shown. We now have only five sets of frequencies to compare. The calculation of $Q$ in the last three columns proceeds exactly as before, and we find that

$$Q = .2476 \quad \text{and} \quad df = r - w - 1 = 5 - 1 - 1 = 3.$$

The right-tail $P$-value from Table B is larger than .95, which implies that the data do support the hypothesis of a binomial distribution with $\theta$ estimated by .35, and thus that $\theta$ is the same for each taste test.

**example   2.6**

President Nixon made two appointments to the Supreme Court in 1971. Although this might appear to be a rather rare event, in fact at least two vacancies occurred in 20 of the years since 1789, when the Court was instituted. The history of appointments to the Court provides an interesting example of a Poisson process. The number of justices on the Court was not constant at the beginning, but varied between 5 and 10 between 1789 and 1869. However, for practical purposes the number has been essentially constant at 9 since 1837 (the only exception is from 1863 to 1865, when the Court had 10 members). Thus we use data starting with that year. The frequencies of numbers of appointments made in a calendar year during the 136-year period, 1837–1972, as reported in *The 1973 World Almanac and Book of Facts* (1973 at p. 795), are given in Table 2.11.

**TABLE 2.11**

| Number of Appointments (1837–1972) | Number of Years $f$ |
|:---:|:---:|
| 0 | 78 |
| 1 | 41 |
| 2 | 15 |
| 3 | 2 |
| 4 or more | 0 |
| | 136 |

**SOLUTION**

The null hypothesis of interest is that the distribution of number of appointments can be regarded as Poisson. The parameter $\lambda$ of the Poisson distribution is not given as part of the hypothesis, and hence must be estimated from the data before the expected probabilities and frequencies can be obtained. Table 2.7 indicates that the appropriate estimate is $\bar{x}$, the mean number of appointments per year. This value is calculated as the sum of the products of the number of years for each number of appointments, divided by the total number of years, or

$$\bar{x} = \frac{0(78) + 1(41) + 2(15) + 3(2)}{136} = .57.$$

Now we need the probability of $x$ occurrences of a Poisson variable with parameter $\lambda = .57$; these are the values of $p_o$ that will be multiplied by $N = 136$ to find the corresponding expected frequencies. In Table 2.12 the entries shown in the column labeled $p_o$ were found from *Tables of Terms of Poisson Distribution*.[3] Less extensive tables of the Poisson distribution are found in most sets of mathematical tables (for example, any edition of the *CRC Standard Mathematical Tables*[4]), but only for selected values of $\lambda$. These other tables could have been used to find $p_o$ here if $\lambda$ had been rounded to .60. Exact values of $p_o$ for any $\lambda$ could also be found easily using most programmable calculators.

Since the expected frequency of four or more appointments is very small, this category is combined with that for three appointments. Then $Q$ is calculated in the usual way from the resulting four categories. Since one

**TABLE 2.12**

| $x$ | $f$ | $p_o$ | $e$ | $(f - e)^2$ | $(f - e)^2/e$ |
|---|---|---|---|---|---|
| 0 | 78 | .5655 | 76.91 | 1.1881 | .01 |
| 1 | 41 | .3223 | 43.83 | 8.0089 | .18 |
| 2 | 15 | .0919 | 12.50 | 6.2500 | .50 |
| 3 | 2 ⎫ | .0175 | 2.38 ⎫ | .5776 | .21 |
| 4 or more | 0 ⎭ 2 | .0028 | .38 ⎭ 2.76 | | |
| | 136 | 1.0000 | 136.00 | | .90 |

parameter $\lambda$ was estimated, and two categories were combined to leave $r = 4$, the degrees of freedom is df $= 4 - 1 - 1 = 2$ for the $Q = .90$. The *P*-value from Table B is larger than .50, so the Poisson distribution appears to fit the data remarkably well. For some interesting and related results, see Kinney (1973) and Wallis (1936).

# 3   Kolmogorov-Smirnov Procedures

### 3.1   The Kolmogorov-Smirnov Goodness-of-Fit Test

Another statistical technique for testing the goodness of fit between a set of sample observations and a theoretical distribution is called the

[3]General Electric Company (1962). *Tables of Terms of the Poisson Distribution*. D. Van Nostrand Company, Inc., New York, p. 75.

[4]Chemical Rubber Company, Cleveland, Ohio.

Kolmogorov-Smirnov test. The procedure is named after two Russian mathematicians, A. N. Kolmogorov and N. V. Smirnov, who were primarily responsible for its development. We introduce the method by means of an example involving the distribution of a function of randomly generated numbers.

If numbers are chosen on a continuum by some random mechanism, the mechanism can be programmed or constructed in such a way that the numbers obey any specified continuous probability model. The simplest continuous probability law to work with, and also one of the most useful, is the continuous uniform distribution. A random variable follows this distribution on an interval of length $S$, or a space of area $S$, if the probability that the variable lies in a subset of size $A$ (interval or space) is $A/S$. (Thus this distribution is an extension of the notion of a finite space of possible values that are equally likely — that is, the discrete uniform distribution.) For example, if a number is chosen at random from the interval $(0, 10)$ according to the uniform distribution, the probability that it is smaller than 4 is $(4 - 0)/(10 - 0) = .4$, or that it is between 2 and 6 is $(6 - 2)/(10 - 0) = .4$. The interval $(0, 1)$ is the simplest to work with since then $S = 1$; for example, the probability that a number from the interval $(0, 1)$ is smaller than 0.4 is .4. But more than that, any interval $(a, b)$ can be translated to the interval $(0, 1)$ by the linear function $X' = (X - a)/(b - a)$; that is, if $X$ is uniformly distributed over $(a, b)$, then $X'$ is uniformly distributed over $(0, 1)$.

One might think that if a number has the uniform probability property, the same property would hold for certain functions of the number, for example, its square root or its square. If that were true for a number between 0 and 1, the probability that its square root is smaller than .4 would be .4 again. However, this is not the case, as statistical theory can show. Such techniques are beyond the scope of this book, but we can investigate such a claim using the Kolmogorov-Smirnov goodness-of-fit test as in Example 3.1.

**example   3.1**

The 20 observations shown in Table 3.1, and denoted by $y$, were chosen randomly according to the uniform distribution over $(0, 1)$, recorded only to four significant figures and then arranged in increasing order of size. In order to investigate the claim that taking a square root does not alter the probability distribution of a random variable, we test the null hypothesis that the square roots of these observations represent a distribution that is uniform over $(0, 1)$.

For each number $y$, we first find its square root, which we call $x$ and list in the second column of Table 3.1, and then find the cumulative relative frequency distribution at $x$, which we call $S_N(x)$. For any number $x$, $S_N(x)$ is defined as the proportion of observations not exceeding $x$. Thus, for any

$x < .111$, the smallest number in the array, $S_N(x) = 0$. For any $x$ that is at least .111 but smaller than the next largest observation, that is, $.111 \leq x < .322$, $S_N(x) = 1/20 = .05$. For any $x$ that is at least .322 but smaller than .442, that is, $.322 \leq x < .442$, $S_N(x) = (1 + 1)/20 = 2/20 = .10$. In this case, no two observations are the same, so the values of $S_N(x)$ for increasing values of $x$ are simply $1/20$, $2/20$, and so on. These values are given in the third column of Table 3.1, but it should be remembered that $S_N(x)$ is defined for all $x$, and each value listed in fact applies to an interval of values of $x$. For example, although the table shows only that $S_N(.830) = .750$, this is to be interpreted as meaning that $S_N(x) = .750$ for all $x$ satisfying $.830 \leq x < .873$. The function $S_N(x)$ appears in Figure 3.1 as a step function.

The hypothesized cumulative distribution $F_o(x)$ in this example is the continuous uniform distribution over $(0, 1)$. From the previous discussion, we know that, under this hypothesis, the probability is $x$ that an observation is less than or equal to $x$, that is, $F_o(x) = x$ for $0 \leq x \leq 1$. $F_o(x)$ then is the 45° line in Figure 3.1.

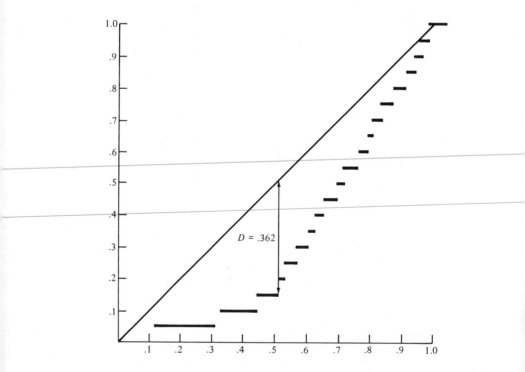

Figure 3.1

**TABLE 3.1\***

| $y$ | $x = \sqrt{y}$ | $S_N(x)$ | $F_o(x)$ |
|---|---|---|---|
| .0123 | .111 | .050 | .111 |
| .1039 | .322 | .100 | .322 |
| .1954 | .442 | .150 | .442 |
| .2621 | .512 | .200 | .512 |
| .2802 | .529 | .250 | .529 |
| .3217 | .567 | .300 | .567 |
| .3645 | .604 | .350 | .604 |
| .3919 | .626 | .400 | .626 |
| .4240 | .651 | .450 | .651 |
| .4814 | .694 | .500 | .694 |
| .5139 | .717 | .550 | .717 |
| .5846 | .765 | .600 | .765 |
| .6275 | .792 | .650 | .792 |
| .6541 | .809 | .700 | .809 |
| .6889 | .830 | .750 | .830 |
| .7621 | .873 | .800 | .873 |
| .8320 | .912 | .850 | .912 |
| .8871 | .942 | .900 | .942 |
| .9249 | .962 | .950 | .962 |
| .9634 | .982 | 1.000 | .982 |

\**H:*   $x = \sqrt{y}$ is uniformly distributed over $(0, 1)$.

Since $F_o(x)$ and $S_N(x)$ are both cumulative relative frequency distributions, their agreement can be used as an indication of the goodness of fit. The Kolmogorov-Smirnov test statistic, written as $D$, measures agreement as the absolute value of the largest vertical difference between the graphs of $F_o(x)$ and $S_N(x)$. (The absolute value is the magnitude, ignoring sign.) Careful inspection of Figure 3.1 shows that the largest absolute difference for these data occurs just to the left of the jump in $S_N(x)$ at $x = .512$ and its value is .362. Thus $D = .362$. If .362 is not too large, we know that none of the absolute differences is too large (since .362 is the largest), and hence the overall fit can be considered satisfactory enough to conclude that $F_o(x)$ is the true distribution. Accordingly, the logical conclusion would be to reject the null hypothesis of fit if $D$ is too large. The sampling distribution of $D$ does not depend on the hypothesized distribution (that is, it is distribution-free), and tables are available. From Table C (as will be explained later) the probability of a $D$ as large as or larger than .362 is less than .01, and therefore it occurs rarely by chance alone. Thus we conclude that the data do not support the null hypothesis of a uniform distribution for the square root of a number that is uniformly distributed.

We now consider the general case for statistical analysis of goodness of fit using the one-sample Kolmogorov-Smirnov test statistic.

*Inference Situation (Two-sided Test)*    The hypothesis set is

   *H:*   The cumulative probability function of the observed random variable is $F_o(x)$

   *A:*   The cumulative probability function of the observed random variable is different from $F_o(x)$.

*Sampling Situation*    The data consist of $N$ independent random observations, which are ordinal data or data measured on at least an interval scale. The data need not be grouped, and in fact usually are not; but if they are, the categories must be numerical intervals.

*Rationale*    For all real numbers $x$, define the quantity

$S_N(x) =$ proportion of sample observations less than or equal to $x$

$$= \frac{\text{number of observations less than or equal to } x}{N}, \qquad (3.1)$$

which is called the *sample* (or *empirical*) *distribution function*. $S_N(x)$ is a non-decreasing function of $x$, which changes value at each *different* observed

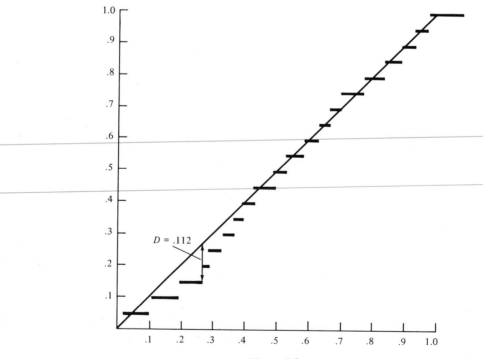

**Figure 3.2**

value of $x$ and is constant in between. Graphically, we can describe $S_N(x)$ as a step function, which takes a jump at each different observed value $x$ and is constant in between jumps (see Figures 3.1 and 3.2, for example). The amount of the jump at any point is the proportional number of observations having that value. (If the only data available are already grouped into numerical interval categories, the only information available about $S_N(x)$ is for $x$ equal to the upper limit of each class interval. Then $S_N(x)$ must be given as the cumulative relative frequency distribution, and the amount of each jump is the frequency of that class. The solid line in Figure 3.3 represents $S_N(x)$ graphically for the grouped data of Example 2.3.)

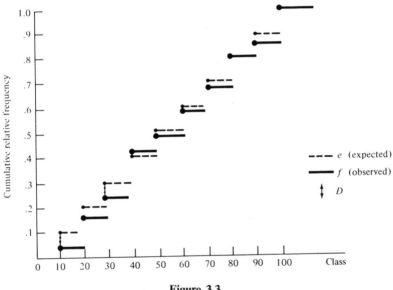

**Figure 3.3**

If $F_o(x)$ is the true cumulative probability function, there should be reasonable agreement between $S_N(x)$ and $F_o(x)$ for all values of $x$, since $S_N(x)$ is the sample image of the true distribution. Equivalently, the deviations (absolute differences) between $S_N(x)$ and $F_o(x)$ should be small for all $x$. The Kolmogorov-Smirnov test statistic is the largest deviation; if the largest deviation is small, it follows that all deviations are small.

It might appear that the largest deviation would be the maximum of the differences $|S_N(x) - F_o(x)|$ for all observed values of $x$. However, this maximum is not necessarily the largest one. Rather, the largest is found by looking at the differences $|S_N(x) - F_o(x)|$ in the *neighborhood* of each observed value of $x$. For example, in Table 3.1 the largest absolute difference between corresponding entries in the last two columns occurs in the fourth row for

$x = .512$ as $F_o(.512) - S_N(.512) = .512 - .200 = .312$. However, from Figure 3.1 it is clear that the largest deviation exceeds .312 by quite a bit. This apparent anomaly follows from the fact that $S_N(x) = .200$ for $x$ equal to or a bit larger than .512, but $S_N(x)$ drops down to .150 for $x$ slightly smaller than .512, while $F_o(x)$ remains very close to .512 on either side of $x = .512$. In summary, the absolute differences between $S_N(x)$ and $F_o(x)$ in the neighborhood of .512 are

$$.512 - .150 = .362 \text{ just to the left of } x = .512$$
$$.512 - .200 = .312 \text{ at or just to the right of } x = .512.$$

Thus the correct value of the test statistic is the larger of these two differences, or .362. This largest difference is called the *supremum*, or sometimes the *least upper bound*, of the differences.

*Test Statistic*    The Kolmogorov-Smirnov goodness-of-fit test statistic (sometimes called simply the Kolmogorov test) is defined as

$$D = \underset{x}{\text{supremum}} |S_N(x) - F_o(x)|, \tag{3.2}$$

where $|c|$ denotes the absolute value of $c$, and the supremum over $x$ means the largest of the differences $|S_N(x) - F_o(x)|$ in the neighborhood of each observed value of $x$.

The value of $D$ can be determined graphically by plotting $S_N(x)$ and $F_o(x)$ on the same graph. The graph of any step function is completely determined once the values of the function are known at the jump points. Thus the graph of $S_N(x)$ is easily found by cumulating the observed frequencies and dividing by $N$. If the hypothesized cumulative distribution function $F_o(x)$ is continuous, its graph is a smooth curve and is plotted just like any mathematical function. As explained above and illustrated in Figure 3.1 for Example 3.1, the differences between $S_N(x)$ and $F_o(x)$ must be inspected on both sides of any jump point in $S_N(x)$ to find the supremum.

This same procedure can be followed to determine $D$ graphically when $F_o(x)$ is a discrete c.d.f. Then $F_o(x)$ is also a step function, and its values at the jump points can be obtained by cumulating the hypothesized probabilities $p_{io}$ for successive outcomes. For instance, in Example 2.3 we tested the null hypothesis that digit pairs of random numbers follow the discrete uniform distribution for each decade class; that is, $p_{io} = 1/10$ for each of the 10 classes. Then the value of $F_o(x)$ for $x$ representing the upper limit of the class 10–19, for instance, is $1/10 + 1/10 = .20$; the corresponding observed cumulative relative frequency is $(2 + 6)/50 = .16$. These distributions are both plotted in Figure 3.3, with each class represented by its upper limit. As shown, the value of $D$ is .06, which occurs for both the classes 00–09 and 20–29. (Note the difference between the graph of the cumulative discrete uniform here and the dashed line in Figure 2.1, which reflects the uncumulated expected frequencies for each class individually.) In general, since a jump in $S_N(x)$ can occur only when a jump occurs in $F_o(x)$, the deviations between $S_N(x)$ and $F_o(x)$ are equal to the differences between the heights of the steps and need be inspected only at the jump points in $F_o(x)$.

The value of $D$ can also be found algebraically, although the graphic procedure is usually easier, especially when $F_o(x)$ is continuous. If $F_o(x)$ is discrete, it is sufficient to list the possible values of $x$, the $S_N(x)$ and $F_o(x)$ for those $x$, and compute the corresponding differences $|S_N(x) - F_o(x)|$. Since the supremum must occur at a jump point and these $x$ are the jump points, the largest absolute difference computed is the value of $D$.

However, if $F_o(x)$ is continuous, the supremum may occur either at a jump point in $S_N(x)$ or slightly to the left of it, and hence it is not sufficient to look at $|S_N(x) - F_o(x)|$ for only the observed values of $x$. Suppose there are $r$ different values of $x$ for some $r \leq N$, and the arrayed values are $x_1 < x_2 < \cdots < x_r$, each observed with frequency at least 1. We know that $S_N(x)$ is constant for all intervals $x_{i-1} \leq x < x_i$, while $F_o(x)$ is nondecreasing for all $x$. Further, for all practical purposes $F_o(x)$ is the same whether $x$ equals, is slightly smaller than, or is slightly larger than, any $x_i$, while $S_N(x)$ takes a jump at each $x_i$. In the interval $x_{i-1} \leq x < x_i$: (a) if $S_N(x)$ is smaller than $F_o(x)$, the largest difference occurs at the right endpoint, that is, at a point slightly smaller than $x_i$, and equals $F_o(x_i) - S_N(x_{i-1})$; (b) if $S_N(x)$ is larger than $F_o(x)$, the largest difference occurs at the left endpoint as $S_N(x_{i-1}) - F_o(x_{i-1})$; and (c) if $S_N(x)$ intersects $F_o(x)$, the largest absolute difference may occur at either endpoint. Thus, in general, it is necessary to look at both these absolute differences for $i = 1, 2, \ldots, r + 1$, where $x_o = -\infty$ and $x_{r+1} = \infty$. However, this is the same as looking at the $2r$ absolute differences $|S_N(x_i) - F_o(x_i)|$ and $|S_N(x_{i-1}) - F_o(x_i)|$ for $i = 1, 2, \ldots, r$, where $S_N(x_o) = 0$. In other words, $D$ is the maximum over all observations of the larger of the two possible deviations for each different observation. Equivalently, we define

$$D = \underset{1 \leq i \leq r}{\text{maximum}} \{\text{maximum} \, [|S_N(x_i) - F_o(x_i)|, |S_N(x_{i-1}) - F_o(x_i)|]\}. \quad (3.3)$$

If all observations are different, so that $r = N$, we have $S_N(x_k) = k/N$ for $1 \leq k \leq N$, and (3.3) reduces to

$$D = \max_{1 \leq i \leq N} \left\{ \max \left[ \left| \frac{i}{N} - F_o(x_i) \right|, \left| \frac{i-1}{N} - F_o(x_i) \right| \right] \right\}. \quad (3.4)$$

It should be noted that the definitions of $D$ in (3.3) and (3.4) are needed only when $F_o(x)$ is continuous. However, this situation arises frequently, since the test based on the $D$ statistic is exact only when $F_o(x)$ is continuous.

*Finding the P-value*    The sampling distribution of the $D$ test statistic is known exactly if the hypothesized probability distribution $F_o(x)$ is continuous and no parameters need be estimated. If $F_o(x)$ is discrete and/or one or more parameters are estimated, the same sampling distribution can be used, but then the test is approximate and $P$-values based on this distribution are conservative. Since the calculated value of $D$ is a measure of the agreement between the hypothesized distribution and the true distribution as reflected by the sample data, a large value of $D$ tends to discredit $H$. Hence the appropriate $P$-value is a right tail probability from the sampling distribution of $D$.

The exact quantile values of the $D$ statistic for some selected right-tail probabilities are given in Table C for $N \leq 40$, under the heading "two-sided tests." For larger values of $N$, the asymptotic quantile values given at the bottom of the table and labeled "two-sided tests" can be used.

**example  3.2**

We use the data of Example 3.1 to give another illustration of the application of the Kolmogorov-Smirnov test. This time we use the test to check whether the 20 original observations, generated by computer and purported to be chosen randomly from the uniform distribution, do in fact follow that distribution. This example will further emphasize the difference between the distribution of a number and its square root, and also show that these data provide no basis for a claim that the random number generator was not operating properly.

**SOLUTION**

The observed values, listed as $y$ in Table 3.1, are repeated in Table 3.2. The variable of interest in this example is $y$, and its empirical distribution function $S_N(y)$ is the same as the $S_N(x)$ (which was given in Table 3.1) since $y = x^2$ increases as $x$ increases. The null hypothesis here is

$$H: \quad y \text{ is uniformly distributed over } (0, 1),$$

and therefore the theoretical c.d.f. is $F_o(y) = y$. The functions $S_N(y)$ and $F_o(y)$ are graphed in Figure 3.2. The largest vertical distance here is $D = .112$, which occurs just to the left of the jump in $S_N(y)$ at $y = .2621$. Algebraically, this difference is found as $S_N(y_4) - F_o(y_3) = .2621 - .1500 = .1121$. From Table C with $N = 20$, we find that the $P$-value is $P > .20$. With a value this large, we can conclude that the hypothesis concerning the distribution of $y$ is not discredited, and that the numbers generated are indeed uniformly distributed.

*Additional Notes on Procedure*    As in the chi-square test, if $F_o(x)$ is not completely specified by $H$, it is generally necessary to estimate some parameters from the data in order to determine $D$. No adjustment is made in the sampling distribution of $D$, however. The maximum likelihood estimates should be used, that is, the estimates listed before in Table 2.7. The estimates should be calculated using the form of the data to be analyzed, whether grouped or ungrouped.

TABLE 3.2*

| $y$ | $S_N(y)$ | $F_o(y)$ |
|---|---|---|
| .0123 | .050 | .0123 |
| .1039 | .100 | .1039 |
| .1954 | .150 | .1954 |
| .2621 | .200 | .2621 |
| .2802 | .250 | .2802 |
| .3217 | .300 | .3217 |
| .3645 | .350 | .3645 |
| .3919 | .400 | .3919 |
| .4240 | .450 | .4240 |
| .4814 | .500 | .4814 |
| .5139 | .550 | .5139 |
| .5846 | .600 | .5846 |
| .6275 | .650 | .6275 |
| .6541 | .700 | .6541 |
| .6889 | .750 | .6889 |
| .7621 | .800 | .7621 |
| .8320 | .850 | .8320 |
| .8871 | .900 | .8871 |
| .9249 | .950 | .9249 |
| .9634 | 1.000 | .9634 |

*$H$:  $y$ is uniformly distributed over $(0, 1)$

**example  3.3**

In formulating a queuing model the operations researcher is interested not only in the basic structure of the queuing system, but also in the arrival or input process, the service discipline, and the service or output process. A widely utilized model is the single-channel/single-phase queue, in which inputs are Poisson distributed, the service discipline is first come/first served, and the service time is exponentially distributed. With such a model, descriptive quantities such as expected time in the queue, expected time in the system, and expected queue length are easily computed.

Consider an operations researcher concerned with the queue length of parts waiting to be machined on a lathe. The researcher postulates Poisson arrivals, a first-come/first-served service discipline, and exponential service times with a mean service rate of 20 minutes. Assume that the model has already been substantiated except for this distribution of service times, and data are collected at random times over a one-week period. Perform a goodness-of-fit test in order to qualify the model.

**TABLE 3.3**

| $x_i$ | $S_N(x_i)$ | $F_o(x_i)$ | $|S_N(x_i) - F_o(x_i)|$ | $|S_N(x_{i-1}) - F_o(x_i)|$ |
|---|---|---|---|---|
| .6 | .0333 | .0295 | .0038 | .0295 |
| 1.6 | .0667 | .0769 | .0102 | .0436 |
| 3.2 | .1000 | .1479 | .0479 | .0812 |
| 3.3 | .1333 | .1521 | .0188 | .0521 |
| 3.5 | .1667 | .1605 | .0062 | .0272 |
| 3.8 | .2000 | .1730 | .0270 | .0063 |
| 5.9 | .2333 | .2555 | .0222 | .0555 |
| 6.2 | .2667 | .2665 | .0002 | .0332 |
| 6.8 | .3000 | .2882 | .0118 | .0215 |
| 11.3 | .3333 | .4316 | .0983 | .1316 |
| 11.9 | .3667 | .4484 | .0817 | .1151 |
| 13.5 | .4000 | .4908 | .0908 | .1241 |
| 15.1 | .4333 | .5300 | .0967 | .1300 |
| 15.7 | .4667 | .5439 | .0772 | .1106 |
| 16.2 | .5000 | .5551 | .0551 | .0884 |
| 16.3 | .5333 | .5574 | .0241 | .0574 |
| 18.5 | .5667 | .6035 | .0368 | .0702 |
| 18.7 | .6000 | .6074 | .0074 | .0407 |
| 20.7 | .6333 | .6448 | .0115 | .0448 |
| 22.0 | .6667 | .6671 | .0004 | .0338 |
| 23.1 | .7000 | .6849 | .0151 | .0182 |
| 23.8 | .7333 | .6958 | .0375 | .0042 |
| 23.9 | .7667 | .6973 | .0694 | .0360 |
| 26.4 | .8000 | .7329 | .0671 | .0338 |
| 27.9 | .8333 | .7522 | .0811 | .0478 |
| 37.4 | .8667 | .8459 | .0208 | .0126 |
| 39.6 | .9000 | .8619 | .0381 | .0048 |
| 40.0 | .9333 | .8647 | .0686 | .0353 |
| 60.4 | .9667 | .9512 | .0155 | .0179 |
| 63.0 | 1.0000 | .9571 | .0429 | .0096 |

$$D = |.3000 - .4316| = .1316$$

**SOLUTION**
The data on service times are shown in Table 3.3, along with the necessary calculations. Since there are 30 observations, all different, the values of $S_N(x)$ for the arrayed values of $x$ are $1/30 = .0333$, $2/30 = .0667$, and so on, as

shown in the second column. The hypothesis states the exponential distribution with mean 20, and thus the theoretical c.d.f. is written as

$$F_o(x) = 1 - e^{-.05x} \quad \text{for } x > 0.$$

(The .05 comes in as the reciprocal of 20.) Hence the value of $F_o(x)$ for any $x$ is computed by multiplying $x$ by .05, finding $e^{-.05x}$, and subtracting the result from 1. Tables of powers of $e$, called exponentials, are included in almost all sets of mathematical tables; for example, see any edition of the *CRC Standard Mathematical Tables*.[5] These powers can also be found from most programmable calculators, or can be generated by a computer, as was done here. The results for $F_o(x)$ are shown in the third column. Corresponding differences are entered in the fourth and fifth columns. The largest value is $D = .1316$. From Table C with $N = 30$, the two-sided $P$-value is larger than .20. Hence the exponential model for service times is supported by these data.

The next example illustrates the procedure for a goodness-of-fit test when the distribution hypothesized is the normal. Hence we first review briefly this probability distribution.

The normal distribution is a family of continuous probability functions indexed by two parameters, the mean $\mu$ and the standard deviation $\sigma$. A graph of the density function is a bell-shaped curve, symmetric about the mean $\mu$. Since this distribution is continuous, the value of its cumulative distribution function at a point $x$ — that is, $F(x)$ — is the area under that curve from $-\infty$ to $x$. For any $\mu$ and $\sigma$, these areas can be found from a table of areas under the *standard normal curve* — that is, the normal distribution for which $\mu = 0$ and $\sigma = 1$. Suppose that a random variable $X$ follows the normal distribution with mean $\mu$ and standard deviation $\sigma$. Then the *standardized random variable*, which is denoted by $Z$ and formed by subtracting $\mu$ and dividing by $\sigma$, that is $Z = (X - \mu)/\sigma$, follows the standard normal distribution. Since the standardization represents only a relabeling of the variable $X$ according to a new origin and scale, a probability $F(x)$ is equal to $\Phi(z)$, where $\Phi$ represents the c.d.f. of the standard normal distribution.

Table A gives the right-tail probabilities under the standard normal curve, specifically the area from $z$ to $\infty$, or $[1 - \Phi(z)]$, for $z$ positive. A left-tail probabilty for $z$ positive — that is, the c.d.f. $\Phi(z)$ — is then found by subtracting the table entry from one. Since the standard normal density is symmetric about zero, these same table entries are left-tail probabilities for the corresponding negative value of $z$; that is, $\Phi(-z) = 1 - \Phi(z)$. Figure 3.4 illustrates these properties specifically when $z = 1.96$, for which the right-

[5]See footnote 4.

tail probability is given in Table A as .0250. The area in each tail is .0250, and the entire shaded area is $\Phi(1.96) = 1 - .0250 = .9750$. These same areas represent corresponding probabilities for the random variable $X$, where $X = \mu + Z\sigma$ or, equivalently, $Z = (X - \mu)/\sigma$.

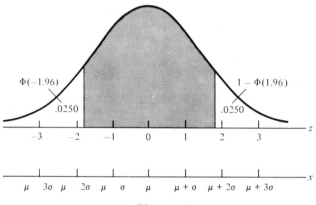

**Figure 3.4**

**example  3.4**

Buses are scheduled to pass a certain corner every 15 minutes. However, the bus arrival times vary considerably because of variations in traffic conditions and number of passengers boarding and leaving the buses.

A thorough study of the adequacy of public transportation facilities is planned, and a probability model for the number of minutes off schedule is needed. It is hoped that a normal model with $\sigma = 3$ can be used, since this would simplify other aspects of the statistical study. Data are collected by observing the arrival times of buses at randomly selected times of day on randomly selected days. The observations shown in Table 3.4 are recorded according to *next largest* integer number of minutes late (positive values) or early (negative). That is, if the bus arrives 1 minute 5 seconds late, it is recorded as $+2$; if 1 minute 5 seconds early, it is recorded as $-1$, so that each integer $x$ represents the interval of all numbers greater than $x - 1$ and not larger than $x$. The $S_N(x)$ in the last column are easily found by dividing the cumulative frequencies for the observed values of $x$ by 20.

**SOLUTION**

Consider any observed time of arrival as representing a "true" time plus an error (positive or negative) that is the result of chance factors operating. Then the normal distribution is a logical model to conjecture for a random variable representing the number of minutes off schedule. Further, if it can be justified, the normal distribution provides a convenient model to use in further studies. Thus we shall test the null hypothesis $H$: $F(x) = F_o(x)$, where $F_o(x)$ is the cumulative normal distribution with $\sigma = 3$. Since $F_o(x)$ is then continuous, the Kolmogorov-Smirnov test is an appropriate procedure for this hypothesis.

TABLE 3.4

| $x$<br>Largest Number of<br>Minutes off Schedule | Observed Frequency | Cumulative Frequency | $S_N(x)$ |
|:---:|:---:|:---:|:---:|
| −5 | 1 | 1 | .05 |
| −3 | 1 | 2 | .10 |
| −1 | 2 | 4 | .20 |
| 0 | 1 | 5 | .25 |
| 1 | 5 | 10 | .50 |
| 2 | 5 | 15 | .75 |
| 4 | 3 | 18 | .90 |
| 7 | 1 | 19 | .95 |
| 8 | 1 | 20 | 1.00 |
|  | 20 |  |  |

Since $\mu$ is not given, it must be estimated from the data before we can evaluate $F_o(x)$. The best estimate is $\bar{x}$, the sample mean, as Table 2.7 indicates. Since the data as given are grouped, $\bar{x}$ is calculated from (2.5) as follows:

$$\bar{x} = [(-5)(1) + (-3)(1) + (-1)(2) + (0)(1) + (1)(5) + (2)(5)$$
$$+ (4)(3) + (7)(1) + (8)(1)]/20$$
$$= 32/20 = 1.6$$

We treat this estimate as if it were the true value of $\mu$. Thus the variable $X$ is standardized here by the relation $Z = (X - 1.6)/3$, and $Z$ follows the standard normal distribution $\Phi(z)$ under $H$.

The calculations necessary to find $F_o(x) = \Phi(z)$ are shown in Table 3.5 in the recommended form. The first three columns are self-explanatory. As already mentioned, the table entry for $z$ is $\Phi(z)$ if $z$ is negative, and the table entries are subtracted from one to find $\Phi(z)$ when $z$ is positive.

**TABLE 3.5**

| $x$ | $z = (x - 1.6)/3$ | Table Entry for $z$ | $\Phi(z) = F_o(x)$ |
|---|---|---|---|
| −5 | −2.20 | .0139 | .0139 |
| −3 | −1.53 | .0630 | .0630 |
| −1 | −.87 | .1922 | .1922 |
| 0 | −.53 | .2981 | .2981 |
| 1 | −.20 | .4207 | .4207 |
| 2 | .13 | .4483 | .5517 |
| 4 | .80 | .2119 | .7881 |
| 7 | 1.80 | .0359 | .9641 |
| 8 | 2.13 | .0166 | .9834 |

Now all necessary calculations have been performed so that we can proceed with the Kolmogorov-Smirnov test in the usual way; see Table 3.6. The largest difference in either of the last two columns is $D = .198$, which for $N = 20$ has a conservative $P$-value of $P > .20$. (The reason *this P-value is conservative is not because of the hypothesized distribution since the normal probability function is continuous*. It is conservative because the parameter $\mu$ was estimated from the data.) The normal model does seem appropriate. However, there may be other probability models that are equally or more appropriate.

**TABLE 3.6**

| $x_i$ | $S_N(x_i)$ | $F_o(x_i)$ | $\|S_N(x_i) - F_o(x_i)\|$ | $\|S_N(x_{i-1}) - F_o(x_i)\|$ |
|---|---|---|---|---|
| −5 | .050 | .014 | .036 | .014 |
| −3 | .100 | .063 | .037 | .013 |
| −1 | .200 | .192 | .008 | .092 |
| 0 | .250 | .298 | .048 | .098 |
| 1 | .500 | .421 | .079 | .171 |
| 2 | .750 | .552 | .198 | .052 |
| 4 | .900 | .788 | .112 | .038 |
| 7 | .950 | .964 | .014 | .064 |
| 8 | 1.000 | .983 | .017 | .033 |

For many kinds of continuous measurements — for example, heights, weights, ages, temperatures, test scores, and diameters — it is very convenient mathematically to assume a normal distribution. Moreover, empirical investigations have shown that such measurements are often approximately normally distributed. This follows theoretically if we consider the measurement as composed of some constant value, which is the true mean for the population, plus a number of independent, additive random values that alter the measurement of each individual member of the population. The theoretical justification for approximate normality of the total measurement rests on the Central Limit Theorem. Nevertheless, in practice it is better to test the assumption of a normal distribution before using it as a model.

*Kolmogorov-Smirnov One-sided Goodness-of-fit Test*   A one-sided alternative should be used with a goodness-of-fit test if a possible directional difference between the hypothesized distribution and the true distribution is of interest. Since the difference could be in either direction, there are two different one-sided alternatives. The null hypothesis is usually changed to an inequality also, so the hypothesis set includes all possibilities.

*Inference Situation*

$$H: \qquad F(x) \le F_o(x) \qquad \text{for all } x$$
$$A_+: \qquad F(x) > F_o(x) \qquad \text{for some } x$$

or

$$H: \qquad F(x) \ge F_o(x) \qquad \text{for all } x$$
$$A_-: \qquad F(x) < F_o(x) \qquad \text{for some } x.$$

For a continuous $F_o(x)$, Figure 3.5(a) illustrates an arbitrary $F(x)$ that never rises above $F_o(x)$ and therefore satisfies the inequality $F(x) \le F_o(x)$ for all $x$ (note that $F(x)$ may touch $F_o(x)$ or coincide with it over an interval). Figure 3.5(b) illustrates the contradiction, namely $F(x) > F_o(x)$ for some $x$, as expressed in the alternative $A_+$. In this figure, $F(x)$ is sometimes above and sometimes below $F_o(x)$, since the cumulative distribution functions cross. It is also possible, but not necessary, for $F(x)$ to lie entirely above $F_o(x)$ under $A_+$.

The sampling situation and rationale are the same as with the Kolmogorov-Smirnov two-sided test, but one-sided statistics must now be defined.

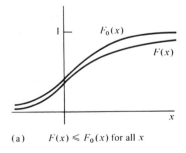

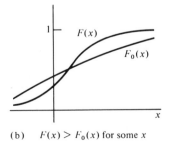

(a)   $F(x) \le F_0(x)$ for all $x$      (b)   $F(x) > F_0(x)$ for some $x$

**Figure 3.5**

*Test Statistics and P-value*    The Kolmogorov-Smirnov one-sided statistics are

$$D_+ = \text{supremum } [S_N(x) - F_o(x)], \tag{3.5}$$
$$D_- = \text{supremum } [F_o(x) - S_N(x)]. \tag{3.6}$$

These suprema are also found by looking at all differences in the neighborhood of each observed value $x$, but now only in one direction. For $D_+$, we consider only those $x$ for which $S_N(x)$ is larger than (above on a graph) $F_o(x)$, and for $D_-$, those $x$ for which $S_N(x)$ is smaller than $F_o(x)$.

Since $S_N(x)$ is the statistical image of the true but unknown distribution $F(x)$, a large value of $D_+$ would support the alternative $A_+$ and consequently discredit its converse in the null hypothesis. Similarly for $D_-$ and $A_-$. The appropriate $P$-values then are right-tail probabilities of the respective sampling distributions of $D_+$ and $D_-$. Because of symmetry, the Kolmogorov-Smirnov one-sided statistics are identically distributed. Thus the same table can be used for either statistic. The entries in the columns labeled "one-sided test" at the bottom of the first part of Table C are the quantile points for the indicated right-tail probabilities for either $D_+$ or $D_-$ when $N \le 40$, and the asymptotic quantile values labeled "one-sided test" can be used for $N > 40$.

## 3.2    Other Inferences Based on the Kolmogorov-Smirnov Statistics

Two useful applications of the Kolmogorov-Smirnov two-sided statistic relate to point and interval estimation of the unknown cumulative distribution function $F(x)$. In the case of a point estimate, suppose that we are planning to use the sample distribution $S_N(x)$ to estimate $F(x)$, but that the size of the sample is yet to be determined. We want a sample large enough to provide a "good" estimate, but no larger than necessary since the cost of the sample increases with the number of observations. The $D$ statistic can be used to determine the minimum sample size required in order to state with a certain probability that the estimate has a certain degree of accuracy.

For example, suppose we consider an estimate that is always within .20 as sufficient precision for the purpose at hand. This says that the value of the $D$ (two-sided) statistic is not to exceed .20, since the maximum error is the value of $D = \sup |S_N(x) - F(x)|$. When using a sample to estimate a population quantity, there is no way to *ensure* this, or any other, degree of precision short of obtaining a total sample, that is, data for the entire population. However, if we take a sample of sufficient size, we can have some specified degree of confidence, say .90, that our estimate has this precision. The problem then is to find $N$ such that the $P$-value of the $D$ statistic is .10 when the observed value of $D$ is no larger than .20. In Table C, looking down the column for a probability of .10 for a two-sided test, we find that the value of

$D$ is .202 when $N = 35$, and .199 when $N = 36$. Since .199 is the first number smaller than .20, the sample size needed is $N = 36$.

Now suppose that the problem had been to find the sample size for a value of $D$ no larger than .15, and the degree of confidence is kept at .90. There is no entry in the .10 column that is as small as .15. Hence the sample size must be larger than 40, and $N$ is found by setting the asymptotic distribution value of $D$ equal to .15. For a two-sided $P$-value of .100 this is $1.22/\sqrt{N}$, so we obtain $N$ by the following manipulation:

$$1.22/\sqrt{N} = .15$$
$$1.22/.15 = \sqrt{N}$$
$$8.13 = \sqrt{N}$$
$$N = (8.13)^2 = 66.09.$$

To be on the safe side, we would round upward and conclude that a sample of 67 observations is required.

For the general case, let $d$ be the maximum error to be allowed and $1 - \alpha$ the selected probability. Then the problem is to find $N$ such that

$$P(D < d) = 1 - \alpha.$$

Scanning the column for a right-tail probability of $\alpha$ with a two-sided test, we find the correct value of $N$ as that number which corresponds to the first table entry equal to or smaller than $d$. If all table entries in that column exceed $d$, set $d$ equal to the entry for the asymptotic distribution that corresponds to probability $\alpha$ for a two-sided test, and solve the resulting equation for $N$. Table 3.7 shows some typical sample sizes found using such a method.

TABLE 3.7
MINIMUM SAMPLE SIZE REQUIRED TO ESTIMATE $F(x)$
WITHIN PRECISION $d$ WITH CONFIDENCE $1 - \alpha$

| $d$ | $1 - \alpha$ | | | | |
|---|---|---|---|---|---|
| | .80 | .90 | .95 | .98 | .99 |
| .05 | 458 | 596 | 740 | 937 | 1063 |
| .10 | 115 | 149 | 185 | 231 | 266 |
| .15 | 51 | 67 | 83 | 105 | 119 |

The other useful and important application of the $D$ statistic is for finding confidence bands for $F(x)$, the unknown cumulative distribution function of the population. Since we are estimating a function $F(x)$ for all $x$, as opposed to a single number, the region is called a confidence band, and its bounds are also functions of $x$. This is in contrast to an ordinary confidence interval for a single parameter, where the lower and upper confidence bounds are each a single number. The point estimate of $F(x)$ is $S_N(x)$, which appears graphically as a step function, and the bounds of the confidence band are also step functions, even for a continuous population. It should be emphasized that the level of confidence applies to $F(x)$ for all values of $x$; that is, it applies to the entire region between the two bounds. For a confidence level $1 - \alpha$, we can find from Table C the number $d$ such that

$$P(D < d) = 1 - \alpha,$$

since $d$ then is the table entry corresponding to a right-tail probability of $\alpha$ for a two-sided test. Since $D$ is the largest difference between $S_N(x)$ and $F(x)$, all differences are less than $d$ if $D$ is less than $d$. Hence the bounds should be at a vertical distance $d$ above and below $S_N(x)$. However, $S_N(x) + d$ can produce values larger than 1 for some $x$, and $S_N(x) - d$ can be negative for some $x$. Since we know that $0 \le F(x) \le 1$ for all $x$, any lower bound that is negative should be increased to 0, and/or any upper bound that is larger than 1 should be reduced to 1. Accordingly, we define the lower and upper confidence bands respectively as

$$L_N(x) = \max\,[S_N(x) - d,\, 0]$$
$$U_N(x) = \min\,[S_N(x) + d,\, 1].$$

The region between $L_N(x)$ and $U_N(x)$ is a confidence band for $F(x)$ with (overall) confidence level $1 - \alpha$.

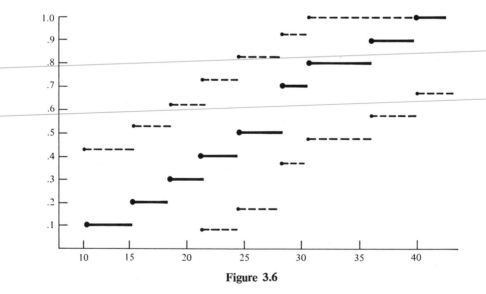

**Figure 3.6**

For example, suppose that in a sample of size 10 the arrayed data are

10.1, 15.2, 18.6, 21.2, 24.5, 27.0, 28.1, 30.5, 36.0, 39.8.

For a confidence level of .80 when $N = 10$, Table C gives $d = .323$. The values of $S_N(x)$, and the corresponding $L_N(x)$ and $U_N(x)$, are shown in Table 3.8 and are graphed in Figure 3.6. The entire region between the broken lines constitutes the confidence region for $F(x)$ at level .80.

**TABLE 3.8**

| $x$ | $S_N(x)$ | $S_N(x) - d$ | $S_N(x) + d$ | $L_N(x)$ | $U_N(x)$ |
|---|---|---|---|---|---|
| 10.1 | .1 | $-.223$ | .423 | 0 | .423 |
| 15.2 | .2 | $-.123$ | .523 | 0 | .523 |
| 18.6 | .3 | $-.023$ | .623 | 0 | .623 |
| 21.2 | .4 | .077 | .723 | .077 | .723 |
| 24.5 | .5 | .177 | .823 | .177 | .823 |
| 27.0 | .6 | .277 | .923 | .277 | .923 |
| 28.1 | .7 | .377 | 1.023 | .377 | 1 |
| 30.5 | .8 | .477 | 1.123 | .477 | 1 |
| 36.0 | .9 | .577 | 1.223 | .577 | 1 |
| 39.8 | 1.0 | .677 | 1.323 | .677 | 1 |

# 4   Comparison of the Chi-Square and Kolmogorov-Smirnov Tests for Goodness of Fit

The chi-square test is specifically designed for use with enumeration data or count data, whereas tests based on the Kolmogorov-Smirnov (K-S) statistics are for random samples from continuous populations, which usually consist of measurements on at least an interval scale. However, when the measurement scale is at least ordinal, the data are such that either of these two goodness-of-fit tests might be used. Accordingly, comparisons of their properties are relevant.

The basic difference between the two tests is that the measure of incompatibility with the chi-square test is based on vertical deviations between the observed and expected (hypothesized) histograms, whereas K-S procedures are based on vertical deviations between the observed and expected cumulative distribution functions. (These deviations, which determine the values of the respective test statistics, are represented graphically for the same data in Figures 2.1 and 3.3, for example.) Both types of deviations are useful and informative in determining goodness of fit, but they are not equivalent for any set of data. Another obvious difference is that the chi-square test *requires* that the data be grouped into categories, either as collected or before analysis, whereas the K-S test does not. Therefore, with interval scale data, the K-S test permits an examination of the goodness of fit for each different value of the $N$ observations instead of only for $r$ groups, where $r \leq N$. In this sense, the investigator using a K-S test can make more complete use of the available data. Further, when the categories are arbitrary and are chosen by the investigator, the chi-square statistic is unstable, since its value is affected by the number and widths of the categories selected. (Of course, in a practical sense, any quantitative measurement could be considered to be grouped data since it represents all values to the nearest unit chosen for recording the measurement. Ungrouped data then refer to data grouped only at the time of measurement, with the width of the groups depending on the precision of measurement. Grouped data are data that are grouped once again for analysis.)

One of the primary advantages of K-S is that its exact sampling distribution is known and tabulated for population distributions that are continuous and completely specified by the null hypothesis, whereas the sampling distribution of $Q$ is only approximately chi-square for any finite sample size. Theoretically then, we might say that K-S requires an infinite number of groups and chi-square requires an infinite number of observations. Neither of these situations is possible in practice, but the former is easier to conceive. The accuracy of the chi-square approximation to the distribution of $Q$ is not known when $N$ is small and/or when the expected frequencies are small. If categories are combined for analysis, the calculated value of $Q$ is not unique, since it is affected by the scheme of combination.

On the other hand, if $N$ is very large, thereby justifying the use of the chi-square distribution for $Q$, the test based on $Q$ is almost certain to lead to rejection of the null hypothesis. The test may detect even a minuscule departure from the hypothesized distribution. This is a case of "too much precision." In most goodness-of-fit applications, the investigator hopes to be able to accept the null hypothesis, even when it is only nearly true.

The K-S procedures are also more flexible than those based on the chi-square test, since the one-sided statistics can be used to test for deviations in a particular direction, whereas chi-square is always concerned equally with differences in either direction, since the deviations are squared. Further, the K-S statistics have inference applications besides goodness of fit, such as, for example, estimation of confidence bands for the true population distribution and estimation of minimum sample size required for a specified precision. In most cases the arithmetic with K-S is somewhat easier than with chi-square, but this advantage is minor if there are computers available to perform the calculations.

Chi-square also has some advantages over K-S. Data measured on a nominal scale and hypothesized distributions that are discrete occur frequently in experimentation. These applications present no problems for chi-square, but the exact properties of K-S are violated by the absence of a continuous population. The *P*-values are overestimated, but it is unknown by how much. Perhaps the main advantage of chi-square is that when unknown parameters must be estimated from the data in order to calculate expected frequencies, a correction factor can be introduced in the sampling distribution by simply reducing the number of degrees of freedom. If parameters must be estimated to apply K-S, no general adjustment is available for the sampling distribution. The *P*-value is conservative, but exactly how conservative is generally not known.

Several studies have been made on relative power of the two tests for specific alternatives, but they do not seem to provide a definitive choice in general. The basic difficulty with comparisons is that their power functions depend on different quantities.

The choice of a test when either could be applied, and neither is exact, would seem to be largely a matter of personal preference. However, it should be remembered that chi-square is *always* approximate, whereas K-S is exact for a hypothesis test concerning a continuous population with all needed parameters specified. When these assumptions are satisfied, K-S would be the preferable statistic, especially when the sample size is small.

## PROBLEMS

2.1 A market research firm was hired by a soap manufacturer to conduct research on packaging materials. From past clients the firm knew that the color of packaging material was an important determinant of consumer selection. In order to test which of four possible colors should be used (red, white, blue, green), 200 randomly selected housewives in the market area were each given four soap packages, one in each of the four colors. The subjects were told that the soap packages were made by different formulas, distinguishable by the color of the package, but in reality all were the some formula. A month later, each housewife was given a free case of soap of her choice. Their selections are as follows:

| Color of Package | Number of Housewives |
|------------------|----------------------|
| Red | 50 |
| White | 75 |
| Blue | 30 |
| Green | 45 |

Is there a significant difference between the colors preferred? Discuss the significance of these results to the manufacturer.

2.2   Suppose it is claimed that anyone can list numbers in such a manner that they could be considered statistically random numbers. In order to test this claim, make a list of at least 60 digits, selected from the digits 0–9, aiming consciously at a sequence that could be considered random. Perform a goodness-of-fit test on the list composed, by comparing the observed and expected frequencies of (a) even and odd digits, and (b) occurrence of each digit 0–9.

2.3   Two types of corn (golden and green-striped) carry recessive genes. When these were crossed, a first generation was obtained that was consistently normal (neither golden nor green-striped). When this generation was allowed to self-fertilize, four distinct types of plants were produced: normal, golden, green-striped, and golden-green-striped. In 1200 plants this process produced the following distribution:

| Type | Number |
|---|---|
| Normal | 670 |
| Golden | 230 |
| Green-striped | 238 |
| Golden-green-striped | 62 |

A monk named Mendel wrote an article theorizing that in a second generation of such hybrids, the distribution of plant types should be in a 9:3:3:1 ratio. Are the above data consistent with the good monk's theory?

2.4   A group of four coins is tossed 160 times, and the following data are obtained.

| Number of Heads | Frequency |
|---|---|
| 0 | 16 |
| 1 | 48 |
| 2 | 55 |
| 3 | 33 |
| 4 | 8 |

Do you think the four coins are balanced?

2.5   The constant $\pi$ occurs frequently in mathematics, for example in computing the area or circumference of a circle. Although the value of $\pi$ is commonly taken to be either 22/7 or 3.1416, it is actually an infinite decimal, which has been evaluated to considerable precision. The data below are a frequency distribution of the digits that occur in the first 4000 decimal places of $\pi$. Is this distribution approximately uniform, as it would be if $\pi$ were a random number instead of a fixed constant?

| Integer | Frequency | Integer | Frequency |
|---------|-----------|---------|-----------|
| 0 | 368 | 5 | 415 |
| 1 | 426 | 6 | 398 |
| 2 | 408 | 7 | 376 |
| 3 | 374 | 8 | 400 |
| 4 | 405 | 9 | 430 |

2.6   The works known to be written by a certain famous author have been thoroughly analyzed as to sentence length, and the distribution of length has been determined by enumerating these results. A newly found manuscript is claimed to have been written by this same author. The data below on sentence length are taken from a sample of 2000 sentences from this manuscript. Compare these data with the known distribution and make a judgment as to whether the new manuscript is by the same author.

| | Proportion of Sentences | |
|---|---|---|
| Number of Words | Known Author | New Manuscript |
| 3 or less | .010 | .007 |
| 4–5 | .030 | .024 |
| 6–8 | .041 | .031 |
| 9–12 | .102 | .034 |
| 13–16 | .263 | .250 |
| 17–20 | .279 | .203 |
| 21–24 | .118 | .198 |
| 25–27 | .105 | .156 |
| 28–29 | .042 | .081 |
| 30 or more | .010 | .016 |

2.7   Either sucrose or glucose can be used as a sweetener for fruit jam. Sucrose is known to be sweeter; a subjective assessment rates sucrose as between 1.2 and 2 times as sweet as glucose. However, glucose has certain properties that make it a desirable ingredient for jam if it does not affect the taste or appeal of the product to the consumer. Gridgeman (1956, p. 108) reported a taste test conducted in Ottawa to compare two basic recipes: (a) the regular jam using only sucrose, and (b) another jam made by replacing 25 percent of the sucrose by weight with glucose. Seven pairs of jams, (a) and (b), of different flavors or made by different manufacturers, were presented to each of 48 subjects. Each subject was asked to rank the jam within each pair according to the criteria of (1) sweetness and (2) their taste preference. In spite of instructions to the contrary, some subjects expressed no preference between (a) and (b) in one or more pairs. Of the 336 paired comparisons made (48 subjects tasted 7 pairs each), the ratings on sweetness were 188 for sucrose only, 138 for the glucose recipe, and 10 neutral; and the taste preferences were 177 for sucrose only, 140 for the glucose recipe, and 19 neutral. The study showed no evidence of association between sweetness and taste preference for the group as a whole, although this is not true for some individual subjects.

Excluding 9 subjects who were neutral as regards at least one pair of jams, a frequency distribution was made for the 39 remaining subjects according to the number of paired comparisons (maximum 7) in which the jam judged sweeter was also the one preferred for taste (which occurred 145 times in the 273 comparisons). The null hypothesis of interest for these data is that the association between taste preference and sweetness is a random variable with probability of agreement estimated by $\theta = 145/273 = .531$ for each subject. This hypothesis can be tested by seeing whether the binomial distribution with $n = 7$ and $\theta = .531$ fits the data. Using the observed and expected frequencies shown below, perform the test and draw an appropriate inference.

| Number of Agreements between Sweetness and Taste Preference | $f$ | $e$ |
|---|---|---|
| 0 | 4 | .20 |
| 1 | 1 | 1.54 |
| 2 | 6 | 5.24 |
| 3 | 8 | 9.89 |
| 4 | 4 | 11.20 |
| 5 | 7 | 7.61 |
| 6 | 6 | 2.87 |
| 7 | 3 | 0.46 |
| | 39 | 39.01 |

2.8   In the same experiment described in Problem 2.7, Gridgeman (1956, p. 109) tallied the number of subjects who judged the glucose recipe jam as sweeter than the regular (all sucrose) recipe jam, and also the number who preferred the glucose jam to the regular jam (those subjects expressing no preference were omitted for this study). The results are shown below, along with the expected frequencies based on the binomial distribution with parameter estimated by $\Sigma fx/7\Sigma f$. The estimate is $132/308 = .429$ for the sweetness data, and $121/280 = .432$ for the taste preference data. Complete the calculations necessary to perform the chi-square test on each set of data.

| Number of Votes | Sweetness | | Taste Preference | |
|---|---|---|---|---|
| for Glucose Jam | f | e | f | e |
| 0 | 1 | .87 | 1 | 0.76 |
| 1 | 6 | 4.58 | 4 | 4.06 |
| 2 | 7 | 10.32 | 9 | 9.27 |
| 3 | 14 | 12.92 | 12 | 11.75 |
| 4 | 11 | 9.71 | 8 | 8.94 |
| 5 | 4 | 4.38 | 5 | 4.08 |
| 6 | 1 | 1.10 | 1 | 1.03 |
| 7 | 0 | 0.12 | 0 | 0.11 |
| | 44 | 44.00 | 40 | 40.00 |

2.9   In one of a series of papers on mental ability of children in Great Britain, Roberts and Griffiths (1937, at p. 39) studied the use of a Binet scale shortened as follows to save time. If a child passes all but one of the tests administered during a complete year of tests, he is credited with passes in all years below. The scale is shortened similarly in the upward direction. The test was extended downward for those 131 children who passed all but one to determine the number of additional tests failed by these children in lower years. For the observed and expected frequencies reported, verify that the distribution of additional failures conforms well to the Poisson distribution, with parameter $\lambda$ estimated from the data as $\bar{x} = 53/131 = .4016$. A similar analysis of data obtained for the upward extension showed no conformance with the Poisson distribution.

| Number of Tests Failed | f | e |
|---|---|---|
| 0 | 88 | 87.41 |
| 1 | 34 | 35.37 |
| 2 | 8 | 7.16 |
| 3 | 1 | 0.97 |
| 4 or more | 0 | .11 |
| | 131 | 131.02 |

2.10  Buses are scheduled to pass a certain corner every 15 minutes, but the arrival times vary about the scheduled times. It is known that passengers arrive at the bus stop according to a Poisson process, with a mean rate of 4 persons per hour. A model is needed for the distribution of the number of empty seats on the buses, and the Poisson distribution with mean 2.5 has been postulated. The point probabilities for this distribution are shown below in the column labeled $p$. Twenty buses are observed at randomly selected times during the day, and the number of empty seats is counted. For the observed frequencies shown below, test the compatibility of the model postulated.

| Number of Empty Seats | Observed Frequency | $p$ |
|:---:|:---:|:---:|
| 0 | 1 | .0821 |
| 1 | 3 | .2052 |
| 2 | 5 | .2565 |
| 3 | 6 | .2138 |
| 4 | 2 | .1336 |
| 5 | 2 | .0668 |
| 6 | 1 | .0278 |
| 7 or more | 0 | .0142 |

2.11 Linguists have been studying word frequency in works of famous authors for over a hundred years. In the late nineteenth century, several linguists discovered the "rank-size law," sometimes called Zipf's law, which says that the number of words occurring $n$ or more times in a text is $K/n$, where $K$ is the total number of distinct words in that text. Then the number of words occurring exactly $n$ times is the number occurring $(n + 1)$ or more times, minus the number occurring $n$ or more times, or

$$\frac{K}{n} - \frac{K}{n + 1} = \frac{K}{n(n + 1)}.$$

This law holds for almost all texts for which it has been investigated, regardless of the language of the text. The data below, from Tanur (1972, p. 198), are for numbers of rarely occurring words in James Joyce's *Ulysses*, which contains $K = 29,899$ distinct words in its text. Show that Zipf's law holds for these data.

| Number of Occurrences | Number of Words Observed | Expected |
|:---:|:---:|:---:|
| 1 | 16,432 | 14,949 |
| 2 | 4,776 | 4,983 |
| 3 | 2,194 | 2,491 |
| 4 | 1,285 | 1,495 |
| 5 | 906 | 997 |
| 6 | 637 | 712 |
| 7 | 483 | 534 |
| 8 | 371 | 415 |
| 9 | 298 | 332 |
| 10 | 222 | 272 |

2.12 The manager of a foundry is of the opinion that the proportion of defective castings produced varies throughout an 8-hour shift, with more being produced during the latter hours of the shift than during the former hours. He attributes this increase to fatigue and a lackadaisical attitude of workers toward the end of their shift. The manager has decided to test his opinion and collected the following data during a randomly chosen 8-hour shift.

| Time | Number of Defectives | Number Produced |
|------|----------------------|-----------------|
| First 2 hours | 25 | 390 |
| Second 2 hours | 33 | 402 |
| Third 2 hours | 49 | 408 |
| Fourth 2 hours | 53 | 400 |

Perform a statistical analysis on these data and state whether you think more supervision during certain hours of each shift would be advisable.

2.13 In order to study frequency of limestone pebbles on a beach near Evanston, Illinois, 100 samples of 10 pebbles each were collected. Krumbein (1954, p. 59) reported this experiment and the following data. Test the null hypothesis that number of limestone pebbles per sample of 10 follows the binomial distribution with parameter $\theta = .50$.

| Number of Limestone Pebbles | Observed Frequency |
|-----------------------------|--------------------|
| 0 | 0 |
| 1 | 1 |
| 2 | 6 |
| 3 | 7 |
| 4 | 23 |
| 5 | 26 |
| 6 | 21 |
| 7 | 12 |
| 8 | 3 |
| 9 | 1 |
| 10 | 0 |

2.14 Edwards and Fraccaro (1960) reported data on the number of males among the first seven children fathered by 1334 Swedish ministers. Test the null hypothesis that the number of boys is binomially distributed, with the parameter $\theta = .50$.

| Number of Boys | Number of Families |
|----------------|--------------------|
| 0 | 6 |
| 1 | 57 |
| 2 | 206 |
| 3 | 362 |
| 4 | 365 |
| 5 | 256 |
| 6 | 69 |
| 7 | 13 |
|   | 1334 |

2.15 Persons called for jury duty are usually selected at random from a list of regis-
tered voters in the community. Thus, for a large number of jury panels, the
representation of various groups in the community should conform to the
makeup of the voting population in that community (but not necessarily to the
entire population of that community). The last census determined that in a
certain community the population is 55 percent female and 30 percent black.
The records of jury panels were collected for a certain period of time and tallied
as to number of females and number of blacks on panels of 12 persons, the typical size. Do these distributions conform to the population of the community?

| Number of Female Jurors | Number of Panels | Number of Black Jurors | Number of Panels |
|---|---|---|---|
| 0 | 1 | 0 | 1 |
| 1 | 4 | 1 | 13 |
| 2 | 10 | 2 | 20 |
| 3 | 14 | 3 | 18 |
| 4 | 24 | 4 | 8 |
| 5 | 7 | 5 or more | 0 |
| 6 or more | 0 | | 60 |
| | 60 | | |

2.16 A company recently initiated a new 3 percent discount policy for customers
who purchase a sufficient amount and pay in full within 6 days of the invoice
date. They are interested in finding a probability model for the distribution of
the number of customers per week who take advantage of the discount; the
manager feels that the binomial distribution might be appropriate. Suppose
there are exactly 10 regular customers whose weekly purchase is of sufficient
size for the discount, and the following data are collected over a 50-week period.
Do the data support the use of the binomial distribution as a model for the
probability function?

| Number of Customers Per Week Taking Discount | Frequency |
|---|---|
| 0 | 3 |
| 1 | 0 |
| 2 | 0 |
| 3 | 2 |
| 4 | 3 |
| 5 | 8 |
| 6 | 10 |
| 7 | 9 |
| 8 | 8 |
| 9 | 6 |
| 10 | 1 |

2.17 During a 50-week period, demand for a certain kind of replacement part for television sets was distributed as shown below. Find the theoretical distribution of weekly demand for a Poisson model with the same mean as the given data and perform a goodness-of-fit test.

| Weekly Demand | Number of Weeks |
|:---:|:---:|
| 0 | 28 |
| 1 | 15 |
| 2 | 6 |
| 3 | 1 |
| More than 3 | 0 |
| | 50 |

2.18 Samples of identical size are taken at random from 50 bolts of cloth, and the number of imperfections is counted for each specimen. Group the data below into a frequency distribution and test the null hypothesis that the number of imperfections follows the Poisson distribution:
(a) With parameter $\lambda = 1.5$.
(b) With parameter $\lambda$ estimated from the data.

| | | | | | | | | | |
|:-:|:-:|:-:|:-:|:-:|:-:|:-:|:-:|:-:|:-:|
| 2 | 1 | 0 | 1 | 1 | 2 | 0 | 5 | 1 | 1 |
| 3 | 0 | 1 | 1 | 2 | 0 | 1 | 1 | 0 | 0 |
| 1 | 3 | 4 | 0 | 0 | 1 | 1 | 4 | 1 | 2 |
| 5 | 2 | 6 | 2 | 4 | 1 | 5 | 1 | 2 | 2 |
| 0 | 1 | 1 | 0 | 3 | 2 | 0 | 2 | 3 | 3 |

2.19 Suppose that monthly collections for home delivery of the *New York Times* in a large suburb are approximately normally distributed, with mean \$150 and standard deviation \$20. A random sample of 10 delivery boys in a nearby suburb is taken, and the arrayed data for monthly collections are as follows:

| | |
|:---:|:---:|
| \$ 90 | \$145 |
| 106 | 156 |
| 109 | 170 |
| 117 | 174 |
| 130 | 190 |

Test the null hypothesis that the same normal distribution model applies to this suburb.

2.20 Krumbein and Graybill (1965, p. 123) reported the actual data below on moisture content of sand on the upper part of a beach foreshore along Lake Michigan near Evanston, Illinois. The variable $X$ is percent moisture by dry weight.
(a) Test the null hypothesis that $X$ is normally distributed.
(b) Test the null hypothesis that $X$ is normally distributed with mean 3.50 and standard deviation 1.

| X Class | Observed Frequency |
|---------|--------------------|
| 0.00–0.99 | 4 |
| 1.00–1.99 | 14 |
| 2.00–2.99 | 38 |
| 3.00–3.99 | 42 |
| 4.00–4.99 | 23 |
| 5.00–5.99 | 6 |

2.21 Most reliability studies are based on specific distribution assumptions about time to failure. Since departures from these assumptions can invalidate the conclusions, a goodness-of-fit test is an important preliminary to any reliability study. The Kolmogorov-Smirnov test is an especially useful aid in determining what assumptions are reasonable, since the length of time that an article will operate without failure is a continuous variable. Foster (1962) gives two sets of data on time to failure to illustrate the K-S test.

(a) In the first example (p. 7) three different theoretical distributions are fitted to a set of 10 observations, whose empirical distribution $S_N(x)$ is shown in the tabulation below. The corresponding values of the three cumulative distributions fitted are in the next three columns; $F_1(x)$ is the exponential distribution, with mean 1500, that is, $F_1(x) = 1 - e^{-x/1500}$; $F_2(x)$ is the normal distribution, with mean 300 and standard deviation 100; $F_3(x)$ is the uniform distribution over the interval (100, 700). Verify the expected frequencies and perform the test for the appropriate three hypothesis sets.

| $x$ | $S_N(x)$ | $F_1(x)$ | $F_2(x)$ | $F_3(x)$ |
|-----|----------|----------|----------|----------|
| 100 | .1 | .0644 | .023 | .000 |
| 150 | .2 | .0952 | .067 | .083 |
| 179 | .3 | .1125 | .113 | .132 |
| 200 | .4 | .1248 | .159 | .167 |
| 250 | .5 | .1535 | .309 | .250 |
| 290 | .7 | .1758 | .460 | .317 |
| 300 | .8 | .1813 | .500 | .333 |
| 310 | .9 | .1867 | .540 | .350 |
| 450 | 1.0 | .2592 | .933 | .583 |

(b) The second example gives the following 60 arrayed observations (p. 8).

| | | | | | | | | | |
|---|---|---|---|---|---|---|---|---|---|
| 1 | 34 | 62 | 102 | 118 | 171 | 193 | 219 | 250 | 343 |
| 9 | 43 | 63 | 111 | 133 | 174 | 204 | 220 | 310 | 356 |
| 18 | 48 | 67 | 114 | 135 | 174 | 209 | 225 | 314 | 376 |
| 21 | 48 | 67 | 116 | 139 | 175 | 210 | 230 | 327 | 384 |
| 23 | 50 | 84 | 116 | 163 | 175 | 212 | 247 | 327 | 409 |
| 29 | 60 | 100 | 117 | 171 | 176 | 213 | 248 | 339 | 410 |

For these data, test the following two null hypotheses:
 (i) The distribution is exponential, with mean 170, that is, $F(x) = 1 - e^{-x/170}$.
 (ii) The distribution is normal, with mean 170 and standard deviation 100.

2.22 A practical application of analysis of numbers arises in estimating equipment acquisitions. Goodman (1954, p. 110) reported a study to estimate total number of equipment acquisitions of the Division of Social Sciences at the University of Chicago. This equipment had been purchased many years previously, and data on number of items purchased are not available. However, each item had been given a serial number. A sample of 31 items of equipment was taken; their arrayed serial numbers are as follows:

| 83  | 895  | 1210 | 1668 | 1880 | 2157 |
|-----|------|------|------|------|------|
| 135 | 955  | 1344 | 1689 | 1936 | 2220 |
| 274 | 964  | 1387 | 1756 | 2005 | 2224 |
| 380 | 1113 | 1414 | 1865 | 2006 | 2396 |
| 668 | 1174 | 1610 | 1874 | 2065 | 2543 |
|     |      |      |      |      | 2787 |

If these data, which have a range of $2787 - 83 = 2704$, can be considered a random sample from the population of continuous, uniformly distributed serial numbers with unknown initial and final values, then the 29 serial numbers in the above listing which are left after excluding 83 and 2787, the minimum and maximum, are uniformly distributed on the interval (83, 2787). Further, if we subtract 83 and divide by 2704 for each of these 29 numbers, the resulting data are uniformly distributed over (0, 1).
 (a) Use these facts to perform an appropriate goodness-of-fit test. (You should obtain $D = .16$, for which $P = .3172$ when $N = 29$.)
 (b) Since the null hypothesis is accepted, we can estimate the range of the population from which the 31 serial numbers were drawn as follows: A group of 29 uniformly distributed numbers divides the interval (83, 2787) into 30 equal parts, each of width 2704/30. Hence the expected width of the interval for 31 serial numbers is $2704 + 2(2704/30) = 2884.26$.

2.23 The "gap test" may be used as an objective criterion in investigating the randomness of a sequence of digits. A gap of length $k$ exists in a sequence of $N$ digits if the digit $x$ is followed by $k$ digits other than $x$ before $x$ occurs again. If an $x$ follows an $x$, the gap length is zero. If $x$ does not occur again, no gap exists. In the following sequence of digits,

$$0, 1, 3, 4, 6, 5, 0, 3, 7, 8, 2, 6, 5, 9, 3, 9, 6, 2, 1, 5,$$
$$7, 4, 0, 1, 8, 9, 5, 3, 1, 4, 2, 0, 1, 6, 3, 8, 7, 8, 9, 2, 5, 6, 4,$$

the digit 1 occurs five times and thus there are four gaps to be measured. If this is done for each digit 0 through 9, the gaps may be tabulated as shown in the table on page 88. The total number of gaps in a sequence of $N$ digits is $N - 10$, since each digit must occur a final time. Assuming that the digits are randomly ordered, the theoretical cumulative distribution function for the length $K$ of the gaps is given by

$$P(K \leq k) = \sum_{i=0}^{k} (0.1)(0.9)^i$$
$$= 1 - (0.9)^{k+1}$$

| Gap Length | Frequency |
|:---:|:---:|
| 0– 2 | 2 |
| 3– 5 | 6 |
| 6– 8 | 11 |
| 9–11 | 3 |
| 12–14 | 6 |
| 15–17 | 5 |
| 18–20 | 0 |
| 21–23 | 0 |
| 24–26 | 0 |
| | 33 |

Compute the theoretical distribution function for these data and perform a Kolmogorov-Smirnov goodness-of-fit test of the null hypothesis that the sequence is random.

# Inferences Concerning Location
# Based on One Sample or
# Paired Samples

## 1 Introduction

In Chapter 2 we discussed how the empirical distribution function can be used to describe a set of data. Since this function does not provide a summary of any particular aspect of the data, it is usually difficult to interpret except as a general descriptor. An especially useful particular descriptor of data is a measure of central tendency — that is, a figure that is typical or representative of the central value of the magnitudes of quantitative observations. A measure of central tendency in the population is a *location parameter*. In this chapter we consider some procedures appropriate for inferences concerning such parameters.

The best-known location parameters are the mean and median. They both represent a "typical" or "average" value of a quantitative variable, or a central value of a distribution, but they typify different concepts of centrality. For a finite population, the mean is determined by summing all the values and dividing by the number of elements, while the median is the middle value in magnitude (the value such that half the elements are smaller and half are larger). For an infinite population, the mean is the center of gravity of the density function, and the median is the value that divides the area under the density function into two equal parts. By definition, then, the probability that a variable exceeds its median is equal to the probability that it is exceeded by it. For each density in Figure 1.1, the area to the left of the vertical dashed line is equal to the area to the right.

Both of these parameters can be interpreted visually in terms of a piece of cardboard cut to represent the plane area bounded above by the probability function $f(x)$ and below by the $x$ axis. There is exactly one value of $x$ for which the cardboard can be balanced, and this is where the mass is in equilibrium; such a point is the mean. In mathematical terms, the mean is the centroid of the area for $f(x)$. There is also a value of $x$ at which the cardboard can be cut vertically into two pieces of equal area; this value is the median. The values of the mean and median are indicated in Figure 1.1 for the two arbitrary density functions.

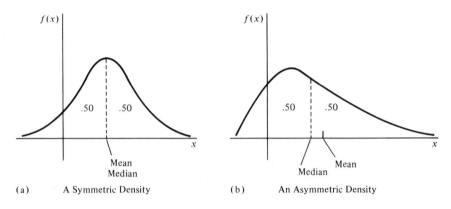

(a)        A Symmetric Density                (b)        An Asymmetric Density

**Figure 1.1**    Reproduced from Figure 3.1, p. 118, of Jean D. Gibbons (1972), *Nonparametric Statistical Inference*, McGraw-Hill Book Company, New York, with permission.

If the area has a center of symmetry, the two pieces of equal area are identical in shape and in weight, and thus the center of symmetry is equal to both the mean and the median, as in Figure 1.1(a). In general, however, the center of mass is not the same as the center of area. The mean is more influenced by extreme values than is the median, since each possible value directly affects the mean while only the *relative* magnitudes affect the median. For example, the mean of a distribution of personal incomes is usually larger than the median of those incomes, since exceptionally large incomes have greater influence on the mean than on the median. Such a distribution is said to be skewed to the right, as in Figure 1.1(b). For any given distribution, the choice between the mean and median as a description of central location may depend on one's judgment as to whether extremes should or should not have much influence on the typical value.

The statistics that correspond to these location parameters are the sample mean and the sample median. The sample mean is the best known sample descriptor of central tendency. It is the value obtained by summing the values observed and dividing by the sample size. Although the sample mean can be computed for any set of numbers, its value can be no more meaningful than the numbers entering its calculation. This factor can be a serious limitation on the proper interpretation of a mean calculated from data measured on a scale lower than interval. For example, if the level of measurement of the sample data is ordinal, but the variable is continuous and therefore capable of more precise measurement, the mean may have little significance.

The sample median is the number occupying the central position in an (ordered) array — that is, a number such that half of the observations are smaller than it and half are larger. For a sample of size $n$, then, the median is defined as the $(n + 1)/2$th from the smallest if $n$ is odd. If $n$ is even, any number between the two middle values is a sample median. Some arbitrary convention may be followed to arrive at a specific value for "the" median, as for example the mean of the two middle values, but it is really "a" median, as is every other value in that interval. Since the median is simply a positional measure, its interpretation is essentially the same whether the data are measured on an ordinal or an interval scale. Further, as in the population, the sample median is less affected by extreme values than the mean. With particularly small sample sizes, the sample median is often a better indicator of central tendency than the mean.

A sample measure of location is frequently needed not only to describe the data, but also for the purpose of statistical inference about a "typical" value in the population. The inference may concern the population mean or median, or just location in general. If we know which parameter is of more interest, or is more representative of the variable, the corresponding sample descriptor is usually the best estimator and basis for a hypothesis test or confidence interval. However, it is also important that the sample measure is representative of the data, that it does not imply a higher level of measurement than was employed, and that it has useful distribution properties.

The sample mean has enjoyed a position of high prominence in classical statistical inference for many reasons. It is an unbiased and consistent estimator of the mean of any population, and it has nice distribution properties, for example, the property defined by the Central Limit Theorem. (This theorem states that the probability function of the standardized mean of a random sample drawn from any population whose variance exists approaches the standard normal distribution as the sample size approaches infinity.) Most of the "good" statistical properties of the sample mean are relevant primarily for inferences about the population mean. Further, the traditional inference procedures are based on samples from normal populations and are used primarily with interval scale data. In nonparametric statistics, the median is often the sample measure of central tendency that has desirable distribution properties. Since the procedures are frequently based on ordinal data, the median may be considered a more appropriate sample descriptor.

The population median is usually the relevant parameter for non-parametric inferences concerning location. However, the inferences are analogous to inferences concerning means in parametric statistics, and are equally important and valuable. Of course, for a symmetric distribution, inferences concerning the median can be considered equivalent to inferences concerning the mean. When the two parameters are not equal, sometimes the median is actually (or should be) of more interest than the mean, either because of the nature of the variable or the nature of the distribution.

Our study of inferences concerning location parameters is divided into four types in this book. In the first case the hypothesis or confidence interval asserts values for the population median and the procedure is based on data from a single sample. An equivalent situation results when the relevant parameter is the median of a population of differences and the analysis employs differences between paired observations. The second case relates to tests of the hypothesis that two population distributions are identical when the data arise from two mutually independent samples. The test presented is particularly sensitive to differences in location, and the corresponding confidence procedure provides an interval estimate of the difference in locations. The third case extends this test to the problem of $k$ independent samples, in which the hypothesis states that the $k$ distributions are equal; again, the test is particularly sensitive to differences in location. The fourth case concerns equality of location parameters when there are $k$ related samples. Only the first case is discussed in this chapter. The second and third cases are treated in Chapter 4. The fourth case is covered in Chapter 7, for reasons to be explained there.

In the one-sample location problem, the inference concerns the value of a single location parameter, specified to be the population median. When the sample data consist of paired observations, the differences within pairs are used to make inferences about the median of the population of differences. The statistical analysis with paired data is exactly the same as in the one-sample model, since once the pairs of observations are replaced by their differences, the data consist of a single sample (of differences). The tests appropriate for these situations and covered in this chapter are the ordinary sign test, which can be applied to nominal data as well as to data measured on higher scales, and the Wilcoxon signed rank test, which is useful for more refined levels of measurement. We also explain the corresponding methods that can be used to find confidence intervals for the population median or median difference. Except for the difference in level of measurement required and the fact that the inferences concern the median rather than the mean, both the sign test and signed rank test procedures can be considered nonparametric analogues of the normal theory (variance known) and the Student's $t$ (variance unknown) tests for means in parametric statistics.

Before proceeding with a discussion of the methodology of these tests, some general remarks about paired samples are appropriate. When observations occur in pairs, the pairs themselves may be independent, but the observations within each pair are related or dependent. Such data arise primarily in comparative experiments — that is, experiments that are planned to discover and evaluate *differences* between effects, treatments, conditions, and the like, rather than the individual effects. When differences are the focus of the investigation, it may be necessary or desirable to have two nonindependent samples, because either (a) the same experimental unit (subject or object) is a member of the two populations under study, such as before and after some treatment, or (b) it is not possible to make a meaningful comparison of responses, as in the case of treatment effects, unless the subjects are as alike as possible in other respects.

In the former case, for example, suppose that two measurements are made on the same experimental unit, but at two different times, say one before and one after some treatment. Then a single random sample of units is drawn, but a pair of observations is obtained for each unit, and each unit acts as its own "control." Under the assumption (not to be treated lightly) that there is no time-related change other than the treatment, the effect of the treatment can be estimated, and with smaller sampling variability than if the controls were chosen independently of the set of units to be treated.

In the second situation, units are matched in some way, either by nature or according to some relevant criterion, in order to form pairs. Then observations are made on both members of the pair for some response or characteristic. The pairs here might be siblings, littermates, married couples, two different sides of a leaf, two different schools, and so on. If the matching of units is effected by the investigator as a design technique, the problem of matching requires careful study. The most important consideration is that the matching be effective in such a way that the members of a pair can be presumed to respond similarly if treated alike. Either treatment-control experiments or comparisons of two treatments can easily be performed here. Within each pair, the two treatments (where one may be a control) are assigned randomly between members of the pair. This reduces any random variation and makes the nonrandom variation in response easier to observe. If a difference between responses is observed, the difference can more readily be attributed to the effect of the treatments than to chance differences between units. The greater the similarity between matched units, the fewer observations are needed to adduce information about the difference in treatment effects. The random assignment of treatments within each pair, or vice versa, is necessary to balance out random differences.

In an analysis of paired observations, it is technically improper and ordinarily disadvantageous to disregard the pairing. The most convenient way to take advantage of the pairing is to subtract the two measurements within each pair and use only the values of these differences for the analysis. This procedure essentially reduces a paired sample problem to a one sample problem. The differences are treated as a single sample, and the random assignment of treatments to units within each pair justifies the assumption of a random sample of (differences of) observations (see Section 5).

## 2    Sign Test Procedures

Since procedures based on the sign test are quite versatile, the presentation in this section is divided into four parts. In Section 2.1 we discuss the ordinary sign test, where the inference concerns a comparison between the probabilities of two different types of outcomes. In the more specific inference situation of Section 2.2, the same technique is used to test a hypothesis about the median of a single population (one-sample case), or the median difference in a population of paired differences (paired-sample case). In Section 2.3 we describe a procedure that leads to confidence intervals for the median or median difference. The tests of Sections 2.1 and 2.2 are both special cases of more general procedures, called the binomial test and the quantile test, respectively. These techniques are discussed briefly in Section 2.4.

### 2.1    Ordinary Sign Test

We introduce the ordinary sign test by the following simple example.

**example   2.1**
Suppose that a political scientist is investigating the theory that women tend to be more conservative than men in their political views, and specifically is interested in the premise that when a husband and wife do not have similar views, the wife is more likely to be the more conservative. A random sample of 50 married couples is taken, and each person is asked individually a series of questions designed to classify political leanings. Only 10 couples differed significantly in opinion. Of these 10, the more conservative individual was indeed the wife in 9 couples.

SOLUTION

We define a "success" as a wife being classified as more conservative than her husband, and a "failure" as her being classified as less conservative. Let $\theta$ represent the probability of a success in the population of couples for which a difference exists. We formulate the null hypothesis that a success and a failure are equally likely in the population, that is, $H:$ $\theta = .5$, and the alternative that a wife is more likely to be more conservative, that is, $A_+:$ $\theta > .5$. Under the null hypothesis, the number of successes in 10 couples follows the binomial distribution with $\theta = .5$, and the expected number of successes is then 5. There are 9 successes in this study. To what extent does this evidence support the null hypothesis? From Table F, the probability of obtaining 9 or more successes in 10 trials with $\theta = .5$ is .0107, which is quite small. Thus we conclude that the data do not support the null hypothesis, and therefore do corroborate the premise that it is more likely that a wife is more conservative than her husband when a difference exists.

*Inference and Sampling Situation*  Suppose that we have $n$ observations, which are independent, and each is classified into one of two categories, such as yes and no, or black and white (nominal level of measurement). Alternatively, we might have quantitative data, that is, $n$ independent observations measured on at least an ordinal scale, which are later classified into two categories according to their relative magnitudes. In either case, after classification the data consist of $n$ dichotomous observations. We call the two categories "success" and "failure" for convenience. (These terms are completely arbitrary; for example, a success could be defined as contracting an illness.) The data to be analyzed are the frequencies in the two categories, that is, count data.

The occurrences of success or failure are assumed to follow a Bernoulli process. Then, however the data for analysis arise, they constitute a random sample from a dichotomous population. Let the symbol $+$ denote success and the symbol $-$ denote failure. The only population parameters relevant are the probability of success, denoted by $p_+$, and the probability of failure, $p_-$. The ordinary sign test procedure can be used for the following types of inference concerning these parameters:

*Two-sided alternative*
$H:$  $p_+ = p_-$     $A:$  $p_+ \neq p_-$

*One-sided alternatives*
$H:$  $p_+ = p_-$     $A_+:$  $p_+ > p_-$
$H:$  $p_+ = p_-$     $A_-:$  $p_+ < p_-$.

Note that in the designation for one-sided alternatives, the subscript on $A$ denotes the kind of sign that predominates; that is, $A_+$ is used to state that

$p_+$ is the larger and $A_-$ says that $p_-$ is the larger. $A_+$ and $A_-$ are called one-sided alternatives and $A$ is called a two-sided alternative. Correspondingly, a test of $H$ versus $A_+$, or $H$ versus $A_-$, is a one-sided test, and a test of $H$ versus $A$ is a two-sided test.

*Rationale*   If the null hypothesis $p_+ = p_-$ is true, the number of plus signs among the $n$ observations should be approximately equal to the number of minus signs. Since the population is dichotomous, we have $p_+ + p_- = 1$. Thus $p_+ = p_-$ can occur only when $p_+ = p_- = .5$. Under $H$, the expected probability of a plus (or minus) sign is then .5, and the expected *number* of plus (or minus) signs is $.5n$. If the observed number of plus signs is much greater than $.5n$, that is, if there are too many plus signs (or, equivalently, too few minus signs), the alternative hypothesis $A_+$ is supported; similarly, too few plus signs (or, equivalently, too many minus signs) supports $A_-$. If the number of signs of either kind is too large, the null hypothesis is discredited in favor of the two-sided alternative.

The number of successes $(+)$ in $n$ independent observations from a dichotomous population follows the binomial probability distribution, with parameter $\theta$ denoting the probability of success, here $\theta = p_+$. Likewise, the number of failures $(-)$ follows the binomial distribution with parameter $1 - \theta = p_-$. Since $p_+ = p_- = .5$ under the null hypothesis, the null distributions of the number of plus signs and the number of minus signs are each binomial, with parameter $\theta = .5$. Hence the number of plus signs, or the number of minus signs, can be used as the basis for a distribution-free, and therefore nonparametric, inference procedure.

*Test Statistic*   The test statistic for the ordinary sign test is defined as either of the following:

$$S_+ = \text{number of plus signs observed}$$
$$S_- = \text{number of minus signs observed.}$$

Note that since the total number of signs, whether plus or minus, is $n$, we have the relation

$$S_+ + S_- = n. \qquad (2.1)$$

*Finding the P-value*   Since the sampling distributions of both $S_+$ and $S_-$ are each binomial with parameter $\theta = .5$ if $H$ is true, the P-values could be found from the appropriate column of Table E. However, the binomial distribution is symmetric when $\theta = .5$, and thus a left-tail cumulative probability for $S_+$ is equal to a right-tail cumulative probability for its complementary value $S_- = n - S_+$, and vice versa. Table F makes use of this property and is more convenient for the sign test, since it gives directly the cumulative probabilities not exceeding .5 in either tail, that is, left tail for the smaller of $S_+$ and $S_-$ and right tail for the larger. For example, suppose that $n = 10$,

$S_+ = 9$, and $S_- = 1$, as in Example 2.1. The left-tail probability for $S_- = 1$ is read from Table F as .0107, and this entry is also the right-tail probability for $S_+ = 10 - S_- = 9$.

Whether the left- or right-tail probability is appropriate depends on which alternative is of interest and which test statistic is used. Suppose that the test statistic is $S_+$. Then if $S_+$ is too much larger than its expected value $.5n$, we tend to believe that $p_+$ is larger than .5. The appropriate $P$-value for the alternative $A_+$: $p_+ > p_-$ is then a right-tail probability, namely the probability of obtaining a sample result as large or larger than $S_+$ in the binomial distribution, with $n$ trials and $\theta = .5$. On the other hand, an $S_+$ that is extremely small supports the alternative $A_-$: $p_+ < p_-$, and thus the appropriate $P$-value is found from Table F as a left-tail probability for the observed value of $S_+$. With the two-sided alternative $A$: $p_+ \neq p_-$, a value of $S_+$ that is either too small or too large does not support $H$. Thus the $P$-value should be the sum of two tail probabilities. However, since the binomial distribution is symmetric, it is reasonable to define the two-sided $P$-value as twice the smaller tail probability, which will be left tail if $S_+ < .5n$ and right tail if $S_+ > .5n$.[1]

In summary, when using $S_+$ as the test statistic, the correct procedure for finding $P$-values from Table F is as follows:

| Alternative | $P$-value (Table F) |
|---|---|
| $A_+$: $p_+ > p_-$ | Right-tail probability for $S_+$ |
| $A_-$: $p_+ < p_-$ | Left-tail probability for $S_+$ |
| $A$: $p_+ \neq p_-$ | 2 (smaller tail probability for $S_+$) |

Note that when using $S_+$ as the statistic for a one-sided test, the right- or left-tail probability is used depending on whether $p_+$ is the larger or smaller of the two parameters, $p_+$ and $p_-$, under the alternatives.

The test could also be based on $S_-$. If $S_-$ is too much larger than its expected value, $.5n$, the hypothesis $H$ should be rejected in favor of the alterna-

---

[1]Throughout this book, the rule for finding $P$-values with a two-sided alternative is given as two times the smaller of the right- and left-tail probabilities, that is, the tail probability that is smaller than .5. The two tail probabilities can be equal; then each must be greater than or equal to .5 and doubling the one tail probability may give a $P$-value which exceeds 1 (because the central value of the test statistic is counted twice). (In the sign test procedure, for example, if $S+ = S- = .5n$, each tail probability exceeds .5.) Since no probability can exceed one, we define any two-sided $P$-value as the smaller of 1.0 and the value calculated from the rule. This will not be mentioned again, since the situation seldom arises and it is intuitively clear that $H$ is supported whenever the test statistic equals its expected value under $H$ or the median of its sampling distribution.

tive that minus signs predominate, that is, $A_-$: $p_+ < p_-$. The appropriate
$P$-value is then a right-tail probability from Table F for the observed value
of $S_-$. Similarly, the appropriate $P$-value for the alternative $A_+$: $p_+ > p_-$ is
a left-tail probability for $S_-$. Thus we could just as well present the following
guide for finding $P$-values.

| Alternative | $P$-value (Table F) |
|---|---|
| $A_+$: $p_+ > p_-$ | Left-tail probability for $S_-$ |
| $A_-$: $p_+ < p_-$ | Right-tail probability for $S_-$ |
| $A$: $p_+ \neq p_-$ | 2 (smaller tail probability for $S_-$) |

Note that for the one-sided tests the tails used with $S_-$ are simply the reverse
of those given for $S_+$. This follows because $S_-$ is too small whenever $S_+$ is too
large, and vice versa.

These guides may be combined in such a way that $P$-values are always
found as right-tail probabilities, or always as left-tail probabilities, as
follows:

| Alternative | $P$-value (Table F) | |
|---|---|---|
| $A_+$: $p_+ > p_-$ | Right tail for $S_+$ | Left tail for $S_-$ |
| $A_-$: $p_+ < p_-$ | Right tail for $S_-$ | Left tail for $S_+$ |
| $A$: $p_+ \neq p_-$ | 2 (right tail for the larger of $S_+$ and $S_-$) | 2 (left tail for the smaller of $S_+$ and $S_-$) |

Right-tail probabilities are somewhat more familiar to most users of statistics, and therefore we shall use only the boxed part of the guide above. Note
then that the test statistic always corresponds to the sign that is stated as
*more* frequent in the alternative, that is, $S_+$ is used for $A_+$ and $S_-$ is used for
$A_-$. If the one-sided alternative is set up in the proper direction, the test
statistic then corresponds to the sign that predominates in the sample, that
is, the larger of $S_+$ and $S_-$. However, it is generally more convenient to count
the smaller number of signs, whether plus or minus, and find the other,
larger number from (2.1).

Table F gives probabilities for the binomial distribution only for
$n \leq 20$, since that distribution is well approximated by the normal distribution for larger $n$. Table A gives tail probabilities of the standard normal
distribution in terms of the standardized random variable $z$; the table entries
are right-tail probabilities for $z$ positive and these are equal to left-tail

probabilities for the negative of $z$ because the distribution of $z$ is symmetric about zero. Recall from Section 3 of Chapter 2 that any variable is standardized by subtracting its mean and dividing that difference by its standard deviation. A binomial variable with $\theta = .5$ has mean $.5n$ and standard deviation $.5\sqrt{n}$. Since we are going to use a continuous distribution to approximate a discrete probability distribution, the approximation is usually improved by considering $S_+$ (or $S_-$) also as continuous in the sense that an integer value of $S_+$ actually represents all values in the interval $(S_+ - 0.5, S_+ + 0.5)$. Since a right-tail probability for the integer $S_+$ includes probabilities for all values greater than or equal to $S_+$, in a continuous distribution we represent this interval as $(S_+ - 0.5, \infty)$. For example, in order to approximate $P(S_+ \geq 9)$, we find $P(S_+ \geq 8.5)$. The quantity 0.5 is called a *correction for continuity*. Because of the manner in which normal curve probabilities are given in Table A, it is more convenient to deal only with right-tail probabilities. Then the appropriate test statistic is always the larger of $S_+$ and $S_-$, and the correction for continuity is effected by always subtracting .5 from that value. Thus we define the large sample test statistic as one of the following standardized random variables:

$$z_{+,R} = \frac{S_+ - 0.5 - .5n}{.5\sqrt{n}} \qquad z_{-,R} = \frac{S_- - 0.5 - .5n}{.5\sqrt{n}}. \qquad (2.2)$$

The statistic $z_{+,R}$ should be used with Table A if $S_+$ is appropriate for use with Table F, and $z_{-,R}$ otherwise. Thus we have the following guide for finding asymptotic approximate $P$-values for the ordinary sign test:

| Alternative | $P$-value (Table A) |
|---|---|
| $A_+$:  $p_+ > p_-$ | Right-tail probability for $z_{+,R}$ |
| $A_-$:  $p_+ < p_-$ | Right-tail probability for $z_{-,R}$ |
| $H_1$:  $p_+ \neq p_-$ | 2 (right-tail probability for the larger[2] of $z_{+,R}$ and $z_{-,R}$ |

Note the complete analogy between the $z$ and $S$ test statistics, and between the procedures for finding $P$-values in terms of right-tail probabilities.

### example   2.1 (continued)

Applying the notation developed here to Example 2.1, we let a plus sign designate a wife who is classified as more conservative than her husband.

---

[2]Since $z_{+,R}$ and $z_{-,R}$ cannot both be positive, the larger of the two is always the nonnegative value, which in turn corresponds to the larger of $S_+$ and $S_-$. If $z_{+,R} = z_{-,R}$, as when $S_+ = S_-$, the two-sided $P$-value is defined to be 1.0, as explained in footnote 1.

Since the political scientist is interested in detecting a predominance of plus signs, a one-sided alternative in that direction is appropriate; specifically, the hypothesis set is

$$H: \quad p_+ = p_- \qquad A_+: \quad p_+ > p_-.$$

The exact $P$-value is a right-tail probability for $S_+ = 9$ when $n = 10$, which from Table F is .0107.

We now use these same data to illustrate the procedure for finding a $P$-value using the asymptotic distribution. The appropriate statistic is $z_{+,R}$, which we calculate from (2.2) as

$$z_{+,R} = \frac{9 - 0.5 - 5}{.5\sqrt{10}} = .7\sqrt{10} = 2.21.$$

The asymptotic approximate $P$-value is the right-tail probability for 2.21 from Table A, which is .0136. Note that this result is in close agreement with the exact $P$-value. The normal approximation to the binomial with $\theta = .5$ is surprisingly accurate, even for $n$ as small as 10. However, as long as $n$ is within the range of a readily available binomial table, here $n \le 20$, we might as well give the exact $P$-value.

Recall that the initial sample in this example consisted of 50 couples, and yet the analysis is based on only those 10 couples for which degree of conservatism is rated as different for the wife and husband. As a result, data on 40 couples are ignored. It may appear that these 40 couples support $H$, and the omission biases the conclusion toward rejection of $H$. However, this is not really the case; we justify the omission by the following argument. The sign test as described here is based on the assumption of a dichotomous population, that is, every member of the population can be classified as either a success or a failure; "indifference" is not allowed. Those couples who were not rated as different in their political views are neither successes nor failures and do not belong to the population under consideration; since they are neither plus nor minus, they provide no information about either $p_+$ or $p_-$, and are not relevant for a test concerning the value of $p_+$. Thus, when using the sign test, the hypothesis formulated applies only to that population in which a difference exists, and accordingly, the conclusions also apply only to this population. The investigator must bear this in mind when he uses the result of such an analysis. In Example 2.1, any conclusion based on the $P$-value found may or may not apply to the population of all couples.

The reader may wonder whether a chi-square goodness-of-fit test could have been applied in the situation we have been discussing in this section. That is, why not consider the data as $n$ repetitions of a one-trial experiment and test the null hypothesis that the data fit the binomial distribution with

$\theta = .5$? When this is done, it is always true that $Q$, as defined in (2.2) of Chapter 2, is equal to both

$$\left(\frac{S_+ - .5n}{.5\sqrt{n}}\right)^2 \quad \text{and} \quad \left(\frac{S_- - .5n}{.5\sqrt{n}}\right)^2.$$

Thus, if we omit the continuity correction in the $z$ statistics, we find that $Q$ is equal to the square of the $z$ statistic used for a two-sided sign test. For example, the calculations needed to find $Q$ from (2.4) of Chapter 2 using the data of Example 2.1 are as follows:

| Number of successes | $f$ | $p$ | $e = 10p$ | $f^2$ |
|:---:|:---:|:---:|:---:|:---:|
| 0 | 1 | .5 | 5 | 1 |
| 1 | 9 | .5 | 5 | 81 |
| | 10 | | | 82 |

$$Q = [2(82)/10] - 10 = 6.4 \qquad \text{df} = 2 - 1 = 1 \qquad .01 < P < .02$$

Suppose that the alternative in Example 2.1 had been $A: \ p_+ \neq p_-$. Then the larger $z$ statistic is $z_{+,R}$, and its value (omitting the 0.5 in the numerator) is

$$z_{+,R} = \frac{9-5}{.5\sqrt{10}} = .8\sqrt{10} = 2.53.$$

The $P$-value for a two-sided test (without continuity correction) from Table A is then $2(.0057) = .0114$, which agrees with the $P$-value based on $Q$. This is always the case, since the square root of a chi-square variable with df = 1 follows the standard normal distribution. Further, we note that

$$(z_{+,R})^2 = (.8\sqrt{10})^2 = 6.4,$$

which agrees exactly with the value of $Q$ above.

Thus, when $n$ is so large that the sign test procedure would be based on the $z$ statistics *and* the alternative is two-sided, a chi-square goodness-of-fit test may be used instead because the results obtained are always equivalent. However, it is not possible to use this goodness-of-fit test when the alternative is one-sided, because the direction of the deviations between observed and expected frequencies are not retained in the calculation of $Q$.

*Additional Notes on One-sided Tests*   Given that the research hypothesis in Example 2.1 is that a wife is *more* likely to be more conservative than her husband when a difference exists, the appropriate alternative is indeed

$A_+$: $p_+ > p_-$. Thus, the reader may wonder why the null hypothesis was not given as a "less than or equals" type of statement, specifically as

$$H': \quad p_+ \leq p_-.$$

Although contrary to the expectation of the political scientist, it is conceivable that wives are *less* likely to be more conservative than their husbands when a difference exists. If this possibility seems reasonable enough to consider, why not broaden $H$ to $H'$ so that the hypothesis set includes every possible relationship between $p_+$ and $p_-$? There is a good statistical answer to this question, which can be explained as follows.

In any hypothesis testing problem, the *statistical* function of the statement in the null hypothesis is to determine the null sampling distribution, which is then used to find the $P$-value and thereby perform the test. The statistical function of the statement in the alternative is to determine the direction in which a sample result is to be considered extreme and thereby show what conclusion is to be reached if the data do not support the null hypothesis. A statistical test, whether distribution-free or not, simply cannot be performed without reference to a specific probability function, either exactly or approximately, as the sampling distribution of the test statistic. In most statistical tests, a unique and specific sampling distribution is determined only when the null hypothesis states an equality relationship of some kind.

In Example 2.1, or any other problem where the sign test is applicable, the number of plus signs follows the binomial distribution with parameter $\theta = p_+$. Making use of the fact that $p_+ + p_- = 1$, the statement $H$: $p_+ = p_-$ uniquely determines the value of $\theta$ to be .5. However, the statement $H'$: $p_+ \leq p_-$ is satisfied by all $\theta$ such that $\theta \leq .5$, and does not provide the information needed to specify the null sampling distribution uniquely. In spite of this arbitrariness of the null sampling distribution, whenever a one-sided alternative is clearly appropriate and the investigator wishes to include all possibilities in the hypothesis set, he can make the null hypothesis a complete negation of the alternative without causing any procedural or practical difficulty. The fact is that the statistical procedure and conclusion are *exactly* the same, whether the null hypothesis is stated as an equality, or as a combination of that equality and the inequality that is in the opposite direction of the inequality given in the alternative. That is, a test of the null hypothesis $p_+ \leq p_-$ is the same as a test of $p_+ = p_-$ as long as the alternative is $p_+ > p_-$.

In order to show that this is true, we again consider Example 2.1, but now with the modified one-sided hypothesis set

$$H': \quad p_+ \leq p_- \qquad A_+: \quad p_+ > p_-,$$

which can be written equivalently in terms of the parameter $\theta = p_+$ as

$$H': \quad \theta \leq .5 \qquad A_+: \quad \theta > .5.$$

The $P$-value for the observed data with $n = 10$, $S_+ = 9$, is no longer a single number; rather, it is a continuous function of $\theta$ for all $\theta \leq .5$, namely the sum

$$\sum_{s=9}^{10} (^{10}_{s})\theta^s(1 - \theta)^{10-s}.$$

The values of this sum can be found from Table E; the function is graphed in Figure 2.1. Note that it increases as $\theta$ increases from 0 to .5, with a unique maximum occurring at the endpoint $\theta = .5$. Hence, the $P$-value found using $\theta = .5$ is always greater than or equal to any $P$-value that would result for $\theta \leq .5$. Consequently, if the $P$-value with $\theta = .5$ is small enough to conclude that the data do not support $H$: $\theta = .5$, then the data certainly do not support $H'$: $\theta \leq .5$; the conclusion then is $A_+$: $\theta > .5$. On the other hand, if the $P$-value is large enough to conclude that the data do support $H$: $\theta = .5$, the statement $H'$: $\theta \leq .5$ is automatically supported since it says that $\theta$ is less than *or equal* to .5. Thus, even though the $P$-value is not unique under $H'$, the statistical conclusion always remains the same, and the $P$-value computed under $H$ is the largest $P$-value that could result.

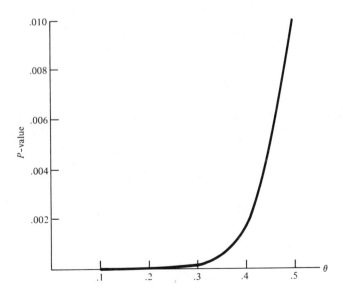

**Figure 2.1**

In summary, it is really unnecessary to distinguish between $H$ and $H'$ in practice. Since it is easier to think of a $P$-value as a unique number, $H$ is always stated as an equality in this book. This means that the method for finding a $P$-value must always be presented according to the statement in the alternative, rather than the statement in $H$. Given a one-sided test and an alternative in a specific direction, the investigator may change the statement given here as $H$ to that $H'$ which is a complete negation of the alternative without affecting the validity of the procedure. This property of one-sided tests holds for all procedures covered in this book.

*Paired Samples*    The ordinary sign test can be applied to any data consisting of a total of $n$ independently produced signs, each either plus or minus, as long as $p_+$ is the same for each sign under $H$. If these signs arise from a set of comparisons of two observations on the same experimental unit, or observations on each of two matched units (as in treatment-control experiments), or of two observations where each is taken under a different condition, the analysis proceeds in exactly the same way as explained before, using $S_+$ or $S_-$ and Table F, or the $z$ statistics of (2.2) and Table A. In fact, most applications of the ordinary sign test are of this type.

In Example 2.1 we essentially had a paired sample — that is, $n$ pairs, where each pair consists of a husband and his wife. Each sign resulted from a comparison between the degrees of conservatism of a husband and his wife. Specifically, a plus could be interpreted there as the sign of the difference, degree of conservatism of the wife minus degree of conservatism of the husband, and $\theta$ is then the probability of a positive difference.

Note that with paired samples, the level of measurement *within* each pair must be ordinal so that the comparison is meaningful, but the ordinal scale need not be the same for all pairs because no comparisons are made between units in different pairs. Since only the direction of the difference is used in the analysis, it is not necessary to make any actual quantitative measurements of units. This is one of the primary advantages of the ordinary sign test, and the property that makes it unique among statistical tests. If more precise measurements are available for each member of each pair, the Wilcoxon signed rank test (Section 3 of this chapter) makes better use of the available information. In the following example with paired observations, a simple order comparison of the elements within each pair is assumed to be the only level of measurement possible.

### example  2.2

An oil company is considering the following two alternative procedures for training prospective service station managers:

1. On-the-job training under actual working conditions for three months.
2. Training in a company-run school program concentrated over one month.

They plan to compare the two procedures by an experiment using 30 current applicants for management training.

No training program can be the sole determinant of the success of a manager; success is also affected by age, intelligence, previous experience or training, and a number of other factors that are completely separate from the training program. In order to eliminate the effects of these other factors as much as possible, personal data cards are prepared for each trainee. A card is selected, and a search is made through the remaining cards to find a trainee whose description closely resembles that of the first one selected; these two are then designated to be a pair. The process is continued, with re-matching where appropriate, until all cards have been paired as well as possible. (If a good match does not exist for a trainee, he is ignored in the analysis.) Once the pairs are determined, one member of each pair is arbitrarily selected to receive on-the-job training, while the other is assigned to the company school.

Assume that the degree of success of the graduates of these two training programs can be measured only to the extent that the personnel manager can judge which member of each pair has done a better job of managing his service station. This evaluation is performed six months after 13 pairs have completed the training programs, with the finding that 10 of the "better managers" received school training. Is this sufficient evidence to claim that school-oriented training programs are more effective?

**SOLUTION**

Since the measurements available are only ordinal within pairs, the differences are observed only as "signs." Thus only the sign test is appropriate for analysis. Denote by plus the event that a product of the school training program is judged a more successful manager than his match who is trained on the job. Then the appropriate hypothesis set for a one-sided test is

$$H: \quad p_+ = p_- \qquad A_+: \quad p_+ > p_-.$$

For these data we have $n = 13$ and $S_+ = 10$. The right-tail probability for $S_+$ from Table F is .0461, and this is the exact $P$-value.

If the question of interest is whether the data provide sufficient evidence to conclude that there is a difference between the two types of training programs, the alternative should be two-sided, as

$$H: \quad p_+ = p_- \qquad A: \quad p_+ \neq p_-.$$

Then the $P$-value is twice the right-tail probability from Table F for $S_+ = 10$, and we obtain $P = 2(.0461) = .0922$.

The $P$-values are now available for both the one-sided and two-sided hypothesis sets, but what about an answer to the question asked? In all examples up to this point, the author has made any requested decisions by simply noting whether the $P$-value is large enough to be attributed to chance so that the null hypothesis is supported, or that the $P$-value is so small that the data do not support $H$. In a sense, the author had no *right* to do this; the *privilege* was taken in order to illustrate the manner in which a purely statistical decision may be reached when there is no information available other than the data at hand. Now that the reader has a more sophisticated knowledge of statistical techniques, we can introduce the idea that a statistical result does not provide a definitive answer to many practical questions. Rather, the statistical analysis should be viewed as only one of the considerations relevant to a practical decision.

Let us look more closely at the situation described in this example. The data consist of 13 pairs of subjects selected from a population of 30 persons; the selection is based first on whether a good mate can be found and then on whether the training program is completed. Four persons were eliminated for one of these reasons; this elimination may produce bias in that the 26 persons observed may not be representative of the original group of 30. The experiment as described is reasonably well designed as regards the matching process and the random assignment to type of training. The observations give no information regarding the degree of success of the managers, either in pairs or individually. Without more precise measurements, we cannot perform a more sensitive test. In fact, the sign test is the only applicable procedure.

What are the economic and practical consequences of a decision? This depends on the answers to some other relevant questions. Would a company school training program be more expensive than on-the-job-training? What sort of physical facilities, instructional materials and personnel are needed for an organized school program? With what frequency would a school program be offered? Do the trainees receive salary during either training program? All these questions are relevant to the practical decision.

The statistical analysis performed in this example is really not sufficient evidence on which to base a decision, or even a recommendation, about the type of training programs to be offered in the future. The statistical job should be considered complete once the appropriate $P$-value is reported. The $P$-value here is no more than a tool that should be helpful to the manager in reaching *his* decision about the type of training program to implement. We do not have sufficient information to offer any more specific advice.

## 2.2    Sign Test for Location

We now consider the problem of using a set of observations from a single sample to make an inference about the value of a population parameter of location, namely the median. The median is a relevant parameter as long as the measurement scale of the underlying population variable is at least ordinal. We will see that the ordinary sign test procedure of Section 2.1 is applicable to this situation as long as the sample data are measured in such a way that an ordinal comparison can be made between each observation and the hypothesized value of the median.

*Inference Situation*   Let $M$ denote the true population median and $M_o$ some particular number. The hypothesis set to be tested is one of the following:

*Two-sided alternative*
$$H: \quad M = M_o \qquad A: \quad M \neq M_o$$

*One-sided alternatives*
$$H: \quad M = M_o \qquad A_+: \quad M > M_o$$
$$H: \quad M = M_o \qquad A_-: \quad M < M_o$$

If the question of interest is whether the true median is smaller than the specified number $M_o$, the hypothesis set with the one-sided alternative $A_-:\quad M < M_o$ should be used. For example, suppose a manufacturer is using some process for which the median time to completion is $M_o$. He will change to another process if its time to completion is smaller than $M_o$, and will continue the original process if the times to completion are the same. Similarly, the alternative $A_+:\quad M > M_o$ is appropriate if the question of interest is whether the true median is larger than $M_o$, or if the predicted difference is in this direction.

Sometimes there is no predicted direction of difference, and the only concern is whether the true median value differs from $M_o$. Then a two-sided alternative should be used.

*Sampling Situation*   Assume a random sample of $n$ observations from a population that is unspecified except for the proviso that the probability is zero that any observation equals $M_o$ (the population is continuous in the vicinity of $M_o$). This assumption implies that no observation is equal to $M_o$; that is, that each observation is either larger than or smaller than $M_o$. As long as each observation is measured on an ordinal scale with respect to $M_o$, the magnitude of each observation relative to $M_o$ can be determined. (It is not necessary to know the magnitude of any one observation relative to the others.) The data can then be represented by $n$ signs, each either plus or minus, where a plus indicates that a sample observation is larger than $M_o$. Since the numbers of signs of each type constitute the information used by the sign test, the data for analysis are count data.

The data as collected, $X_1, X_2, \ldots, X_n$, may be either (a) measured on at least an ordinal scale; (b) measured on an ordinal scale, but ordered only with respect to the value of $M_o$; or (c) nominal, as either plus or minus, where the sign represents an ordinal comparison with $M_o$.

*Rationale*   For sample data of types (a) and (b) above, it is possible to compare each observation with the number $M_o$ and record a plus or minus sign according to whether the observation is greater than or less than $M_o$. The *derived* sample data then are $n$ algebraic signs, each either plus or minus, which are the signs of the $n$ differences $D_i = X_i - M_o$. These derived data are of exactly the same form as the data analyzed for the ordinary sign test in Section 2.1, and can be regarded as a random sample drawn from a dichotomous population of plus and minus signs. Therefore, once the signs are determined, the procedure used in performing the sign test for location is exactly the same as that used in the ordinary sign test. The test statistics are now defined as follows:

$$S_+ = \text{number of plus signs among the } (X_i - M_o)$$
$$S_- = \text{number of minus signs among the } (X_i - M_o).$$

The *P*-values are found in the manner described in Section 2.1, from Table F for $n \le 20$ and the normal approximation for $n > 20$.

*Zeros (or Ties)*   If any sample observation is equal to the hypothesized median, a zero is said to occur because that difference is equal to zero. Even with the assumption that the probability of a zero is zero, a zero is not impossible and must be dealt with in practice. The sign of a zero cannot be determined, because it is neither plus nor minus. Therefore, the sign test procedure developed here does not provide any means of considering zeros in the statistical analysis. As long as the number of zeros is small relative to the size of the sample, the recommended procedure is to ignore the zeros and reduce $n$ accordingly; in other words, the test is performed using only those observations that differ from $M_o$.[3]

### example   2.3

Steel rods produced by a certain company have a median length of 10 meters when the process is operating properly. A sample of 10 rods, randomly selected from the production line, yields the following results:

$$9.8, \ 10.1, \ 9.7, \ 9.9, \ 10.0, \ 10.0, \ 9.8, \ 9.7, \ 9.8, \ 9.9.$$

Does the process need adjustment?

### SOLUTION

Since the median is the location parameter of interest, the process needs adjustment if the true median is either smaller than or larger than 10. Thus we form the hypothesis set with a two-sided alternative as

$$H: \ M = 10 \qquad A: \ M \ne 10.$$

---

[3] An alternative procedure is to allot arbitrarily half the zeros to the plus group and half to the minus group, or to assign a sign according to the outcome of a coin flip; these methods introduce an additional element of randomness, which is objectionable to most investigators.

Using the methods of the ordinary sign test, the first step is to determine whether each observation is greater than, smaller than, or equal to $M_o = 10$. Although it is easy enough just to count the number of signs of each type, the method shown in Table 2.1 can also be used. We have $S_+ = 1$, $S_- = 7$, and $n = 8$ after eliminating zeros. Since the alternative is two-sided, the $P$-value is twice the right-tail probability for the larger number of signs, which here is 7. From Table F the exact $P$-value is then $2(.0352) = .0704$.

TABLE 2.1

| X Lengths | X − 10 | Sign |
|:---:|:---:|:---:|
| 9.8 | −.2 | − |
| 10.1 | .1 | + |
| 9.7 | −.3 | − |
| 9.9 | −.1 | − |
| 10.0 | 0 | |
| 10.0 | 0 | |
| 9.8 | −.2 | − |
| 9.7 | −.3 | − |
| 9.8 | −.2 | − |
| 9.9 | −.1 | − |

As with Example 2.2, the author prefers not to make even a statistical decision here. Only two zeros were eliminated, but these constitute a large proportion — in fact, 20 percent — of the total number of observations made. (If one zero is called plus and the other minus, we have $S_- = 8$ and $n = 10$, so $P = 2(.0547) = .1094$). The problem of zeros could probably have been avoided here by using a more refined measuring device that records rod length to hundredths or even thousandths of meters. If more precise measurements can be obtained within the relevant time frame for analysis, the company should take a new sample and use a test that takes into account the magnitudes of the deviations from the specified median length, as the Wilcoxon signed rank test of the following section permits (see Example 3.1). Whether this is possible or not, the answer to the question of whether the process needs adjustment should depend not only on the $P$-value found, but also on the cost and time required for adjustment, the consequences of producing rods that are too short or too long, and so on. Thus we conclude the solution here with a report of the $P$-value.

**example  2.4**
The technical definition of the word "fatigue" has been given by the American Society for Testing and Materials as the process of localized permanent structural change in a material after subjection to stress condi-

tions. The fatigue life is the amount of stress tolerated before failure. The two main purposes of fatigue testing are (a) to estimate the fatigue properties of a material, or (b) to compare the fatigue properties of two or more sets of materials, and these goals are commonly met by performing an appropriate statistical analysis on sample data. The median is usually preferred to the arithmetic mean as a measure of average fatigue life. In fact, when the term "fatigue life" appears in the literature, it usually refers to the value of the median. (See the ASTM Technical Report (1963) for further details.) Suppose that 10 randomly chosen test specimens of a certain material are subjected to a particular stress level and the fatigue lives, as measured in kilocycles, are 612, 619, 628, 631, 640, 643, 649, 655, 663, 670. A similar standard material has a fatigue life of 625 at this same stress level. Does this material appear to have a greater tolerance for stress?

**SOLUTION**
The material under study here has a greater tolerance for stress than the standard material if its median fatigue life exceeds 625. Thus the appropriate hypothesis set is one-sided, as

$$H: \quad M = 625 \qquad A_+: \quad M > 625.$$

The observations are given in increasing order of magnitude; thus we find that $S_- = 2$ by inspection. Since no observation is equal to 625, we have $n = 10$ and $S_+ = 10 - 2 = 8$. The appropriate P-value is right tail for $S_+$, which from Table F is .0547. As with Example 2.2, we consider the problem solved at this point, primarily because other techniques of analysis are more appropriate (see Problem 3.15).

*Paired Samples*   The sign test for location can be applied to paired sample data by simply taking the difference of each pair of observations and proceeding as in the one-sample case.

*Inference Situation*   Let $M_D$ denote the median[4] of the population of differences between pairs of random variables, and let $M_o$ denote some particular number. The hypothesis sets are exactly the same as those given in the one-sample case, but with $M$ replaced by $M_D$. They are repeated here for convenience.

----

[4] In general, the median of a set of sample differences is not equal to the difference of the medians of the separate samples. Further, it is not always true that the median of a population of differences is equal to the difference of the individual population medians. (On the other hand, the mean of differences is equal to the difference of means, both for samples and populations.) Equivalence does hold in some important cases, though; for example, if $X$ and $Y$ are both normally distributed, the difference of their medians is equal to the median of the distribution of $X - Y$.

$$
\begin{array}{llll}
H: & M_D = M_o & A: & M_D \neq M_o \\
H: & M_D = M_o & A_+: & M_D > M_o \\
H: & M_D = M_o & A_-: & M_D < M_o
\end{array}
$$

In many applications, $M_o$ is zero because the null hypothesis of interest is "no difference."

*Sampling Situation*   The data as collected consist of either (a) a random sample of $n$ pairs of observations, $(X_1, Y_1), (X_2, Y_2), \ldots, (X_n, Y_n)$, where both $X$ and $Y$ are measured on at least an interval scale, or measured on an ordinal scale that is the same for both samples and also for the number $M_o$; or (b) a random sample of $n$ differences of paired observations (which may not have been observed individually), $D_1, D_2, \ldots, D_n$, where the magnitudes of the differences relative to $M_o$ can be classified as either plus or minus.

The minimal requirement for analysis by the sign test is that the sign of the difference within a pair, $D_i = X_i - Y_i$, can be determined relative to $M_o$ for each sample pair. The data entering the analysis are the numbers of signs of each type. We assume that the probability of a zero difference is zero (the population of differences is continuous in the vicinity of $M_o$).

*Test Statistic*   The procedure with paired samples is the same as in the one-sample sign test for location, except that the signs considered are now those of the differences

$$
D_i - M_o = X_i - Y_i - M_o,
$$

and the test statistics are defined as

$$
\begin{aligned}
S_+ &= \text{number of plus signs among the } (D_i - M_o) \\
S_- &= \text{number of minus signs among the } (D_i - M_o).
\end{aligned}
$$

If any differences $D_i - M_o$ are equal to 0, those observations are ignored and $n$ is reduced to correspond to the actual number of signs being analyzed. $P$-values are found exactly as before, from Table F with $S_+$ or $S_-$ or from Table A with the $z$ statistics given in (2.2).

**example 2.5**

In order to investigate the research hypothesis that leadership is a trainable quality, an army psychologist set up matched pairs of individuals from the Officer Candidate School. The matching considered characteristics such as intelligence, personality, and educational background. One member of each pair, called $T$, is randomly selected to receive special training while the other member, $C$, receives no special training. Following the training program, all individuals are given tests designed to evaluate their leadership qualities. The scores are shown in Table 2.2. Perform a suitable test assuming that the rating scale is crude in the sense that the scores reported have no precise quantitative properties except relative to each other.

**TABLE 2.2**

| Pair | *T* Score | *C* Score | Sign of *T − C* |
|------|-----------|-----------|-----------------|
| 1 | 13 | 10 | + |
| 2 | 19 | 7 | + |
| 3 | 34 | 20 | + |
| 4 | 24 | 38 | − |
| 5 | 40 | 22 | + |
| 6 | 39 | 15 | + |

**SOLUTION**

With such measurement of scores, only the signs of the differences are indicative of relative leadership. The pairs of observations are ordinal data relative to each other only, and no test other than the sign test is appropriate. If leadership is trainable, members of the special training group can be expected to have higher scores than their respective mates. Thus the alternative should be one-sided in this direction. If $D$ is the variable representing the population of differences, $T$ scores minus $C$ scores, the appropriate hypothesis set is

$$H: \quad M_D = 0 \qquad A_+: \quad M_D > 0.$$

From the signs of the sample differences $T - C$ given with the data in Table 2.2, we count $S_+ = 5, S_- = 1, n = 6$. The exact $P$-value is then the right-tail probability from Table F for $S_+ = 5$ when $n = 6$, and this is .1094.

**example  2.6**

Many studies have been conducted in an effort to establish a relationship between order of birth, intelligence, and scientific achievement in school age children. The empirical results have supported the conclusion that scientific achievement decreases with birth rank whereas intelligence increases with birth rank for children. Little research seems to have been performed in an effort to determine whether these same relationships hold for adults. De Lint (1967) reported the results of a study conducted to investigate this question. Suppose that the data recorded in Table 2.3 are collected for 13 randomly chosen pairs of adult siblings. Can we conclude that the relationship between IQ and birth rank is the same for adults as for children?

**TABLE 2.3**

| X<br>Firstborn | Y<br>Later Born | X − Y<br>Sign |
|:---:|:---:|:---:|
| 86 | 82 | + |
| 90 | 94 | − |
| 91 | 96 | − |
| 101 | 106 | − |
| 93 | 92 | + |
| 85 | 90 | − |
| 92 | 98 | − |
| 115 | 120 | − |
| 72 | 74 | − |
| 75 | 80 | − |
| 120 | 130 | − |
| 106 | 110 | − |
| 104 | 109 | − |

**SOLUTION**

If the relationship for adults is the same as for children, the firstborn sibling can be expected to have a lower IQ score than the later-born sibling on the average. Accordingly, a one-sided alternative is appropriate. Define the difference $D$ as firstborn minus later born. Since the data appear to be ordinal, we base the decision on an inference about the median difference. Thus the hypothesis set is

$$H: \quad M_D = 0 \qquad A_-: \quad M_D < 0.$$

If we label the firstborn column as $X$ and the later born as $Y$, then $D = X - Y$. Since $M_o = 0$, the signs entering the analysis are simply the signs of $D$ as given with the data in Table 2.3.

We find that $S_+ = 2$, $S_- = 11$ and $n = 13$. The exact $P$-value is a right-tail probability on $S_-$, which is found from Table F as .0112. This means that as many as 11 negative differences occur rarely for variables whose median difference is equal to 0. Accordingly, we can conclude that the data do not support the null hypothesis; rather, the data suggest that intelligence does tend to increase with birth rank.

We use these same data to illustrate the sign test procedure for large samples. The test statistic $z_{-,R}$ is calculated from (2.2) as

$$z_{-,R} = \frac{11 - 0.5 - .5(13)}{.5\sqrt{13}} = \frac{8\sqrt{13}}{13} = 2.22.$$

The right-tail probability from Table A is .0132.

## 2.3   Confidence Intervals for the Median or Median Difference

The statistical procedure of the sign test for location can also be used for estimation. Specifically, a confidence interval can be constructed for the population median in the one-sample case, or the median difference in the paired-sample case. Since the endpoints of these confidence regions are order statistics of the sample, and order statistics have many useful applications in nonparametric statistics, we first define these quantities and introduce some of their general properties.

Suppose we have a set of five observations given as $X_1 = -1, X_2 = 7, X_3 = -6, X_4 = 4$, and $X_5 = 2$. The arrayed data are then

$$-6, -1, 2, 4, 7.$$

The previous subscripts on the $X$'s have no significance regarding relative magnitude; they may indicate the order in which the observations are taken, or the number of the sample element. The usual notation to show relative magnitude is to introduce parenthetical subscripts. The arrayed data are then written as $X_{(1)} = -6, X_{(2)} = -1, X_{(3)} = 2, X_{(4)} = 4, X_{(5)} = 7$; these variables are called the order statistics for the data.

In the general case, we define $X_{(r)}$ as the $r$th order statistic in a set of $n$ variables if $X_{(r)}$ is the $r$th from the smallest in the array. Then for a sample of $n$ different observations $X_1, X_2, \ldots, X_n$, the order statistics of the sample are $X_{(1)} < X_{(2)} < \cdots < X_{(n)}$, that is, the original sample after arrangement in increasing order of magnitude.

The order statistics do not have the same probability function as the original variables, and in fact are not even independent even if the original variables are, as would be true for a random sample. This nonindependence property is intuitively clear since the position of any variable in the array depends on its magnitude relative to all the other variables. However, the order statistics do have nice distribution properties; in particular, the c.d.f. of the $r$th order statistic of a random sample follows the binomial distribution, and a certain transformation of the $r$th order statistic produces a variable that follows the continuous uniform distribution over $(0, 1)$. A discussion of these and other properties is beyond the scope of this book. The interested reader with mathematical background is referred to Gibbons (1971, Chapter 2) and Wilks (1948).

We now return to the problem of constructing a confidence interval for the population median. The general discussion of nonparametric confidence intervals in Chapter 1 introduced the notion that a confidence region consists of those values of a parameter that, if stated in a null hypothesis, would be supported with a certain probability by a test performed on the sample data. Therefore, for any test procedure for inferences about the value of a parameter, there is a corresponding confidence procedure.

Since the present discussion is limited to confidence intervals both of whose endpoints are finite numbers (and are therefore two-sided), the corresponding test must be for a hypothesis set with a two-sided alternative.

Accordingly, the interval, with confidence coefficient $\gamma$, which corresponds to the sign test procedure, should consist of these values of $M$ that, if stated in a null hypothesis, would be supported by the sample data with probability $\gamma$ (when the test is appropriate for a two-sided alternative). The number $\gamma$ is chosen by the investigator. However, if $\gamma$ is to be an *exact* confidence coefficient, then $1 - \gamma$ must be equal to two times any of the probabilities (which are less than .5) given in Table F, since these are the only possible exact $P$-values that could result when the corresponding two-sided sign test is performed. In other words, $(1 - \gamma)/2$ must equal one of the probabilities given in Table F for the appropriate value of $n$. (The restriction $P < .5$ is necessary to ensure that $0 < \gamma < 1$.) Values chosen for $\gamma$ are typically in the range from .85 to .99, but smaller or larger values are justifiable in certain situations. The higher the confidence coefficient, the wider the interval; however, the wider the interval, the less information it provides about the parameter. Hence in choosing $\gamma$, the investigator must weigh the relative importance of precision in the estimate versus confidence in the estimate for that particular application.

For the data of Example 2.3 concerning the lengths of steel rods, the sample size is $n = 10$. From Table F, the choices of $\gamma$ are then as follows:

| *P*-value | $\gamma = 1 - 2(P\text{-value})$ |
|---|---|
| .0010 | .9980 |
| .0107 | .9786 |
| .0547 | .8906 |
| .1719 | .6562 |
| .3770 | .2460 |

Suppose we feel that $\gamma = .8906$ is a sufficient level of confidence for this situation. What values of $M_o$ would be supported by the data with a two-sided $P$-value at least .1094? Let us try a few values of $M_o$ and see. The 10 lengths are listed in Table 2.4 in an array, and various arbitrarily selected values of $M_o$ are subtracted to see what signs result so that $S_+$ or $S_-$ can be counted and the $P$-value obtained. The $P$-values for the respective numbers $S_+$ or $S_-$ in the next to last row are found as twice the corresponding right-tail probabilities from Table F. We see that if any number smaller than 9.8 is subtracted, there will be at least 8 plus signs and then the $P$-value is at most .1094, and if any number larger than 9.8 is subtracted there will be less than 8 plus signs and the $P$-value is greater than .1094. Similarly, if any number smaller than 10.0 is subtracted, there will be 7 or less minus signs, and the $P$-value is greater than .1094; however, subtracting any number larger than 10.0 gives at least 9 minus signs, and the $P$-value is smaller than .1094. Hence the $P$-value is at least .1094 when $M_o$ is any number between 9.8 and 10.0; the confidence interval for $M$ is then $9.8 \leq M \leq 10.0$.

**TABLE 2.4**

| X | X − 9.75 | X − 9.79 | X − 9.81 | X − 9.99 | X − 10.01 |
|---|---|---|---|---|---|
| 9.7 | − | − | − | − | − |
| 9.7 | − | − | − | − | − |
| 9.8 | + | + | − | − | − |
| 9.8 | + | + | − | − | − |
| 9.8 | + | + | − | − | − |
| 9.9 | + | + | + | − | − |
| 9.9 | + | + | + | − | − |
| 10.0 | + | + | + | + | − |
| 10.0 | + | + | + | + | − |
| 10.1 | + | + | + | + | + |
| Larger of $S_+$, $S_-$ | 8 | 8 | 5 | 7 | 9 |
| $P$-value | .1094 | .1094 | 1.0000* | .3438 | .0214 |

*Since $S_+ = S_-$ here, the $P$-value equals 1.00 by definition.

For any set of data, we can always proceed by trial and error in such a manner, testing various arbitrary values of $M$ in hopes of eventually delineating the endpoints of the interval of acceptable values of $M$ and therefore the endpoints of the confidence interval. However, a much simpler and more efficient method is possible using order statistics. Let $X_{(r)}$ and $X_{(u)}$ be the $r$th and $u$th order statistics for the data for some $r < u$. Then the median is larger than the $r$th order statistic — that is, $X_{(r)} < M$ — if and only if at *least* $r$ observations are smaller than the median. Similarly, the median is smaller than the $u$th order statistic, that is $M < X_{(u)}$, if and only if at *most* $u - 1$ observations are smaller than the median. Combining these conditions, we have $X_{(r)} < M < X_{(u)}$ if and only if the number of observations smaller than $M$ is at least $r$ and at most $u - 1$. Hence if the confidence coefficient is $\gamma$ and the corresponding two-tailed test at level $1 - \gamma$ for the original sample size $n$ has a left-tail critical value equal to $k$ for the number of plus or minus signs, the confidence interval endpoints are those numbers that are in the $(k + 1)$th positions from either end in an array of the data, that is, the order statistics $X_{(k+1)}$ and $X_{(n-k)}$.

The value of $k$ can be found from Table F if $n \leq 20$. If $n > 20$, the approximate value of $k$ based on the normal distribution with continuity correction is

$$k = .5(n - 1 - z\sqrt{n}), \tag{2.3}$$

where $z$ is found from Table A as the positive number that corresponds to a right-tail probability of $(1 - \gamma)/2$. The result for $k$ from (2.3) is usually not an integer. If we always round down to the next lower integer, a conservative result is obtained.

The discussion above did not consider the possibility that an observation may be equal to the true median. Since this can occur in practice, it is recommended that the endpoints be included in the confidence interval so that the statement is always conservative. That is, the confidence interval should be given as

$$X_{(k+1)} \leq M \leq X_{(n-k)}.$$

Note, however, that $n$ should not be reduced for the confidence interval procedure since we do not know the true value of $M$.

We illustrate this simple procedure based on order statistics using the data of Example 2.3 again. When $n = 10$, a two-sided test at exact level .1094 would reject if the number of plus signs is less than or equal to 2 or greater than or equal to 8. Hence $k = 2$ and the confidence interval at level $1 - .1094 = .8906$ is $X_{(3)} \leq M \leq X_{(8)}$, or $9.8 \leq M \leq 10.0$. As a verification of this result, we note that if 9.99 is subtracted from each observation, there are 3 plus signs and therefore a null hypothesis $M = 9.99$ would be accepted; however, if 10.01 is subtracted, there is only 1 plus sign and consequently a null hypothesis $M = 10.01$ would be rejected. Thus 10.0 is indeed the upper borderline between acceptance and rejection. Similarly for the lower borderline. Note that this interval is exactly the same as that found by trial and error, but it can be determined with considerably less effort using order statistics.

The procedure using order statistics also gives one-sided confidence regions with confidence coefficient $\gamma$ that correspond to one-sided tests at level $1 - \gamma$. A region giving only a lower confidence bound corresponds to a one-sided test against $A_+$ at level $1 - \gamma$, whereas an upper bound corresponds to a one-sided test against $A_-$. Suppose that $k$ is the left-tail critical value at level $1 - \gamma$, that is, the value of $S_+$ or $S_-$ that has a left-tail $P$-value of $1 - \gamma$. Then $X_{(k+1)}$ is a lower confidence bound with confidence coefficient $\gamma$, whereas $X_{(n-k)}$ is an upper confidence bound with coefficient $\gamma$. Consider the data of Example 2.3 again. At level .0547 the left-tail critical value is $k = 2$; thus $X_{(3)} = 9.8$ is the corresponding lower confidence bound, and we write $M \geq 9.8$ with confidence coefficient $1 - .0547 = .9453$. Similarly, since $n - k = 8, X_{(8)} = 10.0$ is the corresponding upper confidence bound, and we write $M \leq 10.0$ with confidence coefficient .9453.

If the data consist of paired samples, confidence regions for the median difference $M_D$ are found in exactly the same way using the order statistics for the differences of paired observations, that is, using $D_{(1)} \leq D_{(2)} \leq \cdots \leq D_{(n)}$, the arrayed values of $D_1, D_2, \ldots, D_n$, where $D = X - Y$.

### 2.4   The Binomial Test and Quantile Test

In the ordinary sign test the null hypothesis is written as $H: p_+ = p_-$. With data from a dichotomous population, we have $p_+ + p_- = 1$, and thus $H$ can be stated equivalently in terms of the binomial parameter $\theta = p_+ = .5$. In this context, the ordinary sign test can be regarded as a special case of a more general inference procedure called the binomial test. This test is applicable to sample data from any dichotomous population and tests of the null hypothesis $H: \theta = \theta_o$, where $\theta_o$ is any specified number between 0 and 1. For example, suppose that a certain drug is advertised as at least 90 percent effective against a certain condition. In order to investigate this claim we might draw a random sample of $n$ persons who have the condition, administer the drug to them for some appropriate length of time, and then count the number of persons showing significant improvement. This number follows the binomial distribution with parameter $\theta$ representing the probability of significant improvement. Thus it can be used as a test statistic for the hypothesis set of interest, $H: \theta = .90$ versus $A_+: \theta > .90$.

In the general case, the hypothesis set is one of the following for some specified $\theta_o$, $0 < \theta_o < 1$.

*Two-sided alternative*
   $H: \theta = \theta_o$    $A: \theta \neq \theta_o$

*One-sided alternatives*
   $H: \theta = \theta_o$    $A_+: \theta > \theta_o$
   $H: \theta = \theta_o$    $A_-: \theta < \theta_o$

The sampling model is a random sample of size $n$, or a set of $n$ independent trials of the same experiment, or a set of comparisons of elements within $n$ pairs, where each observation is classified as either success or failure; the probability of that outcome designated as success is $\theta$. As before, let $S_+$ denote the observed number of successes, and $S_-$ the number of failures. Under the null hypothesis, the expected number of successes is $n\theta_o$, and the expected number of failures is $n(1 - \theta_o)$. $S_+$ follows the binomial distribution with parameter $\theta_o$, and $S_-$ has the same distribution but with parameter $1 - \theta_o$. (Note that $\theta_o = 1 - \theta_o$ only if $\theta_o = .5$, so that $S_+$ and $S_-$ are not identically distributed in general here.)

Either $S_+$ or $S_-$ can be used as a test statistic. For any value of $\theta_o$, the alternative $A_+$ is supported if $S_+$ is much larger than its expected value $n\theta_o$, and $A_-$ is supported if $S_+$ is much smaller than $n\theta_o$. Thus, in terms of $S_+$, the correct $P$-values are a right-tail probability for $A_+$ and a left-tail probability for $A_-$, using a table of the binomial distribution with parameter $\theta_o$. We summarize these results as follows:

| Alternative | $P$-value (Table E) |
|---|---|
| $A_+$:  $\theta > \theta_o$ | Right-tail probability for $S_+$ with $\theta_o$ |
| $A_-$:  $\theta < \theta_o$ | Left-tail probability for $S_+$ with $\theta_o$ |
| $A$:  $\theta \neq \theta_o$ | 2 (smaller tail probability for $S_+$ with $\theta_o$) |

A similar guide can be given in terms of $S_-$, but then the parameter must be $1 - \theta_o$. Further, as we explained in Section 2.1, the tails must be reversed, since $S_+$ too much larger than $n\theta_o$ can occur only if $S_-$ is too much *smaller* than $n(1 - \theta_o)$. For $S_-$ then, the $P$-values are found as follows:

| Alternative | $P$-value (Table E) |
|---|---|
| $A_+$:  $\theta > \theta_o$ | Left-tail probability for $S_-$ with $1 - \theta_o$ |
| $A_-$:  $\theta < \theta_o$ | Right-tail probability for $S_-$ with $1 - \theta_o$ |
| $A$:  $\theta \neq \theta_o$ | 2 (smaller tail probability for $S_-$ with $1 - \theta_o$) |

Recall that Table E gives the c.d.f. of the binomial distribution for selected values of $\theta$ and $n \leq 20$. Thus it can be used with either $S_+$ or $S_-$ as long as it is entered with the correct parameter value, $\theta_o$ and $1 - \theta_o$ respectively. However, the table gives only left-tail probabilities directly. Hence, the easiest way to find $P$-values is always to use left-tail probabilities. We combine the two guides above to obtain the following recommended test procedures:

| Alternative | $P$-value (Table E) |
|---|---|
| $A_+$:  $\theta > \theta_o$ | Left-tail probability for $S_-$ with $1 - \theta_o$ |
| $A_-$:  $\theta < \theta_o$ | Left-tail probability for $S_+$ with $\theta_o$ |
| $A$:  $\theta \neq \theta_o$ | 2 (smaller left-tail probability, $S_+$ with $\theta_o$ and $S_-$ with $1 - \theta_o$) |

Thus here we enter Table E with $\theta_o$ or $1 - \theta_o$, depending on whether $S_+$ or $S_-$ is the test statistic.

For large samples, asymptotic approximate $P$-values are approximated by the normal curve in the same way, but using right-tail probabilities and reversing the test statistics. The standardized variables are found by subtracting the appropriate mean, namely $n\theta_o$ for $S_+$ and $n(1 - \theta_o)$ for $S_-$, and dividing by the standard deviation $\sqrt{n\theta_o(1 - \theta_o)}$. Specifically, the $z$ statistics, with a continuity correction of 0.5, are:

$$z_{+,R} = \frac{S_+ - 0.5 - n\theta_o}{\sqrt{n\theta_o(1 - \theta_o)}} \qquad z_{-,R} = \frac{S_- - 0.5 - n(1 - \theta_o)}{\sqrt{n\theta_o(1 - \theta_o)}}.$$

Then the $P$-values are obtained from Table A as follows:

| Alternative | $P$-value (Table A) |
|---|---|
| $A_+$: $\theta > \theta_o$ | Right-tail probability for $z_{+,R}$ |
| $A_-$: $\theta < \theta_o$ | Right-tail probability for $z_{-,R}$ |
| $A$: $\theta \neq \theta_o$ | 2 (right-tail probability for the larger of $z_{+,R}$ and $z_{-,R}$) |

### example 2.7

A supermarket is presently closed all day Sunday. They have made an extensive study of the local situation and determined that it will be worthwhile to start opening on Sunday afternoons only if at least 25 percent of their present and potential customers state that they would do their regular grocery shopping there on Sunday if the market were open. A random sample of 50 households in the immediate vicinity are interviewed with regard to their shopping habits and preferences. The interviewers felt that only 18 of these households could be considered regular or potential customers. Each of these 18 is asked about the proposed availability of Sunday shopping, and 7 respond favorably. Should the market open on Sundays?

### SOLUTION

We define a success as a household customer who prefers or needs the availability of Sunday shopping at this market. Since we need to determine whether the data support the conclusion that the probability of success is at least .25, the appropriate hypothesis set is

$$H: \quad \theta = .25 \qquad A_+: \quad \theta > .25.$$

The data results are $S_+ = 7$, $S_- = 11$, $n = 18$. The guide indicates that the appropriate $P$-value is left-tail for $S_- = 11$ with parameter .75, which from Table E is $P = .1390$. This result is not small enough to conclude that the data contradict $H$ in the direction of $A_+$, and hence we accept $H$. These data do not indicate sufficient demand for Sunday shopping to warrant opening.

The sign test for location can also be generalized for inferences concerning quantile parameters other than the median. The median is a quantile of order .50, since the probability is .50 that the variable is smaller than its median. A null hypothesis concerning the value of a quantile of any order other than .50 can be performed using the binomial test, in the same way as the ordinary sign test is used for inferences about the median.

The null hypothesis states a specific value $Q_o$ for a quantile $Q$ of particular order $p_-$ in a population of variables (or differences of variables). We assume that the probability is zero that any variable in the relevant population equals $Q_o$. For a random sample of $n$ observations (or differences of paired observations), we can determine $n$ signs, each either plus or minus, according to the magnitudes relative to $Q_o$ (observations equal to $Q_o$ are ignored and $n$ is reduced accordingly). Let $S_-$ be the number of values smaller than $Q_o$, and $S_+$ the number larger. Recall that a quantile $Q$ is of order $p_-$ if the probability below $Q$ equals $p_-$. Then under the null hypothesis $S_-$ follows the binomial distribution with parameter $p_-$, and $S_+$ follows the binomial distribution with parameter $p_+ = 1 - p_-$. As in the binomial test, let $\theta_o$ denote $p_+$, that is, the probability of a plus sign under the null hypothesis $H: \quad Q = Q_o$. Then $S_+$ follows the binomial distribution with parameter $\theta_o$, and its expected value under $H$ is $n\theta_o$. Similarly, $S_-$ has expected value $n(1 - \theta_o)$. If the quantile of order $p_- = 1 - \theta_o$ is really larger than $Q_o$, we expect fewer than $n(1 - \theta_o)$ observations to be smaller than $Q_o$. Since $S_-$ is the number smaller than $Q_o$, the appropriate $P$-value is then left tail for $S_-$ with parameter $1 - \theta_o$. On the other hand, if this quantile is really smaller than $Q_o$, we expect fewer than $n\theta_o$ observations to be larger than $Q_o$; since $S_+$ is the number larger than $Q_o$, this situation calls for a $P$-value that is right tail for $S_+$ with parameter $\theta_o$. The correct procedures in terms of left-tail probabilities are summarized as follows:

| Alternative | $P$-value (Table E) |
|---|---|
| $A_+: \quad Q > Q_o$ | Left-tail probability for $S_-$ with $1 - \theta_o$ |
| $A_-: \quad Q < Q_o$ | Left-tail probability for $S_+$ with $\theta_o$ |
| $A: \quad Q \neq Q_o$ | 2 (smaller left-tail probability, $S_+$ with $\theta_o$ and $S_-$ with $1 - \theta_o$) |

For sample sizes outside the range of Table E, we compute

$$z_{+,R} = \frac{S_+ - 0.5 - n\theta_o}{\sqrt{n\theta_o(1 - \theta_o)}} \qquad z_{-,R} = \frac{S_- - 0.5 - n(1 - \theta_o)}{\sqrt{n\theta_o(1 - \theta_o)}}.$$

The $P$-values are obtained from Table A according to the following guide:

| Alternative | P-value (Table A) |
|---|---|
| $A_+$: $Q > Q_o$ | Right-tail probability for $z_{+,R}$ |
| $A_-$: $Q < Q_o$ | Right-tail probability for $z_{-,R}$ |
| $A$: $Q \neq Q_o$ | 2 (right-tail probability for the larger of $z_{+,R}$ and $z_{-,R}$) |

Notice that these guides are exactly the same as those given for the binomial test. This follows because $Q < Q_o$ implies that $\theta < \theta_o$, and $Q > Q_o$ implies that $1 - \theta < 1 - \theta_o$; that is, $\theta > \theta_o$.

**example 2.8**
Consider again the data of Example 2.6, but suppose that the question of interest is whether the value of the .25th quantile of difference scores is smaller than $-3$. The hypothesis set then is

$$H: \quad Q = -3 \qquad A: \quad Q < -3,$$

where $Q$ denotes the quantile of order $p_- = .25$. If $H$ is not supported, then we can conclude that the 25th percentile point of the differences, firstborn minus later born, is smaller than $-3$, which is the same as saying that more than 25 percent of these differences are smaller than $-3$, or that $-3$ is a quantile point of order larger than .25. In order to perform the test, we need the signs of $X - Y - (-3) = X - Y + 3$, where $X$ is firstborn and $Y$ is later born.

**TABLE 2.5**

| $X - Y$ | Sign of $X - Y + 3$ | $X - Y$ | Sign of $X - Y + 3$ |
|---|---|---|---|
| 4 | + | $-5$ | $-$ |
| $-4$ | $-$ | $-2$ | + |
| $-5$ | $-$ | $-5$ | $-$ |
| $-5$ | $-$ | $-10$ | $-$ |
| 1 | + | $-4$ | $-$ |
| $-5$ | $-$ | $-5$ | $-$ |
| $-6$ | $-$ | | |

The data in Table 2.5 give $S_+ = 3$, $S_- = 10$, $n = 13$. The guide indicates that the appropriate P-value is a left-tail probability for $S_+$ with $\theta_o = p_+ = .75$, which from Table E is .0001. Hence we conclude that the null hypothesis is not supported.

# 3   Wilcoxon Signed Rank Procedures

## 3.1   Wilcoxon Signed Rank Test for Location

Since the sign test for location utilizes only the signs of the differences between each observation (or the difference of each pair of observations) and the hypothesized median $M_o$, the magnitudes of the differences are not considered. If this information is available because the data are measured on a scale higher than nominal, a test procedure that takes into account the size of these relative magnitudes might be expected to give better performance. The Wilcoxon signed rank test statistic permits both the signs and the magnitudes of the differences to influence the inference. The only additional population assumption required is that of symmetry about the true median or median difference. This test also has a corresponding procedure for finding confidence interval estimates of the median or median difference. We begin the exposition by considering again the situation in Example 2.3.

**example   3.1**
    Assume that the company in Example 2.3 was able to measure the lengths of the steel rods with more precision, and obtained the following results:

$$9.83, \ 10.09, \ 9.72, \ 9.87, \ 10.04, \ 9.95, \ 9.82, \ 9.73, \ 9.79, \ 9.90.$$

With the additional assumption that the lengths are symmetrically distributed about their median, we can use the Wilcoxon signed rank procedure to test the same hypothesis set

$$H: \quad M = 10 \qquad A: \quad M \neq 10.$$

**TABLE 3.1**

| $X$ | $X - 10$ | $|X - 10|$ | Rank of $|X - 10|$ | Sign of $(X - 10)$ |
|---|---|---|---|---|
| 9.83 | $-.17$ | .17 | 6 | $-$ |
| 10.09 | .09 | .09 | 3 | $+$ |
| 9.72 | $-.28$ | .28 | 10 | $-$ |
| 9.87 | $-.13$ | .13 | 5 | $-$ |
| 10.04 | .04 | .04 | 1 | $+$ |
| 9.95 | $-.05$ | .05 | 2 | $-$ |
| 9.82 | $-.18$ | .18 | 7 | $-$ |
| 9.73 | $-.27$ | .27 | 9 | $-$ |
| 9.79 | $-.21$ | .21 | 8 | $-$ |
| 9.90 | $-.10$ | .10 | 4 | $-$ |

The first step is to obtain the differences between each observation and the hypothesized median — that is, the $X - 10$ values in the second column of Table 3.1. Then we temporarily ignore the signs of these differences and rank their absolute values according to relative magnitude, as shown in the third and fourth columns.

If the lengths are symmetrically distributed about the median value 10, the sum of the ranks assigned to positive differences should be approximately equal to the sum of the ranks assigned to negative differences. Suppose that we add those ranks in the fourth column that also have a plus sign in the fifth column. The result, called the "positive rank sum," is

$$T_+ = 3 + 1 = 4.$$

Similarly, the "negative rank sum" is

$$T_- = 6 + 10 + 5 + 2 + 7 + 9 + 8 + 4 = 51.$$

Thus, in this example not only are considerably more of the differences negative, but also more of the negative differences are larger in absolute value. In order to see whether the negative rank sum $T_- = 51$ is too extreme (too much larger than $T_+$) to have occurred by chance under the null hypothesis, we need to find the probability of obtaining such a result through chance alone. As will be explained later, this probability is found by doubling the value .007, which is the entry in Table G for $n = 10$, $T_- = 51$. The exact $P$-value is then quite small, namely $2(.007) = .014$. We conclude that the data do not support the null hypothesis that the steel rods have a median length of 10.

The procedure of statistical analysis using the Wilcoxon signed rank statistic is now described in general.

*Inference Situation* (*One Sample*)   For $M$, the median of a symmetric and continuous population, the hypothesis set to be tested is one of the following:

*Two-sided alternative*
$$H: \quad M = M_o \qquad A: \quad M \neq M_o$$

*One-sided alternatives*
$$H: \quad M = M_o \qquad A_+: \quad M > M_o$$
$$H: \quad M = M_o \qquad A_-: \quad M < M_o$$

*Sampling Situation*   The data consist of a random sample of $n$ observations that are drawn from a symmetric and continuous population and measured on either (a) at least an interval scale, or (b) an ordinal scale in such a way that the ranks and signs needed for the test statistic can be determined (see below).

*Rationale*   Denote the sample observations by $X_1, X_2, \ldots, X_n$. If $H$ is true, these observations are symmetrically distributed about $M_o$ and the $n$ differences between each observation and $M_o$; that is,

$$D_i = X_i - M_o$$

are symmetrically distributed about zero. As a result, positive and negative differences of similar absolute magnitude should be approximately balanced. In order to compare the absolute magnitude of these differences, we temporarily ignore their signs and rank the absolute values $|D_1|, |D_2|, \ldots, |D_n|$ according to relative magnitude. In other words, rank 1 is given to the smallest absolute difference $|D_i|$, rank 2 to the second smallest, $\ldots$, and rank $n$ to the largest. The sum of the ranks assigned to the absolute values of those differences whose original sign is plus, called positive ranks, should be approximately equal to the sum of the ranks of those absolute differences that are originally minus, called negative ranks (even though the ranks themselves are positive). If the sum of the positive ranks is much larger than the sum of the negative ranks, most of the large ranks belong to positive differences, and the data support the alternative $A_+$: $M > M_o$. Similarly, a large sum of negative ranks reflects the situation in which large ranks are associated primarily with negative differences, and the data support $A_-$: $M < M_o$. If either sum is too large, the two-sided alternative is supported. Since the sum of the positive ranks and the negative ranks is a constant value, namely

$$1 + 2 + \cdots + n = \frac{n(n+1)}{2},$$

only one of these sums need be computed, and the inference can be based on either sum.

*Test Statistic*   The Wilcoxon signed rank test statistic is defined as either

$$T_+ = \text{sum of positive ranks, or}$$
$$T_- = \text{sum of negative ranks.}$$

Note that $T_+$ and $T_-$ are both defined as nonnegative numbers, and that

$$T_+ + T_- = \frac{n(n+1)}{2}. \tag{3.1}$$

*Finding the P-value*   The null sampling distributions of $T_+$ and $T_-$ are identical. Each is symmetric about $n(n+1)/4$, the common mean value. $T_+$ and $T_-$ must be integer-valued; they range from a minimum value of zero to a maximum of $n(n+1)/2$. Table G gives the cumulative probabilities for both the left and right tails of the common sampling distribution. However, because of the symmetry, plus the fact that $T_+ + T_- = n(n+1)/2$, any

$P$-value can be defined "au naturel" in terms of a right-tail probability for one of $T_+$ and $T_-$, and analogous to those specified for the ordinary sign test in the preceding section. In Table G, $T$ is interpreted as either $T_+$ or $T_-$.

According to the rationale above, a value of $T_+$ that is much larger than $T_-$ supports $A_+$, and this implies that $T_+$ is much larger than its expected value. Thus the $P$-value for the one-sided hypothesis set with alternative $A_+$ is a right-tail probability for $T_+$. Similarly, a right-tail probability for $T_-$ is the appropriate $P$-value for the alternative $A_-$. The $P$-value for a two-sided alternative is twice the probability that the larger of $T_+$ or $T_-$ is too large. These $P$-values are summarized as follows:

| Alternative | $P$-value (Table G) |
|---|---|
| $A_+$:  $M > M_o$ | Right-tail probability for $T_+$ |
| $A_-$:  $M < M_o$ | Right-tail probability for $T_-$ |
| $A$:  $M \neq M_o$ | 2 (right-tail probability for the larger of $T_+$ and $T_-$) |

Note that the test procedure outlined here calls for the larger of $T_+$ and $T_-$. Since it is usually easier to calculate the smaller statistic, it may be more convenient to find the smaller sum directly and then obtain the other, larger value from Eq. (3.1).

Table G gives $P$-values only for $n \leq 15$, since the sampling distribution of a Wilcoxon signed rank statistic is well approximated by its asymptotic distribution for larger $n$. This asymptotic approximation is the normal curve for the standardized value of $T_+$ or $T_-$, which is found by subtracting the mean $n(n + 1)/4$ and dividing by the standard deviation

$$\sqrt{\frac{n(n + 1)(2n + 1)}{24}}.$$

Incorporating a continuity correction usually improves the approximation. Since $T_+$ and $T_-$ can take on only successive integer values, the continuity correction is 0.5. The procedure for finding asymptotic approximate $P$-values as right-tail probabilities is outlined as follows:

$$z_{+,R} = \frac{T_+ - 0.5 - n(n + 1)/4}{\sqrt{n(n + 1)(2n + 1)/24}} \qquad z_{-,R} = \frac{T_- - 0.5 - n(n + 1)/4}{\sqrt{n(n + 1)(2n + 1)/24}}$$

$$(3.2)$$

| Alternative | $P$-value (Table A) |
|---|---|
| $A_+$:  $M > M_o$ | Right-tail probability for $z_{+,R}$ |
| $A_-$:  $M < M_o$ | Right-tail probability for $z_{-,R}$ |
| $A$:  $M \neq M_o$ | 2 (right-tail probability for the larger of $z_{+,R}$ and $z_{-,R}$) |

*Zeros and Ties*   With the assumption that the population is continuous, it is permissible in theory to ignore the problems that result when the absolute values of two differences are equal, say $|D_i| = |D_j|$, called a tie, and also when a difference is zero, say $D_i = 0$, called a zero. (Note that a tie can occur among the $|D_i|$ even if none of the $X_i$ are tied.) Although the probability is zero that either of these situations occurs, both zeros and ties can occur in application and must be dealt with. A tie presents problems for ranking. Since a zero is neither plus nor minus, the sign associated with the rank of a zero difference is not determined. Zeros may occur with or without ties, and ties may occur in either zero differences or nonzero differences. The simplest procedure to follow, and that usually recommended, is to ignore any zeros, reduce $n$ accordingly, rank the absolute values of the remaining observations, and handle any nonzero ties by a procedure called the midrank method.

The midrank procedure in general assigns the same rank to all observations in a single tied set; the rank assigned is the simple average of the ranks those observations would have if they were not tied. For example, with ordered observations 2, 5, 5, 9, 14, 14, 14, 16, the assigned ranks respectively are 1, 2.5, 2.5, 4, 6, 6, 6, 8. Since the two 5s occupy the positions of ranks 2 and 3, each is assigned rank 2.5. Since the three 14s occupy rank positions 5, 6, and 7, their average rank is 6, and this value is assigned to each. Note that the average rank of ties occurring in rank positions $k$ to $l$ inclusive is always the midpoint of these rank positions, or $(k + l)/2$. Using the midrank method to rank $n$ observations with ties preserves the constant value $n(n + 1)/2$ for the sum of all ranks assigned. Since it is easy to make an error in assigning ranks by the midrank method, a good check on accuracy is provided by verifying that the sum of all ranks assigned equals $n(n + 1)/2$.

Thus, in order to perform the Wilcoxon signed rank test, the recommended procedure is to ignore the zero differences and rank the remaining differences according to absolute value, applying the midrank method to any nonzero ties. This gives a unique ranking of the absolute values of all nonzero differences. Then the quantity $n$ must be interpreted as the number of observations not equal to $M_o$, that is, the reduced sample size. The example of the previous paragraph serves to illustrate this procedure as long as we interpret the numbers 2, 5, 5, 9, 14, 14, 14, 16 as representing absolute values of differences rather than original observations. That is, the possible ranks are averaged for sets of tied absolute differences that are nonzero.

Although the exact sampling distribution of the Wilcoxon signed rank statistic computed for zeros ignored and ties handled by the midrank method is no longer given by Table G even for the reduced $n$, the effect of the nonzero ties is not very significant in most cases. Hence Table G can be used with the reduced $n$ to find an approximate $P$-value. When $T_+$ or $T_-$ is not an integer, it is always halfway between two integers. Then an approximate $P$-value can be given in several different ways. The conservative approach is to report the smaller of the $P$-values that correspond to the integer values on either side of the calculated value of $T_+$ or $T_-$. When using only right-tail probabilities, this is equivalent to rounding down. Alternative approaches are to report simply that $P$ lies between these two $P$-values, or to use linear interpolation to find a single number for $P$ as halfway between. For example, suppose $n = 8$ and we compute $T_+ = 7.5$, $T_- = 28.5$. From Table G the right-tail probabilities are .098 for $T_- = 28$ and .074 for $T_- = 29$. We could report either $.074 < P < .098$, or $P = (.074 + .098)/2 = .086$.

When using the large sample approximation to the distribution of $T_+$ or $T_-$ for $n$ (reduced) nonzero differences with ties among the absolute values, a correction factor can be introduced in the denominator of the standardized $z$ statistics. Specifically, the standard deviation in the denominator of $z_{+,R}$ or $z_{-,R}$ given in (3.2) is replaced by

$$\sqrt{\frac{n(n + 1)(2n + 1)}{24} - \frac{\Sigma u^3 - \Sigma u}{48}}, \tag{3.3}$$

where $u$ is the number of absolute differences that are tied for a given non-zero rank, and the sum is over all sets of $u$ tied ranks. For the above sample of eight nonzero absolute values of differences, 2, 5, 5, 9, 14, 14, 14, 16, there are two sets of tied ranks: the 5s make a set of two tied ranks, or $u_1 = 2$; and the 14s produce a set of three tied ranks, or $u_2 = 3$. These are the values of $u$, and we sum over the two sets to get the correction factor

$$\frac{\Sigma u^3 - \Sigma u}{48} = \frac{(2^3 + 3^3) - (2 + 3)}{48} = \frac{30}{48} = \frac{15}{24}.$$

The adjusted standard deviation is then

$$\sqrt{\frac{8(9)(16 + 1)}{24} - \frac{15}{24}} = \sqrt{50.375}.$$

When there are many ties, the normal approximation may be very inaccurate, even when the standard deviation is corrected and a continuity correction is used in the numerator of $z$.

### example 3.2

A steel company orders a certain size casting in large quantities. Before the castings can be used, they must be machined to a specified tolerance. The machining is either done by the company or is subcontracted, according to the following decision rule:

If median weight of the castings exceeds 25 kilograms, subcontract the order for machining.

If median weight of the castings is 25 kilograms or less, do not subcontract.

The company developed this decision rule in an effort to reduce costs, because the weight of a casting is a good indication of the amount of machining that is necessary, whereas the cost of subcontracting the castings is a function of the number of castings to be machined rather than the amount of machining required by each casting. They also consider the median a better descriptor of central tendency than the mean because it is less affected by extreme values.

A random sample of 8 castings are taken from a lot of 100 castings, and the weight is determined for each. Use these data (shown in Table 3.2) to recommend whether this lot should be subcontracted or not.

**SOLUTION**
The problem calls for a definite decision concerning the subcontracting of this lot. Since the ultimate goal here is reduced costs, it would appear that the more serious of the two types of error would be to subcontract the machining when it could easily be done by the company. While one can never fully justify accepting a null hypothesis on the basis of sample data, the probability of this more serious error is well controlled if a small $P$-value can be interpreted as supporting a decision to subcontract. Then this conclusion must be the alternative and the appropriate null hypothesis is the opposite conclusion, or "Do not subcontract." In terms of the company's adopted criterion based on weight, the appropriate hypothesis set is then

$$H: \quad M = 25 \qquad A_+: \quad M > 25.$$

Since the hypothesis set concerns the median, and since the sample size is quite small, we use the Wilcoxon signed rank test. This requires the assumption that casting weights are symmetrically distributed, but such an assumption seems more reasonable than a normal distribution assumption. The sample data are shown in the second column of Table 3.2. The third through fifth columns are found by subtracting $M_o = 25$ from each observation, ignoring the sign and ranking the absolute values of the nonzero differences, and handling ties by the midrank method. The signs of the differences in the third column are repeated in the sixth column for convenience. The reduced $n$ is equal to 7. We calculate the test statistic as

$$T_- = 6 + 2.5 = 8.5$$
$$T_+ = \frac{n(n+1)}{2} - T_- = \frac{7(8)}{2} - 8.5 = 19.5.$$

The appropriate $P$-value with the alternative $A_+$ is a right-tail probability for $T_+$. Table G gives probabilities only for integer values of $T_+$ since no ties occur in the theoretical distribution, but $T_+$ here is 19.5 because of the ties. We estimate the $P$-value as halfway between the probabilities for those integers on either side of 19.5, namely 19 and 20. Since the right-tail probabilities are .234 and .188, respectively, we obtain $P = (.234 + .188)/2 = .221$. Since this large a $P$-value gives support to $H$, the statistical recommendation is "Do not subcontract"; the castings in this lot do not seem to be so heavy that many will require machining, and the company can probably reduce costs by doing its own machining.

**TABLE 3.2**

| Casting | X Weight | $D = X - 25$ | $|D|$ | Rank of $|D|$ | Sign of $D$ |
|---------|----------|--------------|-------|---------------|-------------|
| 1 | 24.3 | −.7 | .7 | 6 | − |
| 2 | 25.8 | .8 | .8 | 7 | + |
| 3 | 25.4 | .4 | .4 | 4 | + |
| 4 | 24.8 | −.2 | .2 | 2.5 | − |
| 5 | 25.2 | .2 | .2 | 2.5 | + |
| 6 | 25.1 | .1 | .1 | 1 | + |
| 7 | 25.0 | .0 | .0 | | |
| 8 | 25.5 | .5 | .5 | 5 | + |

*Construction of Table G*   The exact sampling distribution of $T_+$ or $T_-$, as given in Table G for small sample sizes, can be easily determined by enumeration. We now give an example of how Table G was constructed. Suppose that $n = 3$; irrespective of the absolute values to be ranked, the ranks are 1, 2, and 3. Under the null hypothesis, the numbers to be ranked are symmetrically distributed about zero; thus each number ranked is as likely to be positive as negative. With three numbers, we then have $2^3 = 8$ possible sets of signs associated with the three ranks. Since each set is equally likely to occur, each one has a probability of $1/8$. The eight possibilities of signed ranks, and the corresponding values of $T_+$, are shown in Table 3.3.

**TABLE 3.3**

| Ranks with Positive Signs | Value of $T_+$ |
|---------------------------|----------------|
| None | 0 |
| 1 | 1 |
| 2 | 2 |
| 3 | 3 |
| 1, 2 | 3 |
| 1, 3 | 4 |
| 2, 3 | 5 |
| 1, 2, 3 | 6 |

Since $T_+ = 0$ occurs for only one possible set of signed ranks, that value occurs with probability 1/8. Similarly for $T_+ = 1$ and $T_+ = 2$. However, $T_+ = 3$ occurs twice in the list, and thus its probability is 2/8. The complete set of point probabilities is then as follows:

| $T_+$ | Probability |
|---|---|
| 0 | $1/8 = .125$ |
| 1 | $1/8 = .125$ |
| 2 | $2/8 = .250$ |
| 3 | $1/8 = .125$ |
| 4 | $1/8 = .125$ |

Any left-tail or right-tail probability can be found from this listing. For example, $P(T_+ \leq 1) = .125 + .125 = .250$, which agrees with the entry in Table G for $n = 3$.

*Paired Samples*   We now turn to an explanation of the application of the Wilcoxon signed rank test procedure for paired sample data that result from comparative experiments on matched units.

*Inference Situation*   Let $M_D$ denote the median of the population of *differences* between pairs of random variables. The hypothesis set is one of the following:

$$H: \quad M_D = M_o \qquad A: \quad M_D \neq M_o$$
$$H: \quad M_D = M_o \qquad A_+: \quad M_D > M_o$$
$$H: \quad M_D = M_o \qquad A_-: \quad M_D < M_o$$

*Sampling Situation*   The data as collected consist of either (a) a set of $n$ pairs of observations $(X_1, Y_1), (X_2, Y_2), \ldots, (X_n, Y_n)$, from which the differences $D_i = X_i - Y_i$ can be formed, or (b) a set of $n$ differences of paired observations (whose individual values may not actually have been observed), $D_1, D_2, \ldots, D_n$.

We assume that the differences $D_i$, whether computed from $X_i - Y_i$ or observed directly, are a random sample from a population of differences that is continuous, and also symmetric about its median $M_D$. The differences must be measured on an interval scale, or on an ordinal scale in such a way that the ranks and signs needed for the test statistic can be determined (see below).

*Test Procedure*   When the Wilcoxon signed rank test is to be applied to data from paired samples, the numbers to be ranked are the absolute values of the differences between the observed differences between pairs and $M_o$; that is,

$$|D_i - M_o| = |X_i - Y_i - M_o|.$$

The test statistics are again

$$T_+ = \text{sum of positive ranks},$$
$$T_- = \text{sum of negative ranks},$$

where "positive rank" means the rank assigned to the absolute value of a difference $(D_i - M_o)$ which was originally positive, and similarly for "negative rank." Hence $T_+$ and $T_-$ are both sums of nonnegative numbers. $P$-values are found as before, using Table G with $T_+$ or $T_-$ for $n \leq 15$, and Table A with the $z$ statistics of (3.2) for larger samples. Any differences $D_i - M_o$ that are equal to zero are ignored, and $n$ is reduced accordingly. If any of the resulting nonzero absolute values $|D_i - M_o|$ are tied, the midrank method is used to assign ranks; for large samples, the correction for ties given in (3.3) can be used in the denominator of the $z$ statistics.

**example   3.3**
Hypoglycemia is a condition in which blood sugar is below normal limits. In order to compare two hypoglycemic compounds, $X$ and $Y$, each one is applied to half of the diaphragm of each of nine rats in an experiment reported by Wilcoxon and Wilcox (1964, p. 9). Blood glucose uptake in milligrams per gram of tissue is measured for each half, producing the following results:

| Rat | X | Y | Rat | X | Y |
|-----|-----|-----|-----|------|------|
| 1 | 9.4 | 8.4 | 6 | 5.2 | 5.2 |
| 2 | 8.5 | 8.7 | 7 | 10.2 | 10.0 |
| 3 | 4.7 | 4.1 | 8 | 3.3 | 4.6 |
| 4 | 3.9 | 3.6 | 9 | 7.0 | 6.1 |
| 5 | 4.7 | 5.1 | | | |

The experimenter is interested in determining whether there is any difference between compounds $X$ and $Y$ as regards median glucose uptake.

**SOLUTION**
Since the experimenter did not anticipate a particular direction of difference between the effects of $X$ and $Y$, we formulate the hypothesis set with a two-sided alternative for the median of $D = X - Y$ as

$$H\!: \quad M_D = 0 \qquad A\!: \quad M_D \neq 0.$$

We make the assumption that the differences $D$ are symmetrically distributed about $M_D$.

The observed differences $D_i$ are given in the fourth column of Table 3.4. Since $M_o = 0$, the quantities to be ranked are $|D_i|$. The one zero difference is eliminated; the two differences tied in absolute value are given the average rank. The ranks are shown in the next-to-last column.

TABLE 3.4

| Rat | X | Y | $D = X - Y$ | $|D|$ | Rank of $|D|$ | Sign of D |
|-----|------|------|------|------|------|------|
| 1 | 9.4 | 8.4 | 1.0 | 1.0 | 7 | + |
| 2 | 8.5 | 8.7 | −0.2 | 0.2 | 1.5 | − |
| 3 | 4.7 | 4.1 | 0.6 | 0.6 | 5 | + |
| 4 | 3.9 | 3.6 | 0.3 | 0.3 | 3 | + |
| 5 | 4.7 | 5.1 | −0.4 | 0.4 | 4 | − |
| 6 | 5.2 | 5.2 | 0.0 | | | |
| 7 | 10.2 | 10.0 | 0.2 | 0.2 | 1.5 | + |
| 8 | 3.3 | 4.6 | −1.3 | 1.3 | 8 | − |
| 9 | 7.0 | 6.1 | 0.9 | 0.9 | 6 | + |

We compute the rank sums

$$T_- = 1.5 + 4 + 8 = 13.5$$
$$T_+ = \frac{8(9)}{2} - 13.5 = 22.5.$$

The appropriate $P$-value is twice the right-tail probability for the larger value, $T_+ = 22.5$. From Table G, with $n = 8$, the tail probabilities are given as .320 for $T_+ = 22$ and .273 for $T_+ = 23$. We average these probabilities to get $(.273 + .320)/2 = .296$ and estimate $P$ by $2(.296) = .592$. Thus the experimenter may conclude that these data do support $H$, and there is not sufficient evidence to conclude that compounds $X$ and $Y$ differ with respect to glucose uptake.

### example   3.4

In order to determine whether constant exposure to violence on television affects an individual's tendency to aggressive behavior and possibly crime, rioting, disturbances, and so on, an extensive controlled study is being conducted. In one facet of this investigation, a group of children are matched as well as possible as regards home environment, genetic factors, intelligence, parental attitudes, and so forth, in an effort to minimize factors other than TV that might influence a tendency for aggressive behavior. In each of the resulting 16 pairs, one child is randomly selected to view the most violent shows on TV, while the other watches cartoons, situation comedies, and the like. The children are then subjected to a series of tests designed to produce an ordinal measure of their aggression factors. The overall aggression scores for the 16 subject pairs are shown in Table 3.5; higher scores represent more aggression. Analyze the data to determine what kind of report might be issued for the study.

**SOLUTION**
Since the scores are ordinal, the median difference score provides a good basis for comparison of aggressive behavior of the two groups. Let us define a difference as the aggression score of a child exposed to violence on TV minus the score of his match. If exposure to violence does have some effect on aggression, the median difference score should increase. Therefore, the appropriate hypothesis set is one-sided in the positive direction, or

$$H: \quad M_D = 0 \qquad A_+: \quad M_D > 0.$$

The symmetry assumption appears reasonable, especially because of the matched design and random assignment to groups. Therefore, the Wilcoxon signed rank test is an appropriate nonparametric procedure for analysis of these data.

The data on aggression scores are shown in Table 3.5. We first form the differences between observations in the second and third columns (Violence–Nonviolence), and rank them in absolute value, as shown in the sixth column. Midranks are assigned in the usual way to tied absolute differences.

**TABLE 3.5**

| Child Pair | TV Exposure Violence | TV Exposure Nonviolence | $D$ | $|D|$ | Rank of $|D|$ | Sign of $D$ |
|---|---|---|---|---|---|---|
| 1 | 35 | 26 | 9 | 9 | 12.5 | + |
| 2 | 30 | 28 | 2 | 2 | 4.5 | + |
| 3 | 15 | 16 | −1 | 1 | 1.5 | − |
| 4 | 20 | 16 | 4 | 4 | 8 | + |
| 5 | 25 | 16 | 9 | 9 | 12.5 | + |
| 6 | 14 | 16 | −2 | 2 | 4.5 | − |
| 7 | 37 | 32 | 5 | 5 | 9 | + |
| 8 | 26 | 24 | 2 | 2 | 4.5 | + |
| 9 | 36 | 30 | 6 | 6 | 10 | + |
| 10 | 40 | 33 | 7 | 7 | 11 | + |
| 11 | 35 | 20 | 15 | 15 | 16 | + |
| 12 | 20 | 19 | 1 | 1 | 1.5 | + |
| 13 | 16 | 19 | −3 | 3 | 7 | − |
| 14 | 21 | 10 | 11 | 11 | 15 | + |
| 15 | 17 | 7 | 10 | 10 | 14 | + |
| 16 | 15 | 17 | −2 | 2 | 4.5 | − |

The rank sums are computed as

$$T_- = 1.5 + 4.5 + 7 + 4.5 = 17.5$$
$$T_+ = \frac{16(17)}{2} - 17.5 = 118.5.$$

Since $n = 16$ is outside the range of Table G, we use the normal approximation. If no correction is made for the ties, the appropriate $z$ statistic from (3.2) is

$$z_{+,R} = \frac{118.5 - 0.5 - 16(17)/4}{\sqrt{16(17)(33)/24}} = \frac{50}{\sqrt{374}} = 2.59.$$

The asymptotic approximate $P$-value (based on the normal approximation) is then .0048 from Table A. The statistical conclusion is that the data do not support $H$, and hence that exposure to violence on TV does seem to increase aggression scores on this particular test.

Since the ties are rather extensive for these data, the investigator might be well advised to compute $z_{+,R}$ using the correction for ties from (3.3) in the denominator of the $z$ statistic in (3.2). The data show ties in the absolute values of the differences between Child 3 and 12, Child 2, 6, 8 and 16, Child 1 and 5. That is, there are 3 sets of ties, and the number of absolute differences tied in these respective sets are 2, 4, and 2 (the values of $u$). The correction factor is then computed as

$$\frac{\Sigma u^3 - \Sigma u}{48} = \frac{(2^3 + 4^3 + 2^3) - (2 + 4 + 2)}{48} = \frac{80 - 8}{48} = 1.5,$$

and the corrected $z$ statistic is

$$z_{+,R} = \frac{118.5 - 0.5 - 16(17)/4}{\sqrt{\frac{16(17)(33)}{24} - 1.5}} = \frac{50}{\sqrt{372.5}} = 2.59.$$

To three significant figures, this is equal to the result obtained for $z$ without the correction. Hence the effect of the ties on the standard deviation and the test statistic is negligible and there is no change in the stated $P$-value or conclusions.

## 3.2  Tests for Symmetry

The Wilcoxon signed rank test procedure can also be used as a test for symmetry once we interchange some of the statements made in describing the basic sampling and inference situations. The sampling distribution of $T_+$ (or $T_-$) is generated using only two population distribution properties, namely a symmetric population and median $M_o$. These properties are separate, but both must be either assumed or hypothesized in order for the null distribution to hold. The allocation of these properties between assumptions and the null hypothesis statement affects the test conclusion. In all situations discussed so far, the symmetry part was assumed, and hence the test was for location. The other two possibilities are described as follows.

1. (a) *Sampling situation*. The data to be analyzed are a random sample of observations (or differences of pairs of observations) from a continuous population (of differences) with median $M$ (or $M_D$) known and given as $M_o$.
   (b) *Inference situation*. The hypothesis set is

   > $H$: The population (of differences) is symmetric (about its median)
   > $A$: The population (of differences) is not symmetric.

2. (a) *Sampling situation*. The data to be analyzed are a random sample of observations (or differences of pairs of observations) from a continuous population (of differences).
   (b) *Inference situation*. The hypothesis set is

   > $H$: The population (of differences) is symmetric and has
   >     median $M$ (or $M_D$) $= M_o$
   > $A$: The statement in $H$ is not true.

In either of these two cases, whether with one sample or paired samples, the Wilcoxon signed rank test procedure is applied exactly as before. A one-sided test can be used in case 1 if the alternative states a direction of skewness (lack of symmetry). If $H$ is rejected in case 1, the conclusion is clear. However, if the null hypothesis in case 2 is not supported or is rejected, we could conclude that either the population is symmetric with median not equal to $M_o$, or the population is asymmetric with median equal to $M_o$, or the population is asymmetric and its median is not equal to $M_o$. Unfortunately, this test provides no information as to which of these conclusions is appropriate.

In case 1, where the median is assumed as known, a Kolmogorov-Smirnov type test for symmetry could be used instead of the signed rank test. It is based on the fact that a cumulative distribution $F(x)$, which is defined for all values of $x$, is symmetric about $M$ if

$$F(x + M) + F(M - x) = 1 \qquad \text{for all } x \geq 0,$$

or, equivalently, if

$$F(x + M) + F(M - x) - 1 = 0 \qquad \text{for all } x \geq 0.$$

Since $S_n(x)$ is the sample estimate of $F(x)$, the sample data support symmetry about $M$ if

$$|S_n(x + M) + S_n(M - x) - 1|$$

is small for all nonnegative values of $x$. Since all these quantities are small if the largest is small, the test statistic is defined as

$$D = \sup_{x \geq 0} |S_n(x + M) + S_n(M - x)^- - 1|.$$

As before, $(M - x)^-$ indicates a value slightly smaller than $M - x$; it is used to simplify the distribution theory. The rejection region is large values of $D$.

Tables of the null distribution of $D$ are given in Chatterjee and Sen (1973). This source also includes tables for corresponding one-sided statistics that are appropriate for alternatives specifying a direction of asymmetry (that is, positive skewness or negative skewness). These procedures are quite new, and little is known about power or other performance properties.

When the median is unknown and the desired inference concerns symmetry only, a test proposed by Gupta (1967) can be applied. However, this procedure is only asymptotically distribution free, and the test statistic is relatively complicated to compute even in small samples.

### 3.3   Confidence Intervals for the Median or Median Difference

In many practical situations, the investigator is not interested in determining whether or not the population median (or median difference) is equal to a specified value. Rather, the inference desired is a confidence interval estimate for the unknown median. The nonparametric confidence procedure that corresponds to the Wilcoxon signed rank test provides an interval estimate of those parameters for which the test is applicable. Thus we can estimate the median $M$ of a single continuous and symmetric population, or the median $M_D$ of a continuous and symmetric population of differences between two variables that are linked by some unit of association. For convenience, the estimation procedure is presented here only for the one-sample model; as with the test procedure, the confidence procedure for a paired-sample model is exactly the same once the differences of pairs are obtained and used for analysis.

As explained in Section 2.3, the interval with confidence coefficient $1 - \gamma$ consists of those values of $M$ that, if stated in a null hypothesis with a two-sided alternative, would be supported by the sample data with probability $\gamma$. Since here we are seeking the interval that corresponds to the Wilcoxon signed rank test procedure, the values possible for $1 - \gamma$ are limited to numbers that are twice any of the probabilities (which are less than .5) given in Table G for the original sample size $n$. Once a number $\gamma$ is selected, we could use trial and error to determine the endpoints of the interval of acceptable values of $M$; however, this may be quite tedious. The method described below gives the same results in a more convenient and systematic way.

Let $X_1, X_2, \ldots, X_n$ be sample observations from a symmetric population with median $M$. Then the $n(n-1)/2$ averages of sets of two different observations, that is,

$$\frac{X_1 + X_2}{2}, \frac{X_1 + X_3}{2}, \ldots, \frac{X_{n-1} + X_n}{2},$$

are also symmetrically distributed about the median $M$. Counting both the observations and these averages, we have a total of $n + n(n-1)/2 = n(n+1)/2$ variables that are symmetrically distributed about $M$. These quantities are frequently called Walsh averages. They can all be represented in symbols by

$$U_{ij} = \frac{X_i + X_j}{2} \quad \text{for } 1 \le i \le j \le n.$$

Note that when $j = i$, we have $U_{ij} = 2X_i/2 = X_i$, since we are averaging the value $X_i$ with itself. If all these averages $U_{ij}$ are arranged in order of magnitude from smallest to largest, the $k$th smallest and $k$th largest values are the endpoints of the confidence interval for $M$; the number $k$ is found from Table G by locating the table entry for $T$ that corresponds to a left-tail probability of $(1 - \gamma)/2$. The value of $k$ is then found as the rank of that number $T$ in the table for that particular $n$. For example, suppose that $n = 7$ and we want an interval with $\gamma$ near .90. From Table G, we decide on $P = .055$, for which $\gamma = 1 - 2(.055) = .890$. The left-tail probability is .055 when $T = 4$, and 4 is the fifth smallest value possible for $T$ when $n = 7$. Thus $k = 5$. Some other examples of $k$ are shown in Table 3.6.

**TABLE 3.6**

| $n$ | $P = (1 - \gamma)/2$ | $\gamma = 1 - 2P$ | Left $T$ for $P$ | $k$ = rank of $T$ |
|-----|------|------|------|------|
| 5 | .062 | .876 | 1 | 2 |
| 6 | .031 | .938 | 1 | 2 |
|   | .047 | .906 | 2 | 3 |
| 7 | .039 | .922 | 3 | 4 |
|   | .055 | .890 | 4 | 5 |
| 8 | .020 | .960 | 3 | 4 |
|   | .027 | .946 | 4 | 5 |
|   | .039 | .922 | 5 | 6 |
|   | .055 | .890 | 6 | 7 |

Notice that the rank of $T$ is always one more than the left-tail critical value of $T$, and thus we could define $k$ equivalently as one more than that value of $T$ for which the left-tail probability is equal to $P$.[5] For $n > 15$, and hence outside the range of Table G, the normal approximation can be used. Then $k$ is given by

$$k = 0.5 + \frac{n(n + 1)}{4} - z\sqrt{\frac{n(n + 1)(2n + 1)}{24}}, \qquad (3.4)$$

where $z$ is the number from Table A that corresponds to a right-tail probability of $(1 - \gamma)/2$. To obtain a conservative result, if the value of $k$ from (3.4) is not an integer, it should be rounded down to the next lower integer.

Note that observations that are zero should *not* be omitted for the confidence procedure, and thus $n$ always refers to the original sample size. We are not testing the null hypothesis $M = 0$ here; in fact, we could consider the interval as a simultaneous test of all possible values of $M$, and so there is no justification for throwing away observations with any particular value.

A systematic method of generating all possible averages $U_{ij}$ from the observations is now illustrated for the following data:

$$-1, 2, 3, 4, 5, 6, 9, 13.$$

The observations are listed in increasing order of magnitude, keeping signs, in the first row of a display like the one below. To construct the rest of the first column, we average $-1.0$, the smallest observation, with every observation larger than it (to the right in the first row); that is,

$$\frac{-1.0 + 2.0}{2} = 0.5, \frac{-1.0 + 3.0}{2} = 1.0, \ldots, \frac{-1.0 + 13.0}{2} = 6.0.$$

For the second column, we average $2.0$, the second smallest observation, with every observation larger than it; for example,

$$\frac{2.0 + 3.0}{2} = 2.5, \ldots, \frac{2.0 + 13.0}{2} = 7.5.$$

In general, the entries in any column are the averages of the first number appearing in that column with every number larger than it (to the right) in first row, until all numbers in the first row have been used and the array contains all $n(n + 1)/2$ possible averages. Thus we obtain

---

[5]Note the analogy between these endpoints and those that correspond to the ordinary sign test. In Section 2.3 we found the lower endpoint as the $(k + 1)$th smallest $X_i$, where $k$ is the value of $S_+$ (or $S_-$) that has a left-tail probability of $P$ from Table F. Thus we could say that the endpoint is the $(S_+ + 1)$th smallest $X_i$. Here the lower endpoint is the $(T_+ + 1)$th smallest $U_{ij}$, where $T_+$ is found from Table G in exactly the same way as $S_+$ is found from Table F.

|       |     |     |     |     |     |     |      |
|-------|-----|-----|-----|-----|-----|-----|------|
| −1.0  | 2.0 | 3.0 | 4.0 | 5.0 | 6.0 | 9.0 | 13.0 |
| 0.5   | 2.5 | 3.5 | 4.5 | 5.5 | 7.5 | 11.0 |    |
| 1.0   | 3.0 | 4.0 | 5.0 | 7.0 | 9.5 |     |      |
| 1.5   | 3.5 | 4.5 | 6.5 | 9.0 |     |     |      |
| 2.0   | 4.0 | 6.0 | 8.5 |     |     |     |      |
| 2.5   | 5.5 | 8.0 |     |     |     |     |      |
| 4.0   | 7.5 |     |     |     |     |     |      |
| 6.0   |     |     |     |     |     |     |      |

With the numbers listed in this fashion it is relatively easy to see which averages are the $k$th smallest and the $k$th largest. Tied values should be counted as many times as they occur. If the ordered arrangement is not clear, all of the averages should be listed in order of magnitude. The results here would be

$$-1.0, \ 0.5, \ 1.0, \ 1.5, \ 2.0, \ 2.0, \ \ldots, \ 8.5, \ 9.0, \ 9.0, \ 9.5, \ 11.0, \ 13.0.$$

For a confidence coefficient of .960 with $n = 8$, we find $k = 4$ from Table G, and the confidence interval endpoints are then 1.5 and 9.0. The confidence interval for $M$ with confidence coefficient .960 is written as $1.5 \leq M \leq 9.0$. With some other selected confidence coefficients, the following intervals are obtained for these data:

| Confidence | $k$ | Interval |
|------------|-----|----------|
| .946       | 5   | $2.0 \leq M \leq 9.0$ |
| .922       | 6   | $2.0 \leq M \leq 8.5$ |
| .890       | 7   | $2.5 \leq M \leq 8.0$ |

This process for determining the endpoints of the confidence interval can be facilitated somewhat by a graphical approach. The necessary steps explained here are illustrated in Figure 3.1 for the observations listed above. Any scale may be chosen for the horizontal axis, in accordance with the magnitudes of the data. Then each of the data points is indicated by a dot on the horizontal axis. Label the smallest value $A$ and the largest value $B$. The line $AB$ is the range of the data. Compute the midrange of the data by averaging the smallest and largest observations, in this case $[(-1) + 13]/2 = 6$. Form an isosceles triangle by joining points $A$ and $B$ with a point $C$ lying anywhere on the vertical line that passes through this midrange value. Through each dot on the horizontal axis indicating an observation, draw a line parallel to $AC$, and also a line parallel to $BC$, and mark each intersection with a dot. These dots, plus those on the horizontal axis, represent all of the averages $U_{ij}$. Now draw a vertical line through the $k$th smallest intersection point from the left, and another through the $k$th largest from the right. The respective confidence interval endpoints can then be read off the figure as the values of those two points where these vertical lines cross the horizontal axis. Here we have $k = 4$, and the endpoints are 1.5 and 9.0.

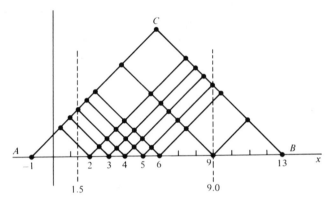

**Figure 3.1**

# 4   Comparison of the Sign Test and Signed Rank Test

In this chapter we have covered two nonparametric test procedures, both applicable to data representing either a single set of observations or differences of paired observations. In each case the variables entering the analysis must be independent. Only the ordinary sign test can be used for strictly dichotomous data, like yes-no responses. However, either the procedure based on the ordinary sign test or the signed rank test can be applied when the inference concerns the value of the population median (or median difference) and the data are at least ordinal, as when rating scales or preferences are used. The assumptions required are minimal: a population that is continuous at $M$ for the sign test and continuous everywhere and symmetric for the signed rank test. In practice, both procedures have the problem of zero differences, and the signed rank test has the additional problem of ties.

The sign test uses only information about the direction of the difference between each observation (or difference between members of a pair) and the hypothesized median, whereas the signed rank test uses this information plus the absolute magnitude of each difference relative to every other such difference. Since this additional information concerning relative magnitude is usually relevant to the question of central location, it seems intuitive that the signed rank test would give better performance in many situations. The reported studies of comparative power support such a view for many continuous and symmetric distributions. For example, with large samples we have the results that the asymptotic efficiency of the sign test relative to the signed rank test is 2/3 for normal distributions and 1/3 for the continuous uniform distribution. Even though there are exceptions, in most situations the signed rank test should be used in preference to the sign test when the assumption of a symmetric distribution appears tenable and the data are measured with a sufficient level of precision.

If the population distribution is symmetric, both these tests are nonparametric counterparts to the Student's $t$ test for means in classical statistics (or the normal theory test if the variance is known). Since the mean and median are then equal, the three tests might be considered direct competitors. However, Student's $t$ test is based on the assumption of a normal distribution, and this is much more stringent than an assumption of symmetry. Since Student's $t$ test is the most powerful procedure possible for normal distributions, clearly it should be used whenever the normal assumption is tenable and the data are measured on an interval scale.

One of these nonparametric tests might be used in preference to the parametric test in any of the following circumstances:

1. The sample size is quite small.
2. The only data available are measured on a nominal or ordinal scale.
3. The sample median seems to be more reliable than the sample mean as a descriptor of central tendency in the data.
4. The median seems to be a more representative location parameter than the mean for the variable of interest.
5. Little or nothing can be assumed about the probability distribution (although note that an assumption of symmetry is needed for the Wilcoxon signed rank test).
6. The distribution is unknown, but thought to bear little resemblance to the normal curve.

In each of these situations, any inferences and conclusions based on the parametric test may be unreliable, or at least of uncertain value.

Comparisons of power can be made between the three tests only under specific distribution assumptions. For any continuous, symmetric distribution, the asymptotic efficiency of the signed rank test relative to Student's $t$ test is never smaller than .864, whereas the same lower bound of the sign test relative to Student's $t$ test is only 1/3. Either ARE can be infinite also. (See Hodges and Lehmann (1956) or Pratt and Gibbons (1975) for further results and limitations.) The ARE of the signed rank test relative to Student's $t$ test is $3/\pi = .955$ for the normal distribution and 1.000 for the continuous uniform distribution. The comparable values for the sign test relative to Student's $t$ test are $2/\pi = .637$ and $1/3$ for the normal and uniform distributions, respectively. Of course, it must be remembered that these values for relative efficiency apply only to large samples. In small samples from nonnormal distributions, one or both of these nonparametric tests may well be more powerful than Student's $t$ test.

# 5   Inference from Nonrandom Samples

In Chapter 1 we stated that the assumption of random sample(s) is required for the validity of all statistical inferences. However, we also said that in many cases the assumptions in nonparametric inference are sufficient, but not necessary, for the validity of procedures. In this section we discuss certain situations in which random sampling is not strictly necessary.

In many experimental studies, particularly in the behavioral, social, and biological sciences, it is almost impossible to obtain a random sample in the true sense of the definition. When performing an experiment with rats to investigate toxicity of pesticides, the psychologist simply purchases a batch of rats to use. They may represent only a few litters, or they may all have been produced in the same laboratory, so that they cannot be considered a random sample drawn from the population of rats. When investigating the cure effectiveness and possible side effects of cancer treatments, the medical researcher generally must use those subjects who are available and willing to be experimented on. Prisoners are often considered good subjects for all types of experimentation; they are always available for observation, they are willing and frequently anxious to volunteer to participate, the external factors are relatively easy to control, and so on. However, such subjects are seldom randomly selected. In all of these situations, we have to deal with a systematic sample, not a probability sample.

Even with a systematic sample, random factors can frequently be introduced into the experiment so as to justify statistical inferences, whether parametric or nonparametric. The justification rests on a technique of experimental design called randomization.

Let us consider the broad class of experiments that focus on inferences about "treatment effects." The treatments may be fertilizers, medicines, tests, learning experiences, or the like. Suppose that a group of experimental units is chosen in a nonrandom fashion, but the treatments are assigned at random to the selected units, or vice versa. Then only random factors determine which unit receives which treatment. This technique of experimental design is regarded as a means of eliminating bias in sampling, that is, the effects of differences between the experimental units that are not attributable to the treatment itself.

However, note that the random assignment essentially provides a random sample of all possible assignments of treatments to the fixed population of experimental units. As a result, we can use the data for statistical inferences *about the treatments* that are valid even though the units are fixed. (The inferences are not about the units themselves.) For example, the sign and signed rank procedures of this chapter are valid for a systematic sample of pairs as long as the treatments are assigned randomly within pairs.

Such inferences are frequently more useful than ones obtained from an ordinary random sample, since then the conclusions must be limited to the population from which that sample is drawn. If these conclusions are to be generalized to some broader group, the population sampled must also be considered a random selection from the same larger group. Suppose that a very large number of prisoners are available for experimentation. Then we can take a random sample from these potential subjects and make inferences about the population of all potential subjects. However, we do not have a random sample of people, and probably not even a random sample of all prisoners.

More generally, it is unfortunate but true that the real target population is frequently unreachable, and the populations actually sampled are of little interest in their own right. The idea of obtaining a true random sample of members of the human or animal population of the world may be theoretically sound, but it is simply not possible. In dealing with measurements of some phenomenon, the population is actually hypothetical as well as infinite, since it consists of all measurements that could be made. In such situations, the investigator should take care that observations are made in some random kind of fashion and also introduce randomization in the design. The statistical inferences apply strictly only to the population actually sampled. Any generalization of these conclusions to the target population must be based on nonstatistical factors.

## PROBLEMS

3.1   A manufacturer of television sets does not make his own tubes; he has purchased them for years from supplier A. Supplier B offers a comparable type of tube at a considerably lower price. The manufacturer is willing to change suppliers unless the less expensive tubes are of inferior quality. A tube is judged of adequate quality if the median lifetime is at least 500 hours. The manufacturer orders one lot of 1000 tubes of this type from supplier B, takes a random sample of 25 tubes from the lot and subjects each of these to a simulated "life test" experiment. In order to reduce the cost of experimentation, the total hours of life are not observed and the experiment is stopped when the tubes are exposed to the equivalent of 500 hours of use. Of the 25 tubes in the sample, 22 survive the test. Perform a statistical analysis of these results to help the manufacturer decide whether to change suppliers.

3.2 Good police officers are expected to exercise authority when needed and not to panic in stressful situations. In order to develop such talents, police officer trainees were traditionally subjected to a high-stress learning program with rigid, authoritarian methods. An assistant sheriff in Los Angeles conducted a controlled experiment at the police academy to investigate the validity of this philosophy of training. A group of 74 candidates were divided into 37 pairs according to age, marital status, race, education, and experience, before beginning a 16-week training course. One member of each pair was randomly selected for a stressful training program while the other received nonstressful training. Both training courses had the same content, but the atmosphere was quite different. The stressful group received instruction in a military atmosphere, while the nonstressful group enjoyed something like a college campus atmosphere. The trainees were evaluated about a year after completing the training programs. A summary of those evaluations stated that 29 of the trainees in the nonstress course outperformed their matched partner in the stress course in all respects, including job knowledge, marksmanship, responsibility, appearance, attitude, adaptability, and so on. As a result of this study, and a similar study performed later on 100 different trainees, the Los Angeles course now emphasizes nonstress training; it even includes courses in sociology, human behavior, and the philosophy of law. Find the $P$-value corresponding to the outcome of the first training experiment.

3.3 A mail survey questionnaire was sent to the 49 persons who completed a particular offering of the Dale Carnegie Course five years ago. The purpose of the questionnaire was to obtain constructive criticisms and evaluations of the course. The final question asked was, "When taking all factors into consideration, do you think that the long-term benefit derived from this course is commensurate with the time, effort, and expense of completing the course?" Of the 49 contacted, only 32 returned the questionnaire, but 25 of these answered yes to the final question. Use the sign test to find a $P$-value for these results. Given only this information, and the fact that the returns were anonymous, could you help someone decide whether a Dale Carnegie Course is worthwhile? Discuss the various problems relevant to such a question, and the value, both statistical and nonstatistical, of the sign test analysis for these data in answering the question posed.

3.4 For more than a century it was believed that oarsmen have a shorter life expectancy than others because of the strain placed on the heart by crew racing. A doctor at Harvard University Health Services recently reported the results of a controlled study on life expectancy of oarsmen. He selected 172 graduates of Harvard and Yale who had rowed at least once in the 4-mile varsity race between 1887 and 1962, and randomly selected a match for each oarsman from his classmates. In spite of the fact that the oarsmen seem to develop a slow-beating "athlete's heart," the oarsmen lived on the average at least 6 years longer than their mates. Further, only half as many rowers died before age 60 as did their classmates.

(a) If 135 of the oarsmen lived at least 10 years longer than their classmates, find the P-value, assuming no difference in life expectancy of the matched groups.

(b) Suppose that exactly 1 member of the pair died before age 60 in a total of 20 pairs, and the oarsman was the dead member in only 5 of these cases. Find the P-value assuming that life expectancy is the same.

3.5    An independent soft drink bottling company is considering the addition of a new "uncola" to their line of products. Since a major competitor has a similar drink that has been quite successful, they decide first to perform a taste preference test. A group of 64 independent tasters are asked to taste both brands in a randomly selected order and to indicate their preference. Make a judgment as to whether the given brand is superior in taste to the competing brand if 38 tasters prefer the given brand, 20 prefer the competing brand, and 6 have no preference.

3.6    A football coach feels that speed is the primary factor for a winning contest in football, and specifically that short reaction times are more important than physical size in a football player (if it is not possible to have both). The team members habitually eat a normal lunch approximately 2 hours before a game. The coach hypothesized that a light snack might improve their speed and alertness during the game. A random sample of 22 players was selected from the varsity team. They were fed a light snack on one day and a normal lunch on another day. The team physician tested the reaction times of all players, 1 hour after the light snack and 2 hours after the normal meal, but the tests were such that he could judge only whether the individual reaction times were slower, faster, or unchanged when eating a normal meal. The results were 17 slower, 3 faster, and 2 unchanged. Do these data support the coach's hypothesis?

3.7    Two different chemical formulas, A and B, are under consideration for dyeing cloth for a new product. The company is primarily interested in a cloth that is highly resistant to fading after exposure to sunlight. Ten specimens of cloth are each cut in half, and each half is dipped in a different dye for a specified period of time. Then the 20 specimens are exposed to intense sunlight for a fixed period of time, and each is measured for color intensity at the end of the period (lower scores indicate less intensity and therefore more fading). Find the P-value for the null hypothesis that there is no difference between the dyes as regards resistance to fading.

| Specimen | Formula A | Formula B | Specimen | Formula A | Formula B |
|---|---|---|---|---|---|
| 1 | 7.2 | 5.1 | 6 | 6.8 | 5.1 |
| 2 | 4.3 | 4.1 | 7 | 6.3 | 5.3 |
| 3 | 5.8 | 5.5 | 8 | 7.0 | 7.3 |
| 4 | 6.5 | 4.1 | 9 | 6.5 | 4.8 |
| 5 | 4.9 | 5.0 | 10 | 6.2 | 5.8 |

3.8 Records on the number of man-hours lost per month due to plant accidents have been kept over the years at each of the six plants of a large company. One year ago, the company instituted an extensive industrial safety program at all plants. The average number of hours lost is shown below for the 12 months prior to the safety program and for the 12 months after its establishment. Do you think the program has been effective in reducing time lost from accidents?

| Plant | Before Program | After Program |
|-------|---------------|---------------|
| 1 | 51.2 | 45.8 |
| 2 | 46.5 | 41.3 |
| 3 | 24.1 | 15.8 |
| 4 | 10.2 | 11.1 |
| 5 | 65.3 | 58.5 |
| 6 | 92.1 | 61.2 |

3.9 A manufacturer of suntan lotion is testing a new formula to see whether it provides more protection against sunburn than the old formula. Out of those employees who volunteer to participate in the test, he chooses 10 persons at random. The two types of lotion are applied to the back of each subject, one on each side, and then each subject is exposed to a controlled but intense amount of sun. Degree of sunburn is measured for each side of each subject, with the results shown below (higher numbers represent more severe sunburn).

| Subject | Old Formula | New Formula |
|---------|-------------|-------------|
| 1 | 41 | 37 |
| 2 | 42 | 39 |
| 3 | 48 | 31 |
| 4 | 38 | 39 |
| 5 | 38 | 34 |
| 6 | 45 | 47 |
| 7 | 21 | 19 |
| 8 | 28 | 30 |
| 9 | 29 | 25 |
| 10 | 14 | 8 |

(a) Assuming that measurements of difference of degree of sunburn (new − old) are symmetric, investigate the theory that the new formula is more effective than the old.

(b) Find a confidence interval for the median difference in degree of sunburn using a confidence coefficient near .90.

3.10 Use the sign test procedure to answer (a) and (b) of Problem 3.9.

3.11  A psychologist was concerned with sampling bias in the selection of rats to be used as subjects for a certain experiment. He was told that if rats were housed two to a cage, the rat placed in the cage first would tend to be more aggressive and get more food than the other rat. The experimenter set up a controlled experiment by placing one rat in each of 10 cages and adding a second rat to each cage one week later. Food sufficient for two rats was placed in the cage for a month. After this period, the weights of the rats are as follows:

| Cage Number | 1 | 2 | 3 | 4 | 5 | 6 | 7 | 8 | 9 | 10 |
|---|---|---|---|---|---|---|---|---|---|---|
| First Rat | 2.49 | 2.42 | 2.31 | 2.15 | 2.83 | 2.10 | 2.43 | 1.88 | 2.48 | 2.28 |
| Second Rat | 2.24 | 2.38 | 2.29 | 1.80 | 1.98 | 2.10 | 2.39 | 2.03 | 2.30 | 2.06 |

Does cage seniority seem to have a significant effect on aggressiveness when measured by total weight? Use both the sign test and the signed rank test to analyze the data.

3.12  A colleague studied the psychologist's results as reported in Problem 3.11. He believes that the theory that aggression is affected by seniority in the cage can be expected to hold true only for male rats. Therefore, he repeats the same experiment but using only male rats. Analyze the data below on male rats.

| Cage Number | 1 | 2 | 3 | 4 | 5 | 6 | 7 | 8 | 9 | 10 |
|---|---|---|---|---|---|---|---|---|---|---|
| First Rat | 2.80 | 3.04 | 3.16 | 2.67 | 3.03 | 3.05 | 2.81 | 2.78 | 3.14 | 2.93 |
| Second Rat | 2.56 | 2.71 | 3.12 | 2.40 | 3.19 | 2.91 | 2.75 | 2.68 | 3.10 | 2.65 |

3.13  A group of 10 students in Typing I are given a 5-minute timed test. Their scores are shown in the table below in the row labeled Preliminary Score. The students are then given a corrective drill, which is designed to increase speed for those who type accurately, and to emphasize accuracy for those who are making too many errors. After this drill, the students are retested; the scores are given in the table below in the row labeled Score after Drill.
(a) Does the drill seem to have a significant effect on improving test scores?
(b) Determine a confidence interval estimate on the median difference of test scores, using a confidence of coefficient of .916 and assuming symmetry.

| Student | A | B | C | D | E | F | G | H | I | J |
|---|---|---|---|---|---|---|---|---|---|---|
| Preliminary Score | 45 | 52 | 34 | 38 | 47 | 41 | 62 | 53 | 51 | 49 |
| Score after Drill | 49 | 56 | 31 | 46 | 54 | 39 | 68 | 55 | 50 | 55 |

3.14  Bender (1958) reports an interesting study of the change in human values over a 15-year period. The subjects are a representative sample of 124 senior men at Dartmouth College in 1939–1940; their values were measured by the Allport-Vernon Study of Values, 1931 edition. A follow-up study was made of these same men in 1955–1958, using the same measure of values. Only 12 of the original men either did not respond, were never located, or had died (a rather remarkable feature of this study). The questionnaire used is designed to measure personal values in six areas, called theoretical, economic, aesthetic, social, political, and religious; the scores are relative, since the sum over all areas is constant for each subject. The only three areas in which values showed a highly significant change over the years are religious, aesthetic, and theoretical. The following artificial data on 12 men simulate the average scores reported for religious and aesthetic values. Use an appropriate one-sided test to draw a research conclusion in each case.

| Subject | Religious Values | | Aesthetic Values | |
|---|---|---|---|---|
| | 1940 | 1956 | 1940 | 1956 |
| 1 | 25.1 | 31.1 | 30.1 | 25.4 |
| 2 | 22.3 | 30.3 | 31.5 | 24.1 |
| 3 | 20.7 | 31.5 | 28.6 | 28.3 |
| 4 | 25.2 | 27.7 | 29.4 | 23.7 |
| 5 | 30.6 | 34.8 | 30.2 | 21.8 |
| 6 | 24.1 | 36.2 | 26.1 | 29.6 |
| 7 | 26.2 | 35.5 | 30.7 | 28.3 |
| 8 | 31.5 | 31.3 | 34.8 | 23.2 |
| 9 | 27.2 | 30.1 | 33.2 | 25.1 |
| 10 | 24.3 | 33.2 | 31.4 | 26.5 |
| 11 | 25.3 | 34.0 | 30.3 | 23.4 |
| 12 | 20.1 | 33.6 | 31.1 | 29.1 |

3.15 Example 2.4 stated that one of the primary purposes of fatigue testing is to estimate the fatigue properties of a material, particularly the median fatigue life. For the sample data given there, find a confidence interval estimate of median fatigue life with a confidence coefficient near .90, (a) without assuming symmetry and (b) assuming symmetry. (c) What is the exact confidence coefficient in (a) and in (b)?

3.16 Prothrombin is one of the clotting factors in the blood. Hill et al (1970, pp. 352, 357) evaluated some technical aspects of the performance of tests of prothrombin times by clinical laboratories. The tables below show their results in a 4-hour stability study of prothrombin time in 10 specimens of patient plasma at 37°C, and in a comparison of prothrombin times on 9 successive days when Extended Control Plasmas are maintained at 37°C and at room temperature. For each set of data, find a confidence interval estimate of the median difference with confidence coefficient near .90.

STABILITY STUDY

| Specimen | Initial Prothrombin Time | 4-Hour Prothrombin Time |
|---|---|---|
| 1 | 14.5 | 19.2 |
| 2 | 13.7 | 17.7 |
| 3 | 15.5 | 18.2 |
| 4 | 14.7 | 19.2 |
| 5 | 16.2 | 18.7 |
| 6 | 22.0 | 25.2 |
| 7 | 25.7 | 30.0 |
| 8 | 17.5 | 19.5 |
| 9 | 24.5 | 26.7 |
| 10 | 19.7 | 23.6 |

TEMPERATURE STUDY

| Day | 37°C | Room Temperature |
|-----|------|------------------|
| 1 | 20.7 | 22.0 |
| 2 | 21.5 | 21.0 |
| 3 | 22.0 | 22.2 |
| 4 | 22.2 | 22.0 |
| 5 | 21.2 | 21.7 |
| 6 | 21.5 | 19.0 |
| 7 | 19.1 | 19.6 |
| 8 | 20.4 | 21.1 |
| 9 | 20.2 | 21.5 |

3.17 Barnett and Youden (1970, p. 456) developed a revised method for evaluating the precision and accuracy of "kits" for chemical analysis. The accuracy may be measured by comparison with the "true" assay value found in analysis by a reference method. In a glucose reproducibility study the results of the test method and reference method analyses of 100 ml of a patient's serum at 10 different times are shown below. Do the data support the conclusion that the test method consistently overestimates the assay value?

| Analysis | Reference | Test | Analysis | Reference | Test |
|----------|-----------|------|----------|-----------|------|
| 1 | 86 | 89 | 6 | 91 | 99 |
| 2 | 92 | 96 | 7 | 92 | 79 |
| 3 | 87 | 92 | 8 | 93 | 101 |
| 4 | 94 | 100 | 9 | 86 | 110 |
| 5 | 90 | 106 | 10 | 91 | 100 |

3.18 The question of whether susceptibility to hypnosis can be learned was investigated by Cooper et al (1967). An objective rating was given to a group of subjects to measure their initial susceptibility. After an extensive period of "hypnotic training," the subjects were rated again. In both cases, higher scores represent greater susceptibility. Do the data below for six subjects support the conclusion that hypnotic susceptibility can be increased through learning?

| Subject | Before | After |
|---------|--------|-------|
| 1 | 10.5 | 18.5 |
| 2 | 19.5 | 24.5 |
| 3 | 7.5 | 11.0 |
| 4 | 4.0 | 2.5 |
| 5 | 4.5 | 5.5 |
| 6 | 2.0 | 3.5 |

3.19 Precision and reliability are quite essential in IQ testing. Some psychologists have hypothesized that individuals can improve their scores on standard IQ tests through preparation. Specifically, they postulate that test questions are fairly standardized in that they require individuals to think logically and in certain mental patterns. If this is so, many persons can study questions of similar nature in order to train themselves to think in certain patterns and thereby answer the questions more readily. In a recent study, twenty psychology students were selected at random and given an IQ test. These same persons were asked to prepare for a second IQ test by studying selected questions which were recommended by the psychologists. The scores on the two tests are as follows:

| Student | 1st | 2nd | Student | 1st | 2nd |
|---------|-----|-----|---------|-----|-----|
| 1 | 101 | 104 | 11 | 110 | 114 |
| 2 | 115 | 117 | 12 | 125 | 128 |
| 3 | 96 | 107 | 13 | 104 | 109 |
| 4 | 93 | 92 | 14 | 110 | 111 |
| 5 | 80 | 90 | 15 | 104 | 104 |
| 6 | 102 | 102 | 16 | 106 | 104 |
| 7 | 95 | 94 | 17 | 100 | 107 |
| 8 | 125 | 131 | 18 | 107 | 115 |
| 9 | 130 | 132 | 19 | 122 | 118 |
| 10 | 111 | 120 | 20 | 114 | 118 |

Do the scores support the psychologists' hypothesis?

3.20 In a study of social interaction and self-concept, Manis (1955) obtained data to test the hypothesis that an individual's self-concept will change over time in the direction of closer agreement with a nonfriend's impression of him when there is sufficient social contact with the nonfriend. The data used in the analysis are differences of paired observations formed by the relationship $X_j = X'_j - X''_j$, where $X'_j$ is the decrease over a period of time in the "distance" between a subject's self-concept and a friend's impression of that subject, and $X''_j$ is the same "distance" between ratings of the subject and a nonfriend. The pairs for observation were formed by matching the subject-nonfriend pair with the subject-friend pair according to "distance" at the beginning of the time period. Since the nonfriends were randomly assigned to be roommates, the design was such that the subjects were expected to have the same amount of contact with nonfriends as with friends over the time period. See whether the data support Manis' hypothesis by performing an appropriate one-sided test on his data for 10 pairs:

$$0.3, -0.8, 0.4, 0.6, -0.2, 1.0, 0.9, 5.8, 2.1, 6.1.$$

3.21 Nine laboratories cooperated in a study to determine the median effective dose of a certain drug. Assuming symmetry, recommend a normal range of effective dosage using a confidence coefficient near .90 and the following medians reported by these individual studies:

$$.41, .52, .91, .45, 1.06, .82, .78, .68, .75.$$

3.22 *The Lancet* (December 16, 1972, pp. 1278–1281) reported a study of the relationship between coffee drinking and acute myocardial infarction (M.I.), more commonly known as coronary thrombosis. Each of 276 patients with acute M.I. was matched with four control patients without that diagnosis; the matching considered the confounding variables of age, sex, and hospital. The M.I. patients were found to drink more coffee than the control group, regardless of age, sex, or hospital. Although this implies some strong association between coffee drinking and coronary thrombosis, it does not imply a causal relationship; for example, personality factors may be the major factor accounting for the association.

As a similar experiment, suppose that 10 matched pairs are obtained and their coffee consumption is observed for 2 weeks. If the average numbers of cups consumed per day over this period are as shown below, see whether these data support the alternative that M.I. patients consume more coffee than the other patients.

| | Average Coffee Consumption | |
| Pair Number | M.I. | Control |
| --- | --- | --- |
| 1 | 4.2 | 3.5 |
| 2 | 1.3 | 1.1 |
| 3 | 1.0 | 1.1 |
| 4 | 3.2 | 0.8 |
| 5 | 5.1 | 2.3 |
| 6 | 6.2 | 2.8 |
| 7 | 5.6 | 4.3 |
| 8 | 2.3 | 1.1 |
| 9 | 2.8 | 2.2 |
| 10 | 1.4 | 1.0 |

3.23 Many medical methods — for example, cauterization, freezing, acid, or surgery — can be used to remove warts from the skin. Surman et al (1973, p. 440) reported success with hypnosis, but they were unable to explain why hypnotic suggestion could succeed in effecting a cure. Twenty-four patients with common or plantar warts were divided into an experimental group of 17 and a control group of 7. The experimental group was treated by hypnosis once a week for 5 weeks and told that the warts would disappear on one side (experimental side) of their bodies, but not on the other (control) side. The control group had no treatment of any kind. The number of warts was counted for each patient before the experiment began and also three months later. The average number of warts at the beginning of the experiment was 29.6 for the treatment group and 29.7 for the control group. Three months later, there was no change in the number of warts for any of the 7 patients in the control group. However, all but seven of the 17 patients in the treatment group experienced a loss of warts on one or both sides of their bodies. The data collected on the treatment group are shown below.

(a) Find the *P*-value for effect of the treatment using the data on number of warts on the experimental side of the body only.

(b) Find the *P*-value for effect of the treatment using the data on total number of warts before and after treatment.

| Patient | Number of Warts Before | | Number of Warts After | |
|:---:|:---:|:---:|:---:|:---:|
| | Exp. Side | Control Side | Exp. Side | Control Side |
| 1 | 80 | 40 | 80 | 40 |
| 2 | 23 | 8 | 2 | 0 |
| 3 | 26 | 30 | 0 | 0 |
| 4 | 4 | 4 | 1 | 1 |
| 5 | 1 | 1 | 1 | 1 |
| 6 | 56 | 38 | 1 | 1 |
| 7 | 9 | 2 | 9 | 2 |
| 8 | 13 | 5 | 0 | 0 |
| 9 | 3 | 2 | 0 | 0 |
| 10 | 4 | 2 | 4 | 4 |
| 11 | 1 | 2 | 5 | 2 |
| 12 | 20 | 21 | 15 | 15 |
| 13 | 26 | 6 | 26 | 6 |
| 14 | 12 | 3 | 12 | 3 |
| 15 | 28 | 20 | 0 | 0 |
| 16 | 2 | 1 | 0 | 1 |
| 17 | 6 | 5 | 0 | 0 |

3.24 Two types of all-purpose glue X and Y, are compared for bonding by measuring the breaking strength of bond produced. Each type is tried on ten different types of materials.

(a) Suppose that the data are as follows, where larger numbers mean greater strength. Which glue seems to be more effective in general?

| | Material | | | | | | | | | |
|:---:|:---:|:---:|:---:|:---:|:---:|:---:|:---:|:---:|:---:|:---:|
| Glue | 1 | 2 | 3 | 4 | 5 | 6 | 7 | 8 | 9 | 10 |
| X | 10 | 9 | 20 | 40 | 14 | 30 | 26 | 60 | 30 | 42 |
| Y | 12 | 10 | 23 | 45 | 12 | 31 | 20 | 65 | 32 | 39 |

(b) Suppose that breaking strength is measured only on a crude scale, and the following table is presented, where a plus sign indicates greater strength. Would your answer be the same? Support your answer with an appropriate analysis.

| | | | | | Material | | | | | |
|---|---|---|---|---|---|---|---|---|---|---|
| **Glue** | **1** | **2** | **3** | **4** | **5** | **6** | **7** | **8** | **9** | **10** |
| X | − | − | − | − | + | − | + | − | − | + |
| Y | + | + | + | + | − | + | − | + | + | − |

3.25 A group of eight niacin-depleted dogs are measured for hemoglobin content (in grams percent) and then given a treatment of 25 mg niacin and measured again. The results reported by Bliss (1967, p. 341) are shown below. Analyze the data to study effectiveness of treatment.

| Dog No. | 1 | 2 | 3 | 4 | 5 | 6 | 7 | 8 |
|---|---|---|---|---|---|---|---|---|
| **Before** | 12.6 | 12.6 | 13.7 | 11.1 | 11.3 | 12.2 | 10.0 | 11.4 |
| **After** | 10.4 | 11.5 | 13.6 | 12.0 | 10.7 | 9.3 | 8.8 | 9.4 |

3.26 The following rates were reported by Rosene (1950) for water influx through the surface of young and old root hairs from eight different radish roots. Find a confidence interval estimate of the effect of age on water influx using a confidence coefficient near .90.

| Root No. | 1 | 2 | 3 | 4 | 5 | 6 | 7 | 8 | 9 | 10 | 11 |
|---|---|---|---|---|---|---|---|---|---|---|---|
| **Young** | 213 | 116 | 260 | 158 | 153 | 170 | 267 | 264 | 219 | 254 | 446 |
| **Old** | 89 | 49 | 91 | 80 | 56 | 79 | 47 | 50 | 108 | 165 | 194 |

3.27 A drug company designed an experiment to evaluate the effectiveness of a tranquilizer in relieving depression. Nine patients diagnosed as having anxiety and depression were rated using the Hamilton Scale Factor IV (the "suicidal" factor) and then given the drug for a fixed period and rated again. The measurement is such that lower factor values indicate improvement. Test the effectiveness of the drug.

| Patient | Before | After |
|---|---|---|
| 1 | 1.83 | .88 |
| 2 | 0.50 | .65 |
| 3 | 1.62 | .60 |
| 4 | 2.48 | 2.05 |
| 5 | 1.68 | 1.06 |
| 6 | 1.88 | 1.29 |
| 7 | 1.55 | 1.06 |
| 8 | 3.06 | 3.14 |
| 9 | 1.30 | 1.29 |

3.28 A lawyer was unhappy with the outcome of a jury trial at which he defended his female client. He claimed a mistrial on the basis that females were under-represented on the jury panel. The population of the community is known to be 65% female, but only three of the nine jurors were female. How likely is such an extreme representation of females in a random selection of nine residents of this community? Set up an appropriate hypothesis set and perform a binomial test.

3.29 Use the binomial test for Problem 2.15 of Chapter 2, for the data on both female jurors and black jurors.

3.30 It is frequently claimed that over the years a husband and wife start to look alike, think alike, and act alike, and that one can predict how the other will respond or react to various situations. Fifty married couples are randomly selected for a study. Each individual is asked to respond to nine true-false questions concerning attitudes and preferences. The husband is told to answer the questions according to his own feelings, and the wife is asked to give the answer that she thinks her husband will choose. For example, the question might be "I prefer the mountains to the beach for a summer vacation." An answer of true from a wife means she predicts her husband will answer true; her own preference is not reflected in the answer. If the wife is only guessing, the probability of an agreement is .5 and the expected number of agreements in all 9 questions is 4.5. If she "knows" her husband well, she should have a considerably higher score. The frequency distribution of scores is shown below, tallied according to number of correct predictions by the wife. Perform a suitable statistical test (a) using the methods of this chapter and (b) using the methods of Chapter 2. Which method do you think is (i) more appropriate, and (ii) more powerful?

| Number of Correct Predictions | Number of Couples |
|---|---|
| 1 or less | 0 |
| 2 | 2 |
| 3 | 8 |
| 4 | 10 |
| 5 | 11 |
| 6 | 12 |
| 7 | 4 |
| 8 | 2 |
| 9 | 1 |

3.31 Two lawyers are discussing the effect of recent changes in the drug laws on drug use among teenagers in their county. One lawyer feels that rate of drug use has increased, and the other feels it has remained the same. The rate of drug use was established to be 7 percent in their county before the drug laws were changed. Suppose that a current study is performed using a random sample of 1000 teenagers from the county, and 78 report drug usage. Which lawyer do you think is correct?

3.32 Suppose it is known that about 70 percent of persons having common colds recover within 2 weeks without any treatment from a physician. A researcher thinks he has found a new cure for the common cold. He chooses 20 persons at random, injects them with viruses that are known to produce the common cold, and then administers the new drug to them. After 2 weeks, only 16 are cured. Is this sufficient evidence to claim that the drug is effective?

3.33 Kephart (1955) reported a study of the incidence of family desertions in the city of Philadelphia. His findings stated that 32 percent of all desertions occurred among semiskilled workers (who comprised 22 percent of the total population), and 4 percent of all desertions occurred among professional people (8 percent of the population). The divorce rate has risen markedly since 1955 for all elements of the population. In order to investigate whether the desertion rate has also risen, or at least changed pattern, suppose that a random sample of 1000 is taken from all family desertions reported this year in Philadelphia. Of these, the man is a semiskilled worker in 280 families, and is a professional in 75 families. Investigate the research hypotheses that desertion rate has (a) decreased among semiskilled workers since 1955 and (b) increased among professionals since 1955.

# Inferences Concerning Location
# Based on Two or More Samples

## 1 Introduction

In the paired-sample situation described in Chapter 3, two characteristics or conditions were compared by taking the differences between matched observations. Since each observation has a particular counterpart, such data are frequently referred to as a single bivariate sample (of pairs). Once the differences between observations on a pair are taken, the situation may be treated as a one-sample problem, and statistical techniques appropriate for one sample are applicable.

Although paired samples are an effective design technique for investigations that involve a comparison of only two variables, the effective sample size is only half of the total number of observations made. Further, in some situations it may be impractical, impossible, or inappropriate to match units for observation. For example, suppose that two different treatments for the same disease are to be compared. Both treatments cannot be applied to the same unit, and appropriate matched units may be very difficult to find. In such a situation, we might take a group of patients who have the disease and divide them randomly into two treatment groups (which need not even be of the same size). If every observation is taken under the same conditions, there is no reason to pair any particular observation from one group with any particular observation of the other group. In fact, the two groups are mutually independent because of the random selection for treatment application.

Effective comparisons can frequently be made using two or more mutually independent groups of random observations. The groups may be formed by drawing a random sample from a single population and assigning units randomly to the groups to be compared (as described above). Alternatively, a group of random samples may be taken, one from each of the populations to be compared. For the groups to be mutually independent, it is not sufficient to have independence only between the elements within each group; it must also be true that every element in one group is independent of every element in the other groups. Recall that independence means that the value of one observation does not influence the value of any other observation. For example, consider a study to compare male and female voting preferences. The groups could be obtained independently by taking a random sample of males and a random sample of females. However, if the females are chosen as the wives of the males chosen, the two groups are not independent. When units are assigned randomly into groups, the units need not be a random sample from some population. As explained in Section 5 of Chapter 3, as long as the units are divided into groups according to a random process, the statistical techniques developed for comparisons of independent random samples remain valid. The only problem lies in generalizing inferences to arbitrary units. We refer to the groups as independent random samples from two or more populations, but this assumption can be relaxed when the design techniques introduce randomness.

When two or more mutually independent groups of units are under study, one of the more common hypotheses in statistical investigations is that the population groups are identical. Under this general null hypothesis, the combined random samples can be considered a single random sample from the common population. In nonparametric applications this common population remains completely unspecified. We do usually make the assumption that the distributions are continuous so that ties do not present a problem.

The type of test procedure appropriate for this general null hypothesis depends on the class of alternatives that is of particular interest — that is, the type of difference between populations that should be detected. Three classes of alternatives are considered in this book. The location alternative is covered in this chapter, the scale alternative in the next chapter, and the general alternative in Chapter 6. A wide variety of nonparametric tests have been proposed for each class of alternatives. Only the best known and most powerful tests are included here.

The subject of this chapter is procedures for comparing two or more populations when measures of central tendency provide the basis for comparison. We call this model the $k$-sample location problem for $k \geq 2$, analogous to the one-sample location problem in the previous chapter. We can justify a comparison based on location parameters either because relative location is of primary interest, or because we have enough information about the $k$ populations to feel that, if a difference exists, it is only in location. We cover first the two-sample model by discussing the Mann-Whitney-Wilcoxon test and its corresponding confidence procedure. The Mann-Whitney-Wilcoxon test procedure was extended to the case of three or more samples by Kruskal and Wallis and is referred to as the Kruskal-Wallis test. This test and corresponding multiple comparison techniques are discussed in Section 3.

## 2   Two Independent Samples: Mann-Whitney-Wilcoxon Procedures

For the two-sample problem we represent the two variables and their corresponding populations by $X$ and $Y$. Then the location model states that the probability functions of the populations are identical in form and shape, including the variance and all other parameters, but are shifted apart by a quantity that represents the difference between some two location parameters, for example, the respective population medians. The theoretical representation of the model is then

$$F_y(u) = F_x(u - \theta) \qquad \text{for all } u, \tag{2.1}$$

where $F_x$ is the cumulative distribution function for the $X$ population and $F_y$ is the cumulative distribution for the $Y$ population, and $\theta$ is a constant representing the amount of the shift. Then $X + \theta$ and $Y$ are identically distributed. If $\theta$ is positive, the variables in the $Y$ population tend to be larger in magnitude that those in the $X$ population; a negative $\theta$ implies the opposite. When $\theta$ is zero, the hypothesis of identical populations is true. If we make the *assumption* that (2.1) holds for some $\theta$, possibly zero, the hypothesis set under the location model can be written in terms of the location shift parameter $\theta$, which can be interpreted as the difference between any two location parameters. It is convenient here to designate $\theta$ as the difference between the two medians, and represent $\theta$ by $M_x - M_y$ where $M_x$ and $M_y$ are the medians of the $X$ and $Y$ populations, respectively.

Figure 2.1 shows three probability functions of the normal distribution family, which are identical in every respect except for the value of the median (which equals the mean and the mode for the normal family). The curve labeled $M = 0$ represents the standard normal density that has median zero (and standard deviation 1). The curve to the right, labeled $M > 0$, represents the same normal density but shifted to the right, whereas the curve to the left of the standard normal has parameter $M < 0$.

The test procedure presented here is valid whether the shift model holds or not. However, since the test is primarily sensitive to differences in location and the corresponding confidence procedure is valid only under the shift model, it will be more convenient to simply assume throughout that the distributions have the relationship stated in (2.1). We do not specify any particular distribution $F$, however, and thus the procedures are still distribution-free.

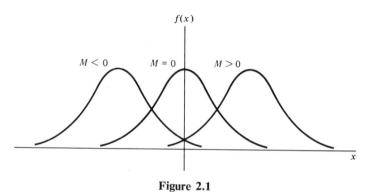

**Figure 2.1**

## 2.1    Mann-Whitney-Wilcoxon Test

We first illustrate the Mann-Whitney-Wilcoxon test procedure by a simple example.

**example    2.1**

In drug screening, a useful experimental technique is to divide animals into two groups that are exposed to the same conditions except that one group is treated with a drug. Then some kind of measurement is made on each animal to see whether the drug is effective. Both groups are frequently quite small; therefore, nonparametric methods are especially appropriate for analysis of the data. For example, suppose that nine mice are to be used in an anticancer drug screening experiment. Cancer cells are implanted in each mouse so that a tumor will grow. Three mice are randomly selected for treatment with a drug. The magnitude of the weight of the tumor is considered indicative of the effectiveness of the drug in curing, arresting, or reducing the cancer. Thus the tumors are removed and weighed for all mice after a fixed period of time. These data are used for screening this particular drug. (The reason the control group is larger than the treatment group is that this same control group may be used as a basis for screening many different drugs. Each drug is used on a different group of three mice, but its effectiveness is measured by comparison with the one group of controls.) The following data on actual tumor weights were reported in the essay by Charles W. Dunnett in Tanur (1972, p. 25).

Treated:    0.96, 1.59, 1.14 grams
Controls:    1.29, 1.60, 2.27, 1.31, 1.81, 2.21 grams

SOLUTION

We call the treated mice the $X$ sample and the controls the $Y$ sample. If the drug is effective, the weights of the treated mice should be smaller than the weights of the controls on the average. Thus we need a hypothesis set with a one-sided alternative. One could argue that the research hypothesis is that the drug is effective, and hence that this should be the statement in the alternative. However, there is an additional reason in this example. In making the decision we must be concerned with two types of errors. We might conclude that the drug is effective when it really is not (a false positive), or we might conclude the drug is not effective when it really is (a false negative). We would like to keep the probability of each error small, but this is not possible with such a small sample size. At best, we can try to control that error which results from rejecting a true hypothesis, that is, the Type I error. Which error should be considered more serious in this problem? The common procedure in drug screening is to perform an initial experiment to eliminate certain drugs from consideration and then do further experimentation using only those drugs found effective by the first screening. While a false positive wastes time and money, a false negative is more consequential to the advancement of medical science. Therefore a false negative should be the Type I error, which means that a negative should be the conclusion asserted in the hypothesis. The alternative then must be a positive, that is, concluding the drug is effective. Since effectiveness if reflected by *decreased* tumor weight, the appropriate alternative in terms of median weights is $M_x - M_y < 0$, where $M_x$ and $M_y$ are the medians of the treatment and control groups, respectively. The hypothesis set is therefore

$$H: \quad M_x - M_y = 0 \qquad A_-: \quad M_x - M_y < 0.$$

Suppose we arrange the two sets of observations in a single array, from smallest to largest, and keep track of which observations are from the treated group by writing them in bold face. The result is as follows:

$$\textbf{0.96}, \textbf{1.14}, 1.29, 1.31, \textbf{1.59}, 1.60, 1.81, 2.21, 2.27.$$

If the drug is effective, we would expect to find the weights of the treated group to be the smaller numbers, that is, at the left end of the array. Hence, if we assign ranks to all the observations, the smaller ranks would be associated with the treated mice ($X$ group) and the larger ranks with the controls ($Y$ group) for an effective drug. In the above array, the treated observations have ranks 1, 2, and 5, with a sum $T_x = 1 + 2 + 5 = 8$. Similarly, the controls have a rank sum $T_y = 3 + 4 + 6 + 7 + 8 + 9 = 37$. Is the rank sum $T_x = 8$ too small, and therefore $T_y = 37$ too large, to have occurred by chance if the drug has no effect? To answer this question, we need the probability of getting a rank sum $T_x$ as small as 8 when the two groups really have the same distribution and therefore the same medians — that is, the

*P*-value. As will be explained later, the *P*-value can be found from Table H as $P = .048$. With such a small *P*-value, we conclude that the sample data do not support the null hypothesis and that the drug does seem to be effective as a cancer treatment.

The general case of hypothesis testing using the Mann-Whitney-Wilcoxon test is now presented following the usual format.

*Inference Situation*    We have two variables, *X* and *Y*, with cumulative distribution functions $F_x$ and $F_y$, respectively. The null hypothesis is that these distributions are the same; that is,

$$F_y(u) = F_x(u) \qquad \text{for all } u.$$

Let $M_x$ denote the median of the *X* population and $M_y$ the median of the *Y* population, and assume that the general location model in (2.1) holds or that the primary interest is whether the populations differ in location. Then the hypothesis set can be written as one of the following:

*Two-sided alternative*
$$H: \quad M_x = M_y \qquad A: \quad M_x \neq M_y$$

*One-sided alternatives*
$$H: \quad M_x = M_y \qquad A_+: \quad M_x > M_y$$
$$H: \quad M_x = M_y \qquad A_-: \quad M_x < M_y$$

*Sampling Situation*    The data consist of two mutually independent random samples of sizes *m* and *n*, denoted by $X_1, X_2, \ldots, X_m$ and $Y_1, Y_2, \ldots, Y_n$, drawn from the continuous populations $F_x$ and $F_y$, respectively. Both sets of data are measured on either (a) at least an interval scale, or (b) an ordinal scale such that the relative magnitude of each element in the pooled sample (both samples combined) can be determined.

*Rationale*    Suppose that we arrange the $m + n = N$ observations in a single ordered sequence, according to relative magnitude from smallest to largest, and keep track of which observations correspond to the *X* sample and which to the *Y*. Under the assumption of continuous populations there is a unique ordering, since $P(X = Y) = 0$ and $P(X_i = X_j) = P(Y_i = Y_j) = 0$ for any two observations from the same or different populations. The data can then be shown as an arrangement of *m* *X*'s and *n* *Y*'s. For example,

$$X\ Y\ Y\ X\ X\ Y\ X\ Y\ Y$$

indicates that for the pooled observations the smallest element is an *X*, the second smallest a *Y*, and so on, and the largest a *Y*.

If the populations are really identical, all *N* observations can be regarded as a single random sample from the common population. Then we would expect the *X*'s and *Y*'s to be well mixed in the ordered sequence that represents the sample data. A sample pattern of arrangement where most of the $Y$'s are greater than most of the $X$'s (or vice versa) would not substantiate random mixing and could be regarded as evidence against the null hypothesis of identical populations. The most extreme cases, which might imply a difference in location, are where all of the $X$'s are smaller than all of the $Y$'s, or vice versa. Then the sequence would be represented as either

$$X\,X\cdots X\,Y\,Y\cdots Y \qquad \text{or} \qquad Y\,Y\cdots Y\,X\,X\cdots X.$$

The positions occupied by the $X$ variables in the combined arrangement are indicative of the magnitudes of the $X$'s in relation to the $Y$'s and hence of the relative locations of the variables. Ranks provide a convenient method of indicating these positions. Therefore, after combining the variables into a single ordered sequence, they are given rank numbers. We assign rank 1 to the smallest, 2 to the second smallest, and so on, and $N$ to the largest. If most of the ranks of the $X$'s are larger than most of the ranks of the $Y$'s, the sum of the $X$ ranks will be larger than its expected value under $H$, supporting the alternative $A_+$. On the other hand, if most of the $X$'s have smaller magnitudes than most of the $Y$'s, the sum of the $X$ ranks will be smaller than its expected value under $H$, supporting $A_-$. If either of these situations occurs, the two-sided alternative $A$ is supported.

Since the sum of all ranks is a constant,

$$1 + 2 + \cdots + N = \frac{N(N+1)}{2}, \tag{2.2}$$

a test could be based on either the sum of the $X$ ranks or the sum of the $Y$ ranks. However, these sums are not identically distributed unless $m = n$. *Test Statistic*   The Mann-Whitney-Wilcoxon test statistic[1] is defined as

$$T_x = \text{sum of the } X \text{ ranks.}$$

We also define the quantity

$$T_y = \text{sum of the } Y \text{ ranks.}$$

From (2.2) we know that $T_x + T_y = N(N+1)/2$. Thus when it is more convenient (as when $T_x$ is larger than $T_y$), the test statistic $T_x$ can be calculated from the equation

$$T_x = \frac{N(N+1)}{2} - T_y. \tag{2.3}$$

---

[1]This test statistic is sometimes called simply the Wilcoxon statistic, or the "sum of ranks" statistic. Since this test is equivalent to one proposed by Mann and Whitney, sometimes called the $U$ test, it is called here the Mann-Whitney-Wilcoxon test. The exact relationship between the $U$ and $T_x$ test statistics is $U = T_x - m(m+1)/2$, where $U$ is defined as the total number of times a $Y$ precedes an $X$ in the configuration of combined samples.

*Finding the P-value*    According to the rationale discussed above, the appropriate P-values are summarized as follows, where smaller tail probability means left or right tail, whichever has the smaller probability.

| Alternative | P-value (Table H) |
|---|---|
| $A_+$: $M_x > M_y$ | Right-tail probability for $T_x$ |
| $A_-$: $M_x < M_y$ | Left-tail probability for $T_x$ |
| $A$: $M_x \neq M_y$ | 2 (smaller tail probability for $T_x$) |

The Mann-Whitney-Wilcoxon statistic $T_x$ takes on only integer values, ranging from $m(m + 1)/2$ to $m(2N - m + 1)/2$, assuming that $m \leq n$. The null sampling distribution of $T_x$ is symmetric about its mean value $m(N + 1)/2$. This means that any left-tail cumulative probability is equal to a corresponding right-tail probability. Table H makes use of this property. It gives left-tail probabilities for $T_x \leq m(N + 1)/2$ and right-tail probabilities for $T_x \geq m(N + 1)/2$ for all $m \leq n \leq 10$. Hence the smaller tail probability is the only one given in this table. Since the arrangement is in terms of increasing values of $n$ for fixed $m$, we first locate the set of entries for the appropriate value of $m$. The fact that the table includes only $m \leq n$ for convenience is not restricting, since the investigator can arbitrarily call the sample with fewer observations the $X$ sample.

For sample sizes outside the range of Table H, the sampling distribution of the Mann-Whitney-Wilcoxon statistic can be approximated by the asymptotic distribution, which is the normal distribution for standardized $T_x$. Incorporating a continuity correction of 0.5, and then subtracting the mean $m(N + 1)/2$ and dividing by the standard deviation $\sqrt{mn(N + 1)/12}$, we obtain the standardized variables $z_{x,L}$ and $z_{x,R}$ as

$$z_{x,L} = \frac{T_x + 0.5 - m(N + 1)/2}{\sqrt{mn(N + 1)/12}} \qquad z_{x,R} = \frac{T_x - 0.5 - m(N + 1)/2}{\sqrt{mn(N + 1)/12}}$$

(2.4)

$$z = \begin{cases} -z_{x,L} & \text{if } T_x < m(N + 1)/2 \\ z_{x,R} & \text{if } T_x > m(N + 1)/2. \end{cases}$$

The following guide for finding P-values from Table A is completely analogous to the one given above for small samples.

| Alternative | P-value (Table A) |
|---|---|
| $A_+$: $M_x > M_y$ | Right-tail probability for $z_{z,R}$ |
| $A_-$: $M_x < M_y$ | Left-tail probability for $z_{z,L}$ |
| $A$: $M_x \neq M_y$ | 2 (right-tail probability for $z$) |

Note that the right-tail probability for $z$ corresponds to the smaller tail probability for $T_x$.

*Ties*   Two or more observations that are equal are called "tied." Ties may occur between observations in the same sample or in different samples or both. Since the populations are assumed continuous, the probability of a tie, either within or across samples, is zero. When ties do occur in practice, the midrank method should be used to obtain a unique value of the test statistic. However, the exact null distribution of $T_x$ is not the same when ties are present. The usual procedure in small samples is to use this distribution anyway, unless there are many ties. When the normal approximation is used for larger sample sizes, a correction for ties can be incorporated. This correction is effected by replacing the standard deviation in the denominator of $z_{x,L}$ or $z_{x,R}$ in (2.4) by the quantity

$$\sqrt{\frac{mn(N+1)}{12} - \frac{mn(\Sigma u^3 - \Sigma u)}{12N(N-1)}}$$

$$= \sqrt{\frac{mn}{12N(N-1)}[N(N^2-1) - (\Sigma u^3 - \Sigma u)]}. \quad (2.5)$$

As before, $u$ is the total number of observations (in either sample) that are tied for a given rank, and the sum is over all sets of $u$ tied ranks. Since this correction tends to make the test anticonservative, it is recommended for use only when the ties are extensive.

To illustrate the calculation, suppose the observations are as follows:

$$X: \quad 1, 2, 4, 4$$
$$Y: \quad 1, 3, 5, 6, 6, 6.$$

Then $m = 4$, $n = 6$, $N = 10$, and ties occur for the observation values 1, 4, 6. There are two observations tied for the value 1, two for 4, and three for 6. Thus $u_1 = 2$, $u_2 = 2$, and $u_3 = 3$, $\Sigma u^3 - \Sigma u = (2^3 + 2^3 + 3^3) - (2 + 2 + 3)$ $= 43 - 7 = 36$, and the corrected standard deviation from (2.5) is

$$\sqrt{\frac{4(6)}{12(10)(9)}[10(99) - 36]} = \sqrt{\frac{954}{45}} = \sqrt{21.20}.$$

If the correction is not used, the standard deviation is

$$\sqrt{\frac{4(6)(11)}{12}} = \sqrt{22}.$$

Thus the correction has little effect even with ties this extensive.

#### example  2.2

Educational psychologists claim that the order in which test questions are asked affects a student's ability to answer correctly, and therefore his total grade. In order to investigate this assertion, a professor randomly divides his class of 20 students into two groups of 10 each. He prepares one set of questions, and ranks them according to his perception of level of difficulty. Based on this ranking he prepares two tests; on Test A the questions are arranged in increasing order of difficulty, that is from easiest to hardest, and on Test B the order is reversed. One group of students takes Test A, and the other group takes Test B; both tests are administered under exactly the same conditions. The test scores are given below. Investigate the claim, measuring ability to answer correctly in terms of median score on these tests.

Test A:    82, 81, 83, 95, 91, 65, 90, 75, 71, 70
Test B:    78, 68, 78, 66, 75, 50, 60, 42, 80, 78.

SOLUTION
The alternative in this example should clearly be one-sided, that scores on Test A are on the average higher than scores on Test B. Since the sample sizes are equal, either set can be labeled the $X$ sample. If we let $X$ represent Set A, the hypothesis set is

$$H: \quad M_x = M_y \qquad A_+: \quad M_x > M_y.$$

The 20 scores are arranged in Table 2.1 in increasing order to magnitude; those scores in boldface type are the $X$ observations. Then the ranks 1 to 20 are assigned, using midranks for tied observations, and the ranks of $X$ observations are in boldface type. Note that when observations are tied across samples it is not clear which one belongs to which sample. However, as long as the midrank is assigned, it does not matter since each observation in any tied set has the same rank. We now calculate the $X$ rank sum as

$$T_x = 4 + 7 + 8 + 9.5 + 15 + 16 + 17 + 18 + 19 + 20 = 133.5.$$

**TABLE 2.1**

| Observation | Rank | Observation | Rank |
|---|---|---|---|
| 42 | 1 | 78 | 12 |
| 50 | 2 | 78 | 12 |
| 60 | 3 | 78 | 12 |
| 65 | 4 | 80 | 14 |
| 66 | 5 | 81 | 15 |
| 68 | 6 | 82 | 16 |
| 70 | 7 | 83 | 17 |
| 71 | 8 | 90 | 18 |
| 75 | 9.5 | 91 | 19 |
| 75 | 9.5 | 95 | 20 |

From Table H with $m = 10$, $n = 10$, the right-tail probabilities for 133 and 134 are .018 and .014, respectively. We estimate the right-tail probability for

133.5 as the average of these probabilities, or $(.018 + .014)/2 = .016$. This estimates the $P$-value. We conclude that the data do not support the null hypothesis, and that these data substantiate the claim. This phenomenon is probably due to the psychological effect of built-up confidence. When hard questions appear first, the student may become more nervous because his initial impression is of a very difficult test. Students who see the easy questions first have an opportunity to build up confidence and thus perform better on the more difficult questions.

**example   2.3**

A psychologist interested in studying the contribution of a drug in controlling aggressive behavior designed the following experiment. Fourteen people, chosen randomly, are given the drug, and 14 different people, also chosen at random, are given a placebo. All subjects are then given a test of aggression, which is scored on an ordinal scale. Suppose that the results are as given in Table 2.2, where higher scores indicate a higher level of aggressive behavior. Does the drug seem to have an effect in reducing aggressive behavior?

**TABLE 2.2**

| Drug Group Scores | | Placebo Group Scores | |
|---|---|---|---|
| 10 | 14 | 13 | 12 |
| 13 | 18 | 10 | 20 |
| 12 | 12 | 19 | 10 |
| 13 | 17 | 14 | 21 |
| 15 | 16 | 9 | 25 |
| 16 | 7 | 18 | 16 |
| 8 | 10 | 14 | 11 |

**SOLUTION**

The research hypothesis is that the drug is effective; we formulate this assertion as the one-sided alternative. If the drug has any effect, the subjects in the drug group would tend to have lower scores on the aggressive behavior test than did the placebo group. Because the measurement level is ordinal, the comparison of scores will be made in terms of medians. Using $X$ to represent the drug group, the appropriate hypothesis set is then

$$H: \quad M_x = M_y \qquad A_-: \quad M_x < M_y.$$

(Since the sample sizes are equal, either group can be labeled the $X$ sample.)

The 28 observations are pooled, arranged in increasing order of magnitude and ranked as shown in Table 2.3; the $X$ scores and ranks are in boldface type. Since both samples are large, the normal approximation can be used. The appropriate $P$-value is a left-tail probability for $z_{x, L}$. We calculate the standard deviation from (2.5) so as to incorporate the correction for ties.

**TABLE 2.3**

| Observation | Rank | Observation | Rank |
|:---:|:---:|:---:|:---:|
| 7 | 1 | 14 | 16 |
| 8 | 2 | 14 | 16 |
| 9 | 3 | 14 | 16 |
| 10 | 5.5 | 15 | 18 |
| 10 | 5.5 | 16 | 20 |
| 10 | 5.5 | 16 | 20 |
| 10 | 5.5 | 16 | 20 |
| 11 | 8 | 17 | 22 |
| 12 | 10 | 18 | 23.5 |
| 12 | 10 | 18 | 23.5 |
| 12 | 10 | 19 | 25 |
| 13 | 13 | 20 | 26 |
| 13 | 13 | 21 | 27 |
| 13 | 13 | 25 | 28 |

The tied observations and their frequencies $u$ are listed in Table 2.4 along with the calculation of $\Sigma u^3 - \Sigma u$. The corrected value of $z_{x, L}$ is then

$$z_{x, L} = \frac{179.5 + 0.5 - 14(29)/2}{\sqrt{\dfrac{14(14)}{12(28)(27)}[(28)(783) - 162]}} = \frac{-23}{\sqrt{470.17}} = -1.06.$$

**TABLE 2.4**

| Tied Observation | $u$ | $u^3$ |
|:---:|:---:|:---:|
| 10 | 4 | 64 |
| 12 | 3 | 27 |
| 13 | 3 | 27 |
| 14 | 3 | 27 |
| 16 | 3 | 27 |
| 18 | 2 | 8 |
| | 18 | 180 |

$$\Sigma u^3 - \Sigma u = 180 - 18 = 162$$

The asymptotic approximate $P$-value, found from Table A, is .1446. The data do not support the research hypothesis; we conclude that the drug does not appear to be effective in reducing aggression score.

This example as described is indeed a valid application of the Mann-Whitney-Wilcoxon test even though the ties endanger the accuracy of the $P$-value. (It is interesting to note that to three significant figures, the value of $z_{x,L}$ is the same when the correction for ties is omitted, and hence the $P$-value is the same.) Let us look at the design of the experiment. Since the primary focus is on whether or not the drug is effective, a control group is necessary. The placebo group serves that function here. However, the scores in either group reflect the individual being measured as well as the effect of the drug. Human aggression is affected by many factors besides whether or not a drug is administered. The two groups were both randomly selected in the hope that they can be expected to have similar aggression scores before the experiment is performed. A pretest might help ascertain whether this assumption is reasonable. Alternatively, the experiment might have been designed such that subjects are matched pairs of individuals, and the treatment number of each pair is selected randomly. Then the differences in aggression scores could be analyzed by the Wilcoxon signed rank test of Section 3 of Chapter 3.
*Construction of Table H*   The exact sampling distribution of $T_x$, as given in Table H for small sample sizes, can be easily found by enumeration. We now give an example to illustrate the construction of Table H. Note that the value of $T_x$ depends only on the configuration of $X$'s and $Y$'s in the combined samples. There are $\binom{N}{m} = N!/m!n!$ distinguishable arrangements of $m$ $X$'s and $n$ $Y$'s, and each is equally likely to occur under the null hypothesis of identical distributions. For $m = 2$, $n = 3$, the $\binom{5}{2} = 10$ possible configurations and their corresponding values of $T_x$ are enumerated as in Table 2.5.

**TABLE 2.5**

| Configuration | Value of $T_x$ |
|---|---|
| XXYYY | 3 |
| XYXYY | 4 |
| XYYXY | 5 |
| YXXYY | 5 |
| XYYYX | 6 |
| YXYXY | 6 |
| YYXXY | 7 |
| YXYYX | 7 |
| YYXYX | 8 |
| YYYXX | 9 |

Since $T_x = 3$ occurs for only one possible configuration, that value occurs with probability $1/10$. Similarly for $T_x = 4$. However, $T_x = 5$ occurs twice, and thus its probability is $2/10$. The complete listing of point probabilities is then as shown in Table 2.6. Any left-tail or right-tail probability can be found from this listing. For example, $P(T_x \leq 4) = .10 + .10 = .20$, the left-tail probability given in Table H for $m = 2$, $n = 3$.

**TABLE 2.6**

| $T_z$ | Probability |
|-------|-------------|
| 3 | $1/10 = .10$ |
| 4 | $1/10 = .10$ |
| 5 | $2/10 = .20$ |
| 6 | $2/10 = .20$ |
| 7 | $2/10 = .20$ |
| 8 | $1/10 = .10$ |
| 9 | $1/10 = .10$ |

*Other Inferences*   In some investigations concerned primarily with location, we are not interested simply in determining whether or not the medians differ, but rather in whether they differ by some predetermined amount. For example, suppose that a production manager is considering a change from Process A to Process B. If the changeover requires a large expenditure of time and money, knowing that Process B is better than Process A may not be sufficient inducement to initiate the change. The research hypothesis of interest might be that Process B is on the average at least 20 units better than Process A.

The Mann-Whitney-Wilcoxon test procedure can also be applied in such problems once the data are adjusted properly. Consider the same sampling situation, but suppose that the null hypothesis states some value other than zero for the difference between the medians of the $X$ and $Y$ populations. Then for some given number $\theta_o$ the hypothesis set is one of the following:

*Two-sided alternative*
$$H: \quad M_x - M_y = \theta_o \qquad A: \quad M_x - M_y \neq \theta_o$$

*One-sided alternatives*
$$H: \quad M_x - M_y = \theta_o \qquad A_+: \quad M_x - M_y > \theta_o$$
$$H: \quad M_x - M_y = \theta_o \qquad A_-: \quad M_x - M_y < \theta_o$$

In each case the null hypothesis can be written equivalently as
$$H: \quad (M_x - \theta_o) - M_y = 0.$$

Now suppose that the variable $X$ is replaced by $X' = X - \theta_o$. Then the median of $X'$ is equal to the median of $X - \theta_o$. But if the variable $X$ is re-duced by $\theta_o$, its median is also reduced by $\theta_o$, and thus $M_{x-\theta_o} = M_x - \theta_o$. If we also assume the shift model (2.1), then $X - \theta_o$ and $Y$ are identically dis-tributed under $H$. The same relationships hold for the sample observations; that is, $X_i - \theta_o$ and $Y_j$.

Therefore, the Mann-Whitney-Wilcoxon procedure can be used with any of these hypothesis sets as long as the test is based on the derived ob-servations $X_1 - \theta_o, X_2 - \theta_o, \ldots, X_m - \theta_o$ and $Y_1, Y_2, \ldots, Y_n$. These $N$ values are pooled into a single array from smallest to largest and assigned ranks $1, 2, \ldots, N$, and the test statistic is

$$T_{x'} = \text{sum of the } X' \text{ ranks.}$$

Otherwise, the test is performed exactly as before.

## 2.2   Confidence Interval Estimation of the Shift Parameter

If the probability functions of $X$ and $Y$ are related by the location model in (2.1), the unknown shift parameter $\theta$ can be estimated by a confidence pro-cedure that corresponds to the Mann-Whitney-Wilcoxon test. Recall that the test is valid under the more general model of identical populations. The corresponding confidence procedure is not. Rather, we must make the spe-cific assumption that two mutually independent random samples are drawn from populations that differ only in location. As in Section 2.1, we call the shift parameter $M_x - M_y$ instead of $\theta$.

The confidence interval for $M_x - M_y$, with confidence coefficient $\gamma$, consists of all values of $M_x - M_y$ that would be supported by the sample data with probability $\gamma$. The number $\gamma$ is chosen by the investigator. How-ever, as in the Wilcoxon signed rank confidence procedure, the *exact* values of $\gamma$ are limited to $\gamma = 1 - 2P$. Here $P$ is one of the tail probabilities in Table H for the relevant $m$ and $n$.

The values of $M_x - M_y$ that belong in the confidence region could be found using the test procedure just described for $H: \quad M_x - M_y = \theta_o$. That is, we could try different values of $\theta_o$ and see which ones lead to acceptance of $H$ when a two-sided test is performed at level $2P$. However, the following method gives the endpoints of the confidence interval in a more convenient and systematic way.

The procedure is to form all of the $mn$ differences

$$U_{ij} = X_i - Y_j \quad \text{for } i = 1, 2, \ldots, m \text{ and } j = 1, 2, \ldots, n,$$

for the observations $X_1, X_2, \ldots, X_m$ and $Y_1, Y_2, \ldots, Y_n$. Arrange these $U_{ij}$ in order of magnitude from smallest to largest. Then the confidence interval endpoints are the $k$th smallest and $k$th largest values of $U_{ij}$. The number $k$ is found from Table H as the rank of that number $T_x$ in the table for that particular $m$, $n$, and left-tail probability $P$, where $\gamma = 1 - 2P$. Some examples of $k$ are shown in Table 2.7. Notice that it is always true that $k$ is one more than the difference between the left $T_x$ for that $P$ and $m(m + 1)/2$, the smallest possible value of $T_x$. (This is analogous to the result for $k$ in the confidence interval procedures described in Chapter 3.)

TABLE 2.7

| $m$ | $n$ | $P = (1 - \gamma)/2$ | $\gamma = 1 - 2P$ | Left $T_z$ for $P$ | $k = $ Rank of $T_z$ |
|---|---|---|---|---|---|
| 4 | 4 | .014 | .972 | 10 | 1 |
| 6 | 8 | .015 | .970 | 28 | 8 |
| 6 | 8 | .041 | .918 | 31 | 11 |
| 8 | 8 | .041 | .918 | 51 | 16 |
| 8 | 8 | .052 | .896 | 52 | 17 |

For sample sizes outside the range of Table H the left-tail critical value of $T_x$ is approximated by

$$-0.5 + \frac{m(N + 1)}{2} - z\sqrt{\frac{mn(N + 1)}{12}}.$$

Since $k$ is the rank of this value $T_x$, $k$ is then approximated by

$$k = -0.5 + \frac{m(N + 1)}{2} - z\sqrt{\frac{mn(N + 1)}{12}} - \frac{m(m + 1)}{2} + 1$$

$$= 0.5 + \frac{mn}{2} - z\sqrt{\frac{mn(N + 1)}{12}}, \tag{2.6}$$

where $z$ is the quantile point of the normal curve, found from Table A, that corresponds to a right-tail probability of $(1 - \gamma)/2$. As usual, if the value of $k$ from (2.6) is not an integer, it should be rounded down to the next lower integer to obtain a conservative result.

As an example, consider the following ordinal observations:

$X$   sample: 1, 6, 7
$Y$   sample: 2, 4, 9, 10, 12.

In order to find the ordered differences $U_{ij}$ systematically, we first list the $X$ values in a column, from smallest to largest, as shown in the first column below. To fill in the second column, subtract from each $X$ observation the smallest value of $Y$. Then we keep subtracting successive increasing values of $Y$ until all $mn$ possible differences $X - Y$ are formed.

| $X$ | $X-2$ | $X-4$ | $X-9$ | $X-10$ | $X-12$ |
|---|---|---|---|---|---|
| 1 | $-1$ | $-3$ | $-8$ | $-9$ | $-11$ |
| 6 | 4 | 2 | $-3$ | $-4$ | $-6$ |
| 7 | 5 | 3 | $-2$ | $-3$ | $-5$ |

We must find the $k$th smallest and $k$th largest difference in this table. Any numbers appearing more than once are counted as many times as they occur. If it seems advisable, a complete array of the differences can be made to avoid any possibility of error. The array here would be $-11$, $-9$, $-8$, $-6$, $-5$, ..., 2, 3, 4, 5. Using Table H with $m = 3$ and $n = 5$, suppose that we decide on $P = .036$, so that $\gamma = .928$. For a left-tail probability of .036, the $T_x$ value is 7, and 7 has rank 2 for that combination of sample sizes; thus $k = 2$. Since $-9$ is the second smallest number and 4 is the second largest number in the array of differences, the confidence interval for $M_x - M_y$ with confidence coefficient .928 is

$$-9 \leq M_x - M_y \leq 4.$$

The confidence interval endpoints can also be determined by a graphical method (Moses, 1965). Each of the $mn$ possible pairs of sample observations is plotted on a graph, the $X$ values on the horizontal scale, and the $Y$ values on the vertical scale; the units are the same on both scales and are chosen in accordance with the magnitudes of the observations. This is illustrated in Figure 2.2 for the data above. Each dot represents one of the $mn$ pairings of an $X$ and a $Y$ observation. Then we draw a 45° line through the $k$th dot counted from the upper left and another 45° line through the $k$th dot counted from the lower right. Since the value of $k$ here is 2, the lines are drawn through the second dot from each direction. The points where these two lines cross the horizontal $X$ axis are the confidence interval endpoints. However, it is really not necessary to extend the lines to meet the $X$ axis, since once they cross either axis the magnitudes of the endpoints are known except for sign. If the point of crossing is observed on the $Y$ axis, the sign is changed for the interval endpoint. Since we observe here a left crossing at $Y = 9$, the lower endpoint is $-9$. The right crossing is $+4$ on the $X$ axis, so that sign is not changed and the upper endpoint is $+4$.

# 3 $k$ Independent Samples: Kruskal-Wallis Procedures

In the introduction to this chapter, the general null hypothesis for mutually independent samples from $k$ populations was defined as the assertion that

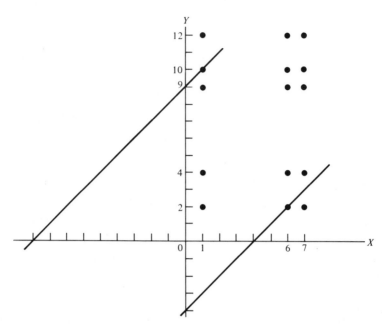

**Figure 2.2**    Reproduced from Figure 5.1, p. 148, of Jean D. Gibbons (1972), *Nonparametric Statistical Inference*, McGraw-Hill Book Company, New York, with permission.

the $k$ populations are identical. In the last section we covered the Mann-Whitney-Wilcoxon procedure, which is applicable for $k = 2$. In this section, we extend this test to an arbitrary number $k$ of populations. The test procedure is known as the Kruskal-Wallis test.

The general null hypothesis for homogeneity of $k$ populations is

$$F_1(x) = F_2(x) = \cdots = F_k(x) \quad \text{for all } x,$$

where $F_1(x)$, $F_2(x)$, $\ldots$, $F_k(x)$ denote the cumulative distribution functions of the $k$ populations. The alternative may be completely general also, stating simply that the distributions are not all equal. The $k$-sample location model asserts that the probability functions of the $k$ populations are identical except for location, usually as described by the respective population medians. If we assume this model, the general null hypothesis is equivalent to a statement of equality between the $k$ population medians. Although assumption of the location model is not necessary for the validity of the Kruskal-Wallis test, the test is primarily sensitive to differences in locations of the $k$ populations. Thus the test is particularly useful when the investigator feels that if a difference exists it is only in location, or when he is primarily interested in detecting a difference in medians if the distributions are not identical. In order to emphasize this aspect, we discuss the Kruskal-Wallis test in terms of inferences about medians. This makes it a natural extension of the Mann-Whitney-Wilcoxon test for two populations. The general null hypothesis of homogeneity is reserved for the more general tests, which are covered in Chapter 6.

### 3.1 Kruskal-Wallis Test

We introduce the methodology of the Kruskal-Wallis test using the following example.

**example 3.1**

   The staff of a mental hospital is concerned with which kind of treatment is most effective for a particular type of mental disorder. A battery of tests administered to all patients delineated a group of 40 patients who were similar as regards diagnosis and also personality, intelligence, and projective and physiological factors. These people were randomly divided into four different groups of 10 each for treatment. For 6 months the respective groups received (1) electroshock, (2) psychotherapy, (3) electroshock plus psychotherapy, and (4) no type of treatment. At the end of this period the battery of tests was repeated on each patient. The only type of measurement possible for these tests is a ranking of all 40 patients on the basis of their relative degree of improvement at the end of the treatment period. In the data shown in Table 3.1, rank 1 indicates the highest level of improvement, rank 2 the second highest, and so forth. On the basis of these data, does there seem to be any difference in effectiveness of the types of treatment?

**TABLE 3.1**

| | Groups | | | |
|---|---|---|---|---|
| | **1** | **2** | **3** | **4** |
| | 19 | 14 | 12 | 38 |
| | 22 | 21 | 1 | 39 |
| | 25 | 2 | 5 | 40 |
| | 24 | 6 | 8 | 30 |
| | 29 | 10 | 4 | 31 |
| | 26 | 16 | 13 | 32 |
| | 37 | 17 | 9 | 33 |
| | 23 | 11 | 15 | 36 |
| | 27 | 18 | 3 | 34 |
| | 28 | 7 | 20 | 35 |
| Sum *R*: | 260 | 122 | 90 | 348 |

**SOLUTION**
Because the data are measured only by scores on an ordinal scale, the only appropriate descriptor of improvement in any group is the median score. Hence we will test the null hypothesis that the median scores are equal for the four types of treatment. With $M_i$ denoting the median of the $i$th group. we write the hypothesis set as

$$H: \quad M_1 = M_2 = M_3 = M_4$$
$$A: \quad \text{At least two of the } M_i\text{'s differ.}$$

If the groups are all the same, the ranks 1, 2, . . . , 40 should be well distributed among the four groups, and the mean of the ranks in any one group should be approximately equal to the mean of the ranks in any other group. The mean rank of each group, denoted by $\bar{R}_j$, is found by dividing the sum $R_j$ of the ranks in that group by $n_j$, the number of units in that group; that is, $\bar{R}_j = R_j/n_j$. For these data the means are

$$\bar{R}_1 = 26.0 \qquad \bar{R}_2 = 12.2 \qquad \bar{R}_3 = 9.0 \qquad \bar{R}_4 = 34.8,$$

and the grand mean is the total sum of ranks divided by 40, or

$$\bar{\bar{R}} = \frac{260 + 122 + 90 + 348}{40} = 20.5.$$

Using the basic idea of ordinary one-way analysis of variance, the groups are compared by computing the sum of the squares of the deviations between the four group means and the grand mean, as weighted by the size of the group, or

$$10(26.0 - 20.5)^2 + 10(12.2 - 20.5)^2 + 10(9.0 - 20.5)^2$$
$$+ 10(34.8 - 20.5)^2 = 4358.8.$$

The Kruskal-Wallis test statistic $H$ is found by multiplying this sum by the constant $12/N(N + 1)$, where $N$ is the total number of observations. The result here is

$$H = \frac{12(4358.8)}{40(41)} = 31.89.$$

As will be explained later, an asymptotic approximate $P$-value can be found by referring the value of $H$ to the chi-square distribution, here with df = 3. Since 31.89 is larger than the largest entry in Table B for df = 3, which is the quantile point for a right-tail probability of .001, the $P$-value is $P < .001$ and $H$ is rejected.

We now describe the general case of the Kruskal-Wallis test following the usual format.

*Inference Situation*   Assume that we have $k$ populations, called populations 1, 2, . . . , $k$, with respective continuous cumulative probability distributions

$F_1, F_2, \ldots, F_k$. The general null hypothesis for which the test is valid is

$$F_1(x) = F_2(x) = \cdots = F_k(x) \qquad \text{for all } x.$$

As explained in the introduction, we emphasize the location model by writing this hypothesis in terms of the respective location parameters, namely, the medians $M_1, M_2, \ldots, M_k$. Then the hypothesis set can be stated as

$$H: \quad M_1 = M_2 = \cdots = M_k$$
$$A: \quad \text{At least two of the } M_i\text{'s differ from each other.}$$

Note that there is no pair of one-sided alternatives that orders the $M_i$'s for $k > 2$, since in fact there are $k! = k(k-1)\cdots(1)$ different ordered arrangements.

*Sampling Situation*   The data consist of $k$ mutually independent random samples, of sizes $n_1, n_2, \ldots, n_k$, which are drawn respectively from the populations with c.d.f.'s $F_1, F_2, \ldots, F_k$. All $k$ sets of observations are measured on either (a) at least an interval scale, or (b) an ordinal scale such that the magnitude of each element relative to every other element can be determined.

*Rationale*   If the null hypothesis is true, the pooled data can be treated as a single random sample of size $N = n_1 + n_2 + \cdots + n_k$ from the common population, and all arrangements of the combined observations in order of magnitude are equally likely. Therefore, when integer ranks $1, 2, \ldots, N$ are assigned to the observations in the combined arrangement, the ranks should be well distributed among the $k$ samples. The average rank for any one of the $N$ observations is

$$\frac{1 + 2 + \cdots + N}{N} = \frac{N(N+1)}{2N} = \frac{N+1}{2}.$$

The expected sum of the ranks for the $j$th sample, which contains $n_j$ observations, is then $n_j$ times the average rank, or

$$\frac{n_j(N+1)}{2},$$

and the expected mean of the ranks for any one of the $j$ samples is $(N+1)/2$. Let $R_j$ denote the actual sum of ranks in the $j$th sample, and $\bar{R}_j$ the actual mean of the ranks in the $j$th sample. The decision regarding random distribution of ranks can be based on the magnitudes of the deviations of the $k$ rank sums from their respective expected values, namely the values of $R_j - n_j(N+1)/2$. (Note that these are $n_j$ times the deviations of the $k$ rank means from their expected value, that is, $\bar{R}_j - (N+1)/2$. It is more convenient to use the deviations of rank sums.) As in the chi-square goodness-of-fit test, a measure of disagreement is provided by squaring these devia-tions of rank sums and reducing them to original units by dividing by the sample size $n_j$. (This division weights the squared deviations inversely ac-cording to sample size.) Except for the multiplication by a constant, this sum is the Kruskal-Wallis test statistic defined below in (3.1). A small value supports $H$. A large value reflects too much disagreement between the actual distribution of ranks over samples and the expected distribution, and thus discredits $H$.

*Test Statistic*    The Kruskal-Wallis $H$ test statistic is defined as

$$H = \frac{12}{N(N+1)} \sum_{j=1}^{k} \frac{[R_j - n_j(N+1)/2]^2}{n_j},  \tag{3.1}$$

where $k$ = number of samples, $R_j$ = sum of the ranks in the $j$th sample, $n_j$ = number of observations in the $j$th sample, and $N$ = total number of observations. The formula for $H$ can be simplified for computation to

$$H = \left( \frac{12}{N(N+1)} \sum_{j=1}^{k} \frac{R_j^2}{n_j} \right) - 3(N+1).  \tag{3.2}$$

This expression is algebraically equivalent to (3.1) and is easier to use, even though it does not show as clearly which samples are the major contributors to disagreement.

*Finding the P-value*    The exact sampling distribution of the test statistic $H$ can be found by enumeration. However, the large number of possible com-binations of $k$, $N$, and $n_1, n_2, \ldots, n_k$ makes tables cumbersome. The asymptotic sampling distribution of $H$ is the chi-square distribution with $k - 1$ degrees of freedom, and this is usually accurate enough for most prac-tical purposes, as long as none of the sample sizes is too small. The asymp-totic approximate $P$-value is then found as a right-tail probability from the chi-square table, Table B, where df = $k - 1$. The general rule of thumb is that this approximation is satisfactory except when $k = 3$ and each $n_j \leq 5$. Since samples this small are seldom taken, tables of the exact distribution are not given in this book. They are available in Kruskal and Wallis (1952), and also in Conover (1971, Table 12), Kraft and Van Eeden (1968, Table R) and Siegel (1956, Table O), among other sources.

*Ties*    Under the assumption of continuous distributions, the probability is zero that two or more observations are equal. In theory then, a unique rank-ing exists. However, ties can occur in practice, either among or between samples. The midrank method is recommended to resolve all ties; the test can then be performed in the usual way. If the ties are extensive, a correction factor might be incorporated for the calculation of $H$. The correction is made by calculating $H$ as in (3.1) or (3.2) and then dividing by the quantity

$$1 - \frac{\Sigma u^3 - \Sigma u}{N(N^2 - 1)},  \tag{3.3}$$

where $u$ is the number of observations in all samples combined that are tied for any rank, and the sum is over all sets of tied ranks. This correction always increases the value of $H$ and therefore decreases the $P$-value estimated by the chi-square distribution; thus the $P$-value is always conservative. The correction has little effect on the value of $H$, even with many ties, as the next example illustrates.

**example 3.2**

It is widely believed that two heads are better than one, and indeed many experimental studies have concluded that problem solving is more effective through team effort than individual effort. However, an extensive study by McConville and Hemphill (1966, p. 270) provided surprising empirical evidence to the contrary. They randomly divided 120 women into 30 teams of four each; no attempt was made to match the subjects within teams. The teams were then allocated to three groups of 10 teams each. Each team independently performed the same task, but teams in different groups had different degrees of constraint on communication. Group I was required to send messages about the solution between trials only to the experimenter. Group II was required to send messages about the solution between trials to each of the other team members; Group III was permitted to send messages of ten words or less to any or all other team members with no restriction on content of message. Trials were repeated until the task was "learned," as evidenced by solving the problem four times in a row. (The task is described fully in the article; we will not describe it here except to say that the task stressed both problem analysis and execution.) The data in Table 3.2 represent the number of the trial on which the last error was made by each team before the task was learned. Thus the measurement level is clearly ordinal, and the medians are appropriate measures of central tendency. Does it appear that median learning time is the same for the three groups?

**TABLE 3.2**

| | Degree of Communication: | |
| None | Limited | Free |
| --- | --- | --- |
| 4 | 9 | 6 |
| 9 | 11 | 4 |
| 3 | 12 | 13 |
| 6 | 10 | 14 |
| 8 | 13 | 5 |
| 2 | 14 | 13 |
| 9 | 10 | 16 |
| 3 | 14 | 4 |
| 5 | 2 | 6 |
| 5 | 4 | 4 |

**SOLUTION**

The null hypothesis to be tested is

$$H: \quad M_1 = M_2 = M_3.$$

We pool the 30 observations into a single array, rank them from smallest to largest, assigning midranks to tied scores, and compute the rank sums for the three groups as shown in Table 3.3. We list the values of $R_j$ and $n_j$ and

**TABLE 3.3**

|  | None | Limited | Free |
|---|---|---|---|
|  | 7 | 18 | 14 |
|  | 18 | 22 | 7 |
|  | 3.5 | 23 | 25 |
|  | 14 | 20.5 | 28 |
|  | 16 | 25 | 11 |
|  | 1.5 | 28 | 25 |
|  | 18 | 20.5 | 30 |
|  | 3.5 | 28 | 7 |
|  | 11 | 1.5 | 14 |
|  | 11 | 7 | 7 |
|  | 103.5 | 193.5 | 168 |

carry out the intermediate calculations needed for $H$. If (3.1) is to be used, the quantities needed are as shown in Table 3.4. Since $k = 3$, we have

**TABLE 3.4**

| Group | $R$ | $n$ | $n(N+1)/2$ | $R - n(N+1)/2$ | $[R - n(N+1)/2]^2$ | $[R - n(N+1)/2]^2/n$ |
|---|---|---|---|---|---|---|
| None | 103.5 | 10 | 155 | −51.5 | 2652.25 | 265.225 |
| Limited | 193.5 | 10 | 155 | 38.5 | 1482.25 | 148.225 |
| Free | 168.0 | 10 | 155 | 13.0 | 169.00 | 16.900 |
|  |  | 30 |  |  |  | 430.350 |

$$H = \frac{12}{30(31)} (430.35) = 5.55$$

df $= 2$; the $P$-value from Table B is a little larger than .05, an asymptotic approximate result. Although this value of $P$ is not extremely small, we could easily justify a decision to reject the null hypothesis and conclude that the groups differ, especially since we know that using the correction for ties will decrease the *P*-value. Before we make a definitive decision, let us see how much the correction will increase the value of *H*. The tied observations and their frequency of occurrence *u* are shown in Table 3.5, along with the

**TABLE 3.5**

| Tied Observation | $u$ | $u^3$ |
|:---:|:---:|:---:|
| 2 | 2 | 8 |
| 3 | 2 | 8 |
| 4 | 5 | 125 |
| 5 | 3 | 27 |
| 6 | 3 | 27 |
| 9 | 3 | 27 |
| 10 | 2 | 8 |
| 13 | 3 | 27 |
| 14 | 3 | 27 |
| | 26 | 284 |

$$\Sigma u^3 - \Sigma u = 258$$

convenient method for calculating the correction for ties. The value of (3.3) for $N = 30$ is then

$$1 - \frac{258}{30(899)} = .9904,$$

and the corrected *H* is

$$H = \frac{5.55}{.9904} = 5.60.$$

The *P*-value is still in the same range, although it is now slightly closer to .05 than before.

Rejection of the null hypothesis implies that there is a difference between the three groups. However, which pairs of groups differ significantly from each other? In order to answer this question, we need to use a statistical technique known as multiple comparisons. These procedures are the subject of the next subsection.

## 3.2   Multiple Comparisons

If the *P*-value from the Kruskal-Wallis test statistic is so small that the null hypothesis is not supported, we conclude that the *k* populations are not all the same. This information may be sufficient, but in many situations the investigator then wants to know *which* populations differ from which others. When the location model is assumed, this question reduces specifically to which populations have different medians. To answer this question, we could use some two-sample location test — for example, the Mann-Whitney-Wilcoxon test — to compare each of the $k(k - 1)/2$ possible combination sets of two samples. If a *P*-value is obtained for each comparison, how can we combine these individual *P*-values into some easily interpreted number? We do not even try.

Since the goal is simultaneous statistical inference, a special technique known as multiple comparisons should be used. Then simultaneous statements of difference can be made with a single overall level of significance $\alpha$. This level is the probability that at least one statement is wrong when the null hypothesis is true. This is equivalent to saying that the probability is $1 - \alpha$ that all statements are correct, and thus $1 - \alpha$ represents a sort of level of confidence in our procedure. The overall level of significance is frequently larger than the numbers that are ordinarily used in an inference involving a single comparison. The values recommended are in the vicinity of .15, .20, or even .25, depending partly on how large $k$ is. As the number of comparisons increases, the overall level of significance is usually increased so that any possible single difference is more likely to be detected.

Several methods of simultaneous multiple comparisons between the locations of all pairs of two of the $k$ populations have been proposed. The procedure described here is attributed to Dunn (1964). It gives conservative results, but it is the simplest method available.

Let $\bar{R}_i = R_i/n_i$ denote the mean of the ranks corresponding to the $i$th sample, and let $\bar{R}_j = R_j/n_j$ be the analogous mean of the ranks for the $j$th sample. Choose a level of significance $\alpha$ that is reasonable, considering that it is the probability of a Type I error when a large number, in fact $k(k - 1)/2$, of comparisons are to be made simultaneously. The next step is to use Table A to find the quantile point of the standard normal distribution that corresponds to a right-tail probability of $\alpha/k(k - 1)$. This quantity is denoted by $z$ and called the *critical z value*. Then the probability is at least $1 - \alpha$ that the following inequality in (3.4) holds for *all* pairs of means, $\bar{R}_i$, $\bar{R}_j$, when the null hypothesis is true, that is,

$$|\bar{R}_i - \bar{R}_j| \leq z\sqrt{\frac{N(N + 1)}{12}\left(\frac{1}{n_i} + \frac{1}{n_j}\right)}. \tag{3.4}$$

If all $k$ sample sizes are equal, we have $n_1 = n_2 = \cdots = n_k = N/k$, and (3.4) can be simplified to

$$|\bar{R}_i - \bar{R}_j| \leq z\sqrt{\frac{k(N + 1)}{6}}. \tag{3.5}$$

The frequently used values of $z$ are given in Table N in terms of a more general quantity $p$, which is defined as the total number of comparisons made. Thus here we have $p = k(k - 1)/2$. For convenience, the appropriate numbers from Table N are repeated in Table 3.6 for all $k \leq 5$.

**TABLE 3.6**
CRITICAL $z$ VALUES FOR MULTIPLE COMPARISONS OF $k$-SAMPLE MEAN RANKS

| | Level of Significance | | | | | |
|---|---|---|---|---|---|---|
| $k$ | .30 | .25 | .20 | .15 | .10 | .05 |
| 2 | 1.036 | 1.150 | 1.282 | 1.440 | 1.645 | 1.960 |
| 3 | 1.645 | 1.732 | 1.834 | 1.960 | 2.128 | 2.394 |
| 4 | 1.960 | 2.037 | 2.128 | 2.241 | 2.394 | 2.638 |
| 5 | 2.170 | 2.241 | 2.326 | 2.432 | 2.576 | 2.807 |

In application of this procedure, the investigator should calculate the right-hand side of (3.4), or (3.5) if all sample sizes are equal, for the given sample sizes and selected level $\alpha$. All differences of means $|\bar{R}_i - \bar{R}_j|$ that are larger than this number are considered significant at level $\alpha$. The direction of each significant difference pair is indicated by which is the larger of $\bar{R}_i$, $\bar{R}_j$. We may find that none of the *pairs* differs significantly (the inequalities in (3.4) are satisfied for all $1 \leq i \neq j \leq k$), even though when taken as a group the populations are found to be different.

**example   3.1 (continued)**
Now that we know how to investigate comparisons of pairs of groups, we can make a simultaneous comparison of all possible treatment pairs for the data in Example 3.1. Since the sample sizes are all equal to 10, the right-hand side of (3.5) with $N = 40$ and $k = 4$ provides the upper limit on non-significant differences between pairs of treatment means. At level .20, the applicable $z$ value from Table 3.6 is 2.128. Hence the treatment pairs that do not differ significantly are those for which the sample means $(\bar{R}_i, \bar{R}_j)$ satisfy the inequality

$$|\bar{R}_i - \bar{R}_j| \leq 2.128\sqrt{\frac{4(41)}{6}} = 2.128(5.23) = 11.12,$$

and the significant differences are those larger than 11.12. A systematic method of generating all differences $|\bar{R}_i - \bar{R}_j|$ is shown in Table 3.7 for Example 3.1, where the $\bar{R}$ values are $R/10$.

TABLE 3.7

| Group $i$ | $\bar{R}_i$ | Group $j$ 1 $\|\bar{R}_i - 26.0\|$ | 2 $\|\bar{R}_i - 12.2\|$ | 3 $\|\bar{R}_i - 9.0\|$ | 4 $\|\bar{R}_i - 12.8\|$ |
|---|---|---|---|---|---|
| 1 | 26.0 | 0 | | | |
| 2 | 12.2 | 13.8* | 0 | | |
| 3 | 9.0 | 17.0* | 3.2 | 0 | |
| 4 | 34.8 | 8.8 | 22.6* | 25.8* | 0 |

*Pairs of sample means whose differences exceed 11.12, and are therefore significant.

The numbers in the first column after the vertical line, labeled $j = 1$, are obtained by subtracting $\bar{R}_1$ from each group mean $\bar{R}_i$, ignoring sign, and the second column is $\|\bar{R}_i - \bar{R}_2\|$, and so on, until all six nonzero values of $\|\bar{R}_i - \bar{R}_j\|$ are given. It is not necessary to complete the entries above the diagonal because the absolute values of the differences are symmetric. The conclusion is that treatment group 2 differs significantly from group 1; group 3 differs significantly from group 1; group 4 differs significantly from both groups 2 and 3. The direction of these significant differences is determined by simply noting for each pair which group has the larger sample mean. The treatments that differ, and their directions of difference, are as follows:

Treatment 2 < treatment 1
Treatment 3 < treatment 1
Treatment 2 < treatment 4
Treatment 3 < treatment 4.

These statements are made with an overall confidence of at least .80, since the level of significance was selected as .20 and the procedure is conservative.

Since for these data rank 1 indicates the *highest* level of improvement, it appears that treatments 2 and 3 are the most effective. Since both of these treatments involve psychotherapy, and electroshock by itself was not determined effective, the experiment presents a strong case for psychotherapy as the best single treatment to use.

**example   3.2 (continued)**

We now return to Example 3.2 in an attempt to derive more information from the experiment. For an overall significance level of .10, the critical $z$ value from Table 3.6 is again 2.128 when $k = 3$. The right-hand side of (3.5) is

$$2.128\sqrt{\frac{3(31)}{6}} = 8.38,$$

to be compared with each of the $|\bar{R}_i - \bar{R}_j|$, where $\bar{R}_1 = 10.35$, $\bar{R}_2 = 19.35$, and $\bar{R}_3 = 16.80$. With only three groups, we need look at only three nonzero differences of paired means, namely

$$\bar{R}_1 - \bar{R}_2 = -9.00$$
$$\bar{R}_1 - \bar{R}_3 = -6.45$$
$$\bar{R}_2 - \bar{R}_3 = \phantom{-}2.55.$$

The absolute values of the second and third differences are both smaller than 8.38, but the absolute value of the first difference is larger. Since this latter difference is negative, we conclude at level .10 that the "no communication" group has a significantly smaller learning time than the "limited communication" group, and hence more effective task performance. Evidently the restraint of opportunity to communicate freely does not affect performance adversely.

**example  3.3**

In an effort to determine whether dependency needs differ among wage earners, salaried personnel, and self-employed individuals, a psychological research institute drew a random sample of 20 individuals from a population in which all three types are well represented. A well-known dependency test was given to each subject. The higher the score on the test, the more independent a person is. The resulting scores for the persons as categorized are as shown in Table 3.8. Is there a significant difference in median dependency needs among these categories? If so, which pairs of categories differ in which direction?

**TABLE 3.8**

| | Self-Employed | Wage Earners | Salaried Personnel |
|---|---|---|---|
| | 42 | 24 | 29 |
| | 35 | 23 | 26 |
| | 41 | 19 | 24 |
| | 38 | 18 | 35 |
| | 37 | 21 | 38 |
| | 28 | 30 | 19 |
| | | 26 | |
| | | 31 | |

**SOLUTION**

The null hypothesis is that the median dependency score is the same for the three populations. The scores of the 20 individuals are ranked, with 1 for the lowest score, indicating the highest level of dependency. These rankings and the calculations needed for the test statistic $H$ are shown in Table 3.9.

**TABLE 3.9**

| (1) Self-Employed | (2) Wage Earners | (3) Salaried Personnel |
|---|---|---|
| 20 | 6.5 | 11 |
| 14.5 | 5 | 8.5 |
| 19 | 2.5 | 6.5 |
| 17.5 | 1 | 14.5 |
| 16 | 4 | 17.5 |
| 10 | 12 | 2.5 |
| | 8.5 | |
| | 13 | |
| 97.0 | 52.5 | 60.5 |

We proceed as usual to calculate $H$ from (3.1); see Table 3.10. The asymptotic approximate $P$-value from Table B is a little larger than .01, so the null hypothesis is rejected.

**TABLE 3.10**

| Group | $R$ | $n$ | $n(N+1)/2$ | $R-n(N+1)2$ | $[R-n(N+1)/2]^2$ | $[R-n(N+1)/2]^2/n$ |
|---|---|---|---|---|---|---|
| 1 | 97.0 | 6 | 63 | 34.0 | 1156.00 | 192.67 |
| 2 | 52.5 | 8 | 84 | −31.5 | 992.25 | 124.03 |
| 3 | 60.5 | 6 | 63 | −2.5 | 6.25 | 1.04 |
| | | 20 | | | | 317.74 |

$$H = \frac{12}{20(21)} (317.74) = 9.08 \qquad df = 3 - 1 = 2$$

The $\bar{R}$ values are as follows:

$$\bar{R}_1 = \frac{97}{6} = 16.2 \qquad \bar{R}_2 = \frac{52.5}{8} = 6.6 \qquad \bar{R}_3 = \frac{60.5}{6} = 10.1.$$

To investigate which particular pairs of groups, if any, differ significantly, we make the multiple comparisons test, with level .15. Then $z = 1.96$. With unequal sample sizes, (3.4) must be used. The maximum nonsignificant differences and corresponding observed differences are calculated as follows:

$$\text{Groups 1, 2:} \quad 1.96 \sqrt{\frac{20(21)}{12}\left(\frac{1}{6} + \frac{1}{8}\right)} = 6.26 \qquad |\bar{R}_1 - \bar{R}_2| = 9.6$$

Groups 1, 3:    $1.96 \sqrt{\frac{20(21)}{12}\left(\frac{1}{6} + \frac{1}{6}\right)} = 6.70$    $|\bar{R}_1 - \bar{R}_3| = 6.1$

Groups 2, 3:    $1.96 \sqrt{\frac{20(21)}{12}\left(\frac{1}{6} + \frac{1}{8}\right)} = 6.26$    $|\bar{R}_2 - \bar{R}_3| = 3.5$

Groups 1 and 2 are the only ones that differ significantly. Since $\bar{R}_1 > \bar{R}_2$, we conclude that self-employed individuals are more independent than are wage earners, but no other differences are judged significant at level .15.

In this example we stated that the three types of individuals are well represented in the population. Therefore, if one simple random sample is taken from the entire population, a representative sample of each type should result. In many situations, this may not be the case. The method of stratified sampling can be used to ensure that representative samples are obtained. The reader is referred to the sampling books listed at the end of Chapter 1 and in the references for a discussion of stratified sampling. The sample sizes obtained with stratification are unequal in most cases; they are affected by the cost of obtaining a sampling unit from each stratum and the variance of the units in each stratum. One advantage of the Kruskal-Wallis test is that it does not require equal sample sizes. An investigator who is familiar with the various sampling methods, especially stratified sampling, can appreciate the fact that this property enhances the usefulness of the Kruskal-Wallis test.

*Other Sets of Contrasts*    In some cases the investigator may not be interested in making all possible $k(k - 1)/2$ paired comparisons between the $k$ populations. For example, if one population represents a control group and the other $(k - 1)$ represent different treatments, the only question of interest might be which treatment groups differ significantly from the one control group. Then a total of only $(k - 1)$ comparisons would be made. In another situation the investigator might wish to pool two or more groups of observations and compare the combined group with each of the others, or with some subset or combination of the other groups. When any number $p$ of contrasts are made between means of groups, whether individual or pooled, the procedure is identical to that described earlier, except for the critical $z$ value. (Of course, then $n_i$ and $n_j$ in (3.4) must be interpreted as the numbers of observations entering into the calculation of the means $\bar{R}_i$ and $\bar{R}_j$ that are being compared.) The quantity $z$ in (3.4) or (3.5) for $p$ contrasts with overall significance level $\alpha$ is always that quantile of the standard normal curve that corresponds to a right-tail probability of $\alpha/2p$. For example, suppose that there are seven groups, one control group, and six different treatment groups, and that each treatment is to be compared only with the control, and not also with other treatments; then $p = 6$. For this multiple comparisons procedure with level $\alpha = .20$, the value of $z$ is found from Table N as

2.128. The overall probability is $1 - \alpha = .80$ that all significance tests lead to correct conclusions when the null hypothesis is true.

The contrasts to be made should be selected by the investigator on the basis of his research hypotheses, and without reference to the data observed, since otherwise the results are biased.

**example   3.4**

In Example 3.1, recall that group 4 had no treatment, group 1 had electroshock only, and groups 2 and 3 both involved psychotherapy. The three comparisons of particular interest in this situation might be group 1 versus group 4, group 1 versus groups 2 and 3 combined, and group 4 versus groups 2 and 3 combined. If the investigator decides to make only these three sets of contrasts, we have $p = 3$. From Table N, the value of $z$ at level .10 is then 2.128. The calculations necessary to perform the inference are shown below:

| Group | R | n | $\bar{R}$ |
|---|---|---|---|
| 1 | 260 | 10 | 26.0 |
| 4 | 348 | 10 | 34.8 |
| 2 and 3 | 212 | 20 | 10.6 |

| Contrast | Right-hand Side of (3.4) or (3.5) | Observed Mean Difference |
|---|---|---|
| 1 vs. 4 | $2.128\sqrt{4(41)/6} = 11.13$ | $-8.8$ |
| 1 vs. 2 and 3 | $2.128\sqrt{\dfrac{40(41)}{12}\left(\dfrac{1}{10} + \dfrac{1}{20}\right)} = 9.64$ | 15.4 |
| 4 vs. 2 and 3 | $2.128\sqrt{\dfrac{40(41)}{12}\left(\dfrac{1}{10} + \dfrac{1}{20}\right)} = 9.64$ | 24.2 |

The first contrast is not significant, whereas the second two are. Thus we make the following statements at overall level .10. There is no significant difference between the electroshock group and control group. Groups receiving any psychotherapy have significantly lower ranks, and thus a significantly higher level of improvement, than those receiving no treatment or only the electroshock treatment. Although this conclusion is essentially the same as when all paired comparisons are made, it is more definitive as regards the effect of psychotherapy and is made with a higher level of confidence.

*Adjustment for Ties*   If contrasts of any kind are to be made for data in which there are extensive ties between or within groups, an adjustment can be made in (3.4) or (3.5) to ensure an overall level which is conservative. The correction is to reduce the quantity under the square root in (3.4) and (3.5) by a factor involving $\Sigma u^3 - \Sigma u$, where, as before, $u$ is the number of observations in all samples combined that are tied at any given rank, and the sum is over all sets of tied ranks. Specifically, the corrected inequality on sample mean differences that holds with probability at least $1 - \alpha$ is

$$|\bar{R}_i - \bar{R}_j| \leq z\sqrt{\frac{[N(N^2 - 1) - (\Sigma u^3 - \Sigma u)]\left[\dfrac{1}{n_i} + \dfrac{1}{n_j}\right]}{12(N - 1)}} \qquad (3.6)$$

where the value of $z$ is obtained as before. When making comparisons between pairs from $k$ groups of equal size, we have $n_i = n_j = N/k$, and (3.6) can be simplified to

$$|\bar{R}_i - \bar{R}_j| \leq z\sqrt{\frac{k[N(N^2 - 1) - (\Sigma u^3 - \Sigma u)]}{6N(N - 1)}}. \qquad (3.7)$$

The correction for ties usually has little effect, but it is simple to use and ensures a conservative result.

**example   3.5**

A study was made of 383 patients judged medically eligible for home care after entering Los Angeles County General Hospital during the period 1959–1961 (Dunn, 1964, p. 244). The patients were divided into three groups according to the assessment of their home situation for satisfactory care. Those patients in group 1 could obtain satisfactory care at home. Groups 2 and 3 could not, but for different reasons. Patients in group 2 had no person responsible for their care, while those in group 3 had a person responsible, but that person was unable or unwilling to provide the care needed. The observations made on each patient so classified concern the prestige of their occupations. Measurements were made on the following ordinal scale, with increasing numbers reflecting decreasing levels of prestige:

10: executives, large proprietors, major professionals
20: business managers, medium proprietors, lesser professionals
30: administrators, small owners, semiprofessionals, farm owners
40: clerical, sales, technical, small business, small farm owner
50: skilled manual, small farmer
60: semiskilled, tenant farmer
70: unskilled, share cropper.

The frequency distribution of observations on all 383 patients is shown in Table 3.11. Note that the table entries are not the observations, but the frequencies of each group in each occupation. For example, the numbers in the column under group 1 indicate that the observations consist of three 10s, twelve 20s, ten 30s, and so on.

**TABLE 3.11**

| Occupation | Group 1 | Group 2 | Group 3 | Total |
|---|---|---|---|---|
| 10 | 3 | 0 | 1 | 4 |
| 20 | 12 | 4 | 2 | 18 |
| 30 | 10 | 7 | 4 | 21 |
| 40 | 20 | 10 | 11 | 41 |
| 50 | 47 | 9 | 10 | 66 |
| 60 | 74 | 12 | 21 | 107 |
| 70 | 62 | 26 | 38 | 126 |
| Total: | 228 | 68 | 87 | 383 |

There are many ties among the observations, but the sum of ranks for each group is easily found as follows from the frequency distribution in Table 3.11. A total of 4 persons are tied at occupation level 10. Thus each of these receives the midrank $(1 + 4)/2 = 2.5$, and the group subtotals of ranks at occupation level 10 are the group frequencies times 2.5, or 7.5, 0, and 2.5, as shown in the first row of Table 3.12. Since 18 persons are tied at the next occupation level 20, these persons would occupy ranks 5 through 22 inclusive if they were not tied, and the midrank value is $(5 + 22)/2 = 13.5$. The subtotals of ranks for occupation group 20 are then the corresponding frequencies multiplied by 13.5, or $(13.5)12 = 162$, $(13.5)4 = 54$, and $(13.5)2 = 27$. Continuing in the same way, we complete Table 3.12 to obtain the rank sums for each of the three groups. The mean ranks for each group are found as usual; see Table 3.13.

**TABLE 3.12**

| | Rank Subtotals for | | |
|---|---|---|---|
| Occupation | Group 1 | Group 2 | Group 3 |
| 10 | 7.5 | 0.0 | 2.5 |
| 20 | 162.0 | 54.0 | 27.0 |
| 30 | 330.0 | 231.0 | 132.0 |
| 40 | 1,280.0 | 640.0 | 704.0 |
| 50 | 5,522.5 | 1,057.5 | 1,175.0 |
| 60 | 15,096.0 | 2.448.0 | 4,284.0 |
| 70 | 19,871.0 | 8,333.0 | 12,179.0 |
| $R$ | 42,249.0 | 12,763.5 | 18,503.5 |

**TABLE 3.13**

| Group | $R$ | $n$ | $\bar{R}$ |
|-------|-----|-----|-----------|
| 1 | 42,249.0 | 228 | 185.4 |
| 2 | 12,763.5 | 68 | 187.7 |
| 3 | 18,503.5 | 87 | 212.7 |

The comparisons of particular interest according to the article are group 1 versus groups 2 and 3 combined, groups 1 and 3 combined versus group 2, and group 1 versus group 3. Then we have $p = 3$ and $z = 1.834$ for significance level .20. Groups 1 and 3 combined have a mean rank of

$$\frac{42,269.0 + 18,503.5}{228 + 87} = \frac{60,772.5}{315} = 192.9,$$

and groups 2 and 3 combined have a mean rank of

$$\frac{12,763.5 + 18,503.5}{68 + 87} = \frac{31,267}{155} = 201.7.$$

The computation of the correction factor for ties within contrasts proceeds as in Table 3.14 for all sets of comparisons.

**TABLE 3.14**

| Occupation Level | $u$ | $u^3$ |
|------------------|-----|-------|
| 10 | 4 | 64 |
| 20 | 18 | 5,832 |
| 30 | 21 | 9,261 |
| 40 | 41 | 68,921 |
| 50 | 66 | 287,496 |
| 60 | 107 | 1,225,043 |
| 70 | 126 | 2,000,376 |
|    | 383 | 3,596,993 |

$$\Sigma u^3 - \Sigma u = 3,596,993 - 383 = 3,596,610$$

For part of the quantity needed in (3.6), we calculate

$$\sqrt{\frac{N(N+1)}{12} - \frac{\Sigma u^3 - \Sigma u}{12(N-1)}} = \sqrt{\frac{383(384)}{12} - \frac{3,596,610}{12(382)}}$$
$$= \sqrt{11,471.4} = 107.1.$$

The relevant contrasts, their observed values and critical values according to (3.6) are as follows:

| Contrast | Observed Value | Critical Value |
|---|---|---|
| 1 vs. 2 and 3 | 16.3 | $1.834(107.1)\sqrt{\dfrac{1}{228} + \dfrac{1}{155}} = 20.4$ |
| 1 and 3 vs. 2 | 5.2 | $1.834(107.1)\sqrt{\dfrac{1}{315} + \dfrac{1}{68}} = 26.3$ |
| 1 vs. 3 | $-27.3$ | $1.834(107.1)\sqrt{\dfrac{1}{228} + \dfrac{1}{87}} = 24.7$ |

The only contrast which is significant at level .20 is group 1 versus group 3, and the difference is negative. The conclusion is that those patients who could obtain satisfactory care at home have lower ranks and therefore occupations with higher prestige than those who have a responsible person at home who could not care for them, and that the other comparisons of interest showed no significant differences.

# 4  Summary

The Kruskal-Wallis test for $k$ samples is simply an extension of the Mann-Whitney-Wilcoxon test for two samples. The rationale and procedure are exactly the same (except for the fact that tables of the exact sampling distribution of $H$ are more tedious to generate and cumbersome to reproduce than tables of $T_x$). Thus we can discuss the two tests together.

These procedures are the most commonly used of the nonparametric techniques available for the location problem. The methods are simple to understand and use, and are quite well known. The only assumption required for validity of either test is that the common probability distribution, whatever it is, is continuous. The Mann-Whitney-Wilcoxon test can be regarded as the nonparametric counterpart of the parametric test known as the Student's $t$ test for means (or the normal theory test if the two variances are known). The $t$ test is exactly valid only if both distributions are normal and their variances are equal (although unspecified). Similarly, the Kruskal-Wallis test is the nonparametric analog of the one-way analysis of variance test (sometimes called the $F$ test). The theory of this parametric test is based on the assumption that all $k$ distributions are normal and their variances are equal (although unspecified). Thus, in each case the assumptions for the parametric test are much more stringent than for the nonparametric.

Both these parametric tests are known to be most powerful as long as the assumptions hold. Although the parametric tests are robust against the assumption of normality, they are sensitive to differences between variances.

The nonparametric tests might be used in preference to the parametric tests in any of the following circumstances:

1. The sample sizes are quite small.
2. The only data available are measured on an ordinal scale.
3. The respective sample medians seem to be more reliable than the sample means as descriptors of central tendency in the data.
4. Medians seem to be a more representative kind of location parameter than means for the comparison of interest.
5. Little or nothing can be assumed about the probability distributions.
6. The distributions are unknown, but thought to bear little resemblance to the normal curve.
7. Homogeneity of variances seems highly questionable.

In each case, inferences and conclusions based on the parametric test may be unreliable, or at least of uncertain value.

Comparisons of the power of the nonparametric and parametric tests can be made only under specific population distribution assumptions. For any continuous distribution, the asymptotic relative efficiency of each of these nonparametric tests relative to the corresponding parametric test is never smaller than .864 and can be infinite. (See Hodges and Lehmann (1956) or Pratt and Gibbons (1975) for further results and limitations.) The ARE is $3/\pi = .955$ for the normal distribution and 1.000 for the continuous uniform distribution.

## PROBLEMS

4.1 Prolonged simulation studies were made in two air-controlled rooms to investigate whether smokers and nonsmokers should be segregated in public places. Ten nonsmokers were placed in a room in which the air was permeated by a controlled amount of the fumes, gases, carbon monoxide, and so forth produced by cigarette smoke. Another group of 10 subjects, all regular smokers, were placed in a different room and told to smoke. The total amount of pollutant particles was kept the same for both rooms. After a period of time, all subjects were given clinical tests designed to measure the effects of the smoke on the blood, cells, organs, and so on. These scores were used to compute the following summary ratings of bodily absorption of potentially dangerous particles from the air:

Smokers:    41, 30, 33, 40, 25, 22, 35, 24, 28, 30
Nonsmokers: 25, 22, 23, 18, 20, 17, 21, 15, 25, 19.

Use a nonparametric test to determine whether there is a significant difference in the amounts of potentially harmful particles absorbed by smokers and nonsmokers.

4.2 An identical block final examination was given to two different sections of the same class. During the semester, section 1 met in the morning and section 2 met in the afternoon. Since the students were assigned randomly to these sections and the same professor lectured to both sections, any differences in student performance on the examination might be attributed at least in part to the hour of the class meeting. Since both classes were quite large, a random sample of eight students was selected from each group. Their scores on the examination are as follows:

Section 1 (morning):   169, 187, 204, 190, 232, 185, 214, 184
Section 2 (afternoon):   181, 145, 190, 167, 174, 158, 184, 221.

Analyze these data to see whether there is any difference in student performance, and find a confidence interval estimate of the shift parameter with a confidence coefficient of .936.

4.3   Suppose that all students in the two sections described in Problem 4.2 were matched as closely as possible as regards factors that could be expected to influenced performance; for example, overall quality point average, scores on the college entrance examination, major field of study, age, and so on. Any students for whom a good match could not be found were eliminated. Of the 36 usable pairs, 25 showed that the student in the morning section had a higher test score than his match in the afternoon section. The professor claims that although he himself is equally alert and clear-minded at all hours of the day, the students in the afternoon section are more prone to mind-wandering and inattentiveness, and therefore are disadvantaged in learning the material he presents in class. Do you agree with the professor?

4.4   An agricultural experiment station ran a controlled experiment to determine whether sex is an important factor in weight gain when young pigs are purchased to fatten and sell for slaughter. If one sex shows higher weight gain on the average, the small-scale farmer could optimize production by purchasing only one sex. Eight young male pigs are placed in one pen, and eight young female pigs are placed in another. Both groups are subjected to identical feeding treatments for a fixed period of time. The initial weights were all between 35 and 50 pounds. The weight gains in pounds are recorded below. Since one of the female pigs died, there are only seven observations in that group. Analyze the data using an appropriate one sided alternative.

Female pigs:   9.31, 9.57, 10.21, 8.86, 8.52, 10.53, 9.21
Male pigs:       9.14, 9.98, 8.46, 8.93, 10.14, 10.17, 11.04, 9.43.

4.5   Adherents of the "encounter group" movement claim that interpersonal attraction is increased by touching. Boderman, Freed, and Kinnucan (1972) report results of an experiment performed to investigate this research hypothesis. Twenty-one female freshman college students were randomly assigned to one of two experimental conditions, called touch and no-touch. Each subject participated in what was represented as an ESP experiment involving blind communication of a card guess, but was actually an attempt to evaluate the effect of touch on interpersonal attraction. The ESP experiment was the same for both groups, except that subjects in the "touch" group were required to have mutual but blind touching with the experimenter as the "message sender" for 110 seconds. Afterwards, all subjects were given a set of four questions designed to evaluate their perception of communication with the message sender. Answers to these questions were measured on an ordinal scale from 0 to 20, where larger numbers represent a higher level of communication perceived.

The total evaluation scores for all four questions averaged 36 for the touch group and 26.2 for the others. A parametric statistical analysis showed that the subjects who touched the experimenter perceived her as a significantly more attractive person than those who did not touch her.

Suppose the total evaluation scores for the subjects in the two conditions are as shown below. (These are artificial data but of comparable magnitudes.) Verify that these data support the premise that a touch condition group feels stronger communication than a no-touch group.

Touch condition:    43, 40, 31, 35, 52, 23, 45, 37, 30, 27
No-touch condition:    40, 23, 21, 31, 19, 27, 15, 36, 33, 26, 22.

4.6    Smokers are commonly thought of as nervous people whose emotionality is at least partly caused by smoking because of the stimulating effect tobacco has on the nervous system. Nesbitt (1972) conducted a study with 300 college students and concluded that smokers are less emotional than nonsmokers, that smokers are better able to tolerate the physiological effects of anxiety, and that, over time, smokers become less emotional than nonsmokers. Subjects of both sexes were drawn from three different colleges, and classified as smokers if they smoked any number of cigarettes on a regular basis. In one aspect of the experiment all subjects were given the Activity Preference Questionnaire (APQ), a test designed to measure the emotionality of the subjects. The APQ is scored using an ordinal scale of 0–33, with lower scores indicating lesser emotionality, that is, greater sociopathy. The mean overall scores were 18 for smokers, and 20.3 for nonsmokers. Suppose that this experiment is repeated using a group of only 8 randomly chosen smokers and 10 randomly chosen nonsmokers, and the (artificial) score results are as shown below. Do these data support the same conclusion concerning emotionality as Dr. Nesbitt's data?

Smokers:    16, 18, 21, 14, 25, 24, 27, 12
Nonsmokers:    17, 15, 28, 31, 30, 26, 27, 20, 21, 19.

4.7    The theory that obese persons eat excessively because of anxiety or other emotional disturbances is widely held. Abramson and Wunderlich (1972) report an experiment to investigate this psychosomatic concept of obesity. Two independent groups of 33 male students were selected to participate; weights of persons in one group were within the normal range for their height, whereas the others where at least 15 percent heavier than the norm. Each of these groups was divided randomly into three experimental groups of 11 persons each. One group was a control; the other two were subjected to conditions of induced interpersonal anxiety and objective fear. The anxiety of each subject was measured and then all subjects were allowed to eat all the crackers they wanted in three different flavors and asked to rank the flavors in order of preference. Thus the experiment was masked as a "taste discrimination" test in order to minimize the subject's self-consciousness about eating. The obese subjects had higher anxiety difference scores than the normal ones as a result of the induced conditions, but the number of crackers eaten was essentially the same for the two sets of groups. The group means reported for number of crackers eaten are as follows:

|                       | Obese | Normal |
| --------------------- | ----- | ------ |
| Control               | 12.18 | 10.82  |
| Interpersonal anxiety | 10.72 | 10.72  |
| Objective fear        | 12.64 | 12.45  |

Suppose that the data (artificial) on number of crackers eaten by the obese and normal subjects in the interpersonal anxiety condition group were as shown below. Use the Mann-Whitney-Wilcoxon test to obtain statistical support to deny the theory that obese people overeat to allay anxiety (one-sided test).

Obese:    10, 11, 9, 15, 8, 18, 11, 6, 12, 10, 9
Normal:   9, 10, 14, 8, 7, 3, 17, 10, 12, 12, 9.

4.8   In reliability engineering, times to failure are frequently assumed to follow a specified probability function (often the exponential distribution) so that usual parametric methods of inference and estimation can be applied. If no specified distribution can reasonably be assumed, whether because of past experience or rejection of the null hypothesis in a goodness-of-fit test (see Problem 2.21 of Chapter 2), nonparametric methods are usually employed. Suppose that a sample of components (for example, capacitors) is randomly divided into two groups. The first group is put to use in a normal operating environment, and the other group is simultaneously subjected to some environmental stress, such as temperature stress or other accelerated conditions. The "normal environment" group is called the control sample, and the other is called the exposure sample. For each component the time to failure is noted; this is called a "life test." Since life test data occur automatically in increasing order of magnitude, that is, as order statistics, nonparametric tests based on ranks are especially simple to apply. Use the Mann-Whitney-Wilcoxon test to see whether the hypothetical data below, representing hours to failure, support the alternative that median time to failure is shorter for the exposure sample than the control sample.

| Control Sample | | Exposure Sample | |
| ---- | ---- | ---- | ---- |
| 5.2  | 17.1 | 1.1  | 7.2  |
| 8.5  | 17.9 | 2.3  | 9.1  |
| 9.8  | 23.7 | 3.2  | 15.2 |
| 12.3 | 29.8 | 6.3  | 18.3 |
|      |      | 7.0  | 21.1 |

4.9   In part of a study reported by Perrotta and Finch (1972, p. 474), the blood films of 16 patients with severe renal anemia and 10 patients with functional heart disease were measured for hematocrit and reticulocyte cell counts. The observations on percentage shift are shown below. Test the null hypothesis that hemolytic anemia and cardiac patients have the same median percentage shift.

Renal:          2.20, 1.52, 1.54, 0.77, 0.34, 0.45, 0.39, 0.29, 0.18,
                0.16, 0.23, 0.24, 0.17, 0.08, 0.02, 0.02

Heart disease:  1.84, 0.44, 0.30, 0.06, 0.20, 0.14, 0.10, 0.09, 0.06,
                0.04

4.10  Glenn and Kehn (1972, p. 327), conducted a study of the side effects induced by giving 1 ml of RhoGAM intramuscularly to postpartum patients. Two groups of normal postpartum women were selected; 13 Rh(D)-positive patients formed a control group and 22 Rh(D)-negative patients were inoculated with Rho-GAM. Serum samples taken within 72 hours of delivery were analyzed for cholesterol and triglyceride levels and lipoprotein anemia. Both cholesterol and triglyceride are measured on an interval scale. The normal ranges for postpartum patients are 150–300 mg per 100 ml cholesterol and 74–172 mg per 100 ml triglyceride. Use their data, given below, to investigate whether Rho-GAM has the effect of (a) raising the average level of cholesterol; (b) raising the average level of triglyceride.

For women not treated with RhoGAM,

Cholesterol:    322, 214, 236, 237, 212, 200, 237, 236, 252,
                237, 250, 178, 195

Triglyceride:   242, 193, 379, 189, 121, 271, 202, 175, 157,
                127, 209, 177, 178.

For women treated with RhoGAM,

Cholesterol:    262, 241, 253, 212, 276, 266, 158, 318, 301, 138,
                220, 320, 256, 270, 370, 335, 250, 255, 300, 245,
                225, 195

Triglyceride:   283, 205, 198, 226, 247, 288, 142, 288, 238, 176,
                184, 424, 335, 251, 371, 380, 181, 279, 103, 301,
                212, 264.

4.11  Kolb (1965) reported a study designed to investigate the research hypothesis that children who are underachievers are also low in motivation and/or concern for achievement. For the purpose of the experiment, underachievers were defined as high school boys with an IQ above 120 but a grade average below C. The subjects were divided randomly into two groups for a summer program. Both groups were given academic training, but one group was also given an Achievement Motivation Training Program. The academic performance of each subject was evaluated at the end of the summer program by administering three tests. Suppose that the overall scores of these two groups were as follows (artificial data). Perform an appropriate statistical analysis to see whether achievement motivation has an effect on academic performance of underachievers.

| Academic Training Only | | | Academic Plus Motivation Training | |
|---|---|---|---|---|
| 66 | 70 | 65 | 81 | 55 |
| 72 | 53 | 70 | 85 | 90 |
| 69 | 48 | 75 | 92 | 93 |
| 80 | 76 | 63 | 71 | 84 |
| 78 | 59 | 68 | 68 | 73 |
| 86 | 61 | 45 | 77 | 65 |

4.12  Baten et al (1958) report a study designed to investigate differences between two methods of preparing fish. Twelve fish are prepared by method A, and twelve by method B. Each item is then scored separately on the criteria of aroma, flavor, texture, and moisture, and also on a composite basis. The composite scores are shown below. Test the null hypothesis that the median composite score is the same for both methods of preparation.

| Method A | | Method B | |
|---|---|---|---|
| 4.05 | 4.18 | 3.31 | 2.35 |
| 5.04 | 4.35 | 3.39 | 2.59 |
| 3.45 | 3.88 | 2.24 | 4.48 |
| 3.57 | 3.02 | 3.93 | 3.93 |
| 4.23 | 4.56 | 3.37 | 3.43 |
| 4.23 | 4.37 | 3.21 | 3.13 |

4.13  A time-and-motion analysis was made in the permanent mold department at the Central Foundry Company, Holt, Alabama. One variable of interest was the length of time required to pour the molten metal into the die and form a casting of a 6- by 4-inch Y-branch. The median time was estimated as 10 seconds for the 1-month period of observation. However, the metallurgical engineer noticed a pattern to the variation of time according to hour of the day, in particular that pouring times before lunch seemed to be shorter than times after lunch. In order to investigate this assertion, 12 independent observations on pouring times in seconds were divided into the following two groups according to whether the observation was made before or after lunch. Perform an appropriate analysis.

| Before Lunch | | After Lunch | |
|---|---|---|---|
| 12.6 | 11.2 | 16.4 | 11.4 |
| 15.4 | 14.0 | 14.0 | 9.4 |
| 13.4 | 12.0 | 13.4 | 11.2 |

4.14  In order to compare two machines which impose stress conditions, 25 test specimens are assigned randomly to the machines, 11 to machine A and 14 to machine B. Since this is a fatigue test, the time to failure in kilocycles is noted for each specimen. Test the null hypothesis that machines A and B give the same median fatigue life, and find a confidence interval, with confidence coefficient near .90, for the difference between median fatigue life.

| Machine A | | Machine B | |
|---|---|---|---|
| 210 | 239 | 205 | 249 |
| 213 | 245 | 216 | 253 |
| 218 | 248 | 224 | 259 |
| 222 | 256 | 225 | 261 |
| 229 | 267 | 228 | 270 |
| 230 |  | 234 | 275 |
|  |  | 238 | 284 |

4.15  In an extension of the two-machine test in Problem 4.14, suppose that the 25 test specimens are assigned randomly to five different machines, giving the fatigue life values below. Test the null hypothesis that the five machines give the same median fatigue life.

| | | Machine | | |
|---|---|---|---|---|
| A | B | C | D | E |
| 206 | 201 | 198 | 200 | 215 |
| 209 | 216 | 202 | 202 | 219 |
| 214 | 238 | 218 | 210 | 226 |
| 231 | 257 | 229 | 214 | 230 |
| 249 | 263 | 243 | 236 | 245 |

4.16  An entering class in law school usually represents a wide variety of undergraduate majors. It is frequently claimed that students with prior study in engineering, mathematics, or similar fields, which train a student in logical reasoning, get better grades in law school than do students from other fields. A random sample of 25 students is chosen from a law school graduating class. Five of them were undergraduate mathematics or engineering majors, 9 were business majors, and 11 were from other fields. Their overall law school class rankings are shown below. Test the null hypothesis that law school performance is the same for all three groups. If this hypothesis is rejected, perform a multiple comparisons test to distinguish those groups that differ significantly.

| Math or Engineering | Business | Other |
|---|---|---|
| 2 | 3 | 1 |
| 5 | 6 | 4 |
| 8 | 11 | 7 |
| 10 | 14 | 9 |
| 13 | 17 | 12 |
|  | 19 | 15 |
|  | 20 | 16 |
|  | 21 | 18 |
|  | 25 | 22 |
|  |  | 23 |
|  |  | 24 |

4.17  An office has three electric typewriters, A, B, and C. They plan to place those machines with higher usage rate under a service contract. Records are kept on each machine for 6 weeks; however, machine A was out for repair for part of 2 weeks. Analyze the data below on weekly machine usage rates to determine whether there is a significant difference in median usage. If the null hypothesis is rejected, perform a multiple comparisons test. Can you make a preliminary recommendation?

| A | B | C |
|---|---|---|
| 15.45 | 10.82 | 41.20 |
| 10.30 | 45.00 | 35.10 |
| 8.00 | 12.36 | 25.00 |
| 16.21 | 8.24 | 8.24 |
|  | 20.15 | 18.42 |
|  | 46.27 | 48.32 |

4.18  Anderson (1962) reported the results of an experiment conducted to see whether (a) fear of failure, and (b) hope of success, in eighth grade students is the same under three degrees of arousal conditions, classified as high, medium, and low.

A teacher talks to each of the groups in such a way that the first group should become quite motivated to success, the second group should be less so, and the third group should have almost no arousal. The variables "fear" and "hope" were each measured by projective tests given to all students; their respective scores represent somewhat crude measures of "hope of success" and "fear of failure." The actual data are not available. Suppose the data below are the scores on "hope of success" in a similar experiment involving 21 students randomly divided into groups of sizes 6, 7, and 8. Test the null hypothesis that the three groups are the same with respect to the hope variable.

| High Arousal | Medium Arousal | Low Arousal |
|---|---|---|
| 77 | 75 | 70 |
| 69 | 81 | 21 |
| 60 | 44 | 39 |
| 89 | 72 | 82 |
| 41 | 79 | 77 |
| 45 | 53 | 67 |
|  | 62 | 48 |
|  |  | 52 |

4.19 A systems analyst is asked to make a recommendation regarding an efficient and productive layout for clerical workers at a work station of a business firm. Suppose that all employees assigned to the work station perform exactly the same task and are homogeneous in work performance, so the primary factor affecting overall productivity at the station is the work-flow layout. Three layouts are developed for consideration and 18 workers are assigned randomly to the three layouts, six per layout. Each worker is given preliminary training in the new layout, and then his productivity is measured as average output per hour over a regular 40-hour workweek. From the results recorded below, which layout do you think the systems analyst will recommend? What degree of statistical justification can you give for your answer?

| | Layout | |
|---|---|---|
| A | B | C |
| 41.3 | 38.2 | 79.1 |
| 50.1 | 61.9 | 63.8 |
| 63.2 | 45.3 | 74.1 |
| 54.9 | 55.6 | 59.4 |
| 69.0 | 42.1 | 85.3 |
| 72.3 | 43.4 | 86.2 |

4.20 Four different persons are hired to conduct household interviews as part of a sample survey on budgeting practices. In a 5-day period, average times spent per completed interview were recorded for each day, but two interviewers are part-time, and worked only 3 days. The data are shown below.

| | Interviewer | | |
| 1 | 2 | 3 | 4 |
| --- | --- | --- | --- |
| 10.0 | 15.0 | 19.1 | 5.1 |
| 25.0 | 5.2 | 25.4 | 9.2 |
| 40.1 | 55.3 | 8.3 | 14.1 |
| 29.2 | 15.1 | | |
| 4.1 | 23.2 | | |

(a) Do the data support the hypothesis that all interviewers take the same amount of time to conduct an interview?

(b) Do you think these interviewers were paid at a certain rate per day or at a certain rate per completed interview? Why?

4.21 Fowler et al (1973) reported the results of two experiments concerning sleep and memory. One aspect of the study considered the effect of two conditions, namely sleep and wakefulness, on memory loss. Subjects were randomly divided into three groups, and each subject was given two different paired-associate tasks. The verbal task was to learn 15 pairs of common words as associates, and the visual forms task was to learn 10 pairs of "nonsense" shapes. The first group learned during the daytime and were tested for recall after 3.5 hours of normal daytime activity. The second and third groups were awakened from sleep for the learning, one during the first half of the night and the other during the second half of the night. They then slept for 3.5 hours more and were awakened for the recall test. The data below on percent loss in memory for the two learning tasks are artificial, but they emulate the summary findings reported.

(a) Test the null hypothesis that median percent loss in memory on the verbal task is the same for the two groups with retention intervals in a sleep condition. Use a one-sided alternative in the appropriate direction.

(b) Do (a) for the visual forms task.

(c) Test the null hypothesis that median percent loss in memory on the verbal task is the same for all three retention conditions. Perform a multiple comparisons test at level .15 to distinguish the best retention conditions.

(d) Do (c) for the visual forms task.

(e) Based on these analyses, what conditions would you recommend as best to facilitate recall for the two tasks?

| Verbal Task | | | Visual Forms Task | | |
|---|---|---|---|---|---|
| Awake | Sleep 1 | Sleep 2 | Awake | Sleep 1 | Sleep 2 |
| 48 | 19 | 36 | 32 | 26 | 22 |
| 45 | 16 | 31 | 35 | 14 | 25 |
| 51 | 14 | 24 | 41 | 10 | 36 |
| 42 | 22 | 39 | 26 | 31 | 10 |
| 55 | 25 | 43 | 19 | 40 | 8 |
| 28 | 23 | 40 | 30 | 24 | 20 |

4.22 A university professor conducted an experiment to compare the effectiveness of three methods of instruction, namely (I) lectures to large groups, (II) lectures to small groups, and (III) seminars or discussion-type sessions with small groups. A population of 50 sophomores were assigned randomly to classes of these three types, all taught in the same semester by the same professor and covering the same subject matter. The students were tested for achievement periodically during the semester, and their average achievement scores (out of a possible 100 points) are as shown below. Test the null hypothesis that the three methods of instruction are equally effective.

| Method I | | | Method II | Method III |
|---|---|---|---|---|
| 60 | 73 | 58 | 75 | 83 |
| 55 | 79 | 38 | 78 | 86 |
| 60 | 89 | 63 | 91 | 98 |
| 84 | 98 | 72 | 86 | 89 |
| 91 | 92 | 78 | 84 | 85 |
| 46 | 51 | 83 | 65 | 81 |
| 49 | 54 | 86 | 73 | 77 |
| 63 | 65 | 92 | 89 | 72 |
| 69 | 86 | 98 | 98 | 70 |
| 71 | 82 | 60 | 81 | 88 |

4.23 Aronson and Mills (1959) collected data to investigate the research hypothesis that a person's liking for, and feeling of affinity with, a group of which he is a member is directly related to the degree of unpleasantness of the initiation procedure he undergoes to join that group. This hypothesis is derived from Festinger's theory of cognitive dissonance. Sixty-three women were randomly divided into three experimental groups of 21 each. Each group had discussions on the psychology of sex but with different initiation procedures. These experimental conditions, in increasing degree of unpleasantness, were (1) Control: No requirements; (2) Mild: Subjects required to read aloud words related to sex but not considered obscene; and (3) Severe: Subjects required to read aloud obscene words and vivid descriptions of sexual activity. After the initiation each subject completed a questionnaire designed to measure her feeling about the discussion and the other participants. The average group scores were Control: 166.7, Mild: 171.1, and Severe: 195.3. The article concluded that the results clearly verify the research hypothesis, since the Severe group scores were significantly higher than both the other two, and no other differences were significant. The data below are artificial but have approximately the same means. Perform a multiple comparisons test at level .10.

| Control | | Mild | | Severe | |
|---|---|---|---|---|---|
| 160 | 185 | 163 | 185 | 203 | 196 |
| 152 | 183 | 170 | 154 | 191 | 189 |
| 188 | 164 | 172 | 151 | 178 | 186 |
| 148 | 178 | 175 | 165 | 195 | 173 |
| 152 | 189 | 155 | 178 | 208 | 198 |
| 163 | 172 | 180 | 179 | 215 | 206 |
| 171 | 165 | 182 | 168 | 182 | 210 |
| 133 | 161 | 176 | 161 | 178 | 205 |
| 149 | 169 | 193 | 169 | 212 | 201 |
| 158 | 175 | 186 | 170 | 190 | 192 |
| 187 | | 158 | | 194 | |

# Inferences Concerning
# Scale Parameters

## 1  Introduction

A measure of central location may be considered the single most representative figure for a population or a sample. However, a much more complete description can be given by reporting, in addition, some measure of the spread or variability of that population or sample. In some studies, variability is an even more relevant characteristic than location — as, for example, when uniformity is the primary factor of concern. In this chapter we study methods of statistical analysis that can be applied when the important difference between two populations is in spread or variability. Since the generic term for a measure of variability in a population is a scale (or dispersion) parameter, the inferences here are concerned with scale parameters.

Whether we are dealing with a population or a sample, variability may be measured as spread around some central value, such as the mean or median, or as spread from some arbitrary point, usually the origin. The standard deviation (or its square, the variance) is a good example of the former, and the range (if it is finite) exemplifies the latter. In either case, the amount of spread is meaningful only on a relative basis. If it is reported that a certain variable or set of data has a standard deviation of $10, there is no way to interpret this figure without some point of comparison. A standard deviation of $10 for annual incomes of public school teachers is absurdly small, whereas a standard deviation of $10 for a daily wage of these teachers is rather large. In order to keep a measure of spread from being devoid of meaning, it is frequently compared with the magnitude of some measure of location. In fact, it is often reported as a proportion or percentage of some number that represents the "typical magnitude" of the variable or data.[1] As a *descriptor* of a population or set of data, some basis for comparison is needed in order to *interpret* a measure of spread.

This problem does not really arise in comparative studies or experiments where the objective is to determine whether two variables have the same spread. As long as spread is measured in the same way for both variables, the values are directly comparable. The spreads may be measured about any location parameter, whether central or not. Further, this parameter need not be the same for both variables; for example, we can compare the scale of, say, $X - \alpha$ with that of $Y - \beta$, where $X$ and $Y$ are the two variables of interest, and $\alpha$ and $\beta$ are any two location parameters (including the origin).

This then is the situation when the general two-sample problem is specialized to the so-called two-sample scale problem. The population model assumed is that the probability functions of the two variables $X$ and $Y$, or adjusted variables $X - \alpha$ and $Y - \beta$, are identical except possibly for the respective values of some scale parameter. This scale parameter is often considered to be the population standard deviation, but this is unnecessary. We use $\sigma$ to denote a scale parameter, and define it as any appropriate measure of spread.

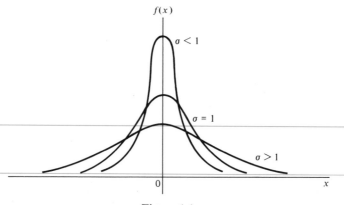

**Figure 1.1**

Figure 1.1 shows three probability density functions of the normal distribution family, which are identical in every respect except for the value of the standard deviation. The probability function of the normal distribution depends on two parameters; the mean (which equals the median and the

---

[1]The most commonly used relative measure in classical statistics is the coefficient of variation, defined as the standard deviation divided by the mean.

mode) describes location, and the standard deviation describes scale. As a result, this probability function is completely specified once these parameters are given. The density labeled $\sigma = 1$ is the "standard normal" distribution, which has mean zero and standard deviation one, that is, the density function for which the probabilities given in Table A represent areas under the curve. The curve with the high peak, labeled $\sigma < 1$, has a smaller standard deviation than the standard normal; this probability function is more concentrated around the mean zero. The flatter curve, labeled $\sigma > 1$, represents a normal distribution where the values of the variable are more spread out from, or less concentrated about, the mean zero than is the standard normal.

In this chapter we study some simple techniques that are especially appropriate for the two-sample scale problem. The tests covered in Section 2 can be used when the variables have, or can be adjusted to have, the same central value. The tests are based on the configuration of $X$'s and $Y$'s in the combined ordered sample. These positions are indicated by scores and the test statistic is the sum of the $X$ scores, but the scores are not assigned in the same way as in Chapter 4. Some corresponding confidence procedures for estimation of the ratio of two scale parameters are discussed in Section 3. When the variables have different location parameters and are not adjusted for this difference, other methods must be used. Section 4 discusses this problem.

# 2   Test Procedures:  Medians Equal or Known

## 2.1   Siegel-Tukey Test

We introduce our discussion of the Siegel-Tukey test for scale with the following example.

### example   2.1

Fix (1949) published results of a study concerning the Old Faithful geyser in Yellowstone National Park. The abstract claims his findings disprove the common beliefs that this geyser "(1) has become irregular and even erratic, (2) does not erupt in winter, (3) erupts to heights much less than formerly, (4) discharges much less water than formerly, and (5) is the only regular large geyser."[2] In fact, data for the period 1870–1947 show that the behavior of the geyser has been amazingly stable as to both height of eruptions and average time interval between eruptions since it was discovered in

---

[2]Fix (1949, p. 246).

1870. For example, the average time between eruptions based on all data available for these years is 65.12 minutes, and A. C. Peale reported 65.10 minutes as the average interval between 97 eruptions observed in 1878. These results seem to justify the assumption that time intervals have the same median in any two different years. On the other hand, the range between the minimum and maximum times between eruptions seems to have been increasing; it was observed as 35 minutes in 1879, 46 minutes in 1937, and 57 minutes in 1947. Suppose the data below are random samples of the measurement on interval between eruptions recorded in 1878 and 1947, six observations from 1878 and seven from 1947. Does it appear that the interval between eruptions had a greater range in 1947 than it did in 1878?

1878 data:   63.1, 67.3, 54.5, 70.8, 71.5, 64.9 minutes
1947 data:   65.2, 50.1, 79.2, 61.3, 47.4, 73.1, 73.6 minutes

**SOLUTION**
We call the 1878 data the $X$ sample, and the 1947 data the $Y$ sample. Represent the range of the interval between all eruptions in 1878 by $\sigma_x$, and the similar range in 1947 by $\sigma_y$. Since the question of research interest is whether $\sigma_x < \sigma_y$, we formulate the hypothesis set as

$$H: \quad \sigma_x = \sigma_y \qquad A_-: \quad \sigma_x < \sigma_y.$$

Because the sample sizes are very small and nothing is known about the distributions, a parametric test for equality of variances is difficult to justify. We perform a nonparametric test of scale in the following way. The first step is to pool the two samples and arrange the resulting 13 observations in increasing order of magnitude, as shown in Table 2.1, where the data from 1878 are in boldface type. Assign to each observation a different integer between 1 and 13, called a "score," following a rule to be explained later.

**TABLE 2.1**

| Observation | Score | Observation | Score |
|---|---|---|---|
| 47.4 | 12 | 65.2 | 1 |
| 50.1 | 11 | **67.3** | **2** |
| **54.5** | **8** | **70.8** | **5** |
| 61.3 | 7 | **71.5** | **6** |
| **63.1** | **4** | 73.1 | 9 |
| **64.9** | **3** | 73.6 | 10 |
| | | 79.2 | 13 |

Notice that the scores that are larger integers correspond to the extreme values in the pooled sample and the smaller integers are in the middle of the array, and that the sums of adjacent pairs of scores equally distant from and on either side of the middle observation in the array are equal; that is, $12 + 11 = 10 + 13$, $11 + 8 = 10 + 9$, and so on. If the median time between eruptions is the same in both 1878 and 1947, but the 1878 intervals have a smaller range, the sample configuration of $X$'s and $Y$'s should have most of the $X$'s in the middle of the array and the $Y$'s at the extremes. Accordingly, the mean of the scores corresponding to the $X$ sample would be smaller than the mean of the $Y$ scores. The test can be based on only the sum of the $X$ scores, which here is

$$T_x = 8 + 4 + 3 + 2 + 5 + 6 = 28.$$

As will be explained later, the $P$-value for $m = 6$ and $n = 7$ is found from Table H to be .026. This value is small enough to conclude that the data do not support the null hypothesis. Thus, although average interval of time between eruptions has remained constant, these data imply that the intervals had a greater range in 1947 than they had in 1878.

We now present the general procedure used to perform this test for scale.

*Inference situation* We have two continuous populations, called the $X$ population, with cumulative distribution function $F_x$, and the $Y$ population, with cumulative distribution function $F_y$. The basic null hypothesis of interest is

$$F_y(u) = F_x(u) \qquad \text{for all } u.$$

Let $\sigma_x$ and $\sigma_y$ denote some appropriate scale parameters of the $X$ and $Y$ populations respectively. If we assume that these populations are identical in every way except possibly for the values of these two parameters, or that the question of primary interest is whether the populations differ in scale, the hypothesis set can be written as one of the following.

*Two-sided alternative*
$$H: \quad \sigma_x = \sigma_y \qquad A: \quad \sigma_x \neq \sigma_y$$

*One-sided alternatives*
$$H: \quad \sigma_x = \sigma_y \qquad A_+: \quad \sigma_x > \sigma_y$$
$$H: \quad \sigma_x = \sigma_y \qquad A_-: \quad \sigma_x < \sigma_y$$

*Sampling situation* The data consist of two mutually independent random samples of sizes $m$ and $n$, denoted by $X_1, X_2, \ldots, X_m$ and $Y_1, Y_2, \ldots, Y_n$, drawn from continuous populations. Both sets of data are measured on either (a) at least an interval scale, or (b) an ordinal scale, such that the relative magnitude of each element in the pooled sample (both samples combined) can be determined.

*Rationale*    Suppose that the $m + n = N$ observations are arranged in a single sequence, according to relative magnitude from smallest to largest, keeping track of which observations correspond to the $X$ sample and which to the $Y$. Under the assumption of continuous populations, we know that $P(X = Y) = 0$ and $P(X_i = X_j) = P(Y_i = Y_j) = 0$ for any two observations from the same or different populations. Hence the probability that any two or more observations are tied is equal to zero, and a unique ordering is theoretically possible.

If the populations are really identical, the mixing of $X$ and $Y$ observations in this arrangement should be completely random. If the populations have the same central value but different scale parameters, the extreme cases which would imply such a difference are

$$X X \ldots X Y Y \ldots Y Y X \ldots X X \quad \text{or}$$
$$Y Y \ldots Y X X \ldots X X Y \ldots Y Y.$$

The first arrangement suggests that the $X$ variables are more spread out than the $Y$'s, and the second reflects a case where the $X$ variables are more concentrated about the common central value.

Suppose that we want a test statistic that is a function of the arrangement of $X$'s and $Y$'s in the combined ordered configuration, and more specifically is a simple sum of some scores representing the positions of one of the variables, say $X$. The simplest scores to use are the positive integers $1, 2, \ldots, N$. However, if these scores are to reflect the relative amount of spread of the $X$'s and $Y$'s, they must be assigned in such a way that the size of the integer assigned to a particular position reflects the distance in either direction between that position and the common central value. For example, we could assign large integers at both extremes of the configuration since those variables are further removed from the central value, and small integers in the middle since those variables are concentrated about the central value. Then if the sum of scores in positions occupied by the $X$ variables is large, the arrangement is such that the $X$ values are more dispersed, or $\sigma_x > \sigma_y$. Similarly, a small sum reflects the opposite, or $\sigma_x < \sigma_y$.

We advocate the following method of assigning these integer scores to the variables in the combined ordered configuration. The score 1 is assigned to the middle observation if $N$ is odd, and to one of the two middle observations if $N$ is even. Then as we move away from the center in both directions, the scores are assigned in sequence until the largest ones are in the tails.

More specifically, suppose first that $N$ is odd, and thus the starting point (the observation with score 1) is unique. Then we decide arbitrarily (without looking at the data) whether to place a 2 to the right or left of that starting point. The remaining scores, starting with 3, are assigned in alternating directions to pairs of observations, always working out from the center, until all $N$ integers are assigned. For example, going right first, we assign score 1 to the starting position and 2 to the observation on its right, then assign scores 3 and 4 to the left of 1, then 5 and 6 to the right of 2, then 7 and 8 to the left of 4, and so on until $N$ is reached.

If $N$ is even, we must first decide arbitrarily which one of the two middle observations is to have a score of 1. If 1 is assigned to the smaller, we go right first and assign the integers 2, 3, ..., $N$ in pairs, as for $N$ odd, except that now the starting point does not count, so the first pair consists of 2 and 3. If 1 is assigned to the larger, the assignment in pairs goes left first.

As this is much easier to see than to explain in words, the proper assignments of scores to the variables in the combined ordered arrangement are shown in Table 2.2 for selected values of $N$.

**TABLE 2.2**

| $N$ | Right First | Left First |
|-----|-------------|------------|
| 4 | 4 1 2 3 | 3 2 1 4 |
| 5 | 4 3 1 2 5 | 5 2 1 3 4 |
| 6 | 5 4 1 2 3 6 | 6 3 2 1 4 5 |
| 7 | 7 4 3 1 2 5 6 | 6 5 2 1 3 4 7 |
| 8 | 8 5 4 1 2 3 6 7 | 7 6 3 2 1 4 5 8 |
| 9 | 8 7 4 3 1 2 5 6 9 | 9 6 5 2 1 3 4 7 8 |

Notice that this method of assignment does not achieve perfect symmetry of scores, as would be desirable, but instead has the property that the sums of any successive pairs of scores are equal for pairs equally distant from the center. For example, when $N = 8$ we have $8 + 5 = 7 + 6$, $5 + 4 = 6 + 3$, $4 + 1 = 3 + 2$.

The scores used in this procedure are frequently called "modified ranks," because we are accustomed to thinking of the integers as ranks. The adjective "modified" is needed to indicate that we are not making an ordinary rank assignment from smallest to largest, where the rank assigned to a variable equals the position of that variable in the combined ordered arrangement (as was true for the Mann-Whitney-Wilcoxon test in Chapter 4).

*Test statistic*   The Siegel-Tukey[3] test statistic is defined as

$$T_x = \text{sum of the } X \text{ scores.}$$

[3]The method presented here for assigning the modified ranks deviates slightly from that originally proposed by Siegel and Tukey (1960); they advocated small ranks in the tails and ignored the middle observation for $N$ odd. The author prefers to place large ranks in the tails so that the prescription for obtaining $P$-values can be consistent with that given for the Mann-Whitney-Wilcoxon test in Chapter 4. Further, this method uses all $N$ observations whether $N$ is even or odd.

We also define the quantity

$$T_y = \text{sum of the } Y \text{ scores.}$$

Since each score is one of the integers $1, 2, \ldots, N$, the sum of all scores is the constant $N(N + 1)/2$. (See (2.2) of Chapter 4.) Thus, when it is more convenient, $T_x$ can be calculated from

$$T_x = \frac{N(N + 1)}{2} - T_y. \tag{2.1}$$

The procedure recommended above for assigning scores is admittedly arbitrary. The test procedure outlined here can be performed with other assignments of integer scores to the positions of $X$'s and $Y$'s. The only requirement is that the large scores be assigned in the two tails, in an approximately symmetric fashion, and small scores in the middle, so that the test statistic is sensitive to differences in scale. For example, if $N = 8$, the scores might be assigned as

$$7\ 5\ 3\ 1\ 2\ 4\ 6\ 8.$$

With this method, we still alternate between left and right of center, but a score is assigned to only one observation before switching direction. As long as the scores are $1, 2, \ldots, N$, the $P$-value for a test based on the sum of the scores in the $X$ sample can be found by the method described here.

*Finding the P-value*    Since the sampling distribution of the test statistic defined here is identical to that of the Mann-Whitney-Wilcoxon test statistic of Chapter 4, the procedure for finding $P$-values is also the same. Table H is used with $T_x$ for small samples where $m \leq n \leq 10$. As before, if $m \neq n$ the sample with fewer observations must be labeled the $X$ sample in order to use this table, and for any $T_x$, only the smaller tail probability is given. For larger sample sizes, Table A is used with the $z$ statistics. The guide for $P$-values is repeated here for convenience.

| Alternative | $P$-value (Table H or A) |
|---|---|
| $A_+$:  $\sigma_x > \sigma_y$ | Right-tail probability for $T_x$ or $z_{x,R}$ |
| $A_-$:  $\sigma_x < \sigma_y$ | Left-tail probability for $T_x$ or $z_{x,L}$ |
| $A$:  $\sigma_x \neq \sigma_y$ | 2(smaller tail probability for $T_x$ or right-tail probability for $z$) |

$$z_{x,L} = \frac{T_x + 0.5 - m(N + 1)/2}{\sqrt{mn(N + 1)/12}} \qquad z_{x,R} = \frac{T_x - 0.5 - m(N + 1)/2}{\sqrt{mn(N + 1)/12}}$$

$$\tag{2.2}$$

$$z = \begin{cases} -z_{x,L} & \text{if } T_x < m(N + 1)/2 \\ z_{x,R} & \text{if } T_x > m(N + 1)/2. \end{cases}$$

*Ties*   Ties are handled by the midrank method, exactly as for the Mann-Whitney-Wilcoxon test. The same correction for ties is also appropriate here; namely, the denominator in the $z$ statistics in (2.2) is replaced by

$$\sqrt{\frac{mn(N+1)}{12} - \frac{mn(\Sigma u^3 - \Sigma u)}{12N(N-1)}}, \qquad (2.3)$$

where $u$ is the number of observations tied for any given position, and the sum is over all sets of ties.

**example   2.2**

In many industrial production processes, each measurable characteristic of any raw material must have some specified average value. However, it may be equally or more important that there be small variability in this characteristic. Variability in the raw material input may result in extremely large variability in some important characteristic of the end product, such as texture, color, operation, efficiency, and the like.

Suppose that samples of lead ingots are taken from two different distributors, A and B, who provide ingots of the same median weight. The following data represent weight in kilograms.

A:   15.7, 16.1, 15.9, 16.2, 15.9, 16.0, 15.8, 16.1, 16.3, 16.5, 15.5
B:   15.4, 16.0, 15.6, 15.7, 16.6, 16.3, 16.4, 16.8, 15.2, 16.9, 15.1

These ingots are to be used as raw material.

In order to ensure uniformity in the end product, the ingots must have small variability in weight. Perform a nonparametric test for variability and make a recommendation as to which distributor should receive the order if there is any preference.

**SOLUTION**

Since the sample sizes are equal, $m = n = 11$, it is immaterial which set of weights is labeled the $X$ sample. Call $X$ the material from distributor A. Since there is no reason at the outset to expect a difference in variability in a particular direction, we use a two-sided alternative with the hypothesis set

$$H:\quad \sigma_x = \sigma_y \qquad A:\quad \sigma_x \neq \sigma_y.$$

The pooled observations are arranged in order of magnitude. Since $N = 22$ here, which is even, suppose we arbitrarily decide to use the larger of the two middle observations as the starting point for assigning the scores. Then the twelfth smallest observation has score 1 and we go left first, always assigning two scores at a time. The resulting arrangement and the respective scores are shown in Table 2.3, with $X$ observations and scores shown in boldface type. Because the assignment of scores requires more thought than an ordinary assignment of (unmodified) ranks, the ties are disregarded at

**TABLE 2.3**

| Obser-vation | Untied Score | Score | Obser-vation | Untied Score | Score |
|---|---|---|---|---|---|
| 15.1 | 22 | 22 | 16.0 | 1 | 1.5 |
| 15.2 | 19 | 19 | 16.1 | 4 | 4.5 |
| 15.4 | 18 | 18 | 16.1 | 5 | 4.5 |
| 15.5 | 15 | 15 | 16.2 | 8 | 8 |
| 15.6 | 14 | 14 | 16.3 | 9 | 10.5 |
| 15.7 | 11 | 10.5 | 16.3 | 12 | 10.5 |
| 15.7 | 10 | 10.5 | 16.4 | 13 | 13 |
| 15.8 | 7 | 7 | 16.5 | 16 | 16 |
| 15.9 | 6 | 4.5 | 16.6 | 17 | 17 |
| 15.9 | 3 | 4.5 | 16.8 | 20 | 20 |
| 16.0 | 2 | 1.5 | 16.9 | 21 | 21 |

first in order to see what scores the tied observations would have if they were not tied; then the midrank (more literally, midscore) method is used on these preliminary scores to get the scores used for the test. It is advisable to check whether the midranks have been assigned correctly by summing all scores given and verifying that their total is $N(N + 1)/2$, which here is $22(23)/2 = 253$. Summing the final scores above, we have

$$22 + 19 + 18 + \cdots + 21 + 21 = 253.$$

The sum of the $X$ (boldface) scores is

$$T_x = 15 + 10.5 + 7 + \cdots + 10.5 + 16 = 86.5.$$

Since the sample sizes here are outside the range of Table H, the normal curve approximation must be used to find the tail probability. Since $T_x = 86.5 < m(N + 1)/2 = 126.5$, the appropriate $z$ statistic to calculate from (2.2) is $-z_{x,L}$. The ties are not extensive; thus the correction for ties should have little effect, and we do not bother to use it.

$$z = -\frac{T_x + 0.5 - m(N + 1)/2}{\sqrt{mn(N + 1)/12}} = -\frac{86.5 + 0.5 - 126.5}{\sqrt{11(11)(23)/12}}$$

$$= \frac{39.5}{11\sqrt{1.917}} = \frac{39.5}{15.23} = 2.59$$

The right-tail probability for 2.59 from Table A is .0048. Since the alternative is two-sided, the (asymptotic approximate) $P$-value is two times this probability, or $2(.0048) = .0096$. The data do not support $H$, and we conclude that the two scale parameters are not the same.

Although the data do support the conclusion that there is a difference in variability, the real question of interest here is which supplier makes ingots with smaller variability in weights. We can make such a decision by comparing the mean values of $T_x$ and $T_y$ — that is, $T_x/m$ and $T_y/n$ — or, equivalently, by simply seeing whether $T_x$ is larger or smaller than $m(N + 1)/2$, its expected value under $H$. For these data, we noted before that $T_x = 86.5 < m(N + 1)/2 = 126.5$. This implies that if $T_x$ is too extreme in one direction, it can only be too small. A left-tail probability on $T_x$ indicates how extreme, and we found this value above to be .0048. Since a small left-tail probability on $T_x$ would be used as support for the one-sided alternative $\sigma_x < \sigma_y$, we can form this conclusion. Thus the raw materials from distributor A have a smaller scale parameter and hence are more uniform in quality than those of distributor B. A recommendation for purchase, however, cannot and should not be made without more information as regards relative cost of the raw materials from the different suppliers, cost of scrapping the end product if it does not meet the quality specifications, and any other relevant factors.

This example illustrates an important general point in statistical analysis that should be mentioned. When the $P$-value for a two-sided test is found by multiplying by two the tail probability for some one-sided test, the $P$-value for that one-sided test is one-half of the two-sided $P$-value. Hence, whenever a two-sided test leads to rejection of the null hypothesis, we can also make the one-sided conclusion that the direction of the difference that is significant is the same as the direction of the difference between the test statistic and its mean. It should be noted, however, that if the null hypothesis is accepted using a two-sided test, the one-sided probability may still be small enough to conclude that a difference exists in a particular direction. Thus, in a situation in which a one-sided conclusion is desirable, a one-sided test in the direction indicated by the sample results is more appropriate than a two-sided test.

**example   2.3**

An institute of microbiology is interested in purchasing microscope slides of uniform thickness. The slides can be purchased from either of two different suppliers; each supplier claims a specified thickness for this product and that the range of difference from that specified value is within acceptable limits. The institute obtains mutually independent samples of 10 slides from each supplier and gauges their thickness with a micrometer. The data below represent differences from specified thickness for each supplier:

$$\text{Supplier A:} \quad .0283, \quad .0291, \quad .0110, \quad -.0293, \quad .0158,$$
$$-.0121, \quad -.0276, \quad -.0176, \quad .0221, \quad -.0231$$

$$\text{Supplier B:} \quad -.0021, \quad .0157, \quad .0047, \quad -.0002, \quad .0003$$
$$.0081, \quad -.0051, \quad -.0091, \quad .0011, \quad -.0187.$$

Assuming that uniformity of thickness is measured by the range of difference from specified thickness, the institute should purchase slides from the supplier whose slides have the smaller range. Which supplier should be used?

**SOLUTION**
Since the data given represent the difference from specified thickness in each case, an assumption that each median (of difference) equals zero is quite tenable. However, any further distribution assumptions would be arbitrary. Thus we apply the Siegel-Tukey test for analysis. Because we are interested in which supplier has the smaller range of difference, a one-sided alternative is appropriate. However, the direction of this alternative cannot be determined until after the test statistic is calculated. The sample sizes are equal; let us call supplier A the $X$ sample. Suppose we arbitrarily decide to assign score 1 to the smaller of the two middle values (the tenth smallest of all). The combined ordered data and their scores are shown in Table 2.4 (there

**TABLE 2.4**

| Observation | Rank | Observation | Rank |
|---|---|---|---|
| −.0293 | 20 | .0003 | 2 |
| −.0276 | 17 | .0011 | 3 |
| −.0231 | 16 | .0047 | 6 |
| −.0187 | 13 | .0081 | 7 |
| −.0176 | 12 | .0110 | 10 |
| −.0121 | 9 | .0157 | 11 |
| −.0091 | 8 | .0158 | 14 |
| −.0051 | 5 | .0221 | 15 |
| −.0021 | 4 | .0283 | 18 |
| −.0002 | 1 | .0291 | 19 |

are no ties here). The observations and scores from the $X$ sample are in bold-face type. It is clear from this array that the sum of the $X$ scores is larger than the $Y$ sum; hence it is easier to compute the value of $T_x$ indirectly, using

$$T_y = 13 + 8 + 5 + \cdots + 11 = 60$$

$$T_x = \frac{N(N + 1)}{2} - T = \frac{20(21)}{2} - 60 = 210 - 60 = 150.$$

Since $T_x$ is the larger observed sum, the appropriate alternative is that $\sigma_x$ is the larger scale parameter, or

$$H: \quad \sigma_x = \sigma_y \qquad A_+: \quad \sigma_x > \sigma_y.$$

The appropriate $P$-value is then a right-tail probability for $T_x = 150$ when $m = n = 10$. From Table H, the exact $P$-value is .000; $H$ is not supported by these data. We conclude that the $X$ population (supplier A) has the larger scale parameter, and that the slides of supplier B seem to be more uniform in thickness. Unless cost and other factors should also be considered, we recommend that the institute purchase from supplier B.

*Other inferences*   The Siegel-Tukey test procedure can also be applied when the null hypothesis asserts that the scale parameters are in some stated ratio. The null hypothesis that we have been considering is $\sigma_x = \sigma_y$, which can be written equivalently as $\sigma_x/\sigma_y = 1$. We now consider the following more general hypothesis sets for some given positive number $\lambda_o$.

*Two-sided alternative*
$$H: \quad \sigma_x/\sigma_y = \lambda_o \qquad A: \quad \sigma_x/\sigma_y \neq \lambda_o$$

*One-sided alternatives*
$$H: \quad \sigma_x/\sigma_y = \lambda_o \qquad A_+: \quad \sigma_x/\sigma_y > \lambda_o$$
$$H: \quad \sigma_x/\sigma_y = \lambda_o \qquad A_-: \quad \sigma_x/\sigma_y < \lambda_o$$

If each $X$ observation is divided by $\lambda_o$, that is, if we form derived observations $X' = X/\lambda_o$, then the scale parameter for the $X'$ observations is $\sigma_{x'} = \sigma_x/\lambda_o$, which equals $\sigma_y$ under the null hypothesis. Thus, if the $X'$ and $Y$ variables have the same medians, the Siegel-Tukey procedure can be used with any of these hypothesis sets as long as the procedure is based on the derived sample observations $X_1', X_2', \ldots, X_m'$ and $Y_1, Y_2, \ldots, Y_n$. These $m + n$ values are pooled in a single array from smallest to largest and assigned scores as before. The test is unchanged except that the test statistic is now defined as

$$T_x = \text{sum of the } X' \text{ scores.}$$

Note, however, that the statement that $X'$ and $Y$ have the same median is not the same as saying that $X$ and $Y$ have the same median. In fact, since $M_{x'} = M_x/\lambda_o$, the relation $M_{x'} = M_y$ implies that $M_y = M_x/\lambda_o$. Of course, if $M_x = M_y = 0$, then $M_{x'} = 0$, also, and $X, X'$, and $Y$ all have the same median.

In general, the Siegel-Tukey test is a poor indicator of differences in scale unless the distributions of the data that are arrayed from smallest to largest have the same central value. Suppose that the central values of the original observations, which for convenience we represent by the medians $M_x$ and $M_y$, are unequal but known. Then the original sample observations can be adjusted for the difference in location by subtracting from each observation the value of its corresponding population median. That is, the derived data $X_1 - M_x, X_2 - M_x, \ldots, X_m - M_x$ and $Y_1 - M_y, Y_2 - M_y, \ldots,$

$Y_n - M_y$, do have equal population medians; in fact, their common central value is zero. Similarly, if the populations have unequal medians with known difference $\theta = M_x - M_y$, then the derived sample observations $X_1 - \theta, X_2 - \theta, \ldots, X_m - \theta$ and $Y_1, Y_2, \ldots, Y_n$ have population medians $M_x - \theta$ and $M_y$, respectively, which are equal. In either case, to test the null hypothesis $\sigma_x = \sigma_y$, the derived observations are pooled and arranged in order of magnitude and the Siegel-Tukey test is applied to these data in the usual way. If the hypothesis is $\sigma_x/\sigma_y = \lambda_o$, the $X - M_x$ or $X - \theta$ observations are further adjusted before pooling by dividing by $\lambda_o$; that is, the data to be analyzed are

$$\frac{X_1 - M_x}{\lambda_o}, \frac{X_2 - M_x}{\lambda_o}, \ldots, \frac{X_m - M_x}{\lambda_o} \quad \text{and}$$
$$Y_1 - M_y, Y_2 - M_y, \ldots, Y_n - M_y,$$

or

$$\frac{X_1 - \theta}{\lambda_o}, \frac{X_2 - \theta}{\lambda_o}, \ldots, \frac{X_m - \theta}{\lambda_o} \quad \text{and} \quad Y_1, Y_2, \ldots, Y_n.$$

Note, however, that in the latter case, the medians of $(X - \theta)/\lambda_o$ and $Y$ cannot then be assumed equal unless their common value is zero.

## 2.2  Other Test Procedures

If the populations are known to have a common median zero, or the two medians are known so that the data can be adjusted to have common median zero by forming $X - M_x$ and $Y - M_y$, inferences concerning relative variability about central values can also be made in other ways. The absolute values of such data are themselves measures of dispersion since they represent deviations from respective central values. We can test whether the average dispersion is the same for both populations against either a one- or two-sided alternative by considering the relative magnitudes of these deviations.

Two different methods can be applied in this situation, depending on whether it is reasonable to assume that the populations are symmetric about their respective medians or not. Under the symmetry assumption, a deviation $X - M_x$ or $Y - M_y$ is as likely to be positive as negative. If, further, the variables have the same scale, then the absolute values of these deviations, $|X - M_x|$ and $|Y - M_y|$, should have the same medians. Since the situation is reduced to a location problem for the absolute values, we can investigate the hypothesis using a test for location on the variables $|X - M_x|$ and $|Y - M_y|$. Applying the Mann-Whitney-Wilcoxon test of Section 2.1 of Chapter 4, we arrange the pooled absolute values $|X_1 - M_x|, |X_2 - M_x|, \ldots,$

$|X_m - M_x|$ and $|Y_1 - M_y|, \ldots, |Y_n - M_y|$ from smallest to largest and assign integer ranks according to relative magnitude, keeping track of which ranks come from which set of sample observations. Then the test statistic is

$$T_x = \text{the sum of the (unmodified) ranks}$$

of the absolute values of the $X$ deviations. $P$-values are found from Table H, right tail for $A_+$, left tail for $A_-$, and twice the smaller tail probability for $A$. As before, if $m \neq n$, the sample labeled $X$ is the one with fewer observations, and ties are handled by the midrank method. For $m$ and $n$ outside the range of Table H the normal approximation is used.

Recall that the data given in Example 2.3 represent differences from respective specified thickness for suppliers A and B. Assuming that deviations from specified thickness are symmetric about zero, we can use these data to illustrate how the Mann-Whitney-Wilcoxon statistic is applied for inferences concerning scale.

**example   2.4**
  The problem treated in Example 2.3 was to test the hypothesis set

$$H: \quad \sigma_x = \sigma_y \qquad A_+: \quad \sigma_x > \sigma_y,$$

using data that represent the differences from specified thickness for suppliers A and B ($X$ and $Y$, respectively). Regarding the absolute values of these differences as measures of spread about the respective medians, we simply form the absolute values and make a pooled ranking from smallest to largest. The absolute values are shown below.

| | Absolute Values for the | | |
|---|---|---|---|
| X Sample | | Y Sample | |
| .0283 | .0121 | .0021 | .0081 |
| .0291 | .0276 | .0157 | .0051 |
| .0110 | .0176 | .0047 | .0091 |
| .0293 | .0221 | .0002 | .0011 |
| .0158 | .0231 | .0003 | .0187 |

The ranking of these data is given in Table 2.5, in which the $X$ observations and ranks are indicated by boldface type. A right-tail probability for 149 is the appropriate $P$-value because the alternative is $A_+$. From Table H the exact $P$-value is .000, and again the conclusion is to reject $H$ in favor of $A_+$.

**TABLE 2.5**
**POOLED ARRANGEMENT**

| Observation | Rank | Observation | Rank |
|---|---|---|---|
| .0002 | 1 | .0157 | 11 |
| .0003 | 2 | **.0158** | **12** |
| .0011 | 3 | **.0176** | **13** |
| .0021 | 4 | .0187 | 14 |
| .0047 | 5 | .0221 | 15 |
| .0051 | 6 | .0231 | 16 |
| .0081 | 7 | .0276 | 17 |
| .0091 | 8 | .0283 | 18 |
| **.0110** | **9** | .0291 | 19 |
| **.0121** | **10** | .0293 | 20 |

$$T_y = 1 + 2 + 3 + \cdots + 14 = 61$$

$$T_x = \frac{N(N+1)}{2} - T_y = \frac{20(21)}{2} - 61 = 210 - 61 = 149$$

It should be noted that the test statistic here, though close to that obtained in Example 2.3, does not have the same value. This will generally be true, since the tests are not equivalent. The observations here are arranged in order of absolute magnitude, and ranks are assigned according to relative position in the pooled arrangement. In Example 2.3, the signs were *not* ignored and scores or *modified* ranks were assigned.

If it is not reasonable to assume that the $X$ and $Y$ populations are symmetric about their known medians $M_x$ and $M_y$, the above procedure based on the Mann-Whitney-Wilcoxon statistic remains valid. However, it is not especially sensitive to differences in scale because it disregards the directions of the deviations. The procedure to be explained now compares separately the magnitudes of those deviations that are positive and those that are negative. That is, we divide all the $X - M_x$ differences into two groups according to sign. Let $m_+$ be the number of positive differences and $m_-$ the number of negative differences. The $Y - M_y$ differences are similarly separated according to sign, $n_+$ positive and $n_-$ negative. (Zero differences are ignored, and $m$ and $n$ are reduced accordingly. Thus we have $m_+ + m_- = m$ and $n_+ + n_- = n$.) Then the $m_+ + n_+$ positive differences are arranged in increasing order of magnitude and assigned ordinary (unmodified) ranks $1, 2, \ldots, m_+ + n_+$. Let $T_+$ be the sum of the ranks for $X$ differences in this positive group. The signs of the negative differences are ignored and these positive values are also ranked from smallest to largest using the integers $1, 2, \ldots, m_- + n_-$. Let $T_-$ be the sum of the ranks corresponding to $X$ deviations in this negative group.

The test statistic is the sum

$$T = T_+ + T_-.$$

If the $X - M_x$ differences are on the average larger in absolute value than those $Y - M_y$ differences that have the same sign, $T$ will be large. Thus the appropriate rejection regions for $T$ are right tail with $A_+$: $\sigma_x > \sigma_y$ and left tail with $A_-$: $\sigma_x < \sigma_y$. Since tables of the exact sampling distribution of $T$ are apparently not available, we give here only the test procedure for large samples, which is based on the normal approximation. The method is to compute one of the following test statistics:

$$z_R = \frac{T - m(N + 2)/4 - 0.5}{\sqrt{mn(N + 7)/48}} \tag{2.4}$$

$$z_L = \frac{T - m(N + 2)/4 + 0.5}{\sqrt{mn(N + 7)/48}}$$

$$z = \begin{cases} -z_L & \text{if } T < m(N + 2)/4 \\ z_R & \text{if } T > m(N + 2)/4. \end{cases}$$

Asymptotic approximate $P$-values are then found from Table A as follows:

| Alternative | $P$-value (Table A) |
|---|---|
| $A_+$: $\sigma_x > \sigma_y$ | Right-tail probability for $z_R$ |
| $A_-$: $\sigma_x < \sigma_y$ | Left-tail probability for $z_L$ |
| $A$: $\sigma_x \neq \sigma_y$ | 2(right-tail probability for $z$) |

$P$-values based on this approximation are generally accurate enough for practical purposes as long as $m$ and $n$ are each at least 10. This test is usually called the Sukhatme (1957) test.[4]

**example   2.5**

Consider again the situation of Example 2.3, with the one-sided alternative $A_+$: $\sigma_x > \sigma_y$. If we do not wish to assume that the measurements of deviations from specified thickness are symmetric, the Sukhatme test may be more appropriate than the test applied in Example 2.4. Again calling supplier A the $X$ variable, we separate the data into four groups (see Table 2.6) and form separate rankings of the positive and negative observations. The procedure for finding the test statistic is shown in Table 2.7, with $X$ ranks

[4]It should be noted that even though $m_+ + m_- = m$ and $n_+ + n_- = n$, the individual values of $m_+$ and $n_+$ depend on the sample results. Thus this test is actually conditional on the observed values of $m_+$ and $n_+$.

**TABLE 2.6**

| X Positive | X Negative | Y Positive | Y Negative |
|:---:|:---:|:---:|:---:|
| .0283 | .0293 | .0157 | .0021 |
| .0291 | .0121 | .0047 | .0002 |
| .0110 | .0276 | .0003 | .0051 |
| .0158 | .0176 | .0081 | .0091 |
| .0221 | .0231 | .0011 | .0187 |

shown in boldface type in each of the pooled sets of rankings. The right-tail probability from Table A is .0021, and thus $H$ is rejected in favor of $A_+$.

**TABLE 2.7**

| Positives | Rank | Negatives | Rank |
|:---:|:---:|:---:|:---:|
| .0003 | 1 | .0002 | 1 |
| .0011 | 2 | .0021 | 2 |
| .0047 | 3 | .0051 | 3 |
| .0081 | 4 | .0091 | 4 |
| **.0110** | **5** | **.0121** | **5** |
| .0157 | 6 | **.0176** | **6** |
| **.0158** | **7** | .0187 | 7 |
| **.0221** | **8** | **.0231** | **8** |
| **.0283** | **9** | .0276 | 9 |
| **.0291** | **10** | .0293 | 10 |

$$m_+ = 5, n_+ = 5, m = 10 \qquad m_- = 5, n_- = 5, n = 10$$
$$T_+ = 5 + 7 + 8 + 9 + 10 = 39 \qquad T_- = 5 + 6 + 8 + 9 + 10 = 38$$
$$T = 39 + 38 = 77$$
$$z_R = \frac{77 - 10(22)/4 - 0.5}{\sqrt{10(10)(27)/48}} = 2.87$$

Examples 2.3, 2.4, and 2.5 illustrate the use of three different statistical procedures for the same set of data and the hypothesis set $H$ versus $A_+$. These methods are not equivalent, but they are all appropriate (although the method in Example 2.4 is less sensitive if the symmetry assumption is not reasonable). The assumption that justifies the use of all three methods is that each set of data analyzed is from a population with median zero. Whenever the sample observations represent differences between actual measurements and some central value, this assumption is frequently tenable.

When it is not reasonable to assume that each population has, or can be adjusted to have, median zero, neither the Mann-Whitney-Wilcoxon (Example 2.4) nor the Sukhatme (Example 2.5) procedure is appropriate for a comparison between variability about some central location value. However, if the medians can be assumed the same, the Siegel-Tukey test can be used even though the common median value is unknown. (If this assumption is not reasonable, the Siegel-Tukey test should not be used; it is not especially sensitive to scale differences when locations are different, as Example 4.1 will show.) The only exception is when the difference of the two medians is known, since then the data can be adjusted to have equal medians. If the medians are unknown and cannot be assumed equal, and the difference of the medians is also unknown, none of these three procedures can be used. This situation will be treated in Section 4.

# 3 Confidence Interval Procedures

If the primary goal of the investigator is estimation of the ratio of two scale parameters, two different simple procedures are available to find a confidence interval. The validity of these procedures rests on the assumption of a certain relationship between the distributions of the variables, which we call a scale model. In terms of a parameter $\lambda$, the scale model can be written as

$$F_y(u) = F_x(\lambda u), \tag{3.1}$$

or, in the more general case, as

$$F_{y-M_y}(u) = F_{x-M_x}(\lambda u), \tag{3.2}$$

where $\lambda$ is any positive number. The model in (3.1) asserts that $X/\lambda$ and $Y$ are identically distributed for some $\lambda$, and model (3.2) says that the distributions are the same except for scale after the variables are corrected for difference in medians; that is, $(X - M_x)/\lambda$ and $(Y - M_y)$ are identically distributed. In either model, the quantity $\lambda$ is interpreted as a scale parameter of $X$ divided by a scale parameter of $Y$. In particular, the variance of $X$ is $\lambda^2$ times the variance of $Y$ in both cases. However, in model (3.1), the expected value of $X$ is $\lambda$ times the expected value of $Y$; hence the means are not the same unless both are zero or the scale parameters are equal. As a result, this model does not correspond to the one we have been discussing for test procedures where the locations were assumed equal but not necessarily zero, or assumed unequal but known. In model (3.2) the medians of $X - M_x$ and $Y - M_y$ are both zero, and hence this model is the same as before if $M_x$ and $M_y$ are known, or if the data are known as deviations from respective central values, as was the case for the data analyzed in Examples 2.3, 2.4, and 2.5. Since model (3.2) is more analogous, we discuss confidence interval techniques for that model first.

Assume that the distributions of the $X$ and $Y$ random variables follow model (3.2). Then there are two confidence interval procedures; one is appropriate for the case where $X$ and $Y$ are symmetric about their respective medians, and the other where they are not. Under the symmetry assumption, we can obtain a confidence interval for $\lambda$ by considering the $mn$ possible ratios, $|X - M_x|/|Y - M_y|$. These are the quotients of the absolute values of the deviations from the respective medians or centers of symmetry, and thus are ratios of measures of spread. These ratios all have the same distribution when $\lambda = 1$. The method is simply to arrange these ratios from smallest to largest, and use as the confidence interval endpoints the $k$th from the smallest and the $k$th from the largest ratios. For $m \le n \le 10$, the value of $k$ is found from Table H as the rank of that value of $T_x$ that has a left-tail probability of $P$ if the confidence coefficient is $\gamma = 1 - 2P$. If $m < n$, the sample with fewer observations must be labeled the $X$ sample. For sample sizes outside the range of Table H, $k$ is found from the following formula, based on the normal approximation:

$$k = 0.5 + \frac{mn}{2} - z\sqrt{\frac{mn(N + 1)}{12}}. \tag{3.3}$$

The value of $z$ is found from Table A as the number that corresponds to a right-tail probability of $P = (1 - \gamma)/2$ if the confidence coefficient is $\gamma$. If this value of $k$ is not an integer, it should be rounded down to the next lower integer. The methodology then is exactly the same as the Mann-Whitney-Wilcoxon procedure to find a confidence interval for location; however, the quantities ranked here are the $mn$ ratios of spread about centrality, and the inference concerns the ratio of the scale parameters. Recall that in Section 2.2 of Chapter 4, the $mn$ differences $X_i - Y_j$ were ranked to find a confidence interval for the difference between medians.

Under model (3.2), if we do not assume that the $X$ and $Y$ populations are both symmetric, the previous procedure remains statistically valid. However, without symmetry, a comparison of two deviations, $X - M_x$ with $Y - M_y$, does not provide information relevant to the relative scales unless both deviations are of the same sign. Thus, here we need to compare the magnitudes of deviations that are in the same direction, rather than absolute values of deviations. Since a ratio $(X - M_x)/(Y - M_y)$ is positive whenever the numerator and denominator have the same sign, the procedure appropriate for model (3.2) without the symmetry assumption is to consider only the positive ratios $(X - M_x)/(Y - M_y)$. There will generally be fewer than $mn$ ratios here. These positive ratios are arranged in order of magnitude from smallest to largest, and the $k$th smallest and $k$th largest are the confidence interval endpoints. Tables of $k$ for small samples are apparently not available, but for $m$ and $n$ sufficiently large, say each at least 10, the value of $k$

can be approximated using the normal distribution. The formula is

$$k = 0.5 + \frac{mn}{4} - z\sqrt{\frac{mn(N+7)}{48}}, \qquad (3.4)$$

where $z$ is the value from Table A that corresponds to a right-tail probability of $P = (1 - \gamma)/2$ if the confidence coefficient is $\gamma$. If the value of $k$ from (3.4) is not an integer, it should be rounded down.

Neither of these two confidence procedures for $\lambda$ under model (3.2) corresponds to the Siegel-Tukey test for scale discussed in Section 2.1, even though the model is analogous. The procedure in which symmetry is assumed corresponds to the Mann-Whitney-Wilcoxon test illustrated in Example 2.4. The second procedure described above, where symmetry is not assumed and $k$ is found from (3.4), considers separately the positive and negative deviations from respective medians. It corresponds to the Sukhatme test illustrated in Example 2.5. In Example 3.1 below we find that this latter method gives a wider interval than the former method for the same level of confidence. One might conjecture that this would generally be the case when the symmetry assumption holds, since more information about the distributions is used.

**example 3.1**

In Example 2.3 we assumed that the data given represent differences from their respective medians. Therefore, model (3.2) applies. The method based on all ratios of absolute values should be used if the symmetry assumption seems appropriate, and the method based on only positive ratios if not. Rather than make a judgment about symmetry here, we illustrate both methods for these same data.

The first step is to form the 100 ratios of an $X$ and a $Y$ observation. Each entry in Table 3.1 is an $X/Y$ value.

For $m = n = 10$ using Table H, a one-sided $P$-value of .045 gives a confidence coefficient of $\gamma = 1 - 2P = .91$; the corresponding left-tail $T_x$ value is 82, and its rank in that table is 28, and hence $k = 28$. Under model (3.2) and assuming symmetry, the confidence interval endpoints are the 28th smallest and 28th largest of the *absolute values* of the 100 numbers listed in Table 3.1, which gives

$$2.16 \leq \lambda \leq 14.36.$$

Now we illustrate finding the interval without assuming symmetry, but again with confidence coefficient $\gamma = .91$ so that $P = .045$. Then the value of $z$ from Table A is 1.70, and the value of $k$ for $m = n = 10$ from (3.4) is

$$k = 0.5 + 25 - 1.7\sqrt{\frac{100(27)}{48}} = 12.75,$$

TABLE 3.1

| X | Y −.0187 | −.0091 | −.0051 | −.0021 | −.0002 | .0003 | .0011 | .0047 | .0081 | .0157 |
|---|---|---|---|---|---|---|---|---|---|---|
| −.0293 | 1.57 | 3.22 | 5.74 | 13.95 | 146.50 | −97.67 | −26.64 | −6.23 | −3.62 | −1.87 |
| −.0276 | 1.48 | 3.03 | 5.41 | 13.14 | 138.00 | −92.00 | −25.09 | −5.87 | −3.41 | −1.76 |
| −.0231 | 1.24 | 2.54 | 4.53 | 11.00 | 115.50 | −77.00 | −21.00 | −4.91 | −2.85 | −1.47 |
| −.0176 | .94 | 1.93 | 3.45 | 8.38 | 88.00 | −58.67 | −16.00 | −3.74 | −2.17 | −1.12 |
| −.0121 | .65 | 1.33 | 2.37 | 5.76 | 60.50 | −40.33 | −11.00 | −2.57 | −1.49 | −.77 |
| .0110 | −.59 | −1.21 | −2.16 | −5.24 | −55.00 | 36.67 | 10.00 | 2.34 | 1.36 | .70 |
| .0158 | −.84 | −1.74 | −3.10 | −7.52 | −79.00 | 52.67 | 14.36 | 3.36 | 1.95 | 1.01 |
| .0221 | −1.18 | −2.43 | −4.33 | −10.52 | −110.50 | 73.67 | 20.00 | 4.70 | 2.73 | 1.41 |
| .0283 | −1.51 | −3.11 | −5.55 | −13.48 | −141.50 | 94.33 | 25.73 | 6.02 | 3.49 | 1.80 |
| .0291 | −1.56 | −3.20 | −5.70 | −13.86 | −145.50 | 97.00 | 26.45 | 6.19 | 3.59 | 1.85 |

which we call 12 in order to ensure a conservative result. The confidence interval endpoints under model (3.2) without symmetry are the 12th smallest and 12th largest of the *positive* numbers listed in Table 3.1, which gives

$$1.85 \leq \lambda \leq 25.73.$$

Up to this point in this chapter, we have been assuming that $X$ and $Y$ have the same central value, or that the dispersions of $X$ and $Y$ are each measured about some location parameter that describes the "typical" magnitude of that variable, that is, some relevant centrality parameter. These parameters have been interpreted as the respective medians for convenience. Comparisons of spread around some noncentral value may also be of interest. For example, the range is a useful descriptor of spread (as long as it is finite), particularly when variability from the origin is of interest. The ranges of two variables can be compared to indicate relative variability without regard for relative locations. If this type of comparison is the one of interest, the scale model in (3.1) is appropriate even if the variables have different centrality parameters. Suppose that each of the two variables has a finite lower limit of possible values (that is, a fixed minimum value or quantile of order zero), and these limits are known to be $\alpha$ and $\beta$ for $X$ and $Y$, respectively. Then the differences $X - \alpha$ and $Y - \beta$ are both positive variables, and the size of $Y - \beta$ relative to $X - \alpha$ as measured from zero is an indicator of the scale of $Y$ relative to the scale of $X$.

In the general case of this situation we have two random variables, which are always positive — either by nature or because they are each measured as a deviation from some finite lower limit. Examples of such variables are ages, weights, or survival times. If we are interested in a comparison of the variability of $X$ and $Y$ from zero, the scale model (3.1) is appropriate. $\lambda$ is still interpreted as the ratio $\sigma_x/\sigma_y$; the only difference is that $\sigma_x$ and $\sigma_y$ now measure spread from the origin, which cannot be a central value for either variable. Thus $\sigma_x^2$ and $\sigma_y^2$ cannot be interpreted as variances here. Given two sets of observations on variables of this type, $X_1, X_2, \ldots, X_m$ and $Y_1, Y_2, \ldots, Y_n$, we can obtain a confidence interval for $\lambda$ by considering all possible ratios $X_i/Y_j$ since these quantities are indicative of relative scale. There are $mn$ ratios, and all are positive. These ratios are arranged in order of magnitude from smallest to largest; the $k$th smallest and $k$th largest are the confidence interval endpoints. When $m < n$, the sample of smaller size is labeled the $X$ sample. For $m \leq n \leq 10$, the value of $k$ is found from Table H as the rank of the value of $T_x$ that has a left-tail probability of $P$ if the confidence coefficient is $\gamma = 1 - 2P$. For sample sizes outside the range of Table H, $k$ is found from the formula

$$k = 0.5 + \frac{mn}{2} - z\sqrt{\frac{mn(N + 1)}{12}},$$

where $z$ is the number from Table A that corresponds to a right-tail probability of $P = (1 - \gamma)/2$. Note that this procedure is identical to that appropriate for model (3.2) under the symmetry assumption, except that absolute values are not needed in forming the ratios since the variables are all positive anyway.

Model (3.1) also has some important practical applications where $\lambda$ is not interpreted strictly as a ratio of two scale parameters. Sen (1963) describes the following situation where this model is especially useful in assay. Suppose that $Y$ is the dosage level of a standard preparation necessary to give a specified response, and $X$ is the dosage required by a test preparation to produce the same response. If the test preparation behaves as a dilution or concentration of the standard preparation, in a diluent that is completely inert in respect to the response used, then the relevant model is $F_y(u) = F_x(\lambda u)$, where $\lambda$ is the relative potency of the test preparation $X$ with respect to standard one $Y$. A confidence interval on $\lambda$, found by the method described above, is an interval estimate of the potency of $X$ relative to $Y$. The reciprocals of these endpoints (in reverse order) provide a confidence interval estimate of the potency of $Y$ relative to $X$.

**example   3.2**

Finney (1952, p. 23) gives data on a study of tolerances of cats for tinctures of strophanthus (a poisonous extract from certain seeds). Fourteen cats were randomly divided into two groups of seven each; one group was given doses of tincture A and the other received tincture B. The doses were recorded as quantities per kg body weight of cat required to kill that cat, and total doses were not available. The data below are on dosages in .01 cc per kg.

A:   1.55, 1.58, 1.71, 1.44, 1.24, 1.89, 2.34
B:   2.42, 1.85, 2.00, 2.27, 1.70, 1.47, 2.20.

Since the data reflect tolerances relative to body weight, a measure of the magnitude of data of one tincture relative to the other is of interest in comparing potency of the tinctures. The problem is to find an appropriate nonparametric confidence interval for the potency of tincture A with respect to tincture B.

**SOLUTION**

The reader may verify that these data do not suggest that average tolerance is the same for both tinctures, whether measured by respective means or medians. However, that presents no problem here since we want to compare potency as measured from the origin. The model in (3.1) is appropriate. Calling tincture A the $X$ sample, $\lambda$ is interpreted as the potency of A relative to B. We arrange each set of observations in order of magnitude and form the 49 ratios $X/Y$ as shown in Table 3.2.

**TABLE 3.2**

| X \ Y | 1.47 | 1.70 | 1.85 | 2.00 | 2.20 | 2.27 | 2.42 |
|-------|------|------|------|------|------|------|------|
| 1.24 | .8435 | .7294 | .6703 | .6200 | .5636 | .5463 | .5124 |
| 1.44 | .9796 | .8471 | .7784 | .7200 | .6545 | .6344 | .5950 |
| 1.55 | 1.0544 | .9118 | .8378 | .7750 | .7045 | .6828 | .6405 |
| 1.58 | 1.0748 | .9294 | .8541 | .7900 | .7182 | .6960 | .6529 |
| 1.71 | 1.1633 | 1.0059 | .9243 | .8550 | .7773 | .7533 | .7066 |
| 1.89 | 1.2857 | 1.1118 | 1.0216 | .9450 | .8591 | .8326 | .7810 |
| 2.34 | 1.5918 | 1.3765 | 1.2649 | 1.1700 | 1.0636 | 1.0308 | .9669 |

From Table H with $m = 7$ and $n = 7$, the left-tail probability is .049 for $T_x = 39$, which has rank 12 in the table. Thus for confidence coefficient $\gamma = 1 - 2(.049) = .902$, we have $k = 12$. The confidence interval for $\lambda$ is then

$$.6960 \leq \lambda \leq 1.0216.$$

This can be interpreted as saying that tincture A is between 69.6 percent and 102 percent as potent as tincture B, or that 1 cc of A is estimated as equivalent to a quantity of B that is between .696 cc and 1.02 cc.

# 4   Inference Procedures: Medians Unequal and Unknown

We now return to the problem of comparing the spread of two variables around some appropriate central location parameters, which for convenience we define as the respective medians. The methods presented in Section 2 are not sensitive to scale differences unless (a) the two populations have equal medians, (b) the two medians are known so that both sets of observations can be adjusted to have zero population medians (by subtracting the value of the corresponding parameter from each set), or (c) the difference between population medians is known so that the observations can be adjusted to have equal but nonzero population medians (by subtracting this difference from the appropriate set). Thus we have not dealt with the situation that arises when the medians are not equal and are not known, and thus their difference is not known. Before discussing possible solutions to this problem, let us look at an example that emphasizes the importance of the assumption that the populations representing the data to be analyzed have equal medians if the Siegel-Tukey method is used.

**example   4.1**

In the situation described in Example 2.3 assume that the thickness specifications of suppliers A and B are 1.250 and 1.220, respectively. The observations given there represented deviations from these specified values. When 1.250 and 1.220 are added to the respective A and B deviations given before, we find the observations on actual thickness as given in Table 4.1.

**TABLE 4.1**

| Supplier A | | Supplier B | |
|---|---|---|---|
| Deviation + 1.250 = Observation | | Deviation + 1.220 = Observation | |
| .0283 | 1.2783 | −.0021 | 1.2179 |
| .0291 | 1.2791 | .0157 | 1.2357 |
| .0110 | 1.2610 | .0047 | 1.2247 |
| −.0293 | 1.2207 | −.0002 | 1.2198 |
| .0158 | 1.2658 | .0003 | 1.2203 |
| −.0121 | 1.2379 | .0081 | 1.2281 |
| −.0276 | 1.2224 | −.0051 | 1.2149 |
| −.0176 | 1.2324 | −.0091 | 1.2109 |
| .0221 | 1.2721 | .0011 | 1.2211 |
| −.0231 | 1.2269 | −.0187 | 1.2023 |

Suppose that these data had been given without the specified values for median thickness. If we assumed the medians to be equal, we might decide to apply the Siegel-Tukey test in the usual way, as shown in Table 4.2. The exact $P$-value from Table H is now .456. The difference in scale, which was supported by the data representing differences from specified thickness, has been completely masked by the fact that the central values are different for the two populations.

**TABLE 4.2**

| Observation | Rank | Observation | Rank |
|---|---|---|---|
| 1.2023 | 20 | **1.2269** | **2** |
| 1.2109 | 17 | 1.2281 | 3 |
| 1.2149 | 16 | **1.2324** | **6** |
| 1.2179 | 13 | 1.2357 | 7 |
| 1.2198 | 12 | **1.2379** | **10** |
| 1.2203 | 9 | **1.2610** | **11** |
| **1.2207** | **8** | **1.2658** | **14** |
| 1.2211 | 5 | **1.2721** | **15** |
| **1.2224** | **4** | **1.2783** | **18** |
| 1.2247 | 1 | **1.2791** | **19** |

$$T_x = 8 + 4 + 2 + \cdots + 19 = 107$$

The foregoing example shows that if the two populations have known medians $M_x$ and $M_y$, respectively, which are not equal, then the data to be analyzed *must* be the respective differences $(X - M_x)$ and $(Y - M_y)$ if we wish to compare spread about those central values (as opposed to variability from zero for positive variables). If the medians are unequal and unknown, is there any way to obtain information about relative scale? One obvious possibility is to estimate the population medians $M_x$ and $M_y$ from the corresponding sets of sample data and subtract these estimates from the corresponding $X$ and $Y$ observations in order to obtain two sets of differences, both of whose central values are likely to be close to zero. With such a procedure, however, the Siegel-Tukey test is no longer exact, and in fact is not even distribution free. A paper by Moses (1963) shows that no test based on relative magnitude of observations (any kind of rank test) can effectively compare spread about unknown central values without some sort of strong restriction on the centrality parameters. Central tendency is simply too interrelated with spread about that central value for rank tests to reflect scale differences alone unless the locations are the same.

Nevertheless, one general type of approach to scale inferences circumvents this interrelationship problem. Strictly speaking, these tests are not rank tests, but they are based on analogous principles. Suppose, as before, that we have two mutually independent random samples of sizes $m$ and $n$ from continuous populations. The scales of the population variables are to be compared, but without assuming any particular scale model. The procedure is to select a positive integer $w \geq 2$ and randomly divide the $X$ observations into some number $m'$ of nonoverlapping subgroups of size $w$. If $m$ is divisible by $w$, we have $m' = m/w$. If not, some $X$ observations will be left over and must be discarded. The $Y$ observations are similarly divided into $n'$ subgroups of size $w$. An appropriate measure of variability is calculated for each of the $m' + n'$ subgroups, ignoring all other observations. If we want to compare spread about some central value, the variance, standard deviation, or average deviation is appropriate. If variability from zero is of interest, the range is more appropriate. (Once an appropriate measure of dispersion is selected, the same quantity must be calculated for each subgroup.) These two sets of calculated measures are identically distributed if the original observations have the same distribution, and their relative magnitudes are indicative of any possible differences in scale between $X$ and $Y$, irrespective of the differences in location. Thus an inference about scale can be based on these $m' + n'$ numbers calculated from the original $m + n$ observations.

If the hypothesis is $H: \lambda = 1$, some appropriate inference procedures have already been explained. The easiest method is to perform a Mann-Whitney-Wilcoxon test for location in the usual way after ranking the calculated numbers from 1 to $m' + n'$. The relevant sample sizes are $m'$ and $n'$, and Table H is entered with these values, where $m' \leq n'$. Either a one- or two-sided alternative can be used. If the null hypothesis of interest is $\lambda = \lambda_o$, that is $\sigma_x/\sigma_y = \lambda_o$, instead of $\sigma_x/\sigma_y = 1$, the same procedure can be used once the original $X$ observations are divided by $\lambda_o$. The method explained in Section 2.2 of Chapter 4 can be used to find a confidence interval for the difference $\sigma_x - \sigma_y$. We take all $m'n'$ differences $U_{ij}$ of $X$ measures of dispersion and $Y$ measures of dispersion; the confidence interval endpoints are the $k$th smallest and $k$th largest differences, where $k$ is found from Table H or approximated by (3.3), with $m$ and $n$ replaced by $m'$ and $n'$, respectively.

Since all calculated measures of spread are positive, we can also find a confidence interval for $\lambda = \sigma_x/\sigma_y$ using the method explained in Section 3 as appropriate for model (3.1) with positive variables. Specifically, we form the $m'n'$ ratios of an $X$ measure of dispersion and a $Y$ measure of dispersion; the confidence interval endpoints are the $k$th smallest and $k$th largest ratios, where $k$ is found from Table H or approximated by (3.3), with $m$ and $n$ replaced by $m'$ and $n'$.

These two procedures are illustrated in Example 4.2, when the subsamples are of size two and the subsample measure of dispersion is the range. A question of interest of course is how to select a value for $w$. If $w$ is a divisor of both $m$ and $n$, no observations will have to be discarded, which is obviously advantageous. The primary criterion for $w$ is that it should not be so large that the resulting $m'$ and $n'$ provide absurdly small sample sizes for the inference procedure. For the validity of any of these procedures, it is necessary that $w$ be chosen independently of the values of the observations (but not of $m$ and $n$), and that the division into subgroups be done by a random process, for example, by using a table of random numbers. Shorack (1969) discusses the choice of $w$, and also the question of which measure of dispersion should be calculated from the subgroups. If there is reason to think that the populations might be normal, the variance (or, equivalently, the standard deviation) would be a logical choice since it is an unbiased estimate of $\sigma^2$. The range is a good choice for the uniform distribution.

### example   4.2

These two methods are illustrated for the data of Example 4.1, where the medians are known to be unequal. Suppose we select $w = 2$. With 10 observations in each sample, each will have 5 subsamples. These can be formed using a table of random digits once the observations are numbered, say 1 to 9 and 0 for 10. The pairs of random digits in the table (from a randomly chosen starting point) indicate which observations are in each subsample (if a digit is repeated, it is discarded). For example, if the random digits are

3 1 5 4 2 1 9 8 0 3 7 6, the $X$ subsamples are $(X_3, X_1)$, $(X_5, X_4)$, $(X_2, X_9)$, $(X_8, X_{10})$, $(X_7, X_6)$, and the $X$ subsample data are as shown below. The $Y$ subsample data are found from another set of random digits.

| X Subsamples | Y Subsamples |
|---|---|
| (1.2610, 1.2783) | (1.2149, 1.2003) |
| (1.2658, 1.2207) | (1.2179, 1.2109) |
| (1.2791, 1.2721) | (1.2211, 1.2357) |
| (1.2324, 1.2269) | (1.2198, 1.2211) |
| (1.2224, 1.2379) | (1.2247, 1.2281) |

For a subsample of size two, say $(X_i, X_j)$, the range is $|X_i - X_j|$, and the average deviation from the mean or median is one-half of the sum

$$\left|X_i - \frac{X_i + X_j}{2}\right| + \left|X_j - \frac{X_i + X_j}{2}\right|,$$

which equals the range, since for any $X_i \leq X_j$ the sum equals

$$\frac{X_i + X_j}{2} - X_i + X_j - \frac{X_i + X_j}{2} = X_j - X_i.$$

Thus, whether the dispersion criterion selected for each subsample is the average deviation or the range, the measures occur in the same combined order and hence have the same ranks and test result. For these data, the ranges, denoted by $U$ for the $X$ subsamples and $V$ for the $Y$ subsamples, are as follows:

| U | V |
|---|---|
| .0173 | .0146 |
| .0451 | .0070 |
| .0070 | .0146 |
| .0055 | .0013 |
| .0155 | .0034 |

Suppose that we want to perform a two-sided test of $H:$ $\lambda = 1$. Using the Mann-Whitney-Wilcoxon test procedure for these measures $U$ and $V$, we arrange the 10 numbers in increasing order, with $U$ in boldface type, assign ranks, and calculate the sum of the ranks assigned to the $U$'s.

| Observation | Rank | Observation | Rank |
|---|---|---|---|
| .0013 | 1 | .0146 | 6.5 |
| .0034 | 2 | .0146 | 6.5 |
| .0055 | 3 | .0155 | 8 |
| .0070 | 4.5 | .0173 | 9 |
| .0070 | 4.5 | .0451 | 10 |

$$T_u = 3 + 4.5 + 8 + 9 + 10 = 34.5$$
$$T_v = 20.5$$

From Table H with $m' = 5$, $n' = 5$ the right-tail probability for 34.5, the larger of $T_u$ and $T_v$, is between .075 and .111, and the $P$-value is between .150 and .222. The null hypothesis should be accepted.

Now to find a confidence interval for $\lambda$, we compute the 25 ratios $U/V$ as follows:

| U \ V | .0013 | .0034 | .0070 | .0146 | .0146 |
|---|---|---|---|---|---|
| .0055 | 4.2 | 1.6 | .8 | .4 | .4 |
| .0070 | 5.4 | 2.1 | 1.0 | .5 | .5 |
| .0155 | 11.9 | 4.6 | 2.2 | 1.1 | 1.1 |
| .0173 | 13.3 | 5.1 | 2.5 | 1.2 | 1.2 |
| .0451 | 34.7 | 13.3 | 6.4 | 3.1 | 3.1 |

From Table H, with $m' = 5$, $n' = 5$, $P = .048$, and $\gamma = .904$, the left-tail critical value is 19, which has rank 5. Hence $k = 5$ and the confidence interval endpoints are the fifth smallest and fifth largest ratios, or

$$.8 \leq \lambda \leq 6.4.$$

# 5   Summary

In this chapter we have considered the problem of comparing two populations with respect to the characteristic of spread, variability, or scale. Unfortunately, the combined ordered arrangement of $X$'s and $Y$'s from two mutually independent random samples reflects differences in location as well as scale. Thus, it was necessary to divide the procedures covered into two types. Those covered in Sections 2 and 3 are appropriate when the medians are either known or can be assumed equal. The methods of Section 4 can be used in the more general case.

Except for the tests of Section 4, all these nonparametric procedures are based on the relative magnitudes of the combined data. Thus it is not surprising that these tests are inadequate unless we invoke some restrictive assumptions about locations.

We summarize the applicability of tests for the equality of scale parameters as follows. When the medians are equal and unknown, the Siegel-Tukey test can be applied to the original data. If the medians are unequal and unknown but their difference is known, the same test can be used after adjusting the data. If the medians are known, three procedures are available. We can either (a) apply the Siegel-Tukey test to the differences of the original data and the respective medians, (b) apply the Mann-Whitney-Wilcoxon test to the absolute values of these differences (preferably when a symmetry assumption is tenable), or (c) separate the positive and negative differences and apply the Sukhatme test. Obviously, these same three methods can be used on the original data when both medians are assumed to be zero. When the medians are unequal and unknown, any of these procedures could be applied using the respective sample medians as estimates of the population medians. Another possibility is to divide the samples into subsamples and apply the "ranklike" tests of Section 4.

In parametric statistics, the variances of two populations are compared using a function of the ratio of the two sample variances as the test statistic. This statistic is sensitive to scale differences irrespective of location differences since sample variances are calculated as functions of deviations from the sample mean. If the two populations are both normal distributions, this statistic follows the $F$ distribution (and hence is called the $F$ test). Unfortunately, the test is not robust against violation of the normality assumption, and hence nonparametric tests are frequently preferable even though they are not as free of assumptions as we would like.

If one of the nonparametric inference procedures is applicable, it might be chosen over the classical $F$ test in any of the following circumstances:

1. The sample sizes are quite small.
2. The only data available are measured on an ordinal scale.
3. Little or nothing can be assumed about the shape of the probability distributions.
4. The distributions are unknown, but thought to bear little resemblance to the normal curve.

Comparisons of the power of the nonparametric and parametric tests can be made only under specific population distribution assumptions. The Siegel-Tukey test and Sukhatme test are asymptotically equivalent, and hence their large sample properties are the same. For either test, the asymptotic efficiency relative to the $F$ test is $6/\pi^2 = .608$ for the normal distribution, and .600 for the continuous uniform distribution. The ARE of the ranklike tests relative to the $F$ test depends on the value of $w$. It ranges from .304 for $w = 2$ to a limit of .955 as $w \to \infty$ when the distributions are normal.

In many two-sample situations, we may be interested in making simultaneous comparisons of location and scale. Then we could use the procedures of this chapter in conjunction with the methods of Section 2 of Chapter 4. If we are interested in any type of difference between populations, whether it is location, scale, or something else, the general two-sample procedures of Chapter 6 are more appropriate.

## PROBLEMS

5.1   A company manufactures and installs central heating and cooling units. Based on a cost analysis, they have decided it is more economical to purchase thermostats than to produce them. The quality of a thermostat is judged by the amount of variation between actual temperature and the reading on the thermostat. The company obtained a random sample of 10 thermostats from each of two prospective suppliers and placed them all in a test chamber with temperature controlled at 72°F. The readings for the thermostats are as follows:

    Company A:   72.0, 72.4, 73.1, 72.5, 71.3, 72.0, 71.5, 72.0, 72.0, 72.5
    Company B:   71.4, 73.5, 72.0, 70.8, 73.2, 72.6, 70.8, 70.9, 73.8, 71.5.

Which company produces thermostats with less variability?

5.2   The psychology departments in two universities have accepted seven and nine applicants respectively for graduate study next fall. The scores of these students on the Graduate Record Examination were reported as follows:

    University X:   1200, 1220, 1300, 1170, 1080, 1110, 1130
    University Y:   1200, 1170, 1000, 1010, 980, 1400, 1430, 1390, 970.

A simple calculation will show that the median scores for the two universities are identical and that the mean scores differ only by one point. Use the Siegel-Tukey test to see if there is any significant difference between variability of the two sets of raw scores.

5.3   Assume that the specified weights in kilograms of the raw materials of distributors A and B in Example 2.2 are both equal to 16.0. Use the Mann-Whitney-Wilcoxon test on the absolute values of the respective differences to test for difference in variability.

5.4   A private investor has decided to invest a fixed sum of money in a mutual fund. He has investigated the various mutual companies, and narrowed the field down to two different funds. One invests heavily in the retailing industry, while the other invests heavily in utilities. The average earnings per share (EPS) ratios for these two industries are approximately equal. Since the investor feels that the EPS ratio provides a valid reference point in terms of anticipated appreciation in the value of the stock as well as dividends paid, the investor wishes to base his decision between the two mutual funds on the variances of the EPS ratios. His reasoning is that the industry with the smaller variability in EPS ratios would be a safer investment for the mutual company (and hence for his funds) because extreme values of EPS on the lower side of the mean in the distribution are more likely to be negative in the distribution with greater spread. *The Fortune Directory* (published annually by *Fortune* magazine) lists the EPS ratios for the 500 largest U.S. industrial corporations as well as the 50 largest for some specific industries. The retailing and utility industries are included; "largest" is measured by net sales for retailing companies and by total assets for utility companies. Using a table of random numbers, random samples of 10 companies were taken from each of these lists for the current year. The EPS ratios for these companies are as follows:

$X$ (Utilities):    1.16, 1.22, 1.48, 1.78, 1.88, 1.97, 2.28, 2.61, 2.80, 2.95
$Y$ (Retailing):    0.97, 1.16, 1.37, 1.45, 1.80, 1.90, 2.07, 3.00, 3.01, 3.08.

The investor feels that if there is a difference in variability, the retailing industry will have the larger spread. See if these data support his view.

5.5    Barnett and Youden (1970) define the precision of a method of chemical analysis as the day-to-day reproducibility of results since patients are usually tested no more frequently than once a day. The precision of a "kit" is as important as its accuracy. Suppose that two different glucose kits have been established as accurate; then we can assume that the median values determined by these test methods are equal to each other and to the true "assay" value. These kits produced the following values of milligrams per 100 milliliters in analyses of 16 samples of a patient's serum. Compare the precision.

Kit A:    85, 97, 94, 86, 91, 88, 96, 87
Kit B:    81, 99, 100, 102, 79, 93, 90, 105.

5.6    A laboratory sample of blood, known to contain 190.0 parts of serum cholesterol, was split; a portion was given to each of two different clinical laboratories. Laboratory A made six separate analyses of serum cholesterol in their sample, and laboratory B made eight. Find a confidence interval estimate at level approximately .90 of the ratio of scale parameters, making whatever assumptions seem reasonable.

Laboratory A:    189.0, 191.3, 188.8, 187.2, 192.5, 190.1
Laboratory B:    186.1, 184.3, 195.1, 193.2, 196.0, 191.8, 185.2, 189.4.

5.7    Jung and Parekh (1970, p. 816) reported an improved method using a new detergent for direct determination of serum iron. The two common methods involve large serum samples plus slow and tedious analysis to avoid loss of iron and production of turbidity. The new method eliminates the deproteinization step and requires smaller samples, and would be quite useful if its precision were found to be comparable to other methods. Hyland Control Sera known to contain 105 micrograms of serum iron per 100 ml were analyzed 20 times by the Jung-Parekh method and 20 times by a commonly used (Ramsay) method. The data below show the amounts of serum iron in micrograms as detected in each analysis. Test the data for precision of method using the null hypothesis that there is no loss in precision and the one-sided alternative that the Jung-Parekh method is a more precise determinator because it produces a smaller spread in the measurements.

| Common | | Jung-Parekh | |
|---|---|---|---|
| 111 | 101 | 96 | 107 |
| 107 | 96 | 108 | 108 |
| 100 | *97 | 103 | 106 |
| 99 | 102 | 104 | 98 |
| 102 | 107 | 114 | 105 |
| 106 | 113 | 114 | 103 |
| 109 | 116 | 113 | 110 |
| 108 | 113 | 108 | 105 |
| 104 | 110 | 106 | 104 |
| 99 | 98 | 99 | 100 |

5.8　Two types of wire are known to have the same average resistance. Low variability of resistance is very important because of its effect on durability. Determine whether any significant difference in variability is exhibited by the data below.

Wire A:　.121, .125, .124, .122, .123, .124, .122, .124
Wire B:　.118, .120, .124, .120, .122, .125, .126, .123.

5.9　Tread wear of tires is affected by many factors besides miles or hours of use. The only way to evaluate wear is to measure the depth of penetration at fixed points in the groove of the tire. In order to compare the precision of two measuring instruments, an inspector is asked to make nine independent measurements with each, all at the same point on the tire. Do the two instruments appear to have the same variation?

| Instrument 1 | | | Instrument 2 | | |
|---|---|---|---|---|---|
| 110 | 115 | 123 | 117 | 121 | 124 |
| 125 | 113 | 126 | 114 | 116 | 121 |
| 131 | 119 | 112 | 119 | 118 | 120 |

5.10　A manufacturer of cloth is concerned about the variation in shrinkage of his polyester-cotton blend fabric. He suspects that the average percent of shrinkage will be the same when washed at any water temperature, but higher temperatures may cause greater variation in shrinkage. Tests were made by washing identical samples of the fabric in an ordinary machine, 8 at the warm-water setting and 10 at the hot-water setting. Do the data below on percent shrinkage support the manufacturer's suspicion?

| Warm Water | | | Hot Water | | | |
|---|---|---|---|---|---|---|
| 1.21 | 1.15 | 1.23 | 1.15 | 1.21 | 1.26 | 1.27 |
| 1.19 | 1.18 | 1.17 | 1.25 | 1.17 | 1.19 | |
| 1.20 | 1.24 | | 1.30 | 1.14 | 1.15 | |

5.11 Finney (1952, p. 23) also reports data (see Example 3.2) on tolerance of cats for a preparation of ouabain, which is not identical to strophanthus but is similar. An independent group of nine cats were given doses of ouabain until they died; the dosages below are recorded in 0.1 mg per kg body weight of cat. Find confidence interval estimates, each with confidence coefficient .910, of the potency of ouabain relative to each of the tinctures of strophanthus.

A:  1.55, 1.58, 1.71, 1.44, 1.24, 1.89, 2.34
B:  2.42, 1.85, 2.00, 2.27, 1.70, 1.47, 2.20
Ouabain:  5.23, 9.91, 4.76, 6.51, 6.68, 5.76, 4.93, 4.58, 6.69

5.12 The article by Fix (1949, p. 250) reports the following annual figures for minimum time interval in minutes between eruptions of Old Faithful. Divide the data into two groups as of the year 1930, that is, "before 1930" and "after 1930." Use the method for positive variables to find a confidence interval estimate, with confidence coefficient .918, of the ratio of minimum time interval "after 1930" relative to "before 1930."

| Year | Minimum Time | Year | Minimum Time |
|---|---|---|---|
| 1872 | 65.00 | 1932 | 38.00 |
| 1873 | 52.08 | 1937 | 42.97 |
| 1877 | 62.00 | 1939 | 40.00 |
| 1878 | 54.07 | 1940 | 39.00 |
| 1928 | 57.50 | 1941 | 38.00 |
| | | 1943 | 41.00 |
| | | 1945 | 39.00 |
| | | 1946 | 34.00 |
| | | 1947 | 39.00 |

5.13 The National Defense Plant at Dallas, a producer of gunpowder charges for large naval guns, was asked to invest gate the consistency of charges used aboard ships in the South Pacific. The guns of interest ranged from 8 to 12 inches; larger guns have larger maximum ranges. The type of 8-inch gun used at that time was set for a range of 1500 meters. However, when 100 rounds were fired, the average distance was 1475 meters, with a standard deviation of 38. With a shell burst radius of, say, 30 meters, many shots would be off target. If the standard deviation were smaller, the shells would fall closer to their mark and thereby produce fewer misses caused by the charges. In another experiment, the 8-inch gun presently used is compared with an 8-inch gun from another manufacturer by firing 20 rounds from each gun. These data are shown below. Assuming that the average distance is the same for both types, would these data result in a recommendation that the navy change suppliers?

| Observed Distance of Exploding Charge for | | | |
|---|---|---|---|
| Present Type | | Other Type | |
| 1401 | 1478 | 1415 | 1476 |
| 1405 | 1480 | 1418 | 1479 |
| 1408 | 1482 | 1423 | 1483 |
| 1410 | 1485 | 1427 | 1487 |
| 1414 | 1491 | 1431 | 1490 |
| 1420 | 1497 | 1439 | 1493 |
| 1428 | 1503 | 1452 | 1498 |
| 1436 | 1521 | 1463 | 1503 |
| 1450 | 1542 | 1469 | 1505 |
| 1470 | 1550 | 1472 | 1520 |

5.14 The median salaries in thousands of dollars of persons in selected professions in 1968 were reported in Schachter, Cohen, and Goldstein (1972, p. 126) as follows:

| | | | |
|---|---|---|---|
| Economics | 15.0 | Biology | 13.0 |
| Statistics | 14.9 | Earth Science | 12.9 |
| Physics | 14.0 | Anthropology | 12.7 |
| Chemistry | 13.5 | Sociology | 12.0 |
| Psychology | 13.2 | Linguistics | 11.5 |
| Mathematics | 13.0 | Agriculture | 11.0 |

Assuming that these data are based on random sampling procedures, do they support the notion that salaries in the natural science professions (include statistics in this group) are less variable than salaries in the other six professions considered?

# 6

General Distribution Tests for
Two or More Independent Samples

## 1 Introduction

In Chapters 4 and 5 we studied methods for testing the null hypothesis that two populations are identical. These tests were described as especially sensitive to differences in location and scale, respectively. In order to emphasize this property, these formulations were called the location and scale problems, and the hypothesis sets were written in terms of the corresponding appropriate parameters. Sometimes there is no particular type of difference between the populations that is of special interest, or the information about the situation is not sufficient to provide any suspicion about what type of difference is likely to exist. As we have already noted and illustrated, a test that is sensitive primarily to location differences may overlook a scale difference, or some other type of difference, and vice versa. Thus, although the Mann-Whitney-Wilcoxon and Siegel-Tukey tests are completely valid for the general alternative of nonidentical distributions, other tests may be more appropriate.

The two-sample tests that we study in Section 2 of this chapter are sensitive not just to location differences, or not just to spread differences, but to differences throughout the distribution. Consequently, these tests compare the entire distributions of the two samples. Accordingly, they are called general two-sample distribution tests. Another appropriate term is two-sample goodness-of-fit tests, since the test statistics measure the compatibility between two sets of relative frequencies or two empirical distribution functions. Since the test statistics are natural extensions of the chi-square and Kolmogorov-Smirnov goodness-of-fit test statistics in Chapter 2, these tests are given the same names. In each case here the second sample plays the role of the hypothesized distribution in the one-sample case. The two-sample chi-square test is appropriate for the null hypothesis of identical distributions if the data to be analyzed are count data, and the Kolmogorov-Smirnov two-sample test applies if the data are measured on at least an ordinal scale. Although the latter test is exactly valid only for populations that are continuous, it provides a conservative test for discrete populations.

In Section 3 of this chapter we consider the hypothesis of $k$ identical populations, for $k > 2$, where the data consist of $k$ mutually independent random samples. The Kruskal-Wallis test presented in Section 3 of Chapter 4 is also valid for this null hypothesis. However, that test is especially sensitive to differences in location and, accordingly, was treated as a location test. The tests covered here are general $k$-sample tests; they compare the entire distributions. Thus they are $k$-sample extensions of the two-sample goodness-of-fit tests. A special case of the $k$-sample problem arises when the samples are drawn independently from $k$ dichotomous populations. Since the number of successes in each sample follows the binomial distribution, the $k$ populations could differ only with respect to the parameter representing the probability of "success." This special case is called a test for the equality of $k$ proportions.

## 2   Two Independent Samples

### 2.1   Chi-Square Two-Sample Test

We introduce the two-sample chi-square goodness-of-fit test using the following simple example.

**example   2.1**

A supplier of television picture tubes stocks two different brands. He has been following the practice of deciding how many tubes to order each quarter and then dividing the order equally between the two brands. He is now making a study of the demand pattern of these brands by quarters in an attempt to plan his inventory better. This study is to be based on a total sample of 550 orders. For each quarter, he takes a random sample of orders with the sample size proportional to quarterly demand. For example, since first-quarter demand ordinarily runs about 20 percent of yearly demand, the sample for the first quarter consists of .20(550) = 110 orders. Suppose that the data in Table 2.1 represent the number of tubes of each brand sold in the quarterly samples. Test the null hypothesis that the distribution of number sold per quarter is the same for the two brands.

TABLE 2.1

| Quarter | Brand A | Brand B | Total |
|---------|---------|---------|-------|
| 1 | 40 | 70 | 110 |
| 2 | 60 | 70 | 130 |
| 3 | 60 | 70 | 130 |
| 4 | 90 | 90 | 180 |
| Total: | 250 | 300 | 550 |

SOLUTION

The total demand in the sample is 550 tubes, with 110 of those in the first quarter. The first quarter then constitutes the proportion $110/550 = .20$ of the total demand. In other words, during this year the probability that a tube was sold during the first quarter is .20. If the demand function is the same for the two brands, .20 should be the probability that a brand A tube was sold during the first quarter, and also for a brand B tube. Since the number of brand A tubes sold in the total sample is 250, the expected number sold during the first quarter is $.20(250) = 50$. Similarly, the expected first quarter demand for brand B is $.20(300) = 60$. Such quantities are calculated for each quarter, as shown in Table 2.2. The total expected demand for each brand must be the same as the observed totals, here 250 and 300. Thus, those expected frequencies indicated by an asterisk could have been found by subtraction. They are included here only for the purpose of illustration. (Note that the four proportions sum to 1.0001, rather than 1.0000. This discrepancy is due entirely to rounding error.)

TABLE 2.2

| Quarter | Proportion of Total Parts Demanded | Expected Number of Parts Demanded | | Total |
|---------|-----------------------------------|-----------------------------------|---|-------|
| | | Brand A | Brand B | |
| 1 | $110/550 = .2000$ | $.2000(250) = 50.0$ | $.2000(300) = 60.0^*$ | 110 |
| 2 | $130/550 = .2364$ | $.2364(250) = 59.1$ | $.2364(300) = 70.9^*$ | 130 |
| 3 | $130/550 = .2364$ | $.2364(250) = 59.1$ | $.2364(300) = 70.9^*$ | 130 |
| 4 | $180/550 = .3273$ | $.3273(250) = 81.8^*$ | $.3273(300) = 98.2^*$ | 180 |
| Total: | 1.0001 | 250.0 | 300.0 | 550 |

We call brand A and brand B samples 1 and 2 respectively. The observed frequencies by quarters are labeled accordingly as $f_1$ and $f_2$, and similarly the respective expected frequencies are $e_1$ and $e_2$. These quantities are then entered in a table, such as Table 2.3. If the null hypothesis of identical distributions is true, the corresponding observed and expected frequencies for each brand in each quarter should be in close agreement. For brand A in quarter 1, 40 parts were demanded, whereas 50 were expected. If the difference $(40 - 50)$ between these quantities is squared and divided by the expected number 50, that is $(40 - 50)^2/50 = 2.00$, we have a measure of the disagreement between the actual demand and the hypothetical demand. Such quantities are calculated for each of the four quarters for brands A and B. Table 2.3 shows all the necessary calculations. Note that the sum of each $(f - e)$ column equals zero. This is always true since $\Sigma f = \Sigma e$ (except possibly because of rounding errors). The eight individual measures of disagreement are then summed to obtain a single overall measure, which we call $Q$.

**TABLE 2.3**

| | Brand A | | | | Brand B | | | |
|---|---|---|---|---|---|---|---|---|
| Quarter | $f_1$ | $e_1$ | $(f_1 - e_1)$ | $\dfrac{(f_1 - e_1)^2}{e_1}$ | $f_2$ | $e_2$ | $(f_2 - e_2)$ | $\dfrac{(f_2 - e_2)^2}{e_2}$ |
| 1 | 40 | 50.0 | −10.0 | 2.00 | 70 | 60.0 | 10.0 | 1.67 |
| 2 | 60 | 59.1 | .9 | .01 | 70 | 70.9 | − .9 | .01 |
| 3 | 60 | 59.1 | .9 | .01 | 70 | 70.9 | − .9 | .01 |
| 4 | 90 | 81.8 | 8.2 | .82 | 90 | 98.2 | − 8.2 | .68 |
| Total: | 250 | 250.0 | 0.0 | 2.84 | 300 | 300.0 | 0.0 | 2.37 |

$$Q = 2.84 + 2.37 = 5.21$$

The $P$-value for $Q$ is found as a right-tail probability from Table B, the chi-square table. The correct number of degrees of freedom here is df $= 3$, because it is necessary to calculate only three of the expected frequencies in order to know all of them. The probability is smaller than .20 but larger than .10. This $P$-value is large enough to conclude that the null hypothesis is supported by these data. Note that we are concluding that the distribution of quarterly demand is the same for brands A and B, not that the demand is the same for the four quarters. It is evident that there is some seasonal variation for this product irrespective of the brand, with peak demand occurring in the fourth quarter.

We now present the chi-square test for a general problem following the usual format.

*Inference situation*   The hypothesis to be tested is that two mutually independent random samples are drawn from identical populations, and the alternative is that the populations are not identical.

> $H$:   The populations are identical
> $A$:   The populations differ in some way.

If we let $F_1(x)$ and $F_2(x)$ denote the respective population distributions, this hypothesis set can be written in symbols as

> $H$:   $F_1(x) = F_2(x)$      for all $x$
> $A$:   $F_1(x) \neq F_2(x)$      for some $x$.

*Sampling situation*   The data consist of two mutually independent random samples, of sizes $n_1$ and $n_2$. Each set of observations is either (a) categorical data, classified into $r$ categories, $r \geq 2$, which do not overlap and cover all classification possibilities, or (b) quantitative data of any sort that are, or can be, grouped into $r$ nonoverlapping numerical categories. These categories may arise naturally due to the nature of the variable, or they may be designated by the investigator.

In either case, the data to be analyzed are two independent sets of count data. The counts represent the numbers of observations from each sample that are classified in each of the $r$ categories. The $r$ categories must be the same for both of the data sets. We denote these counts by $f_{11}, f_{21}, \ldots, f_{r1}$ for the first sample, and $f_{12}, f_{22}, \ldots, f_{r2}$ for the second sample, as shown below.

| Observed Frequencies | |
| :---: | :---: |
| Sample 1 | Sample 2 |
| $f_{11}$ | $f_{12}$ |
| $\cdot$ | $\cdot$ |
| $\cdot$ | $\cdot$ |
| $\cdot$ | $\cdot$ |
| $f_{r1}$ | $f_{r2}$ |

Specifically, $f_{ij}$ is the number of observations in sample $j$ that are classified in the category labeled $i$, for $i = 1, 2, \ldots, r$ and $j = 1, 2$. The total number of observations is $\Sigma f_1 = n_1$ and $\Sigma f_2 = n_2$ for each sample, and the combined total is $N = n_1 + n_2$.

*Rationale*   We test the null hypothesis by comparing each observed frequency with the frequency that would be expected if the populations are identical, that is, if $H$ is true. These expected frequencies are denoted by

$e_{11}, e_{21}, \ldots, e_{r1}$ for the first population and $e_{21}, e_{22}, \ldots, e_{r2}$ for the second population, as shown below.

| Expected Frequencies | |
|:---:|:---:|
| Sample 1 | Sample 2 |
| $e_{11}$ | $e_{12}$ |
| . | . |
| . | . |
| . | . |
| $e_{r1}$ | $e_{r2}$ |

Although the expected frequencies are not specified by $H$, they can be estimated from the data by the following procedure. Let $f_{i.}$ denote the total number of observations in the $i$th category; that is,

$$f_{i.} = f_{i1} + f_{i2}.$$

If $H$ is true, the probability that an observation in the first sample is classified in category $i$ is estimated to be

$$\frac{f_{i.}}{N}. \tag{2.1}$$

Since there are a total of $n_1$ observations in the first sample, the *expected number* of observations in the first sample that are in category $i$ is estimated by

$$e_{i1} = n_1 \frac{f_{i.}}{N}. \tag{2.2}$$

The corresponding quantity in the second sample can be found either by direct calculation or by subtraction, since

$$e_{i2} = n_2 \frac{f_{i.}}{N} = (N - n_1) \frac{f_{i.}}{N} = f_{i.} - e_{i1}.$$

When the data are cast in a table such as Table 2.4, it is easy to see that each expected frequency could be found by taking the product of its respective row total and column total and dividing the result by the combined total sample size $N$. Note also that because the row and column totals remain fixed, only $(r - 1)$ values of $e_{i1}$ need be calculated directly; the other value of $e_1$ can be found by subtraction, as can all the values of $e_2$.

**TABLE 2.4**

| Category | Observed Frequencies | | Total | Expected Frequencies | |
| | $f_1$ | $f_2$ | | $e_1$ | $e_2$ |
|---|---|---|---|---|---|
| 1 | $f_{11}$ | $f_{12}$ | $f_{1.}$ | $n_1 f_{1.}/N$ | $f_{1.} - e_{11}$ |
| 2 | $f_{21}$ | $f_{22}$ | $f_{2.}$ | $n_1 f_{2.}/N$ | $f_{2.} - e_{12}$ |
| . | . | . | . | . | . |
| . | . | . | . | . | . |
| . | . | . | . | . | . |
| $r$ | $f_{r1}$ | $f_{r2}$ | $f_{r.}$ | $n_1 f_{r.}/N$ | $f_{r.} - e_{1r}$ |
| Total: | $n_1$ | $n_2$ | $N$ | $n_1$ | $n_2$ |

If the null hypothesis of identical populations is true, there should be close agreement between the corresponding observed and expected frequencies for each of the two samples. As in Section 2.2 of Chapter 2, we define measures of goodness of fit for the $i$th category as

$$\frac{(f_{i1} - e_{i1})^2}{e_{i1}} \quad \text{and} \quad \frac{(f_{i2} - e_{i2})^2}{e_{i2}}.$$

These individual measures are summed over all categories and added together to obtain the chi-square goodness-of-fit test statistic $Q$ given below in (2.3). A large value of $Q$ reflects an incompatibility between the two population distributions.

*Test statistic*   The chi-square two-sample goodness-of-fit test statistic is

$$Q = \sum_{i=1}^{r} \frac{(f_{i1} - e_{i1})^2}{e_{i1}} + \sum_{i=1}^{r} \frac{(f_{i2} - e_{i2})^2}{e_{i2}}, \tag{2.3}$$

which can also be written as the double sum

$$Q = \sum_{i=1}^{r} \sum_{j=1}^{2} \frac{(f_{ij} - e_{ij})^2}{e_{ij}}. \tag{2.4}$$

In order to perform the calculations needed for $Q$, the tabular presentation shown in Example 2.2, Table 2.6, is recommended.

*Finding the P-value*   The sampling distribution of $Q$ is approximated by the chi-square distribution with $r - 1$ degrees of freedom. The appropriate $P$-value is then a right-tail probability from Table B, the chi-square table, but it is always an asymptotic approximate value. The larger $n_1$ and $n_2$ are in relation to $r$, the better the approximation.

*Additional notes*   As explained in Section 2 of Chapter 2, if any expected frequencies are small, adjacent classes should be combined and the degrees of freedom reduced accordingly. However, in the two-sample case, if classes are combined in one sample, they should also be combined in the other sample so that the categories remain the same for both sets of data.

The curious student may wonder why the number of degrees of freedom here is not $2(r - 1)$ since there are really two sets of $r$ categories in the analysis. The answer is that although we are comparing $(r + r)$ sets of frequencies, a total of $r$ parameters were estimated from the data. These parameters are the probabilities of classification for each category. Since these probabilities must sum to 1, only $r - 1$ parameters are independent of each other. By analogy with the one-sample chi-square test, the degrees of freedom should be reduced by the number of independent parameters estimated. This adjustment leads to the correct number of degrees of freedom as

$$df = 2(r - 1) - (r - 1) = r - 1.$$

**example  2.2**

A consulting firm has a contract from the U.S. Government to conduct studies concerning employment of women in various industries. One facet of the study includes an analysis to determine whether the distribution of number of days absent from work for "married women" differs significantly from the corresponding distribution for "single women." Divorced, separated, or widowed women are considered in the married category if children live at their residence, and in the single category otherwise. Absences are defined as periods missed from work for reasons other than hospitalization; for example, sick leave for medical operations and pregnancy are not included, as such extended absences might bias the results.

Many sampling procedures could be used to obtain data for this study. Assume that the data in Table 2.5 were collected during a one-year period for two mutually independent random samples drawn from the respective target populations.

**TABLE 2.5**

| Number of Days Absent | Frequency of Absences Over the Year | |
|---|---|---|
| | Married Women | Single Women |
| 0–3 | 60 | 130 |
| 4–7 | 21 | 50 |
| 8–11 | 11 | 10 |
| 12–15 | 4 | 6 |
| 16–19 | 3 | 4 |
| 20 or more | 1 | 0 |
| | 100 | 200 |

**SOLUTION**
The problem is to test the following null hypothesis:

$H:$   The distribution of the number of days absent per year
is the same for both married and single women.

The calculations needed to find $Q$ are shown in Table 2.6. The categories $x$ are the intervals of number of days absent. The observed and expected frequencies for married women are denoted by $f_1$ and $e_1$, while the same quantities for single women are shown as $f_2$ and $e_2$. The categories labeled "16–19" and "20 or more" are combined for each sample because the expected frequencies for "20 or more" are, respectively, .33 and .67 for the married and single women.

**TABLE 2.6**

| $x$ | $f_1$ | $f_2$ | Total | $e_1$ | $e_2$ | $(f_1 - e_1)^2/e_1$ | $(f_2 - e_2)^2/e_2$ |
|---|---|---|---|---|---|---|---|
| 0–3 | 60 | 130 | 190 | 63.33 | 126.67 | .175 | .088 |
| 4–7 | 21 | 50 | 71 | 23.67 | 47.33 | .301 | .151 |
| 8–11 | 11 | 10 | 21 | 7.00 | 14.00 | 2.286 | 1.143 |
| 12–15 | 4 | 6 | 10 | 3.33 | 6.67 | .135 | .067 |
| 16–19 | 3 ⎱4 | 4 ⎱4 | 7 ⎱8 | 2.67 | 5.33 | .663 | .332 |
| 20 or more | 1 ⎰ | 0 ⎰ | 1 ⎰ | | | | |
| | 100 | 200 | 300 | | | 3.560 | 1.781 |

$$Q = 3.560 + 1.781 = 5.341 \qquad \text{df} = 5 - 1 = 4$$

From Table B, we find that the asymptotic approximate $P$-value is slightly more than .20 for a $Q$ as small as 5.34, and hence the null hypothesis is supported. The conclusion is that married and single women, as defined here, do not differ with respect to the distributions of number of days absent per year.

*A different rationale*   The rationale for the test statistic $Q$ was explained here in terms of comparisons between observed and expected frequencies for each category and each sample, since this rationale can easily be generalized to the situation of $k$ populations and $k$ samples, where $k > 2$. This generalization will be given in Section 3.1 of this chapter.

In the two-sample case under discussion here, a different rationale can be used. Nevertheless, the test statistic that results is identical to (2.4). Since this rationale provides further insight into the comparisons affecting the value of $Q$, and the logical argument is quite brief, it is included here.

If the null hypothesis of identical populations is true, there should be close agreement between the corresponding relative frequencies, $f_{i1}/n_1$ and $f_{i2}/n_2$, for each of the $r$ categories. If the differences between these proportions are squared, divided by a suitable quantity, and then summed, the result is an overall measure of agreement between the two sets of sample data. The divisor for the $i$th category is an estimate of the probability that an observation from either sample is classified in category $i$ under $H$. Since the two samples are from the same population under $H$, the best estimate of this probability is the combined observed frequency, $f_{i1} + f_{i2}$, divided by the total number of observations $N$. The test statistic, which is equivalent to (2.4), is

$$Q = \frac{n_1 n_2}{N} \sum_{i=1}^{r} \left( \frac{f_{i1}}{n_1} - \frac{f_{i2}}{n_2} \right)^2 \bigg/ \left( \frac{f_{i1} + f_{i2}}{N} \right).$$

### 2.2   Kolmogorov-Smirnov Two-Sample Test

Another test that can be used for the null hypothesis of two identical populations is the Kolmogorov-Smirnov two-sample test. Strictly speaking, the population distributions, though unspecified, should be continuous. If they are not, the test can still be performed, but it is conservative. The data used in this test must be measured on at least an ordinal scale, since the test statistic is a function of the difference between the empirical distribution functions of the two samples.

The Kolmogorov-Smirnov test is not only a good test for the general null hypothesis of identical distributions. It is also especially useful in situations where the Mann-Whitney-Wilcoxon or Siegel-Tukey test would ordinarily be applied, but the ordinal scale of measurement contains so few points that there are many ties (see Example 2.6). For example, if the measurements are simply low, medium, and high, the ties are bound to be extensive. Of course, the two-sample chi-square test could also be used in this situation if the frequencies are large enough. However, with ordinal data, the Kolmogorov-Smirnov test seems to make better use of the available information (note, however, that the assumption of continuous distributions may be questionable).

We introduce this test with the following example, since the situation described meets the requirements for an exactly valid test.

### example 2.3

It is well known that the distribution of a standardized chi-square variable can be approximated by the standard normal distribution if the number of degrees of freedom is large enough. In order to investigate the agreement between these two distributions for a moderate number of degrees of freedom, a computer was used to generate two independent random samples of size 8, one from the standard normal distribution and the other from the chi-square distribution with 18 degrees of freedom. The results are shown below. We use these data to illustrate how the Kolmogorov-Smirnov test can be applied to compare the distributions.

Normal sample:   $-1.91, -1.22, -.96, -.72, .14, .82, 1.45, 1.86$
Chi-square sample:   $4.90, 7.25, 8.06, 14.10, 18.30, 21.21, 23.10, 28.12$

**SOLUTION**
Before a comparison can be made, the values in the chi-square sample must be standardized, since the null hypothesis is that observations on a *standardized* chi-square variable are distributed the same as observations on a standard normal variable. The mean and standard deviation of a chi-square variable with $\nu$ degrees of freedom are $\nu$ and $\sqrt{2\nu}$. Since $\nu = 18$ here, the chi-square sample data are standardized by subtracting 18 and dividing by $\sqrt{2(18)} = 6$. The transformed chi-square data then are

$$-2.18, -1.79, -1.66, -.65, .05, .54, .85, 1.69.$$

The next step is to pool these data with the normal sample and form a single array. For each sample we then compute the value of its empirical distribution function for each of the 16 observations. If the population distributions are the same, corresponding values of the empirical distribution functions should agree reasonably well, and thus all the corresponding differences should be small. The results of these steps are shown in Table 2.7, where $S_1(x)$ and $S_2(x)$ are used to denote the empirical distribution functions of the chi-square and normal samples, respectively, for each observed value $x$.

**TABLE 2.7**

| $x$ | $S_1(x)$ | $S_2(x)$ | $|S_1(x) - S_2(x)|$ |
|---|---|---|---|
| $-2.18$ | 1/8 | 0 | 1/8 |
| $-1.91$ | 1/8 | 1/8 | 0 |
| $-1.79$ | 2/8 | 1/8 | 1/8 |
| $-1.66$ | 3/8 | 1/8 | 2/8 |
| $-1.22$ | 3/8 | 2/8 | 1/8 |
| $-.96$ | 3/8 | 3/8 | 0 |
| $-.72$ | 3/8 | 4/8 | 1/8 |
| $-.65$ | 4/8 | 4/8 | 0 |
| $.05$ | 5/8 | 4/8 | 1/8 |
| $.14$ | 5/8 | 5/8 | 0 |
| $.54$ | 6/8 | 5/8 | 1/8 |
| $.82$ | 6/8 | 6/8 | 0 |
| $.85$ | 7/8 | 6/8 | 1/8 |
| $1.45$ | 7/8 | 7/8 | 0 |
| $1.69$ | 8/8 | 7/8 | 1/8 |
| $1.86$ | 8/8 | 8/8 | 0 |

The absolute values of the deviations between the two empirical distribution functions are shown in the last column. The largest is $2/8$, and this is the value of the Kolmogorov-Smirnov test statistic, which we call $D$. The sampling distribution is given in Table D for small right-tail probabilities only. The tabled values are for the product $mnD$, where $m$ and $n$ are the respective sample sizes. Thus in this example, we enter the table with $(8)(8)(2/8) = 16$ for $m = n = 8$. Since 16 is smaller than the table entry with a probability of .283, we know that $P > .283$ here. The conclusion is that these data do support the null hypothesis of identical distributions.

We now present the general case following the usual format.

*Inference situation* (*two-sided test*)   For samples drawn independently from two populations with cumulative probability distributions $F_1(x)$ and $F_2(x)$, the hypothesis set is

$$H: \quad F_1(x) = F_2(x) \qquad \text{for all } x$$
$$A: \quad F_1(x) \neq F_2(x) \qquad \text{for some } x.$$

*Sampling situation*   The data consist of two mutually independent sets of random observations, of sizes $m$ and $n$, which are both ordinal data or data measured on at least an interval scale. The data need not be grouped, but if they are, the two samples must be grouped into identical categories that represent measurements on at least an ordinal scale.

*Rationale*   Define the two respective sample or empirical distribution functions as

$S_1(x) =$ (number of observations in the first sample that are less
$\qquad\quad$ than or equal to $x$)$/m$

$S_2(x) =$ (number of observations in the second sample that are less
$\qquad\quad$ than or equal to $x$)$/n$.

The "first sample" refers to the sample of size $m$ from $F_1(x)$. Then, for any $x$, $S_1(x)$ is the proportion of observations in the first sample that do not exceed the value $x$, and similarly for $S_2(x)$ in the second sample. If the two samples are drawn from identical populations, there should be reasonable agreement between $S_1(x)$ and $S_2(x)$ for all values of $x$. The absolute values of the differences are a measure of the disagreement. If the maximum absolute difference is small, then all differences are small.

*Test statistic*   The two-sample Kolmogorov-Smirnov goodness-of-fit test statistic (sometimes called simply the Smirnov test) is defined as the maximum, over all observed values of $x$, of the absolute values of the differences between $S_1(x)$ and $S_2(x)$, or

$$D = \text{maximum } |S_1(x) - S_2(x)|. \tag{2.5}$$

Since $S_1(x)$ and $S_2(x)$ are both step functions, it is not necessary to look at the differences in the neighborhood of each observed value, and hence we need not use the word "supremum" here.

*Finding the P-value*   The sampling distribution of the $D$ test statistic is known exactly if both populations are continuous. Since $D$ should be small if $H$ is true, the appropriate $P$-value is a right-tail probability. The first portion of Table D gives exact right-tail probabilities for $2 \leq m \leq n \leq 12$ or $m + n \leq 16$, whichever occurs first, for small $P$.

This table is entered with the value of $mnD$, which is the product of the sample sizes and the test statistic. Since the table includes only $m \leq n$, the smaller size sample should be labeled the "first sample." The second part of Table D gives the smallest values of $mnD$ for which the right-tail probability is less than or equal to .010, .020, .050, .100, and .200 (under the heading "two-sided statistic"), for $9 \leq m = n \leq 20$. For these sample sizes, the exact $P$-value in any problem can be given as an interval. The corresponding method of finding asymptotic approximate $P$-values for large samples is shown in the last part of the table; the entries here are the values of $D$, not $mnD$. Table D can also be used if the population distributions are discrete; the probabilities are then only approximate but conservative.

### example   2.4

Consider again the situation of Example 2.2. The number of days absent could well be considered a continuous random variable, since periods of absence need not be whole days, even though the results for the year would probably be reported after rounding to the nearest whole day; that is, a reported absence of "4 days" represents the interval "at least 3.5, but under 4.5 days." Let us suppose that this is the case, so that the assumptions of the Kolmogorov-Smirnov test are met. The data are analyzed here as grouped. (Problem 6.2 requests an analysis based on ungrouped data.)

### SOLUTION

The sample sizes here are not equal. Since the sample of married women has fewer observations, we call it the first sample; thus $m = 100$ and $n = 200$. The data are grouped, and we do not have information regarding the individual values within any group. However, the comparison between distributions with this test is based on the cumulative frequencies, and we do know that each observation in any group of days does not exceed the upper limit for that group. Hence we represent each category by its upper limit, which we call $x$. The first group is "0–3 days," which we represent by $x = 3.5$. The first step is to sum the consecutive observed frequencies to form the cumulative observed frequencies, denoted by $G_1(x)$ and $G_2(x)$. Then the empirical distribution functions are easily obtained as $S_1(x) = G_1(x)/m$ and $S_2(x) = G_2(x)/n$. The necessary calculations are shown in Table 2.8, from which we find that the maximum absolute difference is $D = .090$.

**TABLE 2.8**

| $x$ | $G_1(x)$ | $G_2(x)$ | $S_1(x)$ | $S_2(x)$ | $|S_1(x) - S_2(x)|$ |
|---|---|---|---|---|---|
| 3.5 | 60 | 130 | .60 | .65 | .05 |
| 7.5 | 81 | 180 | .81 | .90 | .09 |
| 11.5 | 92 | 190 | .92 | .95 | .03 |
| 15.5 | 96 | 196 | .96 | .98 | .02 |
| 19.5 | 99 | 200 | .99 | 1.00 | .01 |
| ∞ | 100 | 200 | 1.00 | 1.00 | .00 |

The sample sizes here are outside the range of Table D. Hence the $P$-value must be found by calculating the quantile points from the large sample approximation at the bottom of Table D. For $m = 100$, $n = 200$, and $N = 300$, we have $\sqrt{N/mn} = \sqrt{300/(200)(100)} = .1225$, and the quantile for $P = .20$ is

$$1.07\sqrt{\frac{N}{mn}} = 1.07(.1225) = .1311.$$

Since the observed value of $D$ is .090, which is smaller than .1311, we know that the asymptotic approximate $P$-value is $P > .20$. When these data were analyzed by the chi-square test in Example 2.2, the $P$-value was found to be .25. Hence the results are in the same range even though the two test statistics compare the distributions in different ways. (This was explained in Section 4 of Chapter 2 for the one-sample goodness-of-fit tests.)

The fact that we assumed continuous populations in this example may bother the reader. If the population distributions are actually discrete, the Kolmogorov-Smirnov test is performed in the same way but the $P$-value is conservative.

*Kolmogorov-Smirnov one-sided test*   A two-sample, but one-sided, Kolmogorov-Smirnov statistic can be defined for the same general hypothesis of identical distributions but with an alternative that states a directional difference between the two population distributions. The one-sided test is particularly useful in situations where a location (or scale) difference is of primary interest, but the data are such that no other test can be used.

*Inference situation*   The hypothesis set with a one-sided alternative is either

$$H: \quad F_1(x) = F_2(x) \qquad \text{for all } x$$
$$A_+: \quad F_1(x) > F_2(x) \qquad \text{for some } x,$$

or

$$H: \quad F_1(x) = F_2(x) \qquad \text{for all } x$$
$$A_-: \quad F_1(x) < F_2(x) \qquad \text{for some } x.$$

*Test statistics and P-values*   The Kolmogorov-Smirnov one-sided statistics are defined analogous to (3.5) and (3.6) of Chapter 2 as

$$D_+ = \text{maximum } [S_1(x) - S_2(x)], \qquad (2.6)$$
$$D_- = \text{maximum } [S_2(x) - S_1(x)]. \qquad (2.7)$$

Since $S_1(x)$ is the sample representation of $F_1(x)$, and $S_2(x)$ for $F_2(x)$, a large value of $D_+$ supports the alternative $A_+$ and a large value of $D_-$ supports $A_-$.

The appropriate $P$-values for either test statistic are right-tail probabilities of the respective sampling distributions, but these distributions are identical. Table D can also be used for either of these one-sided statistics since in the extreme tails, the $P$-value for $mnD_+$ is one-half of that given for $mnD$ in the first portion of the table and, for small $\alpha$, critical values of $mnD_+$ at level $\alpha$ are equal to critical values of $mnD$ of level $2\alpha$. This relationship between the distributions does not hold except in the extreme tails. The last row of the second part of the table gives the correct column labels for selected right-tail probabilities for the one-sided statistics as one-half of those for the two-sided statistic. The corresponding quantile points based on the large sample approximation are given for the probabilities listed on the very last line of the table. The test is exact if $F_1(x)$ and $F_2(x)$ are continuous, and is conservative for discrete populations.

**example   2.5**

The 1970 census statistics for the State of Alabama give data on the total 1970 population and the percentage change in population from 1960 to 1970 for each of the 67 counties. Relative growth rates between rural and nonrural areas are of social, economic, and political interest. The 67 counties were divided into two groups according to 1970 population size. Those with less than 25,000 population were arbitrarily called rural, and those with 25,000 or more were labeled urban. Nine rural and seven urban counties were selected from these groups using a table of random numbers. The reported population percentage changes for those counties selected are as follows:

Rural counties:   1.1, $-21.7$, $-16.3$, $-11.3$, $-10.4$, $-7.0$, $-2.0$,
                  1.9, 6.2
Urban counties:   $-2.4$, 9.9, 14.2, 18.4, 20.1, 23.1, 70.4.

If there is a difference in the distributions of relative growth rates in these two groups, it would be that the urban counties have higher rates than rural counties. We formulate this as the alternative and use a one-sided Kolmogorov-Smirnov goodness-of-fit test.

**SOLUTION**
Since the sample size is smaller for urban counties than for rural counties, the urban observations are labeled the first sample. Correspondingly, the

distribution $F_1(x)$ refers to the population of urban counties. If the urban areas have a higher rate of growth on the average than the rural areas, the proportion of urban areas with growth rates less than or equal to some number would be smaller than the same proportion for rural areas. Hence the population cumulative distribution function for urban areas $F_1(x)$ would be generally below the rural distribution $F_2(x)$; the appropriate one-sided hypothesis set is then

$$H: \quad F_1(x) = F_2(x) \qquad \text{for all } x$$
$$A_-: \quad F_1(x) < F_2(x) \qquad \text{for some } x.$$

The appropriate test statistic is (2.7), where $m = 7$ and $n = 9$.

The sample data are now combined in a single ordered sequence, as shown in Table 2.9, and $S_1(x)$ and $S_2(x)$ are computed for urban and rural counties, respectively, for each different value of $x$ observed.

**TABLE 2.9**

| $x$ | Rural $S_2(x)$ | Urban $S_1(x)$ | $S_2(x) - S_1(x)$ |
|---|---|---|---|
| $-21.7$ | 1/9 | 0 | 7/63 |
| $-16.3$ | 2/9 | 0 | 14/63 |
| $-11.3$ | 3/9 | 0 | 21/63 |
| $-10.4$ | 4/9 | 0 | 28/63 |
| $-7.0$ | 5/9 | 0 | 35/63 |
| $-2.4$ | 5/9 | 1/7 | 26/63 |
| $-2.0$ | 6/9 | 1/7 | 33/63 |
| 1.1 | 7/9 | 1/7 | 40/63 |
| 1.9 | 8/9 | 1/7 | 47/63 |
| 6.2 | 1 | 1/7 | 54/63 |
| 9.9 | 1 | 2/7 | 45/63 |
| 14.2 | 1 | 3/7 | 36/63 |
| 18.4 | 1 | 4/7 | 27/63 |
| 20.1 | 1 | 5/7 | 18/63 |
| 23.1 | 1 | 6/7 | 9/63 |
| 70.4 | 1 | 1 | 0 |

The maximum difference is $D_- = 54/63$, or $mnD_- = 54$. From Table D we find that this value has a right-tail probability of $.003/2 = .0015$ for a one-sided test, and thus the $P$-value here is $.0015$. This value is exact for the situation described. The data do not support the null hypothesis.

**example 2.6**

It is frequently claimed that members of minority groups are discriminated against for top management and other prestige positions or advancement within a profession. A random sample of 400 presently employed persons who are college graduates is taken. Their jobs are classified into the seven occupational categories defined in Example 3.5 of Chapter 4. Recall that these categories, labeled 10, 20, . . . , 70, comprise an ordinal scale with increasing numbers reflecting decreasing levels of prestige. The sample members are also classified according to whether they are members of minority groups (including females and foreigners) or not. For the data in Table 2.10 investigate the claim that members of minority groups tend to have lower prestige occupations than other employed persons, even though they have a comparable level of education.

**TABLE 2.10**

| Occupation | Minority | Other |
|:----------:|:--------:|:-----:|
| 10 | 2 | 14 |
| 20 | 5 | 30 |
| 30 | 13 | 56 |
| 40 | 22 | 73 |
| 50 | 24 | 48 |
| 60 | 25 | 37 |
| 70 | 29 | 22 |
| Total: | 120 | 280 |

**SOLUTION**
Since the total sample is random, all elements are independent, whether combined into subsamples or not. Thus the "minority" and "other" groups comprise two independent samples. The variable occupational prestige is measured on an ordinal scale. We are particularly interested in comparing the location of the two groups, but the Mann-Whitney-Wilcoxon test is not appropriate because the data contain many ties. Thus we use the Kolmogorov-Smirnov test. Since the minority group has fewer elements, we call it the first sample. The claim to be investigated predicts a particular direction of difference, and thus a one-sided test is appropriate. If more minority persons have lower prestige occupations (those indicated by the higher numbers), the minority group will tend to cumulate slower than the other group. Thus the alternative must be

$$A_-: \quad F_1(x) < F_2(x) \quad \text{for some } x,$$

and the appropriate test statistic is $D_-$. The cumulative relative frequency distributions are formed and differences obtained in Table 2.11, using the notation introduced in Example 2.4.

**TABLE 2.11**

| $x$ | $G_1(x)$ | $G_2(x)$ | $S_1(x)$ | $S_2(x)$ | $S_2(x) - S_1(x)$ |
|-----|----------|----------|----------|----------|-------------------|
| 10  | 2        | 14       | .017     | .050     | .033              |
| 20  | 7        | 44       | .058     | .157     | .099              |
| 30  | 20       | 100      | .167     | .357     | .190              |
| 40  | 42       | 173      | .350     | .618     | .268              |
| 50  | 66       | 221      | .550     | .789     | .239              |
| 60  | 91       | 258      | .758     | .922     | .164              |
| 70  | 120      | 280      | 1.000    | 1.000    | .000              |

Thus, for $m = 120$ and $n = 280$, we have $D_- = .268$. Since $m$ and $n$ are outside the range of Table D, we calculate the quantile points of $D_-$ from the asymptotic approximate values given at the end of the table. Here $\sqrt{N/mn} = \sqrt{400/(120)(280)} = .109$, and the quantile point for a one-sided $P$-value of .005 is $1.63(.109) = .178$. Since .268 is larger than .178, we know that $P < .005$ and the null hypothesis should be rejected. These data support the research hypothesis formulated as the alternative. Since it is difficult to conceive of occupational prestige as a continuous variable, this $P$-value should be considered conservative. However, the conclusion is unchanged.

# 3   $k$ Independent Samples

In this section we extend the problem to the situation where $k$ samples are drawn, one from each of $k$ populations, in a manner such that all are mutually independent. The general null hypothesis is again formulated as identical populations; that is,

$$H: \text{ The } k \text{ populations are identical.}$$

Then if $H$ is true, the $k$ random samples constitute a single random sample from the common, but unspecified, population.

The tests covered here are used primarily when the alternative is completely general, asserting simply that the populations differ. If the inference of primary interest is concerned with differences in location, the Kruskal-Wallis test covered in Section 3 of Chapter 4 is considered more sensitive. However, in situations where a location inference is desired but the data are such that the Kruskal-Wallis test should not be used (for example, if ties are very extensive), these tests provide a useful substitute.

### 3.1   Extensions of Chi-Square and Kolmogorov-Smirnov Tests

The two-sample chi-square test has a natural $k$-sample extension. The calculation of $Q$ follows the pattern suggested by (2.4), and the expected frequencies are calculated as in (2.2). Extending the previous notation in the logical way, we have $f_{ij}$ and $e_{ij}$ for the observed and expected frequencies in the $i$th category for the $j$th sample, which is of size $n_j$ for $i = 1, 2, \ldots, r$; $j = 1, 2, \ldots, k$. The formulas here are then

$$e_{ij} = n_j\left(\frac{f_{i\cdot}}{N}\right) \tag{3.1}$$

$$Q = \sum_{i=1}^{r} \sum_{j=1}^{k} \frac{(f_{ij} - e_{ij})^2}{e_{ij}}. \tag{3.2}$$

Again, the sampling distribution of $Q$ is approximately chi-square, with degrees of freedom df $= (k - 1)(r - 1)$. This test will not be discussed further here since it is really just a special case of the chi-square test of independence, which is discussed fully in Chapter 7.

The two-sided Kolmogorov-Smirnov test also has a natural $k$-sample analogue, which is sometimes called the Birnbaum-Hall test. The sample cumulative distribution functions are formed for each of the $k$ samples. The test statistic is defined as the maximum of the absolute differences, at corresponding values of $x$, between the values of all possible pairs of these cumulative distributions; that is,

$$\underset{x, i, j}{\text{maximum}} \, |S_i(x) - S_j(x)|. \tag{3.3}$$

The exact sampling distribution of this test statistic can be determined mathematically if all populations are assumed continuous, but tables have been prepared for few combinations of $k$ and the sample sizes. (See Birnbaum and Hall, 1960, for $k = 3$ and all sample sizes equal.) Because of these practical limitations on usefulness, the test will not be discussed further in this book.

### 3.2 Test for Equality of *k* Proportions

*Inference situation*　Suppose there are $k$ populations, where each element of each population is classified into one of two fixed categories, which we call "success" and "failure" for convenience. In other words, each population is a dichotomous (or Bernoulli) population. Let the parameters $p_1, p_2, \ldots, p_k$ denote the respective probabilities of success in the $k$ populations. The hypothesis set to be tested is

*H:*　$p_1 = p_2 = \cdots = p_k$
*A:*　At least two of the $p_j$'s differ from each other.

A dichotomous population is completely specified by the one parameter that represents the probability of success. Hence, the null hypothesis above, which states that the *k parameters* are all equal, is equivalent to stating that the *k population distributions* are equal.

*Sampling situation*   The data consist of $k$ mutually independent random samples, of sizes $n_1, n_2, \ldots, n_k$, one drawn from each of the $k$ populations. Each sample then is nominal data, classified into the two categories labeled success and failure. The data to be analyzed are enumerative, representing the numbers of successes in each of the $k$ samples; we denote these observed frequencies by $f_1, f_2, \ldots, f_k$.

*Rationale*   If the probability of success is the same in all $k$ populations, the logical sample estimate of this common probability, denoted by $\bar{p}$, is the total number of successes observed divided by the total number of observations, or

$$\bar{p} = \frac{\Sigma f}{\Sigma n}. \tag{3.4}$$

The corresponding estimate of the number of successes in sample number $j$ then is

$$n_j \bar{p}. \tag{3.5}$$

Therefore, if $H$ is true, there should be close agreement between the observed frequency of success $f_j$, and the expected frequency of success $n_j \bar{p}$, for each of the $k$ samples. As with other tests that compare observed and expected frequencies, the respective deviations are squared, divided by some quantity, and summed to get an overall measure of agreement as the test statistic. A large value of this statistic reflects an incompatibility between the $k$ population parameters, and hence between the $k$ populations.

*Test statistic*   The chi-square test statistic for the equality of $k$ proportions is

$$Q = \sum_{j=1}^{k} \frac{(f_j - n_j \bar{p})^2}{n_j \bar{p}(1 - \bar{p})}, \tag{3.6}$$

which can also be written as

$$Q = \frac{1}{\bar{p}(1 - \bar{p})} \sum_{j=1}^{k} n_j \left(\frac{f_j}{n_j} - \bar{p}\right)^2. \tag{3.7}$$

*Finding the P-value*   The sampling distribution of $Q$ is, for large samples, well approximated by the chi-square distribution, with $k - 1$ degrees of freedom. The $P$-value is a right-tail probability from the chi-square table, but it is always asymptotic approximate.

*Additional notes*   As with other tests based on a chi-square statistic, the expected frequencies should not be too small. Samples should be combined in accordance with the rules of thumb discussed in Section 2 of Chapter 2, and the null hypothesis must be modified to reflect the combinations. Some of the parameters in $H$ then are for combined populations, and $H$ may no longer reflect the question of real interest.

This test for the equality of $k$ proportions is simply a special case of the chi-square $k$-sample goodness-of-fit test covered in Section 3.1. Here we have only the two categories of observation, success and failure, so that $r = 2$. The test statistic in (3.6) is equivalent to (3.2) when $r = 2$. To show this, we define a success as the first category and a failure as the second. Then in (3.2) the observed frequencies are $f_{1j} = f_j$, the number of successes, and $f_{2j} = n_j - f_j$, the number of failures. The expected frequencies are $e_{1j} = n_j \bar{p}$ and $e_{2j} = n_j(1 - \bar{p})$, the estimated number of successes and failures, respectively. When these values are substituted in (3.2), we obtain

$$Q = \sum_{j=1}^{k} \left\{ \frac{(f_j - n_j \bar{p})^2}{n_j \bar{p}} + \frac{[(n_j - f_j) - n_j(1 - \bar{p})]^2}{n_j(1 - \bar{p})} \right\}.$$

If the two terms in braces are combined using their least common denominator $n_j \bar{p}(1 - \bar{p})$, the result after simplification is exactly the same as (3.6). The algebraic manipulations are straightforward, but will not be illustrated here.

The reader may also have noticed some similarity between the situation here and the two-sample chi-square goodness-of-fit test in Section 2.1. The analysis is indeed identical except that the roles of categories and samples have been interchanged. The test statistic in (2.3) is equivalent to (3.6). This can be shown by substituting in (2.3) the appropriate values, namely $e_{i1} = n_i \bar{p}$, $f_{i2} = n_i - f_{i1}$, $e_{i2} = n_i - e_{i1}$, and $r = k$.

**example 3.1**

A company is planning to conduct an extensive house-to-house sales campaign in a certain metropolitan area. Since similar campaigns are already under way in other metropolitan areas of the state, funds for this additional project are at a bare minimum. It is therefore considered essential that the campaign be started in the district of the area that can be expected to yield the largest number of sales relative to number of calls — that is, the greatest sale-to-call ratio. The profits from this district could then be utilized to finance the campaign in the other districts of the area. In order to determine where the campaign should be initiated, this area is divided into three sales districts, which are homogeneous with respect to number of dwelling units. One hundred units are randomly selected in each district for a pilot study. The numbers of sales for each district are as follows.

| District: | 1 | 2 | 3 |
|---|---|---|---|
| Sales: | 40 | 25 | 20 |
| Calls: | 100 | 100 | 100 |

Determine whether there is a significant difference among the proportions of sales made.

**SOLUTION**

The null hypothesis is $H$: $p_1 = p_2 = p_3$. The given data are

$$f_1 = 40 \qquad f_2 = 25 \qquad f_3 = 20$$
$$n_1 = 100 \qquad n_2 = 100 \qquad n_3 = 100.$$

From (3.4) we calculate the estimate of the common probability of a sale as

$$\bar{p} = \frac{40 + 25 + 20}{300} = \frac{85}{300} = .283,$$

and from (3.5) the expected numbers of sales in each sample are estimated to be

$$\text{Sample 1:} \quad n_1 \bar{p} = 28.3$$
$$\text{Sample 2:} \quad n_2 \bar{p} = 28.3$$
$$\text{Sample 3:} \quad n_3 \bar{p} = 28.3.$$

The test statistic is calculated from (3.6) as

$$Q = \frac{(40 - 28.3)^2}{(28.3)(.717)} + \frac{(25 - 28.3)^2}{(28.3)(.717)} + \frac{(20 - 28.3)^2}{(28.3)(.717)}$$

$$= \frac{(11.7)^2}{20.3} + \frac{(-3.3)^2}{20.3} + \frac{(-8.3)^2}{20.3}$$

$$= 6.74 + 0.54 + 3.39$$

$$= 10.67.$$

From Table B, with df $= 3 - 1 = 2$, the right-tail probability for 10.67 is between .001 and .010. We report the asymptotic approximate $P$-value as $P < .010$. The conclusion is that the data do not support $H$, and there is a significant difference between the proportions of sales made in the three districts. Since district 1 had the highest proportion of sales, the campaign should be initiated in that district.

# 4   Summary

The two-sample tests covered in Section 2 of this chapter provide an alternative approach to testing the null hypothesis of identical distributions. They are primarily useful when the important difference between populations is not in location, or not in scale, but rather in any kind of difference, or when neither a location model nor a scale model seems to adequately represent the relationship between the forms of the population distributions. Both the chi-square and Kolmogorov-Smirnov tests should be considered general tests that are somewhat sensitive to all kinds of differences between populations and not particularly sensitive to any specific type of difference.

One of these tests might be used in preference to the Mann-Whitney-Wilcoxon test in a location problem, or the Siegel-Tukey test in a scale problem, only if for some reason these latter methods are not appropriate for the situation at hand.

The properties and relative merits of the chi-square and Kolmogorov-Smirnov two-sample tests are essentially the same as in the one-sample case; the reader may wish to review Section 4 of Chapter 2 at this point. The major differences are that chi-square is designed for use with discrete distributions and is always an approximate test, whereas the Kolmogorov-Smirnov test is exactly valid only when the distributions are continuous.

The test of Section 3.2 is applicable only to the situation of dichotomous populations and the null hypothesis that each population has the same unspecified parameter value. It is a special test for a special problem, and has essentially no practical competitors. Nevertheless, it is important to remember that the procedure is always approximate because it is based on an asymptotic distribution. As usual with chi-square tests, the accuracy of the approximation depends on many factors; no single rule of thumb can be given to say when the approximation is adequate for most practical purposes.

## PROBLEMS

6.1  Random samples of orders for refrigerator replacement parts are taken each week over a period of 50 weeks. The demand for two similar parts is distributed as follows:

|  | Number of Weeks | |
| Weekly Demand | Part A | Part B |
| --- | --- | --- |
| 0 | 28 | 22 |
| 1 | 15 | 21 |
| 2 | 6 | 7 |
| 3 | 1 | 0 |
| More than 3 | 0 | 0 |
|  | $\overline{50}$ | $\overline{50}$ |

Test the null hypothesis that the distributions are the same.

6.2  Suppose that the data of Example 2.2 are available in ungrouped form and the results after rounding to the nearest whole day are as shown below. Use the Kolmogorov-Smirnov test for the null hypothesis that the number of days absent is the same for married and single women.

| Number of Days Absent | Frequency for Married Women | Single Women |
|---|---|---|
| 0 | 15 | 35 |
| 1 | 18 | 40 |
| 2 | 15 | 30 |
| 3 | 12 | 25 |
| 4 | 9 | 20 |
| 5 | 7 | 18 |
| 6 | 3 | 6 |
| 7 | 2 | 6 |
| 8 | 3 | 4 |
| 9 | 3 | 3 |
| 10 | 3 | 2 |
| 11 | 2 | 1 |
| 12 | 1 | 2 |
| 13 | 1 | 2 |
| 14 | 2 | 1 |
| 15 | 0 | 1 |
| 16 | 2 | 2 |
| 17 | 1 | 1 |
| 18 | 0 | 0 |
| 19 | 1 | 1 |
| 20 | 0 | 0 |
|  | $\overline{100}$ | $\overline{200}$ |

6.3  Blank (1954, p. 72) reports data on house price indexes in Cleveland and Seattle. The indexes were obtained from 3-year moving averages of prices paid for a new six-room frame house and lot, using 1929 as the base year. Test the null hypothesis that price indexes have the same distribution in these two cities for the period 1920–1930.

|  | Cleveland | Seattle |
|---|---|---|
| 1920 | 86.8 | 104.7 |
| 1921 | 87.8 | 102.9 |
| 1922 | 91.5 | 104.6 |
| 1923 | 96.3 | 101.1 |
| 1924 | 100.0 | 113.1 |
| 1925 | 102.4 | 112.9 |
| 1926 | 103.7 | 114.5 |
| 1927 | 102.4 | 106.1 |
| 1928 | 101.2 | 111.0 |
| 1929 | 100.0 | 100.0 |
| 1930 | 95.1 | 94.3 |

6.4  Bauer et al (1954, p. 49) made a statistical analysis of a study to evaluate therapy for syphilis. A large group of patients with syphilis were treated with medication containing crystalline penicillin G and observed for at least 2 years; one group had intensive follow-up as posttreatment observation while the other had only routine follow-up during this period. The therapy was evaluated by measuring percent of cases requiring retreatment at various observation periods. The data on percents are shown below. Use the Kolmogorov-Smirnov test for the null hypothesis that the distributions of retreated cases are the same for the first 24 months. Assume that there are 1250 retreated cases, with 250 receiving intensive follow-up and 1000 receiving routine follow-up. (The actual number varied according to observation period.)

| Observation Period (Months) | Cumulative Proportion of Retreated Cases | |
|:---:|:---:|:---:|
| | Intensive | Routine |
| 0–1 | .004 | .001 |
| 1–2 | .016 | .003 |
| 2–3 | .016 | .007 |
| 3–4 | .024 | .014 |
| 4–5 | .040 | .026 |
| 5–6 | .064 | .048 |
| 6–7 | .084 | .057 |
| 7–8 | .088 | .067 |
| 8–9 | .109 | .079 |
| 9–10 | .109 | .088 |
| 10–11 | .113 | .099 |
| 11–12 | .117 | .111 |
| 12–15 | .129 | .132 |
| 15–18 | .137 | .139 |
| 18–21 | .150 | .148 |
| 21–24 | .150 | .158 |

6.5  Fatigue test data from an ASTM Technical Report (1963, p. 75) are obtained as millions of revolutions to failure for test specimens in samples from lots obtained from each of two different suppliers. Test the null hypothesis that the fatigue distributions are the same for the two suppliers.

| Lot 1 | Lot 2 |
|:---:|:---:|
| 1.1 | 2.0 |
| 2.3 | 3.7 |
| 4.0 | 5.0 |
| 6.5 | 8.0 |
| 8.6 | 11.5 |
| | 13.0 |
| | 20.0 |
| | 23.5 |

6.6   A gasoline service station manager feels that the customer pattern by time of day is different on an ordinary weekday than on Saturdays. If he can substantiate this belief, he wishes to take advantage of this knowledge and plan his employees' schedules accordingly. From the data shown below, what would you advise?

| Period of Day | Average Number of Customers | |
|---|---|---|
| | Weekdays | Saturdays |
| 6 A.M. to 9 A.M. | 15 | 3 |
| 9 A.M. to 12 noon | 4 | 8 |
| 12 noon to 2 P.M. | 12 | 10 |
| 2 P.M. to 5 P.M. | 10 | 10 |
| 5 P.M. to 8 P.M. | 23 | 12 |
| 8 P.M. to 11 P.M. | 10 | 18 |

6.7   Gynecomastia is excessive development of the male mammary glands, even to the functional state. A group of 351 patients with histologically proven gynecomastia were studied by Bannayan and Hajdu (1972, p. 432) in an attempt to elucidate some clinicopathologic correlations. Of these, 234 had unilateral gynecomastia. The distributions according to age of numbers of these patients having right breast or left breast gynecomastia are shown below. Test the null hypothesis that these two distributions are the same.

| Age Group (Years) | Left Breast | Right Breast |
|---|---|---|
| 1–10 | 1 | 0 |
| 11–20 | 9 | 8 |
| 21–30 | 14 | 7 |
| 31–40 | 17 | 13 |
| 41–50 | 23 | 17 |
| 51–60 | 33 | 17 |
| 61–70 | 32 | 23 |
| 71–80 | 17 | 10 |
| 81–90 | 2 | 1 |
| Total: | 148 | 96 |

6.8   A publication of the National Center for Health Statistics, *Inpatient Health Facilities as Reported from the 1969 MFI Survey* (1972, p. 18), reported statistics from a 1969 survey of inpatient health facilities. One set of data concerned the number of beds in various kinds of nursing home facilities. The data for the two primary types of facilities, nursing care homes and personal care homes, are shown below. Test the null hypothesis that the distribution of number of beds is the same for these two facilities.

| Number of Beds | Nursing Care | Personal Care |
|---|---|---|
| Less than 5 | 81 | 407 |
| 5–9 | 193 | 1,124 |
| 10–14 | 404 | 712 |
| 15–19 | 746 | 568 |
| 20–24 | 1,173 | 311 |
| 25–49 | 3,298 | 494 |
| 50–74 | 2,380 | 111 |
| 75–99 | 1,365 | 29 |
| 100–199 | 1,592 | 32 |
| 200–299 | 176 | 3 |
| 300–499 | 55 | 1 |
| 500 or more | 21 | 0 |
| Total: | 11,484 | 3,792 |

6.9  The NCHS publication described in Problem 6.8 (1972, p. 42) also reported statistics for the distribution of the number of residents (patients) in facilities for the mentally retarded and also in all remaining "other health facilities." The data below include all types of ownership: government, proprietary, and nonprofit. Test the null hypothesis that the distributions are the same.

| Number of Residents | Mentally Retarded | Other |
|---|---|---|
| Less than 25 | 554 | 1,526 |
| 25–49 | 121 | 655 |
| 50–74 | 53 | 345 |
| 75–99 | 31 | 178 |
| 100–199 | 61 | 295 |
| 200–299 | 17 | 102 |
| 300–499 | 18 | 90 |
| 500 or more | 111 | 68 |
|  | 966 | 3,259 |

6.10 An identical aptitude test was given to all seniors in two different high schools, one in a rural area and the other in an urban area. Suppose that the question of interest is whether these two groups are homogeneous as regards test performance; that is, they have the same mean, the same variance, and so on. A table of random numbers is used to select eight students from each high school. Their scores are shown below. Perform an appropriate two-sample goodness-of-fit test to investigate the null hypothesis of identical distributions.

Urban group:  99, 85, 93, 41, 63, 75, 71, 80
Rural group:  82, 96, 75, 49, 75, 61, 63, 50.

6.11 Krumbein and Graybill (1965, p. 201) reported the results of a student project to obtain 26 independent measurements of foreshore slope of a beach on Lake Michigan near Evanston, Illinois. The slopes were measured to the nearest degree with a Brunton compass at randomly chosen locations, 13 on a storm foreshore and 13 on a nonstorm foreshore. Test the null hypothesis that the two sets of slope measurements have the same distribution.

Storm foreshore slope:    3, 5, 6, 6, 6, 6, 7, 7, 7, 7, 8, 8, 14
Nonstorm foreshore slope:    5, 7, 7, 9, 9, 9, 9, 9, 10, 11, 11, 12, 14.

6.12 Linguistics frequently provides some rather esoteric applications of statistics. Suppose that a linguist tabulated the following data on the frequencies of Czech vowel phonemes in random samples taken from two texts.

| Phonemes: | e | a | o | i | í | á | y | u | é | ý | ú | ou |
|---|---|---|---|---|---|---|---|---|---|---|---|---|
| Sample I: | 162 | 127 | 106 | 65 | 38 | 25 | 24 | 29 | 19 | 7 | 5 | 10 |
| Sample II: | 153 | 124 | 119 | 72 | 39 | 32 | 31 | 33 | 19 | 8 | 6 | 6 |

Use the chi-square test for the null hypothesis that there is no difference between the distributions of phonemes for the two samples and give your conclusions.

6.13 In order to compare the occurrence of limestone pebbles in beach gravel on Lake Michigan and Lake Ontario, suppose that 100 samples of 10 pebbles each were collected at each lakeshore. The frequency distributions of number of limestone pebbles per sample of 10 are as shown below. Test the null hypothesis that the distributions are the same.

| Number of Limestone Pebbles | Observed Frequency | |
|---|---|---|
| | Lake Michigan | Lake Ontario |
| 0 | 1 | 2 |
| 1 | 2 | 1 |
| 2 | 4 | 5 |
| 3 | 7 | 9 |
| 4 | 23 | 20 |
| 5 | 26 | 25 |
| 6 | 21 | 23 |
| 7 | 12 | 10 |
| 8 | 3 | 2 |
| 9 | 1 | 2 |
| 10 | 0 | 1 |

6.14 Two drugs are known to be effective in arresting a certain disease and safe when used for short periods of time. The longer-term side effects are still being studied. In order to investigate toxicity, 144 laboratory animals were divided into two experimental groups of 72 each. Drug A was given to the first group and drug B to the second, with equal dosage for each animal. The numbers of *survivors* after nine different intervals of time are recorded below. Test the null hypothesis that the response to the drugs is the same for the two groups, using the .05 level. (*Hint:* Convert the data in each set to a cumulative distribution of the number dying and apply the Kolmogorov-Smirnov two-sample test.)

|  | Number of Survivors | |
| --- | --- | --- |
| Number of Days | Drug A | Drug B |
| 1 | 39 | 45 |
| 2 | 36 | 43 |
| 4 | 29 | 35 |
| 8 | 21 | 25 |
| 16 | 16 | 18 |
| 24 | 12 | 14 |
| 36 | 10 | 12 |
| 48 | 8 | 8 |
| 72 | 5 | 7 |

6.15 It is frequently claimed that smoking dulls the taste buds and thereby reduces the pleasure of eating fine food. In order to compare "taste blindness" of smokers and nonsmokers, random samples of 100 of each were chosen. Each person was blindfolded, given six different foods to taste and asked to identify the primary spice ingredient (all spices used were well known). The distribution of number of correct identifications for each group is shown below. Test the null hypothesis that the distributions are the same. Do the data substantiate the claim?

| Number of Correct Identifications | Smokers | Nonsmokers |
| --- | --- | --- |
| 0 | 0 | 0 |
| 1 | 8 | 4 |
| 2 | 19 | 10 |
| 3 | 31 | 15 |
| 4 | 21 | 29 |
| 5 | 15 | 37 |
| 6 | 6 | 5 |
| | 100 | 100 |

6.16 A random sample of 70 is chosen from the population of married women still living with their husbands. Each woman completes a questionnaire about her marital life and is classified as "satisfied" or "dissatisfied" depending on her score on this questionnaire. These groups are to be compared on the basis of number of years married at the time of the test (recorded to the next full month). Are the distributions the same? Compare the raw data, and also the data after grouping into five intervals of years.

| Duration of Marriage for | | | | | | |
|---|---|---|---|---|---|---|
| Satisfied Group | | | | Dissatisfied Group | | |
| 2 | 38 | 115 | 251 | 8 | 105 | 249 |
| 10 | 41 | 121 | 273 | 23 | 112 | 253 |
| 12 | 42 | 149 | 296 | 40 | 119 | 265 |
| 14 | 49 | 160 | 304 | 45 | 126 | 289 |
| 18 | 55 | 165 | 335 | 58 | 135 | 300 |
| 20 | 63 | 177 | 376 | 69 | 146 | 305 |
| 25 | 68 | 182 | 384 | 78 | 150 | 322 |
| 29 | 71 | 195 | 401 | 85 | 175 | 341 |
| 31 | 75 | 212 | 412 | 96 | 206 | 353 |
| 35 | 99 | 245 | 476 | 98 | 241 | 365 |

6.17 Samples of five different strains of merino sheep were tested for resistance to fleece rot (Dunlop and Hayman, 1958). Test the null hypothesis that the proportions affected are the same for all five strains.

| Strain | Number of Sheep | Number Affected |
|---|---|---|
| FNP | 65 | 11 |
| MNP | 65 | 21 |
| SNP(1) | 74 | 68 |
| MPA(1) | 73 | 35 |
| MPB | 83 | 49 |

6.18 A drug firm needs a study of demand in metropolitan areas for one of its over-the-counter products. A statistician advised them to collect the data through a mail questionnaire, and to use a stratified sample, where stratification is proportional to age group. That is, if a certain age group constitutes 30 percent of the relevant population and a total sample size of 500 is desired, the sample should consist of 150 persons of that age group. The firm decided on the four age groupings shown below, determined the percent of persons in these groups in a certain metropolitan area, and sent questionnaires accordingly to a total of 500 persons. However, as is frequently the case with mail questionnaires, the rate of nonresponse was high, especially in the youngest age group. If the proportion of respondents can be considered the same in each age stratum, the data still represent a proportional stratification of the population. Determine whether such a claim can be made.

| Age Group | Percent of Population | Size of Sample | Number of Respondents |
|---|---|---|---|
| Under 20 | .30 | 150 | 90 |
| 20–34 | .30 | 150 | 60 |
| 35–54 | .30 | 150 | 75 |
| 55 or over | .10 | 50 | 35 |

6.19 An ASTM Technical Report (1963, p. 32) gives the following data from an experiment designed to compare fatigue properties of six lots of phosphor-bronze strip. A random sample of strips is drawn from each lot, with sample size proportional to the size of the lot, and the number of strips surviving $10^8$ cycles of a stress of ±25,000 pounds per square inch is recorded for each sample. Perform an appropriate statistical analysis.

| Lot | Sample Size | Number Surviving |
|---|---|---|
| 1 | 15 | 9 |
| 2 | 20 | 8 |
| 3 | 17 | 10 |
| 4 | 25 | 12 |
| 5 | 19 | 11 |
| 6 | 14 | 7 |

6.20 In a double-blind study of drugs to treat duodenal peptic ulcers, a large number of patients were divided into several groups to compare different treatments, antacid, antipepsin and anticholinergic, both alone and in combination. Antacid has long been considered the major medication for ulcers; the latter two drugs act on different digestive juices. The numbers of patients in three of the groups, and the percent who benefited from that treatment, are shown below. Does there appear to be a difference in beneficial effects of these three treatments?

| Treatment | Number | Percent Benefited |
|---|---|---|
| Antacid | 40 | 55 |
| Anticholinergic alone | 60 | 70 |
| Antipepsin, both alone and with anticholinergic | 75 | 84 |

6.21 Camping was once considered a preoccupation of sportsmen, but it now is a family recreation. As a result, the camping products industry has experienced phenomenal growth. A retail distributor of camping products is planning a spring advertising campaign. If there is some definable group or groups of persons who are more prone to purchase camping equipment than others, the distributor wishes to direct his advertising toward them. They feel that family income would be a key factor influencing purchase patterns, but many persons are reluctant to disclose income data. However, educational training of head of household indirectly groups individuals into homogeneous income classes, and information about education is easy to obtain. The company therefore interviewed 2000 persons regarding their education level and whether or not they had done any camping in the previous year. The data are shown below. Test the null hypothesis that the proportion of persons camping is the same for all three education levels. Which group or groups should the company try to reach through advertising? Discuss the possible advertising media appropriate for these groups.

| Level of Education | Number Interviewed | Number Camping |
| --- | --- | --- |
| High school | 1100 | 743 |
| College | 600 | 287 |
| Professional | 300 | 110 |

6.22  Use the Kolmogorov-Smirnov two-sample test to see whether the chi-square sample data of Example 2.3, when standardized correctly for $v = 8$ degrees of freedom, and the normal sample data come from populations with the same distribution.

6.23  Pick out 30 lines at random from this text and test the null hypothesis that the vowels a, e, i. o, u, and y have an equal probability of occurrence in this book. Could you extend your conclusions to the entire English language? Discuss.

# 7

# Association Analysis

## 1 Introduction

Investigations are frequently designed to study relationships between two or more characteristics — for example, height and weight, or eye color and hair color. In this chapter we present those methods of statistical analysis that are useful for determining whether there exists a relationship, which we call association, between certain designated characteristics. When these characteristics can be considered variables in a multivariate population, association implies a lack of independence.

If the variables are independent, there is no relationship between them and the value of one has no effect on the values of the others. However, if they are not independent, then some kind of association exists. Accordingly, we investigate the research hypothesis that association exists by performing a test of the null hypothesis of independence between the variables.

In addition to studying the *existence* of an association in the population, we need to obtain a descriptive measure of the *degree* of association, or the strength of relationship, between the variables in the sample data. In order to interpret this value, and to facilitate comparisons of degree of association between different sets of samples, or between different sets of characteristics, these measures will all be defined as relative numbers. For consistency, the absolute value of any measure is between zero and one, with zero for independent variables (no association) and one for the highest degree of association. Increasing degrees of association are then represented by increasing absolute values of the measure of association.

Our study of association in this chapter includes four types of problems. In the first type, association between only two variables is of interest, and the sample data measure these variables on at least an ordinal scale. The data then consist of pairs of measurements of observations. The second case involves association between $k$ variables that are measured on an ordinal scale. The same analysis can be applied in a two-way analysis-of-variance type of problem, where the inference concerns the equality of $k$ population means. The third situation covers association between two primary sets of characteristics, each with any number of subgroups. Since the data analyzed are the numbers of observations in each cross classification, the level of measurement of the subgroups is irrelevant. Thus, the characteristics may be entirely qualitative, like political party affiliation, or they may be quantitative. The final problem involves measures of association for two characteristics, of which one is dichotomous and the other is measured on an interval scale.

## 2   Two Related Samples

In Chapter 3 we considered methods of statistical analysis for paired sample data, but only when the inferences were concerned with parameters of location or central tendency. In Sections 2.2 and 3.1 of Chapter 3, respectively, the sign test and signed rank test were applied to the differences of pairs of observations. However, at that point we were not interested in testing for the existence of association. In fact, association is presumably always present in such situations because that is the purpose of the pairing. Since the pairs are linked by some unit of association, either naturally or as an integral part of the design of the experiment, they would presumably behave alike if treated alike. Therefore the experiments were designed to compare the average effects of the treatments, and the inferences concerned the difference in treatment effects.

In this section we consider the problem of describing the association between pairs of variables or characteristics and making inferences about it. The elements under study may be matched units, as two sides of a leaf or brother and sister, or a single unit on which two observations are made, such as height and weight or sex and income. In either case, the data consist of matched pairs of measurements. Thus the original data look the same as in Chapter 3 because of the similarity of the sampling situations, but the description of interest and inference situation are quite different from what they were before. We study here two different measures of association in paired samples, called the coefficient of rank correlation and the Kendall Tau coefficient. These coefficients are descriptive statistics that reflect the degree of association between the sample pairs of observations, but either one can be used to test the null hypothesis of no association in the population. Because of the fact that the descriptive statistics in this chapter are of as much or more interest than the statistical inferences (in fact no inference may be desired at all), we depart from the previous format and defer discussion of the inference situation and test procedure until after the descriptive measure has been discussed in detail. In this section we discuss two different measures of association, called the Spearman rank correlation coefficient and the Kendall Tau coefficient. Since in each case the measure is based on the relative magnitudes of the variables within each of the two sets, we assume throughout that the bivariate population is continuous, or that the probability of a tie within either set is equal to zero.

### 2.1  Rank Correlation

*Sampling situation*  The data consist of a random sample of $n$ pairs of observations, $(X_1, Y_1), (X_2, Y_2), \ldots, (X_n, Y_n)$, where each set (that is, the $X$ set and the $Y$ set) is measured on at least an ordinal scale. For the purpose of analysis, the observations in each set are ranked. That is, the integers $1, 2, \ldots, n$ are assigned to $X_1, X_2, \ldots, X_n$ according to their relative magnitudes within the set of $X$ observations. Let $U_i$ denote the rank of $X_i$, so the set $U_1, U_2, \ldots, U_n$ is some arrangement of the first $n$ positive integers. Similarly, the $Y_1, Y_2, \ldots, Y_n$ are ranked; the results of this ranking are denoted by $V_1, V_2, \ldots, V_n$.

Now the data to be analyzed consist of $n$ pairs of positive integers $(U_1, V_1), (U_2, V_2), \ldots, (U_n, V_n)$, which represent the pairs of ranks of the original observations. In the discussion to follow, we assume for convenience that these pairs are listed in increasing order of the value of the $X$ ranks; that is, $U_i = i$ for $i = 1, 2, \ldots, n$. This has no effect on the general applicability of the methods, since the pairs of sample values can be rearranged in any order after observation. With this assumption, $V_i$ must be interpreted as the rank of that $Y$ which is paired with the $X$ which has rank $i$.

*Rationale for descriptive measure*  The rankings of the two characteristics are in perfect agreement if and only if $U_i = V_i$ for each $i$; that is, the rankings are as follows:

| X-Sample Ranks | Y-Sample Ranks | |
|:---:|:---:|:---:|
| 1 | 1 | |
| 2 | 2 | |
| . | . | |
| . | . | |
| . | . | |
| $n-1$ | $n-1$ | |
| $n$ | $n$ | (2.1) |

The rankings are in perfect disagreement if they are paired in complete reverse orders; that is, we have

| X-Sample Ranks | Y-Sample Ranks | |
|:---:|:---:|---:|
| 1 | $n$ | |
| 2 | $n - 1$ | |
| . | . | |
| . | . | |
| . | . | |
| $n - 1$ | 2 | |
| $n$ | 1 | (2.2) |

The difference between any two paired ranks,

$$D_i = U_i - V_i$$

describes the degree of disagreement between them, in the sense that $D_i$ equals zero when the ranks of $X_i$ and $Y_i$ are equal, and the larger the absolute magnitude of $D_i$ the more the ranks differ in that pair. Thus some function of all the $D_i$ is a logical quantity to consider as a descriptive measure of the overall disagreement. We use the sum of the squares of the differences, or

$$\sum_{i=1}^{n} D_i^2 = \sum_{i=1}^{n} (U_i - V_i)^2. \tag{2.3}$$

The magnitude of this sum is affected by the value of $n$ as well as by the degree of disagreement. A relative measure would facilitate interpretation of its magnitude. Hence we divide the sum in (2.3) by its maximum value. This maximum occurs when the rankings are in perfect disagreement, that is, when the pairings are as in (2.2). Then (2.3) can be evaluated as

$$(n - 1)^2 + [(n - 1) - 2]^2 + \cdots + [2 - (n - 1)]^2 + (1 - n)^2$$

$$= 2[(n - 1)^2 + (n - 3)^2 + \cdots] = \frac{n(n^2 - 1)}{3}. \tag{2.4}$$

The final expression in (2.4) holds for all $n$. It is not obvious, but it can be easily verified for special cases. (The final term in the sum to be multiplied by two is $1^2$ if $n$ is even, and $2^2$ if $n$ is odd.)

Thus a relative measure of the *degree of disagreement* in the sample ranks is the ratio between the observed value of $\Sigma D_i^2$ and its maximum value for complete disagreement, or

$$\frac{\Sigma D_i^2}{n(n^2 - 1)/3} = \frac{3\Sigma D_i^2}{n(n^2 - 1)}. \tag{2.5}$$

This ratio equals 0 for perfect agreement, that is, for samples with the ranks paired as in (2.1); it equals 1 for perfect disagreement, that is, for samples with ranks as in (2.2); and it is always between 0 and 1.

We have found a suitable relative measure of "disagreement," but the stated goal is to define a relative measure of "association." Two characteristics have perfect association if they have either perfect agreement or perfect disagreement, in the sense that if the rank of one variable is given to be $i$, then the rank of the other variable is known as either $i$ (in the case of perfect agreement) or $n - i + 1$ (in the case of perfect disagreement). The value of (2.5) is 0 in the former case and 1 in the latter case. Thus that expression cannot describe association, but a function of it can.

This function is the Spearman coefficient of rank correlation (named after Charles Spearman, an English psychologist), defined as

$$R = 1 - \frac{6\Sigma D_i^2}{n(n^2 - 1)}. \tag{2.6}$$

Note that (2.6) results from subtracting twice the value of (2.5) from 1; this means that $R = -1$ when (2.5) equals 1 (perfect disagreement), and $R = +1$ when (2.5) equals 0 (perfect agreement).

Hence $|R| = 1$ reflects perfect association, and the sign of $R$ indicates the type or direction of association. Perfect agreement means that large values of one variable are associated with large values of the other, and small values are likewise associated. Such an association might be called direct since the variables are moving in the same direction. In an inverse relationship the variables move in opposite directions; that is, large values of one variable are associated with small values of the other, approaching perfect disagreement.

In summary, the value of the coefficient of rank correlation defined in (2.6) is always between $-1$ and $+1$, and it has the following interpretations in the extreme cases:

| Value of $R$ | Type of Association | Type of Agreement |
|---|---|---|
| $R = 1$ | Direct | Perfect agreement |
| $R = -1$ | Inverse | Perfect disagreement |

When $R = 0$, there is no association, and hence neither agreement nor disagreement. Increasing absolute values of $R$ reflect increasing degrees of association, direct if $R > 0$ and inverse if $R < 0$. With respect to degree of association, there is no difference between, say, $R = .8$ and $R = -.8$; the sign indicates only the direction of association, not its strength.

*Computation*    The formula given in (2.6) is frequently the most convenient to use for computation by hand. (Note that $\Sigma D_i = \Sigma U_i - \Sigma V_i = 0$, which provides a useful check on arithmetic.) Other equivalent expressions for $R$ which give exactly the same result are

$$R = \frac{12 \sum_{i=1}^{n} U_i V_i}{n(n^2 - 1)} - \frac{3(n + 1)}{n - 1}, \tag{2.7}$$

or

$$R = \frac{12 \sum_{i=1}^{n} (U_i - \overline{U})(V_i - \overline{V})}{n(n^2 - 1)}. \tag{2.8}$$

The expression in (2.8) is not convenient for computation; it is given here primarily to show that the rank correlation coefficient is equal to the ordinary Pearson product-moment correlation coefficient in parametric statistics if the observations are ranks. The formula in (2.7) is especially convenient for use with desk calculators.

**example 2.1**

Particularly in the behavioral sciences, a widely used experimental technique is to ask panels of "judges" to rank a group of items, which may be products, individuals, actions, competitors, or the like. A question of interest then is how closely do the judges agree among themselves in their ratings? For example, a panel of judges may be asked to rank a number of competitors in a beauty contest. Agreement between ratings (a direct relationship) suggests that the judges are using similar criteria and perceptions to rate the pulchritude of the contestants. Independence of ratings suggests that the judges are either ranking at random or are using different standards of pulchritude and hence supporting the adage that "beauty is in the eye of the beholder."

Suppose that two judges rank eight contestants as follows. Calculate the coefficient of rank correlation.

| Judge | Contestant | | | | | | | |
|---|---|---|---|---|---|---|---|---|
| | A | B | C | D | E | F | G | H |
| 1 | 2 | 1 | 3 | 5 | 4 | 8 | 7 | 6 |
| 2 | 1 | 2 | 4 | 5 | 7 | 6 | 8 | 3 |

**SOLUTION**

The differences between corresponding ranks, $D_i$, are

$$1, -1, -1, 0, -3, 2, -1, 3,$$

and the sum of their squares is

$$\Sigma D_i^2 = (1)^2 + (-1)^2 + (-1)^2 + \cdots + (-1)^2 + (3)^2 = 26.$$

Substituting this result in (2.6), where $n = 8$, gives

$$R = 1 - \frac{6(26)}{8(63)} = \frac{29}{42} = .690.$$

Since $R > 0$, the relationship is direct, and the relative measure of association between rankings of judges is .690.

In order to show that formulas (2.6) and (2.7) are equivalent, we now calculate $R$ again, but this time using the sum of cross products of ranks, which is

$$(2)(1) + (1)(2) + (3)(4) + (5)(5) + (4)(7) + (8)(6) + (7)(8)$$
$$+ (6)(3) = 191.$$

Then, from (2.7),

$$R = \frac{12(191)}{8(63)} - \frac{3(9)}{7} = \frac{764 - 648}{168} = \frac{29}{42} = .690.$$

*Ties*   We have so far ignored the problem that arises in assigning the ranks when some two or more observations in one sample are the same. It is recommended that ties be handled by the midrank method. If the proportion of ties is small, they have little effect on the value of $R$. However, if there are many ties, $R$ may be underestimated when calculated from (2.6) or (2.7). A correction for ties can be incorporated into the value of the descriptive measure by using the following expression:

$$R = \frac{n(n^2 - 1) - 6\Sigma D_i^2 - 6(u' + v')}{\sqrt{n(n^2 - 1) - 12u'}\,\sqrt{n(n^2 - 1) - 12v'}}. \qquad (2.9)$$

Here $u' = (\Sigma u^3 - \Sigma u)/12$ for $u$ the number of observations in the $X$ sample that are tied at a given rank, and the sum is over all sets of $u$ tied ranks; and similarly $v' = (\Sigma v^3 - \Sigma v)/12$ for sets of $v$ tied ranks in the $Y$ sample. An illustration is given in Example 2.3.

*Inference situation*   The coefficient of rank correlation is an important and valuable description of the association of the characteristics as exhibited by the sample data, and hence as an estimate of the association in the population. Thus, it may be desirable to perform some kind of statistical inference based on the sample description. The null hypothesis that can be tested is that the variables that measure the two characteristics in the population are independent. We express this by saying that there exists no association between the $X$ and $Y$ variables.[1] The alternative may be two sided, and hence simply state that an association does exist, or it may specify the direction of association, as either direct or indirect. The hypothesis set is then one of the following:

*Two-sided alternative*
  $H$: No association    $A$:  Association exists

*One-sided alternatives*
  $H$: No association    $A_+$:  Direct association
  $H$: No association    $A_-$:  Inverse association.

*Rationale*   If no association exists, that is, if the null hypothesis is true, then for any given rank order of the $X$ observations the corresponding rank order of the $Y$ observations is equally likely to be any one of all possible arrangements of the $n$ ranks. Each of these arrangements produces a particular value of $R$. If most of the large $X$ ranks are paired with large $Y$ ranks, and accordingly small $X$ ranks with small $Y$ ranks, this supports a conclusion that $X$ and $Y$ have a direct type of relationship; the extreme case of this situation is perfect agreement, as in (2.1), where $R = 1$. Hence large positive values of $R$ call for rejection of $H$ in favor of $A_+$ and the conclusion of a direct type of association. An indirect association is suggested when most of the large $X$ ranks are associated with small $Y$ ranks, and small $X$ ranks with large $Y$ ranks; the extreme situation here is the pairing in (2.2), where there is perfect disagreement and $R = -1$. Hence large negative (small) values of $R$ imply rejection of $H$ in favor of $A_-$. An $R$ that is either too small or too large provides support for the two-sided alternative $A$.

*Test statistic*   The test statistic is the Spearman coefficient of rank correlation, most easily computed from (2.6) or (2.7); that is,

$$R = 1 - \frac{6\Sigma D_i^2}{n(n^2 - 1)}$$

or

$$R = \frac{12 \sum\limits_{i=1}^{n} U_i V_i}{n(n^2 - 1)} - \frac{3(n + 1)}{n - 1}.$$

[1]It should be pointed out that independence is not equivalent to a lack of "correlation" when correlation is defined parametrically as the ratio of the covariance to the product of the standard deviations. Although two variables that are independent are not correlated, it is possible for the parametric correlation coefficient to be zero even though the variables are dependent. Since we are stating the null hypothesis of independence as "no association," the word "association" cannot be interpreted here as being equivalent to parametric correlation. If there is association, there may or may not be parametric correlation.

If there are many ties, the expression in (2.9) should be used. The value of $R$ is always between $-1$ and $+1$ inclusive.

*Finding the P-value*   Under the null hypothesis of no association, the test statistic $R$ is symmetrically distributed about its mean value, 0. As a result, the two-tailed $P$-value is twice the one-tailed value, which is left tail if $R < 0$ and right tail if $R > 0$. According to the rationale discussed above, the appropriate $P$-values are summarized as follows, where smaller tail probability means left or right tail, whichever has the smaller probability.

| Alternative | | P-value (Table I) |
|---|---|---|
| $A_+$: | Direct association | Right-tail probability for $R$ |
| $A_-$: | Inverse association | Left-tail probability for $R$ |
| $A$: | No association | 2(smaller tail probability for $R$) |

Table I can be used to find $P$-values for the calculated value of $R$ for $n \leq 10$; only the smaller tail probability is given as a right-tail probability for $R \geq 0$. The same value is a left-tail probability for the negative of the table entry. For moderate $n$, that is, $10 < n \leq 30$, the second part of the table gives exact quantile values of the null distribution of $R$ for some selected right-tail probabilities given on the top row under the heading "one-sided tests." These also apply to values of $-R$ as left-tail probabilities. The entries here are also the exact quantile values of the null distribution of the absolute value $|R|$ for those tail probabilities given on the bottom row above the heading "two-sided tests." For larger $n$, the quantity

$$z = R\sqrt{n-1} \qquad (2.10)$$

can be calculated and the approximate $P$-value found from Table A as the tail probability on $z$ that corresponds to the appropriate tail probability on $R$. When there are ties, the accuracy of a $P$-value found from either Table I or Table A may be improved when $R$ is calculated from (2.9). A correction for ties could be incorporated in the $z$ statistics; the formula will not be given here.

**example   2.2**

We use Example 2.1 to illustrate the inference procedure for small samples. The null hypothesis of no association between rankings here is equivalent to stating that the judges' evaluations of the contestants are made independently and at random. The appropriate alternative is a direct relationship, since the concern is with agreement between ratings. From Table

I for $n = 8$, the right-tail probability for $R = .690$ is .035, and this is the exact $P$-value.

**example  2.3**

Many countries with a democratic system of government have automatic registration of all voters or no registration procedures at all. On the other hand, the registration process in the United States involves some forethought and, frequently, also inconvenience. The voter turnout in the United States is lower than in many other democratic countries, such as England, France, Norway, and Canada. Some students of politics claim that this is not due to apathy of U.S. citizens, but rather to the obstacles of registration. They claim that there is a very high direct relationship between registration and voting rate, and hence the key to more participation in elections is an increase in registration rate.

The data for 1960 given in Table 2.1 are excerpted from Tufte (1972, pp. 156–157). Both the registration and turnout rates are recorded as a percent of voting age population.

**TABLE 2.1**

| City | Registration | Turnout |
|------|------|------|
| Minneapolis, Minn. | 92.5 | 58.5 |
| Detroit, Mich. | 92.0 | 70.0 |
| Seattle, Wash. | 92.0 | 70.8 |
| Salt Lake City, Utah | 87.0 | 76.6 |
| Topeka, Kan. | 81.9 | 69.3 |
| Pittsburgh, Pa. | 81.2 | 68.3 |
| Oklahoma City, Okla. | 80.4 | 62.7 |
| Philadelphia, Pa. | 77.6 | 69.8 |
| Los Angeles, Calif. | 77.0 | 64.2 |
| Boston, Mass. | 74.0 | 63.3 |
| Charlotte, N.C. | 69.9 | 54.5 |
| Tampa, Fla. | 68.8 | 63.6 |
| Baltimore, Md. | 68.1 | 54.0 |
| San Francisco, Calif. | 68.0 | 64.4 |
| New York, N.Y. | 65.7 | 58.8 |
| Dallas, Tex. | 65.0 | 57.3 |
| Newark, N.J. | 61.4 | 50.4 |
| Honolulu, Hawaii | 60.0 | 54.7 |
| Miami, Fla. | 59.2 | 43.7 |
| New Orleans, La. | 55.6 | 45.9 |
| Richmond, Va. | 46.5 | 31.2 |
| San Antonio, Tex. | 42.6 | 31.4 |
| Birmingham, Ala. | 39.1 | 13.8 |
| Atlanta, Ga. | 33.8 | 25.6 |

**SOLUTION**

We call registration the $X$ characteristic and turnout the $Y$ characteristic. The $X$ and $Y$ data sets are each ranked from 1 to 24, with 1 representing the highest rate in each case. Detroit and Seattle have the same registration rate, so these cities are given the same rank, in this case the average of the ranks 2 and 3. No other ties occur in either set. The resulting 24 pairs of ranks are shown in Table 2.2.

**TABLE 2.2**

| City | X Rank | Y Rank | D | D² |
|------|--------|--------|------|------|
| Minneapolis | 1 | 13 | −12 | 144 |
| Detroit | 2.5 | 3 | − 0.5 | 0.25 |
| Seattle | 2.5 | 2 | 0.5 | 0.25 |
| Salt Lake City | 4 | 1 | 3 | 9 |
| Topeka | 5 | 5 | 0 | 0 |
| Pittsburgh | 6 | 6 | 0 | 0 |
| Oklahoma City | 7 | 11 | − 4 | 16 |
| Philadelphia | 8 | 4 | 4 | 16 |
| Los Angeles | 9 | 8 | 1 | 1 |
| Boston | 10 | 10 | 0 | 0 |
| Charlotte | 11 | 16 | − 5 | 25 |
| Tampa | 12 | 9 | 3 | 9 |
| Baltimore | 13 | 17 | − 4 | 16 |
| San Francisco | 14 | 7 | 7 | 49 |
| New York | 15 | 12 | 3 | 9 |
| Dallas | 16 | 14 | 2 | 4 |
| Newark | 17 | 18 | − 1 | 1 |
| Honolulu | 18 | 15 | 3 | 9 |
| Miami | 19 | 20 | − 1 | 1 |
| New Orleans | 20 | 19 | 1 | 1 |
| Richmond | 21 | 22 | − 1 | 1 |
| San Antonio | 22 | 21 | 1 | 1 |
| Birmingham | 23 | 24 | − 1 | 1 |
| Atlanta | 24 | 23 | 1 | 1 |
| | | | 0 | 314.50 |

The number of ties is small. Using the uncorrected formula in (2.6), we obtain

$$R = 1 - \frac{6(314.5)}{24(575)} = 1 - (.137) = .863.$$

To incorporate the correction for ties, we use the formula (2.9), where $u = 2$, $u^3 - u = 6$, $u' = 6/12$, $v = 0$, $v' = 0$. This result is

$$R = \frac{24(575) - 6(312.5) - 6(6/12) - 0}{\sqrt{13,794}\ \sqrt{24(575)}} = .863.$$

These answers agree to three decimal places, but they are not equal. In practice, we would not ordinarily bother with the latter computation; it is included here solely to illustrate the use of (2.9).

The hypothesis set called for by the problem is one sided, since the claim to be investigated is a significant direct relationship. The alternative then is $A_+$ and the appropriate $P$-value is a right-tail probability on $R$. For $n = 24$, the largest entry in Table I is .608, which corresponds to a right-tail probability of .001. Since the calculated value of $R$ is larger than .608, its tail probability must be smaller than .001. Hence the appropriate conclusion is that the $P$-value is $P < .001$ and the null hypothesis should be rejected; the data support the alternative of a direct relationship between voter registration rate and turnout rate.

For the sake of illustration, we use these data to find an approximation to the $P$-value based on the normal curve. From (2.10), $z = .863\sqrt{23} = 4.14$. The right-tail probability from Table A is .0000 to four decimal places, and this is the asymptotic approximate $P$-value.

## 2.2   Kendall Tau Statistic

The Kendall Tau statistic, named after the well-known English statistician, Maurice Kendall, is another measure of association between two classifications. It is applicable in exactly the same sampling and inference situations as the Spearman coefficient of rank correlation. As we shall see however, the Tau statistic measures association in a slightly different way, and hence in most cases produces a numerical value different from $R$.

*Sampling situation*   The data consist of a random sample of $n$ pairs of observations, $(X_1, Y_1), (X_2, Y_2), \ldots, (X_n, Y_n)$, where each set (that is, the $X$ set and the $Y$ set) is measured on at least an ordinal scale. It is frequently more convenient, though not necessary, to rank the data within each set.

*Rationale for descriptive measure*   Suppose that one of the sets of observations, say the $X$ set, is arranged in natural order from smallest to largest. If the $X$ and $Y$ characteristics are in perfect agreement, the $Y$ set will also be in natural order, as in (2.1). If there is perfect disagreement, the corresponding $Y$ set is in the reverse of the natural order, as in (2.2). The Kendall Tau coefficient is a relative measure of the discrepancy between the actual observed order of the $Y$'s and these two orders that would result from perfect association. The procedure is most easily explained by an example.

**example   2.4**

Suppose that $n = 5$ and the two sets of ranks are paired as follows:

$$X \text{ rank:} \quad 1\ 2\ 3\ 4\ 5$$
$$Y \text{ rank:} \quad 2\ 3\ 1\ 4\ 5$$

Note that the $X$ set has been arranged in natural order. In the resulting arrangement of $Y$ ranks, we consider all possible ordered pairs of $Y$ ranks and score a 1 for each pair of ranks that appear in natural order and $-1$ for those in reverse natural order. We take the pairs in a systematic way, as the 2 paired with each successive rank appearing to its right, then the 3 paired with each to its right, and so on. The first pair of $Y$ ranks, 2 followed by 3, is in natural order, so its score is 1. The second pair, 2 followed by 1, is in reverse order, so $-1$ is scored. The resulting scores for all possible pairs are shown in Table 2.3.

**TABLE 2.3**

| $Y$ Pair | Score | Summary Totals |
|---|---|---|
| 2, 3 | 1 | 8 plus |
| 2, 1 | $-1$ | 2 minus |
| 2, 4 | 1 | |
| 2, 5 | 1 | |
| 3, 1 | $-1$ | |
| 3, 4 | 1 | |
| 3, 5 | 1 | |
| 1, 4 | 1 | |
| 1, 5 | 1 | |
| 4, 5 | 1 | |

If $X$ and $Y$ were in perfect agreement as regards relative order of magnitude, the score would be 1 for each $Y$ pair; perfect disgreement would produce a score of $-1$ for each $Y$ pair. There are $\binom{5}{2} = 10$ distinguishable pairs. The ratio of the total plus score, in this case 8, to the maximum, 10, provides a measure of relative agreement, that is, $8/10$. Similarly, the ratio of the total minus score to the maximum, $2/10$ in this case, measures the relative disagreement. The net relative score of association is then $8/10 - 2/10 = 6/10$, and this is the value of the Kendall Tau statistic.

This same value could have been obtained in a different way. We explain this rationale also since it is completely analogous to the development of the coefficient of rank correlation from the measure of relative disagreement in (2.5). If the proportion of pairs that disagree, in this case $2/10$, is doubled and subtracted from 1, the result is $1 - 2(2/10) = 6/10$, which equals the net relative score found above. (The same result is also obtained by doubling the degree of agreement and subtracting 1 from it, in this case $2(8/10) - 1 = 6/10$.)

*Computation*   To compute Tau in the general case, the first step is to arrange the $(X, Y)$ pairs so that the $X$ values (or ranks) appear in increasing order of magnitude. In this arrangement, let $U$ be the number of pairs of $Y$ values (or ranks) that appear in natural order (that is, the number of plus scores), let $V$ be the number of $Y$ pairs in reverse natural order, and let $S$ be the difference between $U$ and $V$, $S = U - V$. Then the degree of agreement is

$$\frac{U}{\binom{n}{2}} = \frac{2U}{n(n-1)}, \qquad (2.11)$$

and the degree of disagreement is

$$\frac{V}{\binom{n}{2}} = \frac{2V}{n(n-1)}. \qquad (2.12)$$

For perfect agreement, (2.11) equals 1 and (2.12) equals 0; for perfect disagreement, (2.11) equals 0 and (2.12) equals 1. We want a measure that equals $-1$ for perfect disagreement and $+1$ for perfect agreement, so that an absolute value of 1 reflects perfect association. The Tau coefficient $T$ is calculated as some function of (2.11) and (2.12) that has these properties, namely any of the following:

$$T = \frac{4U}{n(n-1)} - 1, \qquad (2.13)$$

$$T = 1 - \frac{4V}{n(n-1)}, \qquad (2.14)$$

$$T = \frac{2S}{n(n-1)}. \qquad (2.15)$$

Note that since $U$, $V$, and $S$ are completely determined by the arrangement of the $Y$ sample that corresponds to the natural order of the $X$ sample, the results are the same whether the observations are ranked or not.

In practice, it is not necessary to list all the pairs to determine the value of $T$, whether it is found from $U$, $V$, or $S$. When the $X$ observations or ranks are written in natural order, simple shortcuts can be used. For instance, we may find the value of $U$ by simply counting the number of $Y$ pairs that also appear in natural order. In Example 2.4, the $Y$ ranks corresponding to the natural order of $X$ ranks were

$$2, 3, 1, 4, 5.$$

We sum the numbers of $Y$ ranks that are greater than and to the right of each successive $Y$ rank. There are 3 ranks greater than and to the right of rank 2, 2 ranks greater than and to the right of rank 3, and so on. Proceeding in this way we find that

$$U = 3 + 2 + 2 + 1 = 8.$$

Similarly, $V$ can be counted as the total number of $Y$ ranks that are smaller than and to the right of each successive $Y$ rank; here $V = 2$. $S$ can be counted directly by summing, for each $Y$ rank, the number of $Y$ ranks to the right that are greater than it minus the number that are less than it.

The value of the Tau coefficient is always between $-1$ and $+1$, and it has the following interpretation.

| Value of $T$ | Type of Association | Type of Agreement |
|---|---|---|
| $T = 1$ | Direct | Perfect agreement |
| $T = -1$ | Inverse | Perfect disagreement |

When $T = 0$ there is no association, since the number of pairs that agree is equal to the number of pairs that disagree. Increasing absolute values of $T$ reflect increasing degrees of association, so that, say, $T = .8$ and $T = -.8$ imply the same degree of association but opposite types of association.

The Tau coefficient can also be interpreted as a coefficient of "disarray," since it can be shown that $V$ is the minimum number of interchanges between consecutive pairs of $Y$ observations or ranks that are needed to transform the given set into the natural order. For example, only two interchanges of the $Y$ ranks 2, 3, 1, 4, 5 are required in Example 2.4. We first interchange 1 and 3 to get 2, 1, 3, 4, 5, and then 1 and 2 to get 1, 2, 3, 4, 5, and thus $V = 2$.

**example 2.5**

We illustrate the computation of $T$ by these methods for the data of Example 2.1. After listing the contestants according to the natural order of one set of ranks, say those for judge 1, the results are as follows.

| | Contestant | | | | | | | |
|---|---|---|---|---|---|---|---|---|
| Judge | B | A | C | E | D | H | G | F |
| 1 | 1 | 2 | 3 | 4 | 5 | 6 | 7 | 8 |
| 2 | 2 | 1 | 4 | 7 | 5 | 3 | 8 | 6 |

In order to count $U$, we note that for judge 2 there are 6 ranks to the right of and larger than 2, 6 ranks to the right of and larger than 1, 4 ranks to the right of and larger than 4, and so on, and thus

$$U = 6 + 6 + 4 + 1 + 2 + 2 + 0 + 0 = 21.$$

From (2.13) we obtain

$$T = \frac{4(21)}{8(7)} - 1 = \frac{28}{56} = .500.$$

For the purpose of illustration, note that $V$ is found as the sum

$$V = 1 + 0 + 1 + 3 + 1 + 0 + 1 + 0 = 7.$$

Now we can use (2.14) to find $T$ as

$$T = 1 - \frac{4(7)}{8(7)} = .500.$$

*Ties*   When ties exist in either or both samples, they are most easily handled by using the midrank method to assign ranks and calculating the score $S$ as the total plus scores minus the total minus scores, except that a score of 0 is given when a tie occurs in *either* an $X$ pair or a $Y$ pair. To avoid making an error, it is better to list not just the $Y$ pairs as we did in Example 2.4, but also the corresponding $X$ pairs. Then $T$ can be calculated from (2.15), or from the formula given in (2.19) below, which incorporates an adjustment for ties.

For a numerical example, consider $n = 5$ and the following sets of ranks, which have ties within both sets of data.

$$\begin{aligned} X \text{ rank:} \quad & 1\ 2.5\ 2.5\ 4.5\ 4.5 \\ Y \text{ rank:} \quad & 2\ 3.5\ 3.5\ 1\ \ \ 5 \end{aligned} \qquad (2.16)$$

The possible $X$ and $Y$ pairs, and their corresponding scores, are as shown in Table 2.4. With 5 plus and 3 minus, the calculated value of $S$ is

**TABLE 2.4**

| X Pair | Y Pair | Score | Summary Totals |
|--------|--------|-------|----------------|
| 1, 2.5 | 2, 3.5 | 1 | 5 plus |
| 1, 2.5 | 2, 3.5 | 1 | 3 minus |
| 1, 4.5 | 2, 1 | −1 | |
| 1, 4.5 | 2, 5 | 1 | |
| 2.5, 2.5 | 3.5, 3.5 | 0 | |
| 2.5, 4.5 | 3.5, 1 | −1 | |
| 2.5, 4.5 | 3.5, 5 | 1 | |
| 2.5, 4.5 | 3.5, 1 | −1 | |
| 2.5, 4.5 | 3.5, 5 | 1 | |
| 4.5, 4.5 | 1, 5 | 0 | |

$S = 5 - 3 = 2$. The total number of pairs is 10, but 10 is not the maximum absolute value of $S$. Given the tied ranks in (2.16), if there were perfect agreement the value of $S$ would be $S = 7$ since the paired ranks would be

$$\begin{aligned} X \text{ rank:} \quad & 1\ 2.5\ 2.5\ 4.5\ 4.5 \\ Y \text{ rank:} \quad & 1\ 2\ \ \ 3.5\ 3.5\ 5, \end{aligned} \qquad (2.17)$$

and for perfect disagreement $S = -8$, since the sets would be

$$X \text{ rank:} \quad 1 \; 2.5 \; 2.5 \; 4.5 \; 4.5$$
$$Y \text{ rank:} \quad 5 \; 3.5 \; 3.5 \; 2 \quad 1. \tag{2.18}$$

In general, the maximum absolute value of $S$ with ties is no longer a constant determined by $n$; it depends on the number of $X$ ranks that are tied, the number of $Y$ ranks that are tied, and also the ranks for which those ties occur. Since it is not possible to take the latter factor into account and obtain any general kind of correction for ties, we consider only the first two factors. Recall that the score is always 0 for any $X$ pair chosen from a group of $u$ tied $X$ ranks, or from any $Y$ pair chosen from a group of $v$ tied $Y$ ranks. Thus the number of $X$ pairs that can contribute to a nonzero score is $\binom{n}{2} - \Sigma\binom{u}{2}$, and similarly for the number of $Y$ pairs. We take a sort of geometric mean of these respective contributions to scores and use it as the denominator of $S$ to form the Kendall Tau coefficient adjusted for ties. The formula is then

$$T = \frac{S}{\sqrt{\binom{n}{2} - u'} \; \sqrt{\binom{n}{2} - v'}}, \tag{2.19}$$

where $u' = \Sigma\binom{u}{2}$ for $u$ the number of observations in the $X$ sample that are tied at a given rank and the sum is over all sets of $u$ tied ranks, and similarly $v' = \Sigma\binom{v}{2}$ for sets of $v$ tied ranks in the $Y$ sample. For the data in (2.16), $u' = \binom{2}{2} + \binom{2}{2} = 2$ and $v' = \binom{2}{2} = 1$, so that $T = 2/\sqrt{8(9)} = .236$.

If there are many ties, it is recommended that (2.19) be used to calculate $T$; it always gives a coefficient whose absolute value is larger than (2.15), although its maximum absolute value may still not be 1, as it is not for the ranks above. With perfect agreement as in (2.17), (2.19) gives .825; for the perfect disagreement in (2.18), (2.19) gives $-.943$.

We should point out that formulas (2.13) and (2.14) are *not* equivalent to (2.15) if there are ties, and should not be used to calculate $T$. This follows since $U + V$ does not equal $n(n - 1)/2$ when ties are present.

*Inference situation*   The Kendall Tau coefficient is useful for exactly the same inference situation as the coefficient of rank correlation. The possible hypothesis sets are then:

*Two-sided alternative*
   $H:$ No association   $A:$ Association exists

*One-sided alternatives*
   $H:$ No association   $A_+:$ Direct association
   $H:$ No association   $A_-:$ Inverse association.

*Rationale*   Assume that the set of $X$ ranks is arranged in natural order from smallest to largest. If no association exists, the corresponding arrangement of $Y$ ranks is equally likely to be any one of those possible. If there is a direct association, most of the pairs of $Y$ ranks should also appear in natural order. Then most of the scores will be $+1$ and $U$ will be close to $\binom{n}{2}$, making $T$ in (2.13) close to 1. Hence large positive values of $T$ call for rejection of $H$ in favor of $A_+$. Similarly, if there is an indirect association, most pairs of $Y$ ranks should appear in reverse natural order. Then many scores will be $-1$, making $V$ close to $\binom{n}{2}$ and the value of (2.14) close to $-1$. Thus large negative values of $T$ are evidence against $H$ in favor of $A_-$. A value of $T$ that is either too small or too large supports the two-sided alternative.

*Test statistic*    The Kendall Tau coefficient $T$, as computed from (2.13), (2.14), or (2.15), or from (2.19) if there are ties, provides a test statistic for the null hypothesis of no association.

*Finding the P-value*    Under the null hypothesis of no association, the test statistic $T$ takes on values between $-1$ and $+1$ and is symmetrically distributed about its mean value 0. By the rationale above, the appropriate $P$-values are summarized as follows, where smaller tail probability means right or left tail, whichever has the smaller probability.

| Alternative | | P-value (Table J) |
|---|---|---|
| $A_+$: | Direct association | Right-tail probability for $T$ |
| $A_-$: | Inverse association | Left-tail probability for $T$ |
| $A$: | No association | 2(smaller tail probability for $T$) |

Because the distribution of $T$ is symmetric, one table can be used for either right- or left-tail probabilities. Table J gives exact $P$-values corresponding to the calculated value of $T$ for $n \leq 10$. The table is entered with the absolute value $|T|$ for the appropriate $n$; the probability given is right tail if $T > 0$ and left tail if $T < 0$. For moderate $n$, that is $10 < n \leq 30$, the second part of the table gives exact quantile values of the null distribution of $T$ for some selected tail probabilities. The entries are the absolute values $|T|$, so that the probability figures given on the top row under the heading one-sided tests are right tail if $T > 0$ and left tail if $T < 0$. The probability values for two-sided tests are given on the bottom row above that heading. For larger $n$, the quantity

$$z = \frac{3T\sqrt{n(n-1)}}{\sqrt{2(2n+5)}} = \frac{S}{\sqrt{n(n-1)(2n+5)/18}} \qquad (2.20)$$

is calculated. The asymptotic approximate $P$-value is then found from Table A as the tail probability on $z$ that corresponds to the appropriate tail probability on $T$.

**example   2.6**

A civic club sponsored an essay contest for high school seniors in their community. The 10 essays submitted were judged by a committee of seven persons, three from the faculty of the local university and four from the business community. Each judge scored all 10 papers independently of the other judges. The scores were summed for each paper, and the winner determined as the paper with the highest total score. After the contest was over, there was some discussion among the judges as to whether or not the business judges and faculty judges were in agreement about their ratings of the papers. In order to resolve the conflict, the scores given by these two groups of judges were summed separately for each paper and ranked within each group, giving the results below. Note that the worst inconsistency is for contestants C and F. Does there appear to be significant agreement between ratings?

| | Contestant | | | | | | | | | |
|---|---|---|---|---|---|---|---|---|---|---|
| | B | G | C | J | A | F | D | H | I | E |
| Business | 1 | 2 | 3 | 4 | 5 | 6 | 7 | 8 | 9 | 10 |
| University | 2 | 1 | 6 | 5 | 4 | 3 | 8 | 7 | 10 | 9 |

**SOLUTION**

Since the question of interest concerns agreement only, we will use the following hypothesis set with a one-sided alternative.

$H$:   There is no association between sets of rankings
$A_+$:   There is a direct association between sets of rankings.

For the Kendall coefficient, we calculate the following:

$$U = 8 + 8 + 4 + 4 + 4 + 4 + 2 + 2 = 36$$

$$T = \frac{4U}{n(n-1)} - 1 = \frac{144}{90} - 1 = \frac{27}{45} = .600$$

From Table J, the $P$-value is .008; we conclude that there is a direct association. Therefore, when all contestants are considered simultaneously, the two groups of judges can be considered consistent in their ratings, even though there is considerable inconsistency in ratings for contestants C and $F$.

The Spearman coefficient of rank correlation could also have been used in this example. The reader may verify that $\Sigma D^2 = 26$ and $R = .8425$. The right-tail probability from Table I is .002; the conclusion is the same, although the $P$-value is smaller here.

**example 2.7**

The problem of Example 2.3 is now treated using the Kendall Tau statistic. We first calculate the value of $S$. Since the only tie is in the $X$ sample, it does not seem necessary to list all the possible pairs. However, then we must remember to score 0 for the $Y$ pair that corresponds to this tied $X$ pair. Taking the 24 $Y$ ranks in the order listed, we obtain

$$U = 11 + 20 + 20 + 20 + 18 + 17 + 12 + 16 + 14 + 12 + 8 + 11$$
$$+ 7 + 10 + 9 + 8 + 6 + 6 + 4 + 4 + 2 + 2 + 0 + 0$$
$$= 237$$

$$V = 12 + 1 + 1 + 0 + 1 + 1 + 5 + 0 + 1 + 2 + 5 + 1$$
$$+ 4 + 0 + 0 + 0 + 1 + 0 + 1 + 0 + 1 + 0 + 1 + 0$$
$$= 38$$

$$S = U - V = 237 - 38 = 199.$$

Using (2.15) we obtain the Kendall Tau coefficient as

$$T = \frac{2(199)}{24(23)} = .721.$$

(Note that (2.13) gives $T = .717$ and (2.14) gives .725, both of which are incorrect.) If (2.19) is used to permit a correction for ties, the result is

$$T = \frac{199}{\sqrt{275}\ \sqrt{276}} = .722.$$

Thus the correction here is negligible to three decimal places.

We now test the null hypothesis of no association against the one-sided alternative $A_+$, a direct relationship. For $n = 24$, from Table J we find that .377 has a right-tail probability of .005. Since the observed value .721 is larger than .377, the exact $P$-value for these data must be smaller than .005. Again, the appropriate conclusion is that the data support the alternative of a direct association between voter registration and turnout rate.

To illustrate the use of the large sample approximation to determine an asymptotic approximate $P$-value, from (2.20) we calculate

$$z = \frac{3(.721)\sqrt{24(23)}}{\sqrt{2(53)}} = 4.93.$$

From Table A the right-tail probability for this value of $z$ is .0000 to four decimals.

**example 2.8**

Schachter, Cohen, and Goldstein (1972, p. 129) reported an extensive study of empirical determinants for the supply of and demand for economists in U.S. institutions of higher learning. Data were collected by sending a questionnaire to approximately 1200 departments of economics in 1970; only 402 departments responded. The authors felt that this group of respondents could be considered a representative sample for their purposes if the respondents were geographically distributed in reasonably close accordance with the regional distribution of all academic institutions. The United States was divided into 10 nonoverlapping geographic regions. Table 2.5 below shows the percent distributions of institutions according to these regions, for all institutions and for those responding. Find a measure of association for these data.

**TABLE 2.5**

| Region | All Institutions | Responding Institutions |
|---|---|---|
| New England | 9.0 | 7.8 |
| Middle Atlantic | 16.6 | 19.3 |
| East No. Central | 16.2 | 20.1 |
| West No. Central | 11.3 | 7.1 |
| South Atlantic | 16.2 | 13.0 |
| East So. Central | 7.1 | 4.8 |
| West So. Central | 7.8 | 8.9 |
| Mountain | 4.0 | 7.4 |
| Pacific | 11.2 | 10.0 |
| Other | .3 | 1.5 |

**SOLUTION**
We calculate both $R$ and $T$ for these data, first listing the pairs of observations so that the "all institutions" percentages ($X$) are in natural order; see Table 2.6.

**TABLE 2.6**

| X | Y | X Ranks | Y Ranks | D | D² |
|---|---|---|---|---|---|
| .3 | 1.5 | 1 | 1 | 0 | 0 |
| 4.0 | 7.4 | 2 | 4 | −2 | 4 |
| 7.1 | 4.8 | 3 | 2 | 1 | 1 |
| 7.8 | 8.9 | 4 | 6 | −2 | 4 |
| 9.0 | 7.8 | 5 | 5 | 0 | 0 |
| 11.2 | 10.0 | 6 | 7 | −1 | 1 |
| 11.3 | 7.1 | 7 | 3 | 4 | 16 |
| 16.2 | 20.1 | 8.5 | 10 | −1.5 | 2.25 |
| 16.2 | 13.0 | 8.5 | 8 | .5 | .25 |
| 16.6 | 19.3 | 10 | 9 | 1 | 1 |
| | | | | | $\overline{29.50}$ |

Without correction for ties,

$$R = 1 - \frac{6(29.50)}{10(99)} = .821.$$

With correction for ties,

$$R = \frac{10(99) - 6(29.50) - 3}{\sqrt{10(99) - 6}\ \sqrt{10(99)}} = .821.$$

Now we calculate

$$S = (9 + 6 + 7 + 4 + 4 + 3 + 3 + 0 + 1)$$
$$- (0 + 2 + 0 + 2 + 1 + 1 + 0 + 1 + 0 + 0) = 37 - 7 = 30.$$

Without correction for ties,

$$T = \frac{2(30)}{10(9)} = .667.$$

With correction for ties,

$$T = \frac{30}{\sqrt{(45 - 1)}\ \sqrt{45}} = .674.$$

Note that although the correction for ties has little effect on the value of either measure (since the ties are not extensive), the difference between the values of $R$ and $T$ seems substantial. No hypothesis test is called for here, but, for the sake of illustration, the one-tailed $P$-values from Tables I and J, respectively, are

$R$:    $P$-value is about .003
$T$:    $P$-value is between .002 and .005.

Thus, from a probability point of view, the difference between the observed values of $R$ and $T$ is almost negligible. This illustrates the danger in trying to give an interpretation to the numerical value of a measure of association without reference to its $P$-value.

### 2.3    Additional Notes

The proper interpretation of a measure of association, whether it is computed as the coefficient of rank correlation, the Kendall Tau, or any other measure, is quite important. Association is simply a description of the relationship between the variables; the existence of a significant association provides no evidence about any kind of a causal relationship between the variables. Any association that exists between two variables may be due to another factor or variable, or even several other factors, or it may not be due to any factor that is identifiable. Attributing an association to some particular factor is a nonstatistical, and usually dangerous, conclusion. The statement of any results of a study of association should be prepared carefully so that opportunities for improper interpretation are kept to a minimum. This is true for any measure of association, whether parametric or nonparametric.

With the nonparametric methods under study here, we have an additional problem of interpretation in that the association being measured is of the *positions* of the variables relative to each other — that is, of the ranks, not of the values of the variables on some more refined scale of measurement. If the distribution of such a variable is such that there is a high correlation between ranks and variate values, the nonparametric measures can be interpreted as reasonable approximations to the association between the variables, but otherwise they simply measure the association between ranks.

Either the coefficient of rank correlation or the Kendall Tau can be used to describe trend or perform a test of trend. Suppose that we have a set of $n$ observations, $Y_1, Y_2, \ldots, Y_n$, measured on at least an ordinal scale, and each measurement is made at some specific point in time or covers some specific time period. Then if we consider each $Y_i$ as paired with an $X_i$ that represents the time of measurement, we again have a bivariate sample such that each variable can be ordered. If there is a definable trend over time, the $Y$ measurements tend to become larger (or smaller) as time goes by. Thus a positive trend is suggested if $X$ and $Y$ have a significant direct association, and a negative trend is implied by a significant indirect association. If we assume that the $Y_i$ are independent, the sampling distributions of $R$ and $T$ are the same as before if we formulate the null hypothesis as

> $H:$ The $Y_i$ are identically distributed.

The one-sided alternatives are then

> $A_+:$ The $Y_i$ exhibit an upward trend
> $A_-:$ The $Y_i$ exhibit a downward trend,

and the two-sided alternative is

> $A:$ There is either an upward or a downward trend.

The $P$-values are found as before. This use of $T$ to test for trend was proposed by Mann (1945), and the corresponding use of $R$ was suggested by Daniels (1950).

It should be noted that the validity of these tests relies on the assumption that the $Y_i$ are independent. This would not be the case for monthly, data, for example, if there is a seasonal pattern by months. Of course, if the seasonal variation can be removed, these procedures are applicable to the adjusted data. For such applications, $R$ and $T$ should be considered nonparametric analogues of the parametric test based on the regression coefficient (which assumes the normal distribution). The asymptotic efficiency of either of these nonparametric tests relative to the regression test is about .98 for normal distributions (Stuart, 1956).

Other procedures which can be used as tests of trend are discussed in Chapter 8.

### 2.4    Comparison of the Rank Correlation Coefficient and Kendall Tau Statistic

For bivariate data such that each sample is measured on at least an ordinal scale, either the rank correlation coefficient or Kendall Tau statistic can be used to describe association between the two variables and/or to test the null hypothesis that the two variables are independent. Hence, it is appropriate to compare the properties, interpretation, and performance of these coefficients. Although $R$ is probably better known, the Tau statistic has certain advantages.

In each case, an exact test is easily performed for small samples. The exact distributions are found by first listing all possible arrangements of the ranks of the $X$'s paired with the ranks of the $Y$'s. Under the null hypothesis of independence, each of the $n!$ distinguishable arrangements of ranks is equally likely. Hence the sampling distributions are found by calculating the value of $R$ or $T$ for each arrangement, counting the number of arrangements that produce each value, and dividing that number by $n!$ to obtain the probability of that value of $R$ or $T$. The asymptotic distribution of the standardized test statistic is the normal distribution in each case. However, $T$ approaches normality more rapidly than $R$. Thus, in samples of moderate size, a $P$-value based on $T$ is more reliable than one based on $R$, and this is a distinct advantage of $T$.

The numerical values of $R$ and $T$ are in general not the same for a given set of rankings, except when there is perfect agreement or perfect disagreement and no ties. This is easily verified by comparing the results obtained in Examples 2.1 and 2.5, or 2.3 and 2.7, or those of Example 2.8. The two coefficients have different scales, even though both vary between $-1$ and $+1$. The value of $R$ is usually larger than the value of $T$ when both are calculated from the same set of data. It can even happen that, say, $T = 0$ while $R = .5$. For most degrees of association that occur in practice (that is, absolute values not too close to 1) $R$ is about 50 percent greater than $T$ in absolute value. Even though numerical values of $R$ and $T$ are not comparable, $P$-values are, and both statistics usually produce the same result when used for inference. This should not be considered an anomaly, since the inference in each case is about the independence of the rank arrangements, or the independence of the underlying variables that produced the ranks, and not about some parameter that measures degree of association.

The two coefficients measure association in somewhat different ways. It is considerably easier to give an interpretation to $T$ than to $R$. In fact, a precise meaning can be given to the value of $T$ in either of two ways. Two observation pairs, say $(X_i, Y_i)$ and $(X_j, Y_j)$, are called concordant if the direction of difference is the same with $X$ as with $Y$; that is, $X_i < X_j$ whenever $Y_i < Y_j$, or $X_i > X_j$ whenever $Y_i > Y_j$. They are called discordant when the direction of difference is not the same. Thus $T$ can then be interpreted as the number of concordant pairs minus the number of discordant pairs, divided by the total number of distinguishable pairs, or equivalently as the excess of the proportion of concordant pairs over the proportion of discordant pairs. Of course, we can interpret $R$ as a relative measure of the sample covariance between rank pairs, or the covariance divided by the square root of the product of the variances of the $X$ ranks and $Y$ ranks. Although covariance has a precise mathematical definition, it is difficult to give it a meaningful interpretation otherwise.

Since $R$ and $T$ are useful descriptive measures for association in paired observations, it would be nice if we could interpret $R$ and $T$ as estimates of some parameter that describes association in the population. The logical parameter is of course the one used in parametric statistics, namely $\rho$, the Pearson product-moment correlation coefficient. However, even if we do consider $R$ and $T$ as estimates of $\rho$, we cannot interpret a test of independence as equivalent to a test that $\rho = 0$. Independence and zero correlation are not equivalent properties of variables in general. It is true that independence always implies zero correlation, but the reverse implication does not necessarily hold. That is, it is possible for two variables to be dependent even though they are not correlated. The bivariate normal is a significant exception, and this distribution is assumed for the classical parametric test. Therefore, in classical statistics, if we reject the null hypothesis of independence we can conclude that $\rho \neq 0$, $\rho > 0$ or $\rho < 0$, as the case may be. However, in nonparametric statistics, where no particular distribution is assumed for the population, we cannot interpret the results of a test for independence in terms of a value of $\rho$ unless the null hypothesis is accepted.

In actuality, we should not think of either $R$ or $T$ as an estimate of $\rho$ anyway. $T$ is an unbiased estimator of a parameter denoted by $\tau$, and defined as the probability of concordance minus the probability of discordance. Concordance and correlation do measure association in the same spirit but they are not the same. Even though $R$ measures the correlation between ranks in a sample of pairs, it cannot be considered the sample analogue of a population coefficient of rank correlation. We assumed the population to be continuous, but continuous variables cannot be ranked because an infinite number of values cannot be ordered. Nevertheless, $R$ is an unbiased estimator of a parameter that does exist for continuous distributions. This parameter is a function of both $\tau$ and a quantity sometimes called the grade correlation. (See Gibbons (1972, pp. 235–240), for further details.)

The above discussion is not meant to imply that tests based on $R$ and $T$ cannot be viewed as the nonparametric counterparts of the parametric test based on the sample correlation coefficient. Because that test is based on the assumption of a bivariate normal distribution, the inferences of independence and zero correlation are equivalent. The asymptotic efficiency of the tests based on both $R$ and $T$ relative to this parametric test is $9/\pi^2 = .912$ for normal distributions, and 1 for uniform distributions.

### 2.5   Partial Correlation

As mentioned earlier, any association that exists between two variables may be due to one or more other factors or variables. If these other variables can be measured, their influence can be eliminated by computing what is called a partial correlation measure. It is partial in the sense that any indirect correlation that can be attributed to other factors is removed from the description of correlation between the two variables of primary interest. We limit the discussion here to three variables. Then we have two primary variables, but some third variable is thought to be strongly associated with both of them. The partial correlation describes the relationship between the other two variables when the third variable is held constant.

Specifically, suppose we have $n$ triplets of observations, $(X_1, Y_1, Z_1)$, $(X_2, Y_2, Z_2)$, . . . , $(X_n, Y_n, Z_n)$, that is, $n$ observations on each of the three characteristics under study. Then we could compute the Kendall Tau coefficient for each of the three different sets of paired observations. Let $T_{xy}$ denote the value of Tau for the set of observations $(X_1, Y_1)$, . . . , $(X_n, Y_n)$, and similarly $T_{xz}$ and $T_{yz}$ are the Tau coefficients for the sets $(X_1, Z_1)$, . . . , $(X_n, Z_n)$ and $(Y_1, Z_1)$, . . . , $(Y_n, Z_n)$, respectively. Each of these coefficients measures the association between the respective characteristics indicated in the subscript of $T$ without regard for the other characteristic.

Now suppose that we are interested in measuring the association between the $X$ and $Y$ characteristics when the $Z$ factor is held constant. This measure is called a *partial Tau coefficient*, denoted by $T_{xy.z}$ to show that $Z$ is the factor held constant. It may be computed from the formula

$$T_{xy.z} = \frac{T_{xy} - T_{xz}T_{yz}}{\sqrt{(1 - T_{xz}^2)(1 - T_{yz}^2)}}. \tag{2.21}$$

This coefficient takes on values between $-1$ and $+1$, and stronger association is reflected by larger absolute values. Interpretations of the sign and magnitude of the partial Tau are the same as for the ordinary Tau, except that the effects of the third factor, in this case $Z$, have been eliminated. However, Table J is *not* applicable for partial Tau coefficients. Since the sampling distribution of the partial Tau has not been tabled in the literature, the value of $T_{xy.z}$ must be considered only a descriptive measure, from which statistical inferences cannot be made.

**example   2.9**

Although postbaccalaureate training or a graduate degree is almost essential for many professions, students interested in actuarial science are frequently advised that on-the-job training will be a more important factor in career advancement and success than the pursuit of extensive formal academic study. Many aspirants to a career in actuarial science therefore take a position in the insurance field right after the bachelor's degree and devote their time occasionally to graduate courses in probability and mathematics, but primarily to independent study for the professional actuary examinations. Suppose we have the figures on age, education, and income of a group of professional actuaries as shown in Table 2.7. Since age is clearly associated with both income and education, find the Tau coefficient of partial correlation between education and income when age is held constant.

**TABLE 2.7**

| Actuary | Age in Years | Course Units of Graduate Education | Income in Thousands |
|---------|--------------|-----------------------------------|---------------------|
| A | 40 | 18 | $25.0 |
| B | 35 | 0 | 20.0 |
| C | 30 | 6 | 15.0 |
| D | 36 | 15 | 24.1 |
| E | 41 | 24 | 30.0 |
| F | 45 | 30 | 28.0 |
| G | 48 | 45 | 29.1 |

**SOLUTION**

We call $Z$ the age characteristic, and $X$ and $Y$ education and income, respectively. The three tabulations below are formed to compute the values of the Tau coefficients from (2.13) for all possible pairs of variables.

$$X: \quad 0 \quad 6 \quad 15 \quad 18 \quad 24 \quad 30 \quad 45$$
$$Z: \quad 35 \quad 30 \quad 36 \quad 40 \quad 41 \quad 45 \quad 48$$

$$U_{xz} = 5 + 5 + 4 + 3 + 2 + 1 + 0 = 20$$
$$T_{xz} = \frac{4(20)}{7(6)} - 1 = \frac{38}{42}$$

$$X: \quad 0 \quad 6 \quad 15 \quad 18 \quad 24 \quad 30 \quad 45$$
$$Y: \quad 20.0 \quad 15.0 \quad 24.1 \quad 25.0 \quad 30.0 \quad 28.0 \quad 29.1$$

$$U_{xy} = 5 + 5 + 4 + 3 + 0 + 1 + 0 = 18$$

$$T_{xy} = \frac{4(18)}{7(6)} - 1 = \frac{30}{42}$$

$$Y: \quad 15.0 \quad 20.0 \quad 24.1 \quad 25.0 \quad 28.0 \quad 29.1 \quad 30.0$$
$$Z: \quad 30 \quad 35 \quad 36 \quad 40 \quad 45 \quad 48 \quad 41$$

$$U_{yz} = 6 + 5 + 4 + 3 + 1 + 0 + 0 = 19$$

$$T_{yz} = \frac{4(19)}{7(6)} - 1 = \frac{34}{42}$$

Substituting these results in (2.21), we obtain

$$T_{xy.z} = \frac{\dfrac{30}{42} - \left(\dfrac{38}{42}\right)\left(\dfrac{34}{42}\right)}{\sqrt{\left(1 - \dfrac{1444}{1764}\right)\left(1 - \dfrac{1156}{1764}\right)}} = \frac{-32}{\sqrt{(320)(608)}} = -.073.$$

From Table J, the one-sided $P$-value for $T_{xy} = 30/42 = .714$ when $n = 7$ is .015. We would conclude that there exists an association between $X$ and $Y$. However, when the effects of age, the $Z$ variable, are removed, the measure of association between income and education is reduced to $-.073$. While we cannot state a $P$-value for this coefficient, $T_{xy.z}$ is quite small. Such data are too limited to draw any major conclusions, but they do seem to imply a lack of association between $X$ and $Y$ except through the $Z$ variable. Since the value of $T_{xy.z}$ is so small, no significance should be attached to the minus sign in this case.

## 3   $k$ Related Samples

We now extend the situation of the preceding section to the problem of defining a descriptive measure of association between some $k$ characteristics, and finding a test statistic that can be used for inference about the existence of association between the $k$ characteristics in the population. The data here consist of $k$ sets of rankings. These sets may arise from measurements on each of $k$ characteristics for a group of experimental units, or from measurements on experimental units that appear in $k$ matched groups. The measurements are related because either they are all made on the same experimental unit, or they are made on units that are matched into groups by design or by nature. For example, we might have test scores measuring $k$ different attributes on each of $n$ individuals, or a set of rankings of $n$ beauty contestants by a panel of $k$ judges.

A measure of the agreement between the $k$ sets of rankings of these $k$ related measurements describes the association between the $k$ characteristics. This quantity, called the Kendall coefficient of concordance, can also be used to test the null hypothesis that the $k$ characteristics are independent. If the agreement appears to be strong (the null hypothesis is rejected), we might also be interested in determining which ranking of the $n$ individuals seems to be the one agreed upon — that is, in estimating the order of preference.

The test of significance based on the coefficient of concordance is equivalent to a well-known nonparametric test called Friedman's two-way analysis of variance by ranks. This test is applicable to *related* samples drawn from $k$ populations, and the inference concerns the homogeneity of the population means or medians. (This sampling model is to be distinguished from the situation in Section 3 of Chapter 4, where we dealt with $k$ *independent* samples and used the Kruskal-Wallis test.) A multiple comparisons procedure can be used to determine which pairs of means differ significantly.

The final topic discussed in this section is a coefficient of concordance for $k$ sets of rankings that are incomplete but balanced in a certain way. Inferences about independence can also be made in this situation.

Throughout this section, we assume that the probability of a tie within any of the $k$ sets of observations is equal to zero, so that unique rankings exist.

### 3.1 Kendall Coefficient of Concordance for Complete Rankings

*Sampling situation*    The data to be analyzed consist of $k$ complete sets of rankings of $n$ objects. Each ranking is made by a comparison of objects within that set only. The data as collected or observed consist of $nk$ ranks or measurements on at least an ordinal scale. Such data can arise in several conceptually different ways. In the simplest case, the same group of $n$ objects is presented to each of $k$ observers; the observers score or rank each of the objects according to some criterion. Alternatively, we might have scores or measurements on $k$ characteristics for each of $n$ objects or subjects, for example, scores on tests in mathematics, English, and history for each of $n$ students. The situation is then the same once the sets of $n$ math scores, $n$ English scores, and $n$ history scores are ranked. Another possibility is economic data on $k$ measurable variables for each of $n$ different countries or groups. Each of these situations is completely analogous to having observations on $n$ experimental units under each of $k$ matching conditions.

The various possibilities for sampling situations will become clearer as examples are presented. The important thing about the data is that the comparisons that are of interest, are relevant, or make sense, are comparisons *within* each of the $k$ groups or characteristics. Hence the rankings are assigned separately within each of the $k$ groups (as opposed to one assignment of $nk$ ranks, which was the type of data analyzed by the Kruskal-Wallis test in Section 3 of Chapter 4).

**example 3.1**

In order to illustrate the data to be analyzed, we consider the extension of the beauty contest situation of Example 2.1 to the case where each member of a panel of three judges is asked to rank the eight contestants. Suppose the results are as shown in Table 3.1. The bottom row of the table shows the sums of the three ranks given to each contestant. The last column shows the sum of all ranks given by each judge; these sums are, of course, equal since each judge ranks each contestant.

**TABLE 3.1**

| Judge | Contestant A | B | C | D | E | F | G | H | Sum |
|---|---|---|---|---|---|---|---|---|---|
| 1 | 2 | 1 | 3 | 5 | 4 | 8 | 7 | 6 | 36 |
| 2 | 1 | 2 | 4 | 5 | 7 | 6 | 8 | 3 | 36 |
| 3 | 3 | 2 | 1 | 7 | 5 | 8 | 6 | 4 | 36 |
| Sum $R_j$: | 6 | 5 | 8 | 17 | 16 | 22 | 21 | 13 | 108 |

Assume that the investigator is interested in measuring the association between the three sets of rankings. If there is *no* agreement between the three sets of rankings, the column totals would be approximately equal for each contestant. If they were exactly equal, each column sum would have to be $108/8 = 13.5$. This value is, of course, impossible since it is not an integer. However, it is the *expected* value of each column sum if there is no association — that is, if the judges are assigning ranks independently. A measure of the departure from equal column sums can be found by taking the sum of squares of deviations between the observed column sums and the expected column sums, which we denote by $S$. In this case

$$S = (6 - 13.5)^2 + (5 - 13.5)^2 + (8 - 13.5)^2 + \cdots + (13 - 13.5)^2 = 306.00.$$

(Note the similarity between this situation and the development of the chi-square goodness-of-fit test in Section 2 of Chapter 2. However, a goodness-of-fit test is not appropriate here, since the data are not counts but ranks.)

Although the number 306.0 is a descriptive measure of the lack of equality between rank sums, it is not a figure that can be interpreted easily until it is reduced to a relative value. To be consistent with the measures developed for association in the paired-sample case, the absolute value of the relative measure should be between 0 and 1, with 1 representing perfect agreement of rankings and 0 representing no association.

If the three sets of rankings by the judges were in perfect agreement, the contestant who is judged most beautiful by all would have been assigned rank 1 by all judges, the next most beautiful would be ranked 2 by all, and so on, so that the column sums would be some rearrangement of the numbers

$$1(3), \ 2(3), \ 3(3), \dots, 8(3).$$

The sum of squares of deviations of these column sums (with perfect agreement) from the column sums expected (with no association) is

$$(3 - 13.5)^2 + (6 - 13.5)^2 + (9 - 13.5)^2 + \cdots + (24 - 13.5)^2 = 378.0.$$

The relative agreement can be measured by the ratio of the actual sum of squares to the sum of squares under perfect agreement, or

$$\frac{306.00}{378.00} = \frac{51}{63} = .810.$$

The ratio formed in this way would clearly equal 1 for perfect agreement and tend toward 0 for no agreement, but it can never be negative. What about a measure of perfect disagreement? The question is not relevant here because there is no such thing as perfect disagreement for more than two sets of rankings. If, say, the first set of ranks is arranged in natural order, and the corresponding ranks of the second set appear in reverse natural order, there is no way that the third set can be in complete disagreement with both of the first two sets of rankings.

*Rationale and Descriptive Measure*  In order to develop a formula for the general case, suppose that the original data are recorded in a table following the format used in the beauty contest example. Then each set of numbers that are to be ranked or are already ranked as a complete set occupies a row in the table. If there are *k* sets of *n* ranks, the table has *k* rows and *n* columns. Since each row is some rearrangement of the first *n* integers, each row sum equals

$$1 + 2 + \cdots + n = \frac{n(n + 1)}{2}.$$

The grand total of all ranks in the table is then $kn(n + 1)/2$, and the average column sum, or the column sum expected if there is no association, is

$$\frac{k(n + 1)}{2}. \tag{3.1}$$

Now let $R_1, R_2, \ldots, R_n$ denote the actual column sums observed. The sum of squares of the deviations between observed and expected column sums is

$$S = \sum_{j=1}^{n} \left[ R_j - \frac{k(n+1)}{2} \right]^2.$$  (3.2)

$S$ measures the departure from lack of agreement in the sense that the sum would be small if there is no agreement between sets of rankings and large if there is association.

For perfect agreement, the column sums are a rearrangement of the numbers

$$1k, 2k, 3k, \ldots, nk.$$

The sum of squares of the deviations of these column sums from the expected column sums is

$$\sum_{j=1}^{n} \left[ jk - \frac{k(n+1)}{2} \right]^2 = k^2 \sum_{j=1}^{n} \left[ j - \frac{(n+1)}{2} \right]^2$$
$$= \frac{k^2 n(n^2 - 1)}{12},$$  (3.3)

and it can be shown that this is the largest possible value of (3.2) for any set of column sums. Hence the relative measure of association, or the degree of communality or agreement, which is called the Kendall coefficient of concordance, is (3.2) divided by (3.3), or

$$W = \frac{12S}{k^2 n(n^2 - 1)}$$  (3.4)

$$= 12 \sum_{j=1}^{n} \frac{[R_j - k(n+1)/2]^2}{k^2 n(n^2 - 1)}.$$  (3.5)

The value of $W$ is always between 0 and 1, and the extreme values are interpreted as follows.

| Value of $W$ | Interpretation |
|---|---|
| 0 | No association |
| 1 | Perfect association |

*Computation*   The computation of $W$ can be simplified in most cases by using the following expression, which is algebraically equivalent to (3.4) and (3.5):

$$W = \frac{12 \sum_{j=1}^{n} R_j^2 - 3k^2 n(n+1)^2}{nk^2(n^2 - 1)}.$$  (3.6)

This formula is usually easier because it involves only a sum of squares of integer values, whereas the deviations that are squared and summed in (3.2) or (3.5) are frequently not integers.

We illustrate the use of (3.6) for the data of Example 3.1. The sum of squares of column totals is

$$6^2 + 5^2 + 8^2 + \cdots + 13^2 = 1764,$$

and $n = 8$, $k = 3$. Substituting in (3.6), we obtain

$$W = \frac{12(1764) - 3(9)(8)(81)}{8(9)(63)} = \frac{3672}{4536} = \frac{51}{63} = .810,$$

which is equal to the result obtained before.

*Ties*  If any of the observations within a set to be ranked are tied, the mid-rank method should be used to assign ranks. The sum of squares of deviations, $S$ in (3.2), is calculated in the usual way, but with tied ranks the maximum value of $S$ is smaller than (3.3). The recommended adjustment is to replace (3.3) by

$$\frac{k^2 n(n^2 - 1) - k(\Sigma t^3 - \Sigma t)}{12},$$

where $t$ is the number of observations tied at any rank in any set of rankings, and the summation is over all sets of $t$ tied ranks and all $k$ sets of rankings. This is equivalent to replacing the denominator of $W$ in (3.6) by

$$k^2 n(n^2 - 1) - k(\Sigma t^3 - \Sigma t). \tag{3.7}$$

As usual, the correction factor has little effect on the value of $W$ unless the ties are very extensive.

*Inference situation*  If the $k$ sets of rankings result from measurements on $k$ characteristics, the Kendall coefficient of concordance can be used to test the null hypothesis that the characteristics are independent — that is, that there exists no association or relationship between the $k$ variables that measure the $k$ characteristics. More generally, for any $k$ sets of rankings the Kendall coefficient can be used to test the null hypothesis of no association between rankings. We write the hypothesis set as

$$H: \quad \text{No association} \qquad A: \quad \text{Association exists.}$$

Note that the alternative does not distinguish any direction of relationship; with more than two sets of rankings, a completely inverse relationship cannot be defined.

*Rationale*  If no association exists (that is, if the null hypothesis is true), then the ranks have been assigned independently to the $n$ objects for each of the $k$ sets of rankings, and the column sums tend toward equality, so $W$ is close to 0. If there were association, then most of the large ranks would appear together in the same columns, and small ranks would also be clustered in the same columns. $S$ becomes larger as this kind of agreement increases, and hence $W$ becomes close to 1. Accordingly, large values of $W$ call for rejection of $H$ in favor of the alternative $A$.

*Test statistic and computation*    The Kendall coefficient of concordance $W$, as defined in (3.4), (3.5), or (3.6), may be used as a test statistic. However, a test based on the quantity $S$, as defined in (3.2), is equivalent, and tables are usually designed for use with $S$. The simplest form for calculation of $S$ is indicated by the numerator of (3.6), that is

$$12S = 12 \sum_{j=1}^{n} R_j{}^2 - 3k^2 n(n + 1)^2. \tag{3.8}$$

For large samples, the test statistic that is easiest to use is $Q$ as calculated from either

$$Q = k(n - 1)W, \tag{3.9}$$

or

$$Q = \frac{12S}{kn(n + 1)}, \tag{3.10}$$

which are equivalent expressions.

*Finding the P-value*    Since the test statistics $S$ and $Q$ are monotonically increasing functions of $W$, and specifically $S$ and $Q$ are large when $W$ is large and zero when $W$ is zero, the appropriate $P$-values are right-tail probabilities no matter which test statistic is used.

Table K gives exact right-tail probabilities for $S$ for $n = 3$, $k \leq 8$, and $n = 4$, $k \leq 5$. For combinations of $n$ and $k$ that are not covered by this table, we use an approximation, which is adequate for most practical purposes unless $P$ is very small. The distribution of the test statistic $Q$ in (3.9) or (3.10) can be approximated by the chi-square distribution with $n - 1$ degrees of freedom. Hence Table B can be used to find asymptotic approximate $P$-values based on the value of $Q$.

In Example 3.1, we had $k = 3$ sets of rankings of $n = 8$ objects, and this $n$, $k$ combination is outside the range of Table K. Hence we use the chi-square distribution, with $n - 1 = 7$ degrees of freedom, to find an asymptotic approximate $P$-value. For the result $W = .810$ found before, the value of $Q$ from (3.9) is

$$Q = 3(7)(.810) = 17.01.$$

Since 17.01 lies between the entries in Table B for right-tail probabilities of .01 and .02 when df $= 7$, the $P$-value is $P < .02$. The data do not support the null hypothesis of independence, and we conclude that there is a relationship between the rankings of the three judges.

*Estimation*   If there are no ties and the value of $W$ equals 1, the sets of ranks are in perfect agreement and there is a consensus as regards the ordering of the objects. We know what that ordering is, since it is the same as the ordering of the sums of ranks assigned to the objects. If $W$ is less than one but large enough that the $P$-value is small or the null hypothesis of no association is rejected, the investigator may wish to conclude that there exists a consensus as regards the ordering of the objects, and that the observed results differ from this unique communal ranking only because of sampling variation. The natural estimate of this ordering is again provided by the ordering of the magnitudes of the column sums in the $k \times n$ table since the smaller the column sum, the more preferred is the object by the judges as a group. For the data of Example 3.1, the contestants are listed below in increasing order of column sums, that is, decreasing order of beauty as assessed by the three judges.

B    A    C    H    E    D    G    F

Most beautiful ↔ least beautiful

Using this basis of estimation, the estimated rank of B is 1, of A is 2, . . . , and of F is 8.

This estimation in accordance with column sums is not only the natural one; it also has some good statistical properties. The property which is perhaps easiest to understand and appreciate is that the average of the rank correlation coefficients between this estimated set and all $k$ observed sets is a maximum. Another property is that this estimated order is the one that minimizes the sum of the squares of the error in the estimate; that is, it is the least-squares estimate.

**example  3.2**

    Coffee sold commercially in the United States is usually a blend of roasted green coffee beans. The blending operation harmonizes the divergent characteristics of the green coffee beans in various ways to produce coffees with distinct flavors, aromas, and tastes. There are some 8 to 10 characteristics defined for each green coffee bean and thus to the blended coffee. A few of these are measured objectively, but most are measured on a rating scale by professional coffee tasters. A coffee manufacturer, unsure of the ability and objectivity of these professional tasters, chose 10 blends of coffees differing by only a small degree in one specific characteristic. The manufacturer then asked four professional tasters to judge the quality of this characteristic for each blend, and to rank the coffees from 1 to 10. The results are presented below.

| | Blend | | | | | | | | | |
|---|---|---|---|---|---|---|---|---|---|---|
| Taster | I | II | III | IV | V | VI | VII | VIII | IX | X |
| 1 | 10 | 9 | 7 | 8 | 6 | 5 | 4 | 1 | 2 | 3 |
| 2 | 10 | 9 | 8 | 6 | 7 | 3 | 5 | 2 | 1 | 4 |
| 3 | 10 | 9 | 8 | 7 | 6 | 5 | 4 | 1 | 2 | 3 |
| 4 | 9 | 10 | 7 | 8 | 6 | 5 | 4 | 1 | 2 | 3 |

What can be said concerning the agreement of these four professionals and their ability to distinguish these 10 blends? Estimate the order of preference for the blends.

**SOLUTION**

We test the null hypothesis of no agreement between the 4 sets of rankings. where $k = 4$, $n = 10$. The respective column sums of ranks and their squares are given in Table 3.2.

**TABLE 3.2**

| Blend | $R_j$ | $R_j^2$ |
|---|---|---|
| I | 39 | 1521 |
| II | 37 | 1369 |
| III | 30 | 900 |
| IV | 29 | 841 |
| V | 25 | 625 |
| VI | 18 | 324 |
| VII | 17 | 289 |
| VIII | 5 | 25 |
| IX | 7 | 49 |
| X | 13 | 169 |
| Sum: | 220 | 6112 |

From (3.8) we find that

$$12S = 12(6112) - 3(4)^2 10(11)^2 = 15,264,$$

and, from (3.6), the measure of association is

$$W = \frac{15,264}{10(4)^2(99)} = .964.$$

In order to find the $P$-value, we calculate $Q$ from (3.10) as

$$Q = \frac{15,264}{4(10)(11)} = 34.7, \quad \text{with df} = 9.$$

From Table B, the asymptotic approximate $P$-value is $P < .001$, so we reject the null hypothesis and conclude that the tasters are not ranking the blends independently and in fact agree reasonably well on a quality ranking of the blends as VIII, IX, X, VII, VI, V, IV, III, II, I.

*Additional notes* For $k$ sets of rankings it is, of course, possible to select any pair of rankings and compute either the coefficient of rank correlation or the Kendall Tau coefficient, as in Sections 2.1 and 2.2, respectively. If a comparison between some *two* sets of rankings is of particular interest, one of these measures is informative for either description or inference. However, when there are $k$ sets of rankings, or $k$ relevant characteristics, all possible comparisons of two sets of rankings may be of interest. There are a total of $k(k-1)/2$ distinguishable sets of pairs of rankings, each producing its own value for the measure of association selected.

Suppose that $R$ is calculated for each pair following the formula in (2.6) or (2.7). How might these $k(k-1)/2$ numbers be combined into a single, overall measure of association? A logical possibility is to take their simple average; that is, sum the values of $R$ and divide by $k(k-1)/2$. When there are no ties, this average is always equal to the value of $(kW-1)/(k-1)$, where $W$ is the Kendall coefficient of concordance for these $k$ sets of rankings.

We illustrate this interesting fact by computing the value of $R$ for each possible pair of rankings, that is, each subset of two judges, in Example 3.1. The computations are shown in Table 3.3.

**TABLE 3.3**

| Judge | Contestant | | | | | | | | Sum |
|---|---|---|---|---|---|---|---|---|---|
| | A | B | C | D | E | F | G | H | |
| 1 | 2 | 1 | 3 | 5 | 4 | 8 | 7 | 6 | |
| 2 | 1 | 2 | 4 | 5 | 7 | 6 | 8 | 3 | |
| $D$ | 1 | $-1$ | $-1$ | 0 | $-3$ | 2 | $-1$ | 3 | |
| $D^2$ | 1 | 1 | 1 | 0 | 9 | 4 | 1 | 9 | 26 |

$$R = 1 - \frac{6(26)}{8(63)} = \frac{29}{42}$$

**TABLE 3.3 Continued**

| Judge | A | B | C | D | E | F | G | H | Sum |
|---|---|---|---|---|---|---|---|---|---|
| | | | | Contestant | | | | | |
| 1 | 2 | 1 | 3 | 5 | 4 | 8 | 7 | 6 | |
| 3 | 3 | 2 | 1 | 7 | 5 | 8 | 6 | 4 | |
| D | −1 | −1 | 2 | −2 | −1 | 0 | 1 | 2 | |
| $D^2$ | 1 | 1 | 4 | 4 | 1 | 0 | 1 | 4 | 16 |

$$R = 1 - \frac{6(16)}{8(63)} = \frac{34}{42}$$

| Judge | A | B | C | D | E | F | G | H | Sum |
|---|---|---|---|---|---|---|---|---|---|
| | | | | Contestant | | | | | |
| 2 | 1 | 2 | 4 | 5 | 7 | 6 | 8 | 3 | |
| 3 | 3 | 2 | 1 | 7 | 5 | 8 | 6 | 4 | |
| D | −2 | 0 | 3 | −2 | 2 | −2 | 2 | −1 | |
| $D^2$ | 4 | 0 | 9 | 4 | 4 | 4 | 4 | 1 | 30 |

$$R = 1 - \frac{6(30)}{8(63)} = \frac{27}{42}$$

$$\text{Average } R = \left(\frac{29}{42} + \frac{34}{42} + \frac{27}{42}\right)/3 = \frac{5}{7}$$

In Example 3.1 we found that $W = 51/63 = 17/21$ for these three sets of rankings. Then, since $k = 3$, we have

$$\frac{kW - 1}{k - 1} = \frac{3(17/21) - 1}{2} = \frac{10/7}{2} = \frac{5}{7},$$

which agrees with the result obtained for average $R$ in Table 3.3.

### 3.2  Friedman Test and Multiple Comparisons

Recall that for procedures based on the coefficient of concordance, the data to be analyzed consist of $k$ sets of rankings. In this subsection we present a procedure, called the Friedman test, that is also applicable to this same kind of data. As we shall see, the methodology is identical, but the inference here is concerned with homogeneity of ranked objects rather than independence of rankings, and the sampling model is somewhat more complicated.

In the situation of the preceding subsection, for example, the data might arise as scores of students on math, English, and history tests, where each student takes each test. For each subject separately, the students are ranked according to their relative performance on that test. If there are, say, four students, we have three sets of rankings of four objects here. The coefficient of concordance describes the consistency of performance on the three tests or the agreement between rankings and can be used to test the hypothesis of independence between rankings. If this hypothesis is rejected, we conclude that there is agreement between rankings, and therefore a communality of performance in math, English, and history.

If we suspect a priori that there is some confounding between students and subjects, the implications of such a conclusion are difficult to assess. For example, if the students consist of one freshman, one sophomore, one junior, and one senior, the ranks in each subject may follow educational level. Then $W = 1$ and the agreement is perfect, but we have gained no information about the general relationship between the three subjects. If the goal is to compare student performance in math, English, and history, a better experiment would involve comparisons separated according to educational level. Suppose we have 12 different students, selected such that there are three freshmen, three sophomores, three juniors, and three seniors. Then there are four matched groups as opposed to four individuals, and valid comparisons can be made even if each student takes only one test. From the three freshmen, we select one at random to take the math test, another to take the English test, and the remaining one takes the history test. A similar random allocation is also made within each of the other three groups of students. Then we do not have to compare the math score of a senior with the math score of a freshman, or even with the history score of a freshman. Comparisons made within the freshman group, within the sophomore group, and so on, provide information relevant to relative performance. If we rank the three test scores within each group of four students, we have four sets of rankings of three objects.

As a somewhat more complicated design, suppose that a college counseling center is studying the effectiveness of three different types of therapy, namely individual therapy, group therapy, and computer simulated therapy, in comparison with each other and also in comparison with no therapy. A group of 12 students request therapy, but their schedules are such that they cannot all come to the center with the same frequency; frequency of therapy is recognized as an instrumental factor in improvement. Suppose that four students can come once a week, four can come twice a week, and four can come three times a week. Within each "frequency of therapy" group, the four students are randomly assigned to a "type of therapy," and the study begins. After a certain period of time, the students are all scored as regards amount of improvement. We are interested in comparing the effectiveness of the three types of therapy without interference from the "frequency of therapy" factor. Then comparisons between students are appropriate only within each frequency group. We rank observations in each frequency group from 1 to 4, and we have three sets of four rankings. These data can be used to test the null hypothesis that there is no difference between the four types of therapy.

We now generalize the sampling model in these examples. There are $nk$ experimental units, which either naturally or by design are separated into $k$ distinct groups (frequently called blocks or matching conditions) of $n$ matched or homogeneous units. Treatments are assigned randomly to the $n$ units in each block and treatment effects are measured in some way. This design is commonly known as a randomized complete block design.

The important aspect of the sampling model is that the relevant comparisons, or the comparisons that make sense, are within each of the $k$ blocks of $n$ matched units. To effect these comparisons, ranks $1, 2, \ldots, n$ are assigned to the observations within each block. If these ranks constitute the data to be analyzed, we have a sampling situation of $k$ sets of rankings of $n$ objects, exactly as in Section 3.1.

For statistical analysis we could proceed to find a measure of association between the $k$ rankings and/or to test the null hypothesis of no association. However, in most experiments or studies a measure of the association between the $k$ sets of rankings would have little meaning, because of the nature of the groups or because the original purpose of imposing the groups was solely to improve the design of the experiment. The same reasoning applies to an inference concerning the existence of an association between the rankings.

However, an important and useful inference can be made concerning the difference between the $n$ treatment effects. Letting $\mu_j$ denote the average effect of the $j$th treatment, the null hypothesis of interest is that the treatment effects are homogeneous, or

$$H: \quad \mu_1 = \mu_2 = \cdots = \mu_n,$$

and the alternative is

$$A: \quad \text{At least two of the } \mu_j\text{'s differ from each other.}$$

Suppose that the $k$ sets of rankings are presented in a table using the format of Example 3.1. If the null hypothesis is true, the $n$ column sums should be approximately equal to their average value $k(n + 1)/2$, and the quantity $S$ defined in (3.2) should be small. If certain treatments have more significant effects, the corresponding columns would tend to have more of the extreme ranks, and hence column sums that differ considerably from the average; the value of $S$ would be inflated. Accordingly, the test here can be based on the statistic $S$ defined in (3.2), or $Q$ in (3.9) and (3.10). Small values of $S$ or $Q$ support $H$ and large values discredit $H$ in favor of $A$. This test of no difference between $n$ treatment effects is the same as the test of no association between $k$ sets of rankings. However, when applied to this inference situation and to data that originate in the manner described here, it is usually referred to as Friedman's two-way analysis of variance by ranks. The sampling and inference situations are not equivalent, but there is absolutely no difference in the statistical methodology.

*Multiple comparisons*   If the $P$-value from the test statistic is so small that the null hypothesis of no difference between treatment effects should be rejected, we might want to know which treatment effects differ significantly from which others. The column sums in the $k \times n$ table, once divided by $k$, are the means of the ranks of the $n$ treatment effects for the data, and accordingly can be regarded as estimates of the ranks of the corresponding treatment effects in the population. That is, $R_j/k$ is an estimate of the rank of $\mu_j$ for $j = 1, 2, \ldots, n$. Multiple comparisons — that is, simultaneous tests with an overall level of significance — can be performed using these estimated treatment ranks or, equivalently, the column totals. The procedure is like that explained in Section 3.2 of Chapter 4 in connection with the Kruskal-Wallis test. We take all possible differences between rank sums of two treatments (column totals). Then the probability is at least $1 - \alpha$ that the following inequality is satisfied by all pairs $(R_i, R_j)$ for $1 \le i \ne j \le n$:

$$|R_i - R_j| \le z\sqrt{\frac{kn(n + 1)}{6}}. \tag{3.11}$$

(This is analogous to (3.4) of Chapter 4.) The constant $z$ is the quantile point of the normal curve that corresponds to a right-tail probability of $\alpha/n(n-1)$, since the total number of comparisons is $n(n - 1)/2$. Thus $z$ can be obtained from Table A for any $n$ and $\alpha$. For small $n$ and the typical values of $\alpha$, $z$ can be read from Table N, with $p = n(n - 1)/2$. Table 3.6 of Chapter 4 can also be used, with $k$ replaced by $n$.

At overall level $\alpha$, all pairs of differences of column sums that are larger than the right-hand side of (3.11) are significantly different pairs, and the direction of difference is determined by the sign of $R_i - R_j$. If some subset of the $n(n - 1)/2$ possible comparisons of column sums is desired, (3.11) may still be used, but the critical $z$ value must correspond to that $p$ which equals the number of comparisons actually made.

### example 3.3

Most parents assert that preprimary children ask questions constantly. Conversely, some recent studies have concluded that preprimary children show a lack in ability to ask questions. As a result of these findings, many researchers have been studying methods to develop children's skills in asking questions. One such study, reported in Torrance (1970), dealt with group size and number of questions asked of a person familiar to them in a classroom atmosphere. The following situation simulates this experiment. Suppose that 46 preprimary children are divided into four groups of fixed size; these group sizes are 24, 12, 6, and 4. Children are assigned randomly to these groups. The number of questions asked by all members of each group combined are recorded for a fixed short period of time on each of 8 days. The data in Table 3.4 represent the total numbers of questions asked by

**TABLE 3.4**

|  | Group Size | | | |
|---|---|---|---|---|
| Day | 24 | 12 | 6 | 4 |
| 1 | 14 | 23 | 26 | 30 |
| 2 | 19 | 25 | 25 | 33 |
| 3 | 17 | 22 | 29 | 28 |
| 4 | 17 | 21 | 28 | 27 |
| 5 | 16 | 24 | 28 | 32 |
| 6 | 15 | 23 | 27 | 36 |
| 7 | 18 | 26 | 27 | 26 |
| 8 | 16 | 22 | 30 | 32 |

each group according to day. Test the null hypothesis that group size has no significant effect on the question-asking performance of preprimary children, and perform a multiple comparisons analysis to determine which group sizes differ in question-asking performance.

**SOLUTION**
Let $\mu_j$ denote the average number of questions asked by the $j$th group size. The null hypothesis of interest is

$$H: \quad \mu_1 = \mu_2 = \mu_3 = \mu_4.$$

The comparisons of interest with these data are across group sizes for different days. Since the environmental and situational conditions are not the same each day, it would not make sense to compare the number of questions asked by a group of size 24 on Monday with the number asked by a group of size 6 on Wednesday. Hence comparisons will not be made simultaneously between all 32 observations, but only across group sizes separately for each different day. Thus we rank across rows, giving rank 1 to the smallest number of questions in each case. The data for analysis are presented in the usual $k \times n$ table of ranks, where $k = 8$ and $n = 4$, and we use the Friedman test to perform the analysis as in Table 3.5.

**TABLE 3.5**

| | Group Size | | | |
|---|---|---|---|---|
| Day | 24 | 12 | 6 | 4 |
| 1 | 1 | 2 | 3 | 4 |
| 2 | 1 | 2.5 | 2.5 | 4 |
| 3 | 1 | 2 | 4 | 3 |
| 4 | 1 | 2 | 4 | 3 |
| 5 | 1 | 2 | 3 | 4 |
| 6 | 1 | 2 | 3 | 4 |
| 7 | 1 | 2.5 | 4 | 2.5 |
| 8 | 1 | 2 | 3 | 4 |
| Sum $R_j$: | 8 | 17 | 26.5 | 28.5 |

From (3.8), we calculate

$$12S = 12(8^2 + 17^2 + 26.5^2 + 28.5^2) - 3(8)^2 4(5)^2$$
$$= 12(1867.5) - 19,200$$
$$= 3210,$$

and, from (3.10),

$$Q = \frac{3210}{8(4)(5)} = 20.1, \quad \text{with df} = n - 1 = 3.$$

From Table B the right-tail probability is found to be smaller than .001. The asymptotic approximate *P*-value is $P < .001$, and the null hypothesis is not supported by these data.

Let us use level .10 for the multiple comparisons analysis. From Table N with $p = n(n - 1)/2 = 6$, the critical $z$ value is 2.394. From (3.11), we have

$$|R_i - R_j| \le 2.394 \sqrt{\frac{8(4)(5)}{6}} = 2.394 \sqrt{26.67} = 12.4.$$

Thus 12.4 is compared with the observed difference between each pair of column sums. The pairs that differ significantly at level .10 are the group sizes 24 and 6, 24 and 4.

From these data, we would estimate the order of effectiveness of group sizes in decreasing order of question-asking performance as 4, 6, 12, 24. Thus it appears that question-asking performance is inversely related to group size, and smaller group size encourages more questions. Since the group size is small in a home situation of parents and their children, the studies are not inconsistent with the assertion by parents.

*Additional notes*   The technique of multiple comparisons between column sums may also be useful in the sampling and inference situation of Section 3.1. If $W$ is found to be significantly different from zero, we might wish to determine which pairs of objects have rank sums that differ significantly.

Consider, for example, the coffee-tasting experiment in Example 3.2, but suppose that the manufacturer is trying to decide which blend to use for his product. All other things being equal, he would select VIII, since that blend was most preferred among the tasters. However, cost or other considerations may alter this decision. Suppose that the coffee will be sold at a fixed price per pound regardless of which blend is used, that the costs are known, and the manufacturer feels that those blends that do not differ significantly in quality would have approximately the same demand. Then if several blends are preferred and they do not differ significantly, profit can be maximized by choosing from these the blend that has the lowest price. For instance, the blends VIII and IX may be much more expensive than blend X. Unless VIII and IX are significantly preferred to X, the profit would be greater using X.

We calculate the right-hand side of (3.11) for $\alpha = .20$ with the data of Example 3.2, where $n = 10$, $k = 4$. Since $p = \binom{10}{2} = 45$ is outside the range of Table N, the critical $z$ value is found from Table A. We get $z = 2.85$ as the value for which the right-tail probability is $\alpha/n(n - 1) = .20/90 = .0022$. The inequality for differences of column sums is

$$|R_i - R_j| \leq 2.85\sqrt{\frac{4(10)(11)}{6}} = 24.41.$$

The only pairs that do differ significantly at level .20 are I with VIII, IX, and X; II with VIII and IX; and III with VIII.

*Replicated designs*   In the previous design we assumed that $nk$ experimental units were divided into $k$ homogeneous blocks, each of size $n$, and that each treatment is applied to only one unit in each block. If there are $n\lambda k$ experimental units but still $n$ treatments and $k$ blocks, each treatment can be observed $\lambda$ times in each block, and the design is said to be replicated. Replicated designs are usually more expensive, but they are quite effective in reducing sampling error. As long as $\lambda$ is the same for each treatment and each block, the Friedman statistic is easily extended to provide a test for the null hypothesis of no difference between treatments.

The data can still be presented in a two-way table with $n$ columns, but now there are $\lambda k$ rows. The first $\lambda$ rows are observations in the first block, rows $\lambda + 1$ through $2\lambda$ constitute the second block, and so forth. The $n\lambda$ observations within each block (set of $\lambda$ rows) are ranked from 1 to $n\lambda$. Under the null hypothesis, the expected rank of any observation is $(n\lambda + 1)/2$, and the expected total of the ranks observed for any one treatment is $\lambda(n\lambda + 1)/2$ in any one block, and $k\lambda(n\lambda + 1)/2$ in all $k$ blocks combined. Let $R_j$ denote the sum of ranks in column $j$, that is, the sum of all ranks for treatment $j$. Then the sum of squares of deviations $S$ between actual column sums and expected column sums is indicative of the degree to which the data support $H$. A large value of $S$ is inconsistent with the expected result under $H$, so the logical rejection region is the right tail of the distribution of $S$. The exact distribution is tedious to evaluate, but a monotonically increasing function of $S$ is approximately chi-square distributed, with $n - 1$ degrees of freedom. This function is

$$Q = \frac{12S}{nk\lambda^2(n\lambda + 1)},$$

where

$$S = \sum_{j=1}^{n} \left[ R_j - \frac{k\lambda(n\lambda + 1)}{2} \right]^2,$$

which can be simplified to

$$S = \sum_{j=1}^{n} R_j^2 - \left[ \frac{k^2 n\lambda^2 (n\lambda + 1)^2}{4} \right].$$

When $Q$ is used as the test statistic, the asymptotic approximate $P$-value is a right-tail probability from Table B, with df $= n - 1$.

### 3.3   Kendall Coefficient of Concordance for Incomplete Rankings

*Sampling situation*   We now consider another design where the data to be analyzed consist of $k$ sets of rankings of $n$ objects, as introduced in Section 3.1. Here we assume that the sets are incomplete in the sense that not all of the objects are ranked by each judge. However, the number of ranks in each set is the same, say $m$ ranks for some $m < n$, and the design is balanced in the sense that every object is ranked the same total number $km/n$ times and every *pair* of objects is ranked the same total number of times, say $\lambda$ times. Note then that $\lambda$ can be interpreted as the number of complete sets of paired comparisons among all judges combined.

In Example 3.1 there were eight contestants and three judges. Since there is no $m < 8$ for which $km/n = 3m/8$ is an integer, no such balanced incomplete design exists. But if there are three contestants and three judges, so that $k = 3$ and $n = 3$, we could have $m = 2$; then every object is ranked a total of $km/n = 2$ times. A possible design then is as follows, where the asterisk indicates that the contestant is to be ranked by that judge.

| Judge | Contestant | | |
|:---:|:---:|:---:|:---:|
|  | A | B | C |
| 1 | * | * |   |
| 2 |   | * | * |
| 3 | * |   | * |

The general restrictions on the choices of the quantities $m$, $n$, $k$, and $\lambda$ are that

$$\lambda n(n - 1) = km(m - 1) \qquad (3.12)$$

is satisfied, and that $km/n$ is an integer. Some possible sets of constants for which these designs can be constructed are shown in Table 3.6. Designs of this type are called Youden squares, or a special case of incomplete Latin squares. Specific layouts have been tabulated, for example, in Cochran and Cox (1957, pp. 520–544).

**TABLE 3.6**

| $\lambda$ | $k$ | $n$ | $m$ |
|:---:|:---:|:---:|:---:|
| 1 | 3 | 3 | 2 |
| 1 | 6 | 4 | 2 |
| 1 | 7 | 7 | 3 |
| 1 | 10 | 5 | 2 |
| 2 | 4 | 4 | 3 |
| 2 | 6 | 3 | 2 |
| 2 | 12 | 4 | 2 |

If every one of the $n$ objects were ranked in each of the $k$ sets, a total of $kn$ observations would be required. With the incomplete design here, only $km$ observations are made. If the cost of the analysis is on a per-observation basis, either actually or in the sense that the cost increases as the number of objects increases, as opposed to a cost per set of rankings, this design allows more objects to be considered for the same cost. Another possible advantage is improved reliability of the results, especially when the ordering of the objects in each set is a (subjective) judgment (as opposed to ranking objective or physical measurements obtained on at least an ordinal scale). Many persons find that comparing and ranking more than two or three objects is an almost impossible task since the decision process involves too many factors. Consider, for example, a tasting experiment where we want to compare $k$ different types or brands of bourbon. The discriminatory powers of the tasters may legitimately be questioned if each one tastes (and swallows) more than three or four types. Particularly with human observers, the ability to rank objects in an effective and reliable way may be a direct function of the number of comparative judgments to be made. In other inference situations, there may be a fixed limit on the number of objects that *can* be ranked at once. For example, only four kinds of tires can be tried simultaneously on a single automobile, and a television set may use only three tubes of a particular type. Because of all these factors, it is important to learn how to analyze data from this type of design.

The simplest design of the type described here is where each incomplete set of rankings involves only two objects; that is, each ranking consists of a single statement of preference between two items, as in the example above. This is frequently called a paired comparison experiment. Since $m = 2$, the expression in (3.12) then requires that $k = \lambda n(n - 1)/2$. In the typical practical situation, the number $k$ of judges or participants in the experiment is easily determined because the total number $n$ of objects to be compared is fixed according to the purpose of the experiment. If $\lambda = 1$, $k = n(n - 1)/2$, and if $\lambda = 2$, $k = n(n - 1)$. For a paired comparison experiment involving 4 objects, 6 judges are required for $\lambda = 1$, and 12 for $\lambda = 2$. Of course, the decision as to which objects are presented to which judges cannot be completely arbitrary, because of the assumptions of balance mentioned earlier. In particular, when $m = 2$, every object must be ranked exactly $2k/n$ times and every pair of objects must be ranked exactly $\lambda$ times.

*Rationale and descriptive measure* Once the design is selected and the observations are made, the data for analysis can be presented in exactly the same format as before — that is, a table with $k$ rows and $n$ columns. However, now there are only $m$ entries in each row and $km/n$ entries in each column. The row sums are then $m(m + 1)/2$, and the average column sum is the total sum of ranks, $km(m + 1)/2$, divided by the number of columns, $n$, or

$$\frac{km(m + 1)}{2n}.$$

Since each object is ranked the same total number of times, the column sums will tend to be all equal to this average value if there is no agreement between rankings and hence no communality of preference between objects. As before, then, a measure of the departure from lack of agreement between rankings is given by the sum of squares of deviations of the actual column sums $R_j$ around the average column sum $km(m + 1)/2n$, or

$$S = \sum_{j=1}^{n} \left[ R_j - \frac{km(m + 1)}{2n} \right]^2. \tag{3.13}$$

For complete agreement between rankings, it can be shown that $S$ becomes

$$\frac{\lambda^2 n(n^2 - 1)}{12}, \tag{3.14}$$

and that this is the largest possible value of $S$. Hence, a measure of the relative agreement between rankings is the ratio of the observed value of $S$ to its maximum, or

$$W = \frac{12S}{\lambda^2 n(n^2 - 1)} \tag{3.15}$$

$$= 12 \sum_{j=1}^{n} \frac{[R_j - km(m + 1)/2n]^2}{\lambda^2 n(n^2 - 1)}. \tag{3.16}$$

(If $m = n$ and $\lambda = k$, so each observer ranks each object, then (3.15) reduces to (3.4), and (3.16) to (3.5).)

The measure $W$ is called the Kendall coefficient of concordance for $k$ incomplete sets of rankings of $n$ objects. The value of $W$ is always between 0 and 1, and the extreme values are interpreted as follows:

| Value of $W$ | Interpretation |
| :---: | :---: |
| 0 | No association |
| 1 | Perfect association |

It should be emphasized that in a consumer preference study, the word association can be replaced by "consistency or communality of preference," and no association implies the conclusion either that the consumers are making independent choices or that there really is no discernible preference between the objects under study.

*Computation*   Computing the value of $S$ in (3.13), and hence the numerator of $W$ in (3.15) or (3.16), is usually simpler when the following equivalent expression is used:

$$12S = 12 \sum_{j=1}^{n} R_j^2 - \frac{3k^2 m^2(m + 1)^2}{n} \tag{3.17}$$

*Ties*   When ties are present in any ranking, they are generally assigned midranks. However, $W$ is still calculated from (3.15) since no simple correction is available to reduce (3.14), and account for the ties. Alternatively, integer ranks may be assigned randomly among the tied observations.

**example 3.3**

A taste-test experiment to compare seven different kinds of wine is to be designed such that no taster should be asked to rank more than three different wines. Thus we have $n = 7$ and $m = 3$. If each pair of wines is compared only once, so that $\lambda = 1$, the required number of tasters is found from (3.12) as

$$k = \frac{\lambda n(n - 1)}{m(m - 1)} = \frac{7(6)}{3(2)} = 7.$$

A design which satisfies the balance requirements is shown in Table 3.7.

**TABLE 3.7**

| | Wine | | | | | | |
|---|---|---|---|---|---|---|---|
| **Taster** | **A** | **B** | **C** | **D** | **E** | **F** | **G** |
| 1 | * | * | | * | | | |
| 2 | | * | * | | * | | |
| 3 | | | * | * | | * | |
| 4 | | | | * | * | | * |
| 5 | * | | | | * | * | |
| 6 | | * | | | | * | * |
| 7 | * | | * | | | | * |

Note that each wine is ranked three times, as it must be since $km/n = 7(3)/7 = 3$. When each taster ranks the three wines presented to him, suppose that the results are as shown in Table 3.8. Calculate the Kendall coefficient of concordance as a measure of the agreement between preferences of the tasters.

**TABLE 3.8**

| | Wine | | | | | | |
|---|---|---|---|---|---|---|---|
| **Taster** | **A** | **B** | **C** | **D** | **E** | **F** | **G** |
| 1 | 1 | 2 | | 3 | | | |
| 2 | | 1 | 3 | | 2 | | |
| 3 | | | 3 | 2 | | 1 | |
| 4 | | | | 2 | 3 | | 1 |
| 5 | 1 | | | | 3 | 2 | |
| 6 | | 2 | | | | 1 | 3 |
| 7 | 1 | | 3 | | | | 2 |
| $R_i$: | $\overline{3}$ | $\overline{5}$ | $\overline{9}$ | $\overline{7}$ | $\overline{8}$ | $\overline{4}$ | $\overline{6}$ |

**SOLUTION**
We compute $S$ using (3.17) as

$$12S = 12(3^2 + 5^2 + 9^2 + 7^2 + 8^2 + 4^2 + 6^2) - \frac{3(7)^2(3)^2(4)^2}{7}$$
$$= 12(280) - 3024$$
$$= 336,$$

and $W$ from (3.15) is

$$W = \frac{336}{7(48)} = 1.$$

For these data, $W$ has taken on its maximum value and the tasters are in perfect agreement.

*Inference situation*    The Kendall coefficient of concordance for incomplete sets of rankings can be used to test the null hypothesis that there is no association between the sets of rankings. This test is also known as the Durbin (1951) test. The hypothesis set is

$$H: \text{ No association} \quad A: \text{ Association exists.}$$

If the $k$ sets of rankings represent measurements on $k$ characteristics, this null hypothesis can be interpreted as saying that the characteristics are independent. If the rankings represent preferences of individuals, the null hypothesis means a lack of agreement, or lack of communality or consistency in preferences.

*Rationale*    If no association exists, that is, if the null hypothesis is true, then the column sums tend toward equality, and thus $S$ and $W$ are small. If there is association, the large ranks would tend to appear in the same columns and the small ranks would also appear together; this increases $S$, and $W$ gets close to its maximum value of 1. Accordingly, large values of $S$ or $W$ call for rejection of $H$.

*Test statistic*    Since tables of the sampling distribution of $S$ or $W$ are not practical, we give the test statistic as a linear function of $S$ or $W$ whose asymptotic approximate distribution is the chi-square distribution. This statistic is

$$Q = \frac{\lambda(n^2 - 1)W}{m + 1} \tag{3.18}$$

$$= \frac{12S}{\lambda n(m + 1)}, \tag{3.19}$$

which has approximately the chi-square distribution with $n - 1$ degrees of freedom. A statistic that permits a more exact but also less convenient test is given in Kendall (1962, p. 105).

*Finding the P-value*   Since $Q$ is a monotonically increasing function of $W$, the appropriate $P$-value is a right-tail probability for $Q$, which can be found from Table B with df $= n - 1$; this $P$-value is always asymptotic approximate.

*Multiple comparisons*   If the $P$-value from the test statistic is so small that the null hypothesis of no association is not supported, the technique of multiple comparisons among the column sums can be used to obtain information about which column sums differ significantly. In the case of $k$ incomplete sets of rankings of $n$ objects, we are then investigating which pairs of objects are consistently given ranks that differ significantly from each other. Since any column sum $R_i$ divided by the number of ranks given, that is, $R_i/(km/n)$, is the average rank given to the $i$th object, such comparisons are frequently of interest. The inequality for the differences of all pairs of column sums, $R_i$, $R_j$, which has probability at least $1 - \alpha$ is

$$|R_i - R_j| \le z\sqrt{\frac{km(m^2 - 1)}{6(n - 1)}} \tag{3.20}$$

or, equivalently,

$$|R_i - R_j| \le z\sqrt{\frac{n\lambda(m + 1)}{6}}. \tag{3.21}$$

The value of $z$ may be obtained from Table A as the quantile point of the normal curve that corresponds to a right-tail probability of $\alpha/n(n - 1)$, or from Table N with $p = n(n - 1)/2$. All pairs of differences that are larger than the right-hand side of (3.20) or (3.21) are significantly different pairs.

**example   3.4**

We illustrate these inference methods using the data and calculations of Example 3.3. For the null hypothesis of no association, the test statistic from (3.18) is

$$Q = \frac{1(48)(1)}{4} = 12 \quad \text{with df} = 6.$$

From Table B the right-tail probability is a little larger than .05, and this is the asymptotic approximate $P$-value. The conclusion is rather optional, but since $W$ equals its maximum possible value, it seems logical to reject the null hypothesis and conclude that there is a communality of preference. Based on the rankings given by these tasters, we could estimate the preferential order of kinds of wines, starting with the most preferred, as A, F, B, G, D, E, C.

To perform the multiple comparisons analysis at, say, level .10, we first find the critical $z$ value for $n = 7$ and $p = 21$, from Table N as 2.823. From

(3.21), the interval of nonsignificant differences of pairs is

$$|R_i - R_j| \leq 2.823\sqrt{\frac{7(4)}{6}} = 2.823\sqrt{4.67} = 6.10.$$

The conclusion is that none of the pairs is significantly different at level .01. The reader may verify that if the level is increased to .30, A is significantly more preferred than C since its rank sum is lower, but no other differences are significant.

*Additional notes*    The statistical analysis of $k$ incomplete sets of rankings is also applicable to balanced randomized incomplete block designs. In such a case the $k$ sets of rankings are the $k$ blocks and observations are made within each block on some subset of the $n$ treatments. The balance requirements are the same; namely, each treatment is observed the same number $m$ of times in each block, $m < n$, a total of $km/n$ (an integer) observations are made on each treatment for all blocks combined, every pair of treatments appears together in $\lambda$ different blocks, and (3.12) is satisfied by $m$, $n$, $k$ and $\lambda$.

The null hypothesis is that the treatment effects are all equal. The test statistic is $Q$, as given in (3.18) or (3.19), and the $P$-value is a right-tail probability from Table B with df $= n - 1$. When used for this sampling situation and inference, the test based on $Q$ is frequently called the Durbin (1951) test. The multiple comparisons analysis in (3.20) or (3.21) can then be used to investigate which pairs of treatment effects differ significantly.

# 4    Count Data from Classificatory Samples

In each of the situations already described in this chapter, the data to be analyzed have been presented as a two-way layout — that is, a rectangular table with a fixed number of rows and columns, and particular designations for each row and each column. Each table entry has represented a measurement or ranking (relative to certain other units) of the experimental unit which has the characteristics of its row and column label. For example, in Section 3.1 the ranking given by judge 2 to object 3 is the entry in the second row and third column of a table for which judge number is the row designation and object number is the column designation. In Section 3.2, the same table entry represents the ranking of the effect of treatment 3 when applied in block 2. In either case, only one set of characteristics (one set of variables), the row or column, was of real statistical interest. The two-way classification resulted from the necessity of distinguishing between objects, or the imposition of blocks or groups, and the inferences concerned relationships between variables representing the other classification set.

The situation we consider now differs in two ways. First, the relationship between *two sets* of characteristics is of statistical interest, Second, each entry in the rectangular table is not a rank; it is the *number* of experimental units having the particular pair of characteristics designated by its row and column. Hence the data for analysis are the counts that represent the numbers of observations classified into each possible cross-category of the two characteristics. The table in which these counts are presented is commonly known as a *contingency table*. The two sets of characteristics that form the subgroup classifications may be quantitative or qualitative in designation.

For example, in a public opinion survey concerning a proposed school bond issue, the results of each interview or questionnaire may be classified according to the two sets of characteristics, opinion and level of educational attainment. If the opinion subgroups are favor, oppose, and undecided, and education level is determined by five classes of highest level of formal schooling completed, the data can be presented in a 3 $\times$ 5 or a 5 $\times$ 3 rectangular table with 15 cells. In each cell we record the number of persons having both characteristics of the two subgroups which specify that cell.

We assume that the question of statistical interest for these data is whether the two main groups (sets or characteristics or broad categories) of classification are independent. A descriptive measure of association between these groups may also be desired. The method of statistical analysis, frequently called contingency table analysis, is quite well known and commonly used. The method can be extended to three- and higher-way contingency tables, but the presentation here will be restricted to two-way tables. Once this methodology is understood, the investigator can easily comprehend an explanation of the general procedure for higher-way layouts. The interested reader is referred to Ostle (1963, p. 131), for example.

## 4.1   Chi-Square Test for Independence

*Inference situation*   Assume there are two primary sets of characteristics, say $A$ and $B$, which we will call categories, and each category has two or more subcategories. The hypothesis set is

> $H:$   The $A$ and $B$ categories are independent
> $A:$   The $A$ and $B$ categories are not independent.

Rejection of the null hypothesis then implies that the $A$ and $B$ categories are associated, but the type of association or its direction is not a part of this conclusion.

*Sampling situation*   We begin by introducing a general notation. Suppose there are $r$ subcategories of the $A$ category, denoted by $A_1, A_2, \ldots, A_r$, and $c$ subcategories of the $B$ category $B_1, B_2, \ldots, B_c$; the subcategories of $A$ exhaust all relevant possibilities and do not overlap, and similarly for the subcategories of $B$. Each observation can then be classified in exactly one of the $A$ subcategories and one of the $B$ subcategories.

Let $f_{ij}$ denote the number of observations that belong to or are classified as belonging to both of the subcategories $A_i$ and $B_j$. Then the data to be analyzed can be cast in a rectangular table of $r$ rows and $c$ columns if the $r$ subcategories of $A$ are the row labels and the $c$ subcategories of $B$ are the column labels. For $N$ observations, the data would appear as follows:

|  | $B_1$ | $B_2$ | $\cdots$ | $B_c$ | Total |
|---|---|---|---|---|---|
| $A_1$ | $f_{11}$ | $f_{12}$ | $\cdots$ | $f_{1c}$ | $f_{1.}$ |
| $A_2$ | $f_{21}$ | $f_{22}$ | $\cdots$ | $f_{2c}$ | $f_{2.}$ |
| $\cdot$ | $\cdot$ | $\cdot$ |  | $\cdot$ | $\cdot$ |
| $\cdot$ | $\cdot$ | $\cdot$ |  | $\cdot$ | $\cdot$ |
| $A_r$ | $f_{r1}$ | $f_{r2}$ | $\cdots$ | $f_{rc}$ | $f_{r.}$ |
| Total | $f_{.1}$ | $f_{.2}$ |  | $f_{.c}$ | $N$ |

The indicated notation is that $f_{1.}, f_{2.}, \ldots, f_{r.}$ are the respective row totals, and $f_{.1}, f_{.2}, \ldots, f_{.c}$ are the column totals. Thus

$$f_{i.} = \sum_{j=1}^{c} f_{ij} \tag{4.1}$$

$$f_{.j} = \sum_{i=1}^{r} f_{ij} \tag{4.2}$$

$$N = \sum_{i=1}^{r} \sum_{j=1}^{c} f_{ij} = \sum_{i=1}^{r} f_{i.} = \sum_{j=1}^{c} f_{.j}. \tag{4.3}$$

Further $f_{i.}/N$ is the observed proportion of units in subcategory $A_i$, while $f_{.j}/N$ is the observed proportion in subcategory $B_j$.

*Rationale*   If the $A$ and $B$ categories are independent, then the probability that a unit is in both subcategories $A_i$ and $B_j$ is the product of the respective probabilities, and the expected number of units so cross-classified is $N$ times the product of these probabilities. These probabilities are unknown, but they can all be estimated from the data by the corresponding observed proportions. Accordingly, the product of these proportions and $N$ is an estimate of the expected number of units classified as belonging to both subcategories $A_i$ and $B_j$. Denoting this number by $e_{ij}$, we have

$$e_{ij} = N\left(\frac{f_{i.}}{N}\right)\left(\frac{f_{.j}}{N}\right) = \frac{f_{i.}f_{.j}}{N} \tag{4.4}$$

for any $i = 1, 2, \ldots, r; j = 1, 2, \ldots, c$. These calculations can be checked using the fact that

$$\sum_{i=1}^{r} \sum_{j=1}^{c} e_{ij} = \sum_{i=1}^{r} \sum_{j=1}^{c} f_{ij} = N. \tag{4.5}$$

Thus, we have an estimated expected frequency $e_{ij}$ corresponding to each observed frequency $f_{ij}$. If the categories are independent, there should be close agreement between these corresponding frequencies for each combination of subcategories. The deviations $f_{ij} - e_{ij}$ measure lack of agreement, but they sum to zero by (4.5). Besides, amount of disagreement is reflected by the absolute magnitude of these deviations. We eliminate the signs by squaring each difference, and reduce that value to original units by dividing by the respective $e_{ij}$. Thus $(f_{ij} - e_{ij})^2/e_{ij}$ measures lack of agreement for the $(i, j)$ table entry. An overall measure of the lack of agreement, for all subcategories simultaneously, is the sum of these individual measures, that is, the statistic $Q$ defined below in (4.6). A small value of $Q$ supports $H$, whereas a large value reflects a general incompatibility between the frequencies observed and those expected under independence. Since $Q = 0$ for independence (no association), the value of $Q$ is itself a measure of the association between categories.

*Test statistic*   The chi-square test statistic is defined as

$$Q = \sum_{i=1}^{r} \sum_{j=1}^{c} \frac{(f_{ij} - e_{ij})^2}{e_{ij}}, \tag{4.6}$$

where

$$e_{ij} = \frac{f_{i.} f_{.j}}{N}$$

and $f_{i.}$ and $f_{.j}$ are respectively the totals of the $i$th row and $j$th column, as seen from (4.1) and (4.2). In practice it is not necessary to compute all of the $e_{ij}$. It can be shown using (4.4) that the row totals and column totals for the estimated frequencies are all equal to the corresponding totals for the observed frequencies. Hence, once $r - 1$ expected frequencies in each column, or $c - 1$ in each row, are calculated, the remaining ones can be found by subtraction. The total number of calculations that need to be performed using (4.4) is the product $(r - 1)(c - 1)$. Of course, calculating all expected frequencies provides an internal check on arithmetic, as does (4.5).
*Computation*   The computation of $Q$ can be simplified in most cases by using the following formula, which is algebraically equivalent to (4.6):

$$Q = N\left[ \sum_{i=1}^{r} \sum_{j=1}^{c} \frac{f_{ij}^2}{f_{i.} f_{.j}} - 1 \right]. \tag{4.7}$$

This requires no calculations from (4.4), and is considerably easier to use than (4.6), but it does not show clearly which subcategories are the main contributors to a large value of $Q$. The equivalence will be illustrated in Example 4.1 below.

If the data to be analyzed are presented in a $2 \times 2$ contingency table, that is, if each category is divided into only two subcategories, the value of $Q$ is most easily computed from the following equation:

$$Q = \frac{N(f_{11} f_{22} - f_{12} f_{21})^2}{f_{.1} f_{.2} f_{1.} f_{2.}}. \tag{4.8}$$

Note that the numerator is $N$ times the square of the difference between the product of the two frequencies along the main diagonal and the product of the two frequencies off the main diagonal, and the denominator is the product of the four row and column totals. The formula in (4.8) is algebraically equivalent to (4.6) and (4.7) whenever $r = 2$ and $c = 2$, but it cannot be used except for $2 \times 2$ tables.

*Finding the P-value*    The exact sampling distribution of the test statistic $Q$ defined in (4.6) or (4.7) is quite complicated, but for large samples it may be approximated by the chi-square distribution, with $(r - 1)(c - 1)$ degrees of freedom. (Note that the number of degrees of freedom equals the minimum number of expected frequencies that must be calculated.)

Since the value of $Q$ measures the lack of agreement between the observed frequencies and those expected under $H$, small values of $Q$ support $H$ and large values support $A$. Hence the $P$-value for lack of agreement is a right-tail probability from Table B, but it is always asymptotic approximate.

As with the goodness-of-fit test in Section 2 of Chapter 2, the chi-square approximation may not be reliable if more than 20 percent of the $e_{ij}$ are less than 5, or if any $e_{ij}$ is less than 1. Subcategories can be combined until these restrictions are satisfied. If the subcategories are ordinal classifications, the logical combination is of adjacent subcategories. With nominal-type subcategories, any combinations may be difficult to interpret. When subcategories are combined, the degrees of freedom should be reduced to correspond to the actual number of rows and columns used for the analysis.

When (4.6) is used to calculate $Q$, it is a simple matter to observe whether subcategories should be combined. If (4.7) or (4.8) is used, expected frequencies are usually not found; the investigator must verify separately that the $e_{ij}$ are large enough for the chi-square approximation to be reliable.

**example   4.1**
A small, private opinion poll was conducted to investigate views on the requirement of celibacy for Roman Catholic priests. A sample of the members of a Catholic church in a small rural community produced the data below on opinions about celibacy requirement. Test the null hypothesis that opinion and sex are independent.

| | Agree | Disagree | No Opinion | Total |
|---|---|---|---|---|
| Male | 15 | 24 | 5 | 44 |
| Female | 7 | 22 | 2 | 31 |
| Total | 22 | 46 | 7 | 75 |

## SOLUTION

Using (4.4), we calculate $e_{11}$ and $e_{12}$; all remaining expected frequencies can be found by subtraction.

$$e_{11} = (22)(44)/75 = 12.9$$
$$e_{12} = (46)(44)/75 = 27.0$$

The table below shows the observed frequencies, with corresponding expected frequencies in parentheses. Since the row and column totals must be

| | Agree | Disagree | No Opinion | Total |
|---|---|---|---|---|
| Male | 15 (12.9) | 24 (27.0) | 5 (4.1) | 44 |
| Female | 7 (9.1) | 22 (19.0) | 2 (2.9) | 31 |
| Total | 22 | 46 | 7 | 75 |

exactly the same for both the observed and expected frequencies, the uncalculated expected frequencies are found easily as that row or column total minus the sum of all the other entries in that row or column. In this case, after the calculations for $e_{11}$ and $e_{12}$ are made and entered in the table, the sequential steps followed to complete the table are as follows:

$$e_{21} = 22.0 - 12.9 = 9.1$$
$$e_{22} = 46.0 - 27.0 = 19.0$$
$$e_{13} = 44 - (12.9 + 27.0) = 4.1$$
$$e_{23} = 31 - (9.1 + 19.0) = 2.9$$

Both expected frequencies in the "no opinion" column are less than 5, and these constitute a third of the total number of entries. Hence some subcategories should be combined; the combination must be some two columns, since there are only two rows. Then only two groups of opinion can be differentiated. It is probably more common in opinion polls to combine "no opinion" with "disagree," although arguments could be made either way for the issue under consideration here. After combining in this way, the 2 × 2 table is as follows:

|  | Agree | Do Not Agree | Total |
|---|---|---|---|
| Male | 15 (12.9) | 29 (31.1) | 44 |
| Female | 7 (9.1) | 24 (21.9) | 31 |
|  | $\overline{22}$ | $\overline{53}$ | $\overline{75}$ |

Calculating $Q$ from (4.6), we obtain

$$Q = \frac{(15 - 12.9)^2}{12.9} + \frac{(7 - 9.1)^2}{9.1} + \frac{(29 - 31.1)^2}{31.1} + \frac{(24 - 21.9)^2}{21.9}$$
$$= .34 + .48 + .14 + .20 = 1.16.$$

In order to illustrate the equivalence of the alternative computation formulas, we now compute $Q$ again using (4.7), and also (4.8) since the reduced table is $2 \times 2$. From (4.7),

$$Q = 75\left[\frac{15^2}{22(44)} + \frac{7^2}{22(31)} + \frac{29^2}{53(44)} + \frac{24^2}{53(31)} - 1\right]$$
$$= 75[.232 + .072 + .361 + .351 - 1] = .75(.016) = 1.20,$$

and, from (4.8),

$$Q = \frac{75[15(24) - 7(29)]^2}{22(53)(44)(31)} = \frac{75(157)^2}{1,590,424} = 1.16.$$

The result from (4.7) disagrees slightly, but only because of rounding. If one more decimal place had been carried within the brackets, $Q$ would equal 1.16 here also.

After combining cells, we have df $= (2 - 1)(2 - 1) = 1$. From Table B, the right-tail probability for 1.16 is between .20 and .30, and hence the asymptotic approximate $P$-value is given as $P > .20$. The data do support the null hypothesis of independence between sex and opinion on celibacy.

## 4.2  Measures of Association in Contingency Tables

In an $r \times c$ contingency table, the chi-square test of Section 4.1 is used to investigate the existence of an association between two categories of classification. Particularly when the $P$-value is so small that the null hypothesis of independence between categories is rejected, a descriptive measure of the degree of association or strength of the relationship is desirable and useful. The test statistic $Q$ is itself a measure of the lack of agreement between the observed frequencies and those estimated under the assumption of independence. Accordingly, $Q$ can be considered a measure of association between categories in the sense that it is approximately equal to zero if the categories are independent and it increases as the deviations between observed and expected frequencies increase in absolute value. However, $Q$ is not really a satisfactory descriptor because its magnitude is considerably affected by the number of items classified ($N$) and also by the number of degrees of freedom (which is determined by the dimensions of the table). A relative measure of association is needed for any interpretation about strength of relationship.

The traditional relative measure, called the *contingency coefficient*, is defined by $\sqrt{Q/(Q + N)}$. This value is zero whenever $Q$ is zero, and it increases as $Q$ increases. However, its maximum is not necessarily equal to one; that maximum depends on the dimensions of the contingency table. Therefore, contingency coefficients of two different tables are not comparable as measures of strength of association unless the tables have the same dimensions.

Since $Q$ does have a known constant maximum value for any contingency table of fixed dimensions, a better relative measure of association is obtained by dividing $Q$ by its maximum value. It can be shown that the largest possible value of $Q$ in any $r \times c$ contingency table is $N(t - 1)$, where $t$ is the smaller of $r$ and $c$. The resulting measure of association, frequently called the *Cramér statistic*, is then defined as

$$C = \sqrt{\frac{Q}{N(t - 1)}}, \tag{4.9}$$

where $t = \min(r, c)$. $C^2$ could also be used to measure association.

$C$ is then interpreted like other measures of association, in that it is a statistic that measures the departure from independence divided by the maximum value of that statistic. If $r = c = 2$, it can be shown that $Q/N$ equals $T^2$, where $T$ is the Kendall Tau statistic adjusted for ties, as computed by the formula given in (2.19). Since (4.9) gives $C = \sqrt{Q/N}$ when $r = c = 2$, $C^2 = T^2$ is this special case (see Example 4.3 below). $C$ is sometimes called the *phi coefficient* (denoted by $\phi$) when $r = c = 2$.

Since $C^2$ is always a linear function of $Q$, any probability concerning a value of $Q$ holds also for the corresponding value of $C^2$. Accordingly, a $P$-value obtained for $Q$, using the methods explained in the preceding subsection, is equal to a $P$-value for the value of $C^2$ computed for the same data.

**example 4.2**

*Economic Report: The State of South Carolina 1972* reported the data in Table 4.1 from the 1970 census for South Carolina residents in 1970 who were at least 25 years of age at that time. Test the null hypothesis that educational attainment is independent of sex and find a measure of the association between the two attributes.

**TABLE 4.1**

| Maximum Number of Years of School Completed | | Numbers in Thousands (Rounded) | | |
|---|---|---|---|---|
| | | Male | Female | Total |
| None | | 18 | 16 | 34 |
| Elementary: | 1–4 | 67 | 55 | 122 |
| | 5 and 6 | 61 | 67 | 128 |
| | 7 | 44 | 49 | 93 |
| | 8 | 52 | 59 | 111 |
| High School: | 1–3 | 130 | 179 | 309 |
| | 4 | 120 | 144 | 264 |
| College: | 1–3 | 47 | 59 | 106 |
| | 4 | 38 | 42 | 80 |
| Postbaccalaureate | | 23 | 12 | 35 |
| Total: | | 600 | 682 | 1282 |

**SOLUTION**

It can be seen by inspection that none of the expected frequencies will be too small. The computation of $Q$ is most easily performed using (4.7) as follows:

$$Q = 1282\left[\frac{18^2}{600(34)} + \frac{67^2}{600(122)} + \frac{61^2}{600(128)} + \cdots + \frac{12^2}{682(35)} - 1\right]$$

$$= 1282(1.00941 - 1) = 12.06$$

$$\text{df} = (10 - 1)(2 - 1) = 9.$$

To test the null hypothesis of independence, we look up 12.06 in Table B for 9 degrees of freedom, and find that the asymptotic approximate $P$-value is between .20 and .30. The null hypothesis is supported by these data. The Cramér statistic to measure the association between sex and level of educational attainment is found from (4.9) as

$$t = \min (10,2) = 2 \qquad N = 1282$$

$$C = \sqrt{\frac{12.06}{1282(1)}} = \sqrt{.00941} = .097.$$

The $P$-value found before for the hypothesis test applies to the Cramér statistic also, namely $P > .20$.

**example  4.3**

Several studies have been undertaken in order to substantiate the proposition that a person who is firstborn (oldest child) or an only child is more prone to become a member of social and recreational clubs than is a later-born child. De Lint (1966, p. 178) reported the following results of an affiliative behavior evaluation of male college students. Analyze the data for association between birth rank and membership, and find a measure of the strength of the relationship.

| | Birth Rank | | |
|---|---|---|---|
| Membership | First | Later | Total |
| Do not belong | 20 | 18 | 38 |
| Belong | 46 | 33 | 79 |
| Total: | 66 | 51 | 117 |

**SOLUTION**
Since there are only two subcategories of each classification, we can calculate $Q$ from the simplified formula in (4.8) as follows:

$$Q = \frac{117[(20)(33) - (46)(18)]^2}{66(51)(38)(79)} = \frac{3,302,208}{10,104,732} = .327.$$

The null hypothesis is

$H:$    Birth rank and membership are independent.

Since $r = 2$ and $c = 2$, we have df $= (r - 1)(c - 1) = 1$. From Table B the right-tail probability for $Q = .327$ with df $= 1$ is larger than .50. Thus the asymptotic approximate $P$-value is $P > .50$; we conclude that these data do support the null hypothesis.

The Cramér measure of association between the two types of classification is found from (4.9) with $t = 2$ to be

$$C = \sqrt{\frac{.327}{117}} = \sqrt{.002795} = .053.$$

We now use these same data to illustrate the fact that $Q/N$ for $2 \times 2$ tables equals the square of the Tau statistic. For this purpose, we must consider the 117 observations as each ranked according to two separate criteria. Designating membership rank by $X$, with 1 as lack of membership, and birth rank by $Y$, with 1 for firstborn, the data from the evaluation study could have been presented as follows:

| | a | b | c | d |
|---|---|---|---|---|
| $X$ (membership rank) | $1 \cdots 1$ | $1 \cdots 1$ | $2 \cdots 2$ | $2 \cdots 2$ |
| $Y$ (birth rank) | $1 \cdots 1$ | $2 \cdots 2$ | $1 \cdots 1$ | $2 \cdots 2$ |
| | 20 pairs | 18 pairs | 46 pairs | 33 pairs |

$T$ is to be calculated from (2.19). The value of $S$, the total plus scores minus the total minus scores, can be counted from the data as listed above. The membership ranks $X$ are already in natural order. We now consider all possible pairs of birth ranks $Y$ for which membership rank is not tied and birth rank would score a plus. Since the $X$ ranks in groups a and b, and in groups c and d, are tied, the only pairs of $Y$ ranks to consider are members of group a or b when paired with members of group c or d. Pairs formed between groups a and c, or between b and d, score zero since the $Y$ ranks are tied. Hence, the only pairs left are between members of a and d, and between b and c. Any $Y$ pair formed between members of a and d gives a score of $+1$, and there are 20(33) of these; pairs from b and c give a score of $-1$, and there are 18(46) of these.

Accordingly, we have

$$S = 20(33) - 18(46) = -168$$

for the numerator of (2.19). Now for the denominator, in the $X$ sample there are $20 + 18 = 38$ observations tied at rank 1 and $46 + 33 = 79$ observations tied at rank 2. Hence we have

$$u' = \binom{38}{2} + \binom{79}{2} = 3784,$$

and similarly

$$v' = \binom{66}{2} + \binom{51}{2} = 3420.$$

Substituting these results and $N = 117$ in (2.19) gives

$$T = \frac{-168}{\sqrt{\binom{117}{2}} - 3784 \ \sqrt{\binom{117}{2}} - 3420} = \frac{-168}{\sqrt{3002(3366)}}$$

$$= \frac{-168}{\sqrt{10,104,732}} = -.0528,$$

or

$$T^2 = \frac{(-168)^2}{10,104,732} = \frac{28,224}{10,104,732} = \frac{Q}{117} = \frac{Q}{N}.$$

This verifies the statement made earlier about the relationship between $T^2$ and $Q$ (or $C$) in $2 \times 2$ tables.

The fact that $T^2 = Q/N$ in $2 \times 2$ contingency tables shows that $T$ can be calculated in another, simpler way when the variables analyzed for association are both dichotomous. Substituting $Q/N$ from (4.8) into $T = \sqrt{Q/N}$, the formula for $T$ becomes

$$T = \frac{f_{11}f_{22} - f_{12}f_{21}}{\sqrt{f_{1.}f_{2.}f_{.1}f_{.2}}} = \frac{ad - bc}{\sqrt{(a+b)(c+d)(a+c)(b+d)}}. \tag{4.10}$$

In the second expression of (4.10), $a$, $b$, $c$, and $d$ designate the number of pairs in the groups designated above by $a$, $b$, $c$, and $d$. That is, $a = f_{11}$, $b = f_{12}$, $c = f_{21}$, and $d = f_{22}$. Of course, the Tau coefficient requires data measured on at least an ordinal scale, while the subcategories in a contingency table need not have any kind of ordinal relationship. The special case of $T$ in (4.10) is usually denoted by $\phi$ and called the *phi coefficient*.

In any $2 \times 2$ table the subcategories effect a dichotomy of the $A$ category and of the $B$ category. If these subcategories are assigned arbitrary numbers and the ordinary Pearson product-moment correlation coefficient is calculated for the resulting $N$ pairs of numbers, the result is equal to $\phi$. Suppose that 0 and 1 are the numbers assigned; then $f_{11}$ or $a$ is the number of observations classified as 0 in both the $A$ and $B$ categories, and $f_{22}$ or $d$ is the number classified as 1 in both categories. If $A$ and $B$ are two different attributes (or traits, conditions, or the like), and the 0 and 1 indicate absence and presence of an attribute, then the sign of $\phi$ can be given an interpretation; a plus sign implies relatively more agreements and a minus sign implies relatively more disagreements between the $A$ and $B$ subcategories. Similarly, if 0 and 1 are indications of ordinal subcategories, such as degree of preference, where, say, 0 denotes the lower level in both categories, the sign of $\phi$ can be given the same interpretation. If the 0 and 1 are completely artificial, the sign of $\phi$ indicates simply that the association, if it exists, is along the $f_{11} f_{22}$ diagonal rather than the $f_{12} f_{21}$ diagonal in the $2 \times 2$ table.

It should be noted that while the value of $\phi$ as defined in (4.10) ranges between $-1$ and $+1$, these extreme values cannot always be attained; the maximum absolute value depends on the marginal totals. Thus an alternative measure of association is sometimes used in $2 \times 2$ tables. This measure is sometimes called the *Yule coefficient;* it is denoted by $\gamma$ and defined as

$$\gamma = \frac{f_{11} f_{22} - f_{12} f_{21}}{f_{11} f_{22} + f_{12} f_{21}} = \frac{ad - bc}{ad + bc}. \qquad (4.11)$$

Note that $\gamma$ has the same numerator as $\phi$ but a different denominator; $\gamma$ can always achieve the absolute value 1.

We can interpret the coefficient $\gamma$ in terms of the comparisons that must be made to compute the Kendall Tau coefficient. The numerator in (4.11) is the total number of agreements minus the total number of disagreements, but the denominator is the total number of comparisons made that produce either an agreement or a disagreement. In terms of $U$ and $V$, the number of plus scores and minus scores for Tau, the $\gamma$ coefficient expresses $S = U - V$ as a proportion of the total number of nonzero scores $U + V$, or

$$\gamma = \frac{U - V}{U + V}. \qquad (4.12)$$

That is, we eliminate those scores which show no association in either direction and consider only the remaining pairs. Accordingly, the value of $\gamma$ relates to the particular set of data scores obtained more than does $\phi$.

For the data of Example 4.3, the value of $\gamma$ is

$$\gamma = \frac{20(33) - 18(46)}{20(33) + 18(46)} = \frac{660 - 828}{660 + 828} = \frac{-168}{1488} = -.113,$$

which is almost twice the value of Tau calculated from (2.19) or (4.10) in Example 4.3. Of the plus and minus comparisons, which total 1488, over one-tenth, or 168, represent an excess of minus over plus comparisons. Note that the number of zero comparisons is

$$20(18) + 20(46) + 18(33) + 46(33) + \binom{20}{2} + \binom{18}{2} + \binom{46}{2} + \binom{33}{2} = 5298,$$

which when added to 1488 gives 6786, which equals $\binom{117}{2}$, the total number of comparisons.

Up to this point, the coefficients $\phi$ and $\gamma$ have been defined only for contingency tables with $r = c = 2$. An advantage of the $\gamma$ coefficient is that it can be generalized to higher-order contingency tables with ordered subcategories, that is, where the subcategory designations $A_1, A_2, \ldots, A_r$ and $B_1, B_2, \ldots, B_c$ are each ordinal measurements. Then $U$ and $V$ can be calculated exactly as they were for the Kendall Tau coefficient and substituted in the formula (4.12). In this generalized case, $\gamma$ is usually called the *Goodman-Kruskal coefficient*. For a contingency table of $N$ observations, the calculation of $U$ and $V$ in the ordinary way would require listing the $N$ pairs of ranks and then making the $\binom{N}{2}$ comparisons between an $X$ (or $A$) rank and a $Y$ (or $B$) rank. If the contingency table is set up so that the subcategories are the ranks — that is, $A_1$ is rank 1, $A_2$ is rank 2, and so on, for $X$, and similarly $B_1, B_2, \ldots$, for $Y$ — then $U$ and $V$ are easily calculated by multiplication. Using the $f_{ij}$ notation of before, the contingency table is now written as follows:

|  |  | Y rank | | | |
|---|---|---|---|---|---|
|  |  | 1 | 2 | $\cdots$ | c |
|  | 1 | $f_{11}$ | $f_{12}$ | $\cdots$ | $f_{1c}$ |
| X rank | 2 | . | . |  | . |
|  | . | . | . |  | . |
|  | . | . | . |  | . |
|  | r | $f_{r1}$ | $f_{r2}$ | $\cdots$ | $f_{rc}$ |

Consider the elements in cell (1, 1) compared with every other cell, going across row 1, then row 2, and so on. Each element in row 1 or column 1 has either $X$ rank 1 or $Y$ rank 1 and hence contributes nothing to the score. However, compare cell (1, 1) with cell (2, 2). Both the $X$ and $Y$ ranks increase so that $+1$ is scored for each pair and there are $f_{11} f_{22}$ pairs. Similarly for cell (1, 1) compared with cell (2, 3), and in fact every other cell for which $i > 1$ and $j > 1$. For each of the $rc$ cells, the frequency is multiplied by the total number of elements in all other cells which have higher rank on both variables; the sum of these products over all cells is $U$. To find $V$, the frequency in each cell is multiplied by the total number of elements in all other cells that have higher $X$ rank but lower $Y$ rank. This procedure is illustrated in Example 4.4.

**example   4.4**

Data from an early study of the relationship between family planning and level of formal education are reported in Goodman and Kruskal (1954, p. 752). The data are adapted from results of a stratified sample of white Protestant married couples living in Indianapolis in the late 1940s and married in 1927, 1928, or 1929 (Whelpton and Kiser, 1950). The educational level categories are (1) less than three years of high school, (2) three or four years of high school, and (3) some college. The couples were rated according to effectiveness of planning and spacing of children, with 1 designating the least effective planning.

| X | Y | | | | |
|---|---|---|---|---|---|
| | **Family Planning Rank** | | | | |
| **Education Rank** | **1** | **2** | **3** | **4** | **Total** |
| 1 | 223 | 168 | 90 | 110 | 591 |
| 2 | 122 | 215 | 80 | 191 | 608 |
| 3 | 34 | 68 | 35 | 102 | 239 |
| Total: | 379 | 451 | 205 | 403 | 1438 |

The systematic calculations of $U$ and $V$, as indicated by the cell designations $(X, Y)$, are as given in Table 4.2. The cells that are not listed all have 0 scores because at least one of the $X$ or $Y$ ranks is tied. Substituting these results in (4.12), the $\gamma$ coefficient is

$$\gamma = \frac{311{,}632 - 168{,}295}{311{,}632 + 168{,}295} = \frac{143{,}337}{479{,}927} = .299.$$

That is, of the 479,927 nonzero comparisons, the excess of the $+1$ scores over the $-1$ scores is 143,337, and the proportional excess is .299.

**TABLE 4.2**

| Cell | $U$ | $V$ |
|---|---|---|
| 1, 1 | $223(215 + 80 + 191 + 68 + 35 + 102) = 154{,}093$ | $168(122 + 34) = 26{,}208$ |
| 1, 2 | $168(80 + 191 + 35 + 102) = 68{,}544$ | $90(122 + 215 + 34 + 68) = 39{,}510$ |
| 1, 3 | $90(191 + 102) = 26{,}370$ | $110(122 + 215 + 80 + 34 + 68 + 35) = 60{,}940$ |
| 1, 4 | | |
| 2, 1 | $122(68 + 35 + 102) = 25{,}010$ | $215(34) = 7{,}310$ |
| 2, 2 | $215(35 + 102) = 29{,}455$ | $80(34 + 68) = 8{,}160$ |
| 2, 3 | $80(102) = 8{,}160$ | $191(34 + 68 + 35) = 26{,}167$ |
| 2, 4 | | |
| Total: | $311{,}632$ | $168{,}295$ |

Thus when the contingency table is of arbitrary order $r \times c$ and the subcategory designations are ordinal, either the Goodman-Kruskal coefficient $\gamma$ or the Kendall Tau coefficient with correction for ties can be used to describe association. In the $f_{ij}$ notation, the numbers of ties in the denominator of (2.19) are

$$u' = \binom{f_{1.}}{2} + \binom{f_{2.}}{2} + \cdots + \binom{f_{r.}}{2}$$
$$v' = \binom{f_{.1}}{2} + \binom{f_{.2}}{2} + \cdots + \binom{f_{.c}}{2}.$$

For the data of Example 4.4, we have

$$u' = \binom{591}{2} + \binom{608}{2} + \binom{239}{2} = 387,314$$
$$v' = \binom{379}{2} + \binom{451}{2} + \binom{205}{2} + \binom{403}{2} = 275,019.$$

Thus $\binom{N}{2} = \binom{1438}{2} = 1,033,203$ and $U - V = 311,632 - 168,295 = 143,337$. From (2.19), $T$ is found as

$$T = \frac{143,337}{\sqrt{(645,889)(758,184)}} = \frac{143,337}{699,787.6} = .2048.$$

We now summarize briefly the measures of association available for data presented in a contingency table. For any kind of subcategories, and tables of any dimension, the Cramér statistic $C$ in (4.9) can be used. When $r = c = 2$, $C$ is usually called the $\phi$ coefficient; it is most easily calculated from (4.10). The result is equal to the Kendall Tau coefficient with correction for ties as defined in (2.19). When the subcategories represent ordinal measurements, the association can also be measured by either the Goodman-Kruskal coefficient $\gamma$ calculated from (4.12), or by the Kendall Tau coefficient with correction for ties in (2.19). When $r = c = 2$, $\gamma$ is sometimes called the Yule coefficient, and it is most easily calculated from (4.11).

# 5 Measures of Association for Special Types of Data

The measures of association in Section 2, namely the rank correlation coefficient and the Kendall Tau, are appropriate for paired data where each set is measured on at least an ordinal scale. The measures in Section 4.2 can be used for any kind of variables once the results are cast in an $r \times c$ table. The data consist of the numbers of observations in cross-categories, where the subcategories of each category may represent a nominal or ordinal scale, or even an interval scale if the subcategories are intervals of numbers. Nonparametric inferences can be performed in each case. If the data on each variable are measured on an interval scale, the methods of classical statistics, usually the Pearson product-moment correlation coefficient, provide appropriate measures of association.

The one remaining situation for which bivariate association might be measured is where one variable is dichotomous and the other is measured on an interval scale and the interval scale data are not grouped into subcategories of numbers in order to form a contingency table. Data of this type arise frequently in studies in the behavioral, social, and human sciences, since many of the interesting characteristics are dichotomous by nature. Some examples are sex (male and female), employment status (employed or unemployed), opinion (favor or opposed, true or false), treatment effect (cured or uncured, survive or die), and test scores (pass or fail). It is convenient to code observations on a dichotomous variable using values such as 0 and 1. However, the numbers are not meant to imply an ordinal scale of measurement; they are simply numerical indicators of the observation.

The problem of describing association between a dichotomous variable and some variable measured on an interval scale is called biserial correlation. Several different measures of association have been proposed. Because of the type of variables involved, the problem of describing their association falls within the definition of nonparametric methods given in Chapter 1. However, the usual inference procedures based on biserial correlation coefficients require rather stringent assumptions about one or both of the underlying populations that the data represent. Thus the inferences are not nonparametric, and we discuss here only the problem of description of biserial correlation. The interested reader is referred to Walker and Lev (1953, Chapter 11), for instance.

The simplest descriptive measure is called the point biserial coefficient of correlation. The observations on the dichotomous variable are assigned arbitrary numbers and the ordinary Pearson product-moment correlation coefficient is calculated. The absolute value of this coefficient is the same irrespective of what numbers are used, but the arithmetic is simplest if 0 and 1 designate the two dichotomous outcomes. Suppose that $Y$ is a dichotomous variable recorded as 0 or 1, and $n_1$ is the number of 1s in $n$ pairs. Let $\bar{X}_1$ be the mean of the $X$'s that are paired with those $Y$'s that are designated as 1s, and $\bar{X}_0$ the corresponding mean for the $X$'s paired with the remaining $Y$'s. Then the ordinary correlation formula reduces to

$$R_{pb} = \sqrt{\frac{n_1 n_0}{n}} \frac{\bar{X}_1 - \bar{X}_0}{\sqrt{\Sigma(X - \bar{X})^2}}. \qquad (5.1)$$

(This formula is applicable only when the $Y$ data are recorded as 0 or 1.)

As a simple example, consider data for scientific aptitude scores $X$ and sex $Y$ for a sample of 10 persons. Using 1 to denote male and 0 for female, the calculations proceed as shown in Table 5.1.

**TABLE 5.1**

| Subject | Test Score | Sex |
|---------|-----------|-----|
| 1  | 92 | 0 |
| 2  | 85 | 1 |
| 3  | 90 | 0 |
| 4  | 80 | 1 |
| 5  | 76 | 1 |
| 6  | 74 | 1 |
| 7  | 88 | 0 |
| 8  | 83 | 0 |
| 9  | 72 | 1 |
| 10 | 98 | 1 |

$$\bar{X}_1 = \frac{85 + 80 + 76 + 74 + 72 + 98}{6} = 80.83 \qquad n_1 = 6$$

$$\bar{X}_0 = \frac{92 + 90 + 88 + 83}{4} = 88.25 \qquad n_0 = 4$$

$$\bar{X} = \frac{838}{10} = 83.8$$

$$\Sigma(X - \bar{X})^2 = 637.6$$

$$R_{pb} = \sqrt{\frac{24}{10}} \left( \frac{80.83 - 88.25}{\sqrt{637.6}} \right) = -0.455.$$

# 6 Summary

In Section 2 we discussed two nonparametric measures and how they can be used to test for independence between two variables. These procedures were compared in Section 2.4.

The remainder of the chapter was devoted to a study of methods of analysis that are appropriate for multivariate data. The partial correlation coefficient of Section 2.5 is used to describe the relationship between two variables when a third variable is held constant. The Kendall coefficient of concordance, whether for complete or incomplete sets of rankings, is like a measure of total correlation when more than two variables are involved. It provides a descriptive measure of the overall agreement between $k$ sets of rankings of the same objects. It has a precise interpretation as the proportional amount of variation between rankings that can be explained by the existence of a relationship between the sets of rankings. More specifically, the coefficient is the sum of squares of observed deviations from average rank divided by the corresponding sum that would result if the rankings were in perfect agreement. The coefficient of concordance can be used as a test statistic for the null hypothesis that the sets of rankings are independent.

When there are only two primary variables, but each variable has subcategories of classification and the data are counts, methods of contingency table analysis in Section 4 are appropriate. Then we can test the null hypothesis of independence between variables using the chi-square test statistic, and can obtain a measure of the strength of relationship between primary variables by computing the contingency coefficient or the Cramér statistic (frequently called the phi coefficient for $2 \times 2$ contingency tables). Unfortunately, this test is always approximate, and its accuracy is difficult to assess in general.

These procedures for multivariate data have no directly comparable parametric analogue, because they apply to count data in the contingency table situation, and to sets of rankings in the coefficient of concordance situation. The one multivariate method covered in this chapter that has a direct parametric analogue is the Friedman test. Here we are comparing the locations of several related samples (the sampling model is essentially an extension of the matched pairs situation covered in Chapter 3). It is reasonable to compare the Friedman test to the classical two-way (without interactions) analysis of variance test, or the analysis-of-variance test for the randomized complete block design. In fact, if the usual $F$ statistic is computed using ranks rather than the actual observations, it is a linear function of the Friedman test statistic. The classical test, based on the $F$ distribution, is completely valid only under the assumption that the "treatment" effects are normally distributed with a common variance. The Friedman test is a realistic alternative procedure to consider in any of the following cases:

1. The sample sizes are quite small.
2. The only data available are measured on an ordinal scale within each block.
3. The respective sample medians seem to be more reliable than sample means for a comparison between treatment effects.
4. The population medians seem to be more representative parameters for the comparison of interest.
5. The assumption of normality and/or equal variances is questionable.
6. Little or nothing can be assumed about the probability distributions.

The asymptotic efficiency of the Friedman test relative to the parametric analysis-of-variance test is always at least .576 and can be infinite. For normal distributions the efficiency is .637 for $k = 2$, .716 for $k = 3$, .764 for $k = 4$, and keeps increasing as $k$ increases. The value for infinite $k$ is $3/\pi = .955$. The corresponding values for uniform distributions range from $2/3$ to 1.00.

# PROBLEMS

7.1  M. S. Bartlett (1957, Tables 1 and 2) collected data that emphasize the relevance of the population, or community size, in the recurrence of measles epidemics. He observed that if the rate of immigration of new infection can be assumed to be proportional to the population of the community, then the average period between epidemics tends to be larger for smaller communities. Show that his data, given below, reflect this phenomenon because there exists a significant inverse association between population and periodicity of measles epidemics.

## MEASLES EPIDEMICS FOR TOWNS IN ENGLAND AND WALES (1940–1956)

| Town | Population (Thousands) | Average Periodicity (Weeks) |
|------|------------------------|------------------------------|
| Birmingham | 1046 | 73 |
| Manchester | 658 | 106 |
| Bristol | 415 | 92 |
| Hull | 269 | 93 |
| Plymouth | 180 | 94 |
| Norwich | 113 | 80 |
| Barrow-in-Furness | 66 | 74 |
| Carlisle | 65 | 75 |
| Bridgewater | 22 | 86 |
| Newbury | 18 | 92 |
| Carmarthen | 12 | 79 |
| Penrith | 11 | 98 |
| Ffestiniog | 7.1 | 199 |
| Brecon | 5.6 | 149 |
| Okehampton | 4.0 | 105 |
| Cardigan | 3.5 | >284 |
| South Molton | 3.1 | 191 |
| Llanrwst | 2.6 | >284 |
| Appleby | 1.7 | 175 |

7.2  Nobel Laureate William Shockley used the term "population pollution" recently to describe the possible decline in intelligence of the general population as a result of the disproportionately higher birthrate among the most disadvantaged. In general, concern has been expressed over the seemingly inverse relationship between measured intelligence of children and the size of their families. Osborne (1970, p. 189) reported the following results of a recent study based on children in a small rural community where racial composition was approximately equal.

| IQ Class | Midpoint | Average Number of Children in Families Represented |
|----------|----------|----------------------------------------------------|
| 130–139  | 135      | 3.15 |
| 120–129  | 125      | 3.28 |
| 110–119  | 115      | 3.04 |
| 100–109  | 105      | 3.49 |
| 90–99    | 95       | 3.75 |
| 80–89    | 85       | 4.25 |
| 70–79    | 75       | 4.82 |
| 60–69    | 65       | 5.04 |
| 50–59    | 55       | 5.17 |
| 40–49    | 45       | 4.84 |

Calculate the Kendall Tau coefficient and make an appropriate inference concerning this community. Is it possible to make population inferences from such a study?

7.3    Research and development are important activities in any industry, but especially in the pharmaceutical industry. Funds applied to R & D increased steadily in the 1960s, at least when the total funds include receipts from Federal sources. However, the employment opportunities for scientists and engineers have not been good. The data below were reported by National Science Foundation (1971, Tables B-2, B-12, B-19, and B-53). Use your judgment to decide what statistical methods are appropriate, and write a short report on your findings.

| Year | Funds for R&D | | Full-Time Equivalent R&D Scientists and Engineers | |
|------|------------------|---------------------|-------|-------------------|
|      | Total (Millions) | Percent of Net Sales | Total | Per 1000 Employees |
| 1960 | $162 | 4.6 | 6000 | 50 |
| 1961 | 180 | 4.3 | 6200 | 42 |
| 1962 | 195 | 4.3 | 6800 | 44 |
| 1963 | 216 | 4.7 | 6900 | 47 |
| 1964 | 238 | 5.9 | 7300 | 54 |
| 1965 | 274 | 5.7 | 7700 | 53 |
| 1966 | 318 | 6.2 | 8000 | 53 |
| 1967 | 356 | 6.3 | 9300 | 51 |
| 1968 | 394 | 6.1 | 10000 | 51 |
| 1969 | 437 | 6.1 | 10300 | 50 |
| 1970 | 484 | 6.0[a] | 11000 | 52[a] |

[a]Estimated.

7.4    A random sample of nine households was chosen in a certain community to participate in a pilot study of the relationship between household food expenditures and gross total household income. The households selected kept careful records for one month of amounts spent on food items, and these totals were divided by that month's take-home pay to determine percent of disposable income spent on food. The data collected are shown below, where $X$ represents gross income and $Y$ represents percent of this income spent on food items. Find the coefficient of rank correlation and the Tau coefficient for these data.

X:   4,700 5,260 8,420 9,000 10,000 10,500 11,500 15,000 20,100
Y:    42    36    33    25    27     26     29     25     20

7.5   It is frequently conjectured that income is one of the primary determinants of social status for an adult male. In a study to investigate this statement, suppose that 10 adult males are chosen at random in a certain community where the residents are primarily professional people; each subject is asked to complete a questionnaire designed to measure social status within the community. The questionnaires are scored on the basis of 100 points, with higher scores corresponding to higher social status. The test scores and gross incomes of the 10 subjects are shown below. State and test a reasonable hypothesis. What can you say about the conjecture that prompted the investigation?

Social status:   92   51   88   65   80   31   38   75   45   72
Income (thousands):   29.9 18.7 32.0 15.0 26.0  9.0 11.3 22.1 16.0 25.0

7.6   Numerous articles have dealt with the subject of risk taking and utility preferences among corporate executives. Vroom and Pahl (1971) used a sample of 1484 male managers employed in over 200 companies to examine the association between a manager's age and his aversion to risk. This aversion was measured by giving each subject a set of questions extracted from the standard Kogan and Wallack choice-dilemma questionnaire. The following data on mean risk scores for 15 different age groups are a representative subset of the results presented graphically in Vroom and Pahl (p. 404). Lower mean risk scores indicate a more conservative attitude toward risk taking. Measure the association between age and risk preference and interpret the result.

| Age | Mean Risk Score | Age | Mean Risk Score |
|-----|-----------------|-----|-----------------|
| 25  | .42             | 37  | .12             |
| 26  | .46             | 38  | .13             |
| 28  | .32             | 41  | .12             |
| 31  | .35             | 43  | .14             |
| 32  | .11             | 46  | .17             |
| 33  | .26             | 55  | .05             |
| 34  | .25             | 57  | .01             |
| 35  | .13             |     |                 |

7.7   Twelve individuals have been assigned the following score ranks on two attitudinal measures. Is there a significant association between the two sets of scores?

| Test | Individual | | | | | | | | | | | |
|---|---|---|---|---|---|---|---|---|---|---|---|---|
| | **1** | **2** | **3** | **4** | **5** | **6** | **7** | **8** | **9** | **10** | **11** | **12** |
| 1 | 12 | 5 | 2 | 1 | 6 | 4 | 3 | 7 | 11 | 9 | 8 | 10 |
| 2 | 9 | 4 | 5 | 3 | 7 | 1 | 2 | 10 | 8 | 12 | 6 | 11 |

7.8  Randomly select 20 lines of text from this book. Ignoring one-letter words, determine the association between word length and number of occurrences in these 20 lines. Is the relationship direct or inverse? Are there benefits to be derived from an inverse relationship as opposed to a direct relationship? (*Hint:* Compare the two types of relationships with respect to the effect they would have on the length of this book.)

7.9  Krumbein and Pettijohn (1938, p. 262) collected the following data to study the relationship between average roundness and average particle size in gravel.

| Sample | Average Roundness | Average Particle Size |
|---|---|---|
| 1 | .62 | 52 |
| 2 | .74 | 43 |
| 3 | .65 | 36 |
| 4 | .71 | 32 |
| 5 | .68 | 27 |
| 6 | .59 | 26 |
| 7 | .49 | 22 |
| 8 | .67 | 37 |
| 9 | .64 | 24 |
| 10 | .56 | 19 |
| 11 | .51 | 13 |

Do these data support the judgment that there is a meaningful relationship between particle size and roundness?

7.10  The data below show the scores obtained by 15 students on a general college entrance examination and a verbal comprehension test. Test for association between the two sets of data.

| Student | Verbal Score | General Exam Score | Student | Verbal Score | General Exam Score |
|---|---|---|---|---|---|
| 1 | 450 | 62 | 9 | 302 | 45 |
| 2 | 350 | 50 | 10 | 490 | 65 |
| 3 | 478 | 63 | 11 | 415 | 57 |
| 4 | 430 | 55 | 12 | 420 | 60 |
| 5 | 319 | 30 | 13 | 476 | 61 |
| 6 | 389 | 57 | 14 | 428 | 53 |
| 7 | 325 | 48 | 15 | 488 | 62 |
| 8 | 460 | 58 | | | |

7.11  The Consumer Price Index (CPI) is a well-known measure of price levels in personal terms for most consumers. The Wholesale Price Index (WPI) is a comparable measure for wholesalers. The WPI has a history of fluctuations, both up and down, whereas the CPI has pushed relentlessly upward over the past several decades. The data below, from the *Statistical Abstract of the United States* (1970, p. 338), are the WPI and CPI from 1951 through 1969, each with a base 1957–1959. Find an appropriate measure of the association between these two measures of different types of price changes. Since purchasing power of a current dollar in terms of 1957–1959 dollars is the reciprocal of the price index with that base, the degree of association between the WPI and the CPI is the same as the degree of association between purchasing power of a consumer and a wholesaler. For the data below on consumer purchasing power (CPP) and wholesale purchasing power (WPP), find the coefficient of rank correlation to measure the association between CPI and WPI.

| Year | WPP | CPP | Year | WPP | CPP | Year | WPP | CPP |
|------|-----|-----|------|-----|-----|------|-----|-----|
| 1951 | 1.034 | 1.106 | 1958 | .996 | .994 | 1965 | .976 | .910 |
| 1952 | 1.064 | 1.081 | 1959 | .994 | .985 | 1966 | .944 | .884 |
| 1953 | 1.079 | 1.072 | 1960 | .993 | .971 | 1967 | .943 | .860 |
| 1954 | 1.076 | 1.069 | 1961 | .997 | .960 | 1968 | .920 | .825 |
| 1955 | 1.073 | 1.071 | 1962 | .994 | .949 | 1969 | .885 | .782 |
| 1956 | 1.040 | 1.056 | 1963 | .997 | .937 | | | |
| 1957 | 1.010 | 1.021 | 1964 | .975 | .925 | | | |

1.12  Krumbein and Graybill (1965, p. 278) report actual data for foreshore slope measurements in degrees at different sites at a selected beach on Lake Michigan. Sand samples obtained at the site are analyzed for average size of sand grain and degree of sand sorting. For 12 samples, the slope $X_1$, the median grain diameter $X_2$, and the degree of sand sorting $X_3$ are shown below.
(a) Evaluate Kendall's Tau as a measure of association between $X_1$ and $X_2$, and also between $X_1$ and $X_3$.
(b) Find the value of Kendall's partial Tau for $X_1$ and $X_2$ given $X_3$, and for $X_1$ and $X_3$ given $X_2$, and interpret the results in view of your answers to (a).

| Sample | $X_1$ | $X_2$ | $X_3$ | Sample | $X_1$ | $X_2$ | $X_3$ |
|--------|-------|-------|-------|--------|-------|-------|-------|
| 1 | 12.0 | .870 | 1.69 | 7 | 6.0 | .205 | 1.14 |
| 2 | 5.5 | .202 | 1.17 | 8 | 4.5 | .670 | 1.92 |
| 3 | 4.5 | .203 | 1.17 | 9 | 5.5 | .205 | 1.22 |
| 4 | 4.5 | .198 | 1.17 | 10 | 6.0 | .271 | 1.71 |
| 5 | 10.5 | .730 | 1.63 | 11 | 4.0 | .203 | 1.16 |
| 6 | 4.5 | .510 | 1.59 | 12 | 6.0 | .264 | 1.37 |

7.13  Blank (1954) has studied the relationship between an index of residential cost construction and house price indexes for a group of 22 cities, and specifically

for Cleveland and Seattle. All indexes use a 1929 base. The data below are from Blank (pp. 72, 75, 76). Find a measure of
(a) The association between cost and "typical city" price.
(b) The association between price in Cleveland and Seattle.
(c) The partial correlation of price in Cleveland and Seattle when the effect of cost is removed.

| Year | Cost Index | Price Index | | |
| | | 22 Cities | Cleveland | Seattle |
|------|-----------|-----------|-----------|---------|
| 1920 | 118.7 | 90.8 | 86.8 | 88.2 |
| 1921 | 95.4 | 90.0 | 87.8 | 86.3 |
| 1922 | 87.7 | 92.5 | 91.5 | 99.8 |
| 1923 | 98.3 | 95.2 | 96.3 | 100.0 |
| 1924 | 96.9 | 96.7 | 100.0 | 117.6 |
| 1925 | 96.2 | 103.1 | 102.4 | 109.8 |
| 1926 | 96.9 | 100.4 | 103.7 | 107.8 |
| 1927 | 95.6 | 97.9 | 102.4 | 99.9 |
| 1928 | 95.9 | 100.7 | 101.2 | 102.0 |
| 1929 | 100.0 | 100.0 | 100.0 | 100.0 |

7.14 Eight graduate students are each given examinations in quantitative reasoning, vocabulary, and reading comprehension. Their scores are listed below. It is frequently claimed that persons who excel in quantitative reasoning are not as capable with verbal, and vice versa, and yet a truly intelligent person must possess all of these abilities. Test these data to see if there is an association between scores, and estimate the rank order of the students' abilities. Does there seem to be an indirect relationship between quantitative and verbal abilities?

| Test | Student | | | | | | | |
| | 1 | 2 | 3 | 4 | 5 | 6 | 7 | 8 |
|------|---|---|---|---|---|---|---|---|
| Quantitative | 90 | 60 | 45 | 48 | 58 | 72 | 25 | 85 |
| Vocabulary | 62 | 81 | 92 | 76 | 70 | 75 | 95 | 72 |
| Reading | 60 | 91 | 85 | 81 | 90 | 76 | 93 | 80 |

7.15 In an effort to improve the equity and consistency of figures used as appraisals of market value of real estate property, a realty company set up guidelines and training sessions for local appraisers. To test the success of this program, four properties in a similar value and use category were shown to each of five appraisers. Their independent appraised values (in thousands of dollars) are shown below. Test the consistency of the appraised values, and estimate the order of value of these properties.

|          | Property |      |      |      |
|----------|----------|------|------|------|
| Appraiser | A       | B    | C    | D    |
| 1        | 25.1     | 18.2 | 31.1 | 29.2 |
| 2        | 10.2     | 15.6 | 19.3 | 21.2 |
| 3        | 23.1     | 20.4 | 40.0 | 27.8 |
| 4        | 24.0     | 15.6 | 35.2 | 28.0 |
| 5        | 24.5     | 21.3 | 26.0 | 26.2 |

7.16 Ten graduate students take identical comprehensive examinations in their major field. The grading procedure is that each professor ranks each student's paper in relation to all others taking the examination. Suppose that four professors give the following ranks, where 1 indicates the best paper and 10 the worst.

|           | Student |   |   |   |   |   |   |   |   |    |
|-----------|---------|---|---|---|---|---|---|---|---|----|
| Professor | 1 | 2 | 3 | 4 | 5 | 6 | 7 | 8 | 9 | 10 |
| 1         | 5 | 3 | 8 | 9 | 2 | 7 | 6 | 1 | 4 | 10 |
| 2         | 7 | 4 | 6 | 2 | 3 | 9 | 8 | 5 | 1 | 10 |
| 3         | 3 | 5 | 7 | 6 | 4 | 10 | 8 | 2 | 1 | 9 |
| 4         | 4 | 5 | 7 | 8 | 3 | 9 | 6 | 1 | 2 | 10 |

(a) Is there evidence of agreement among the four professors?

(b) Give an overall estimate of the relative performance of each student.

(c) Will it be difficult to decide which students should be given a passing grade?

7.17 The following table shows cell counts on a sensitive murine leukemia cell line (L51784) reported by Williams and Path (1970, p. 659). These data were collected after three different periods of time for a control group in tube A, and two different concentrations of L-asparaginase added in tubes B and C. Test the null hypothesis that cell counts are the same for the three tubes.

|          | Tube  |      |      |
|----------|-------|------|------|
| Time     | A     | B    | C    |
| 2 hours  | 5.55  | 5.16 | 5.46 |
| 4 hours  | 5.58  | 5.18 | 5.22 |
| 12 hours | 12.16 | 2.83 | 2.51 |

7.18 In a clinical experiment at the beginning of the century, Cushny and Peebles (1905) investigated the effectiveness of various hydrobromides on sleep. The average number of hours of sleep on 3 to 9 nights was recorded for patients following the four different treatments: A = no hypnotic, B = levo-hyoscyamine, C = levo-hyoscine, and D = racemic hyoscine. The results are shown below.

| | Average Hours of Sleep by Patient Number | | | | | | | | | |
|---|---|---|---|---|---|---|---|---|---|---|
| Drug | 1 | 2 | 3 | 4 | 5 | 6 | 7 | 8 | 9 | 10 |
| A | .6 | 1.1 | 2.5 | 2.8 | 2.9 | 3.0 | 3.2 | 4.7 | 5.5 | 6.2 |
| B | 1.3 | 1.1 | 6.2 | 3.6 | 4.9 | 1.4 | 6.6 | 4.5 | 4.3 | 6.1 |
| C | 2.5 | 5.7 | 8.0 | 4.4 | 6.3 | 3.8 | 7.6 | 5.8 | 5.6 | 6.1 |
| D | 2.1 | 5.8 | 8.2 | 4.3 | 6.4 | 4.4 | 8.3 | 4.7 | 4.8 | 6.7 |

Test the hypothesis that there is no difference in effectiveness of the four drugs. Use multiple comparisons to determine which drugs differ significantly at level .10.

7.19 Cullinan (1963) reported results of a study to investigate stability of adjustment to stimulation, called adaptation, in persons who stutter. Twenty-three stutterers enrolled in the speech clinic at the University of Illinois were each given a 300-word factual passage to read aloud five successive times. This was repeated one week later and then again two weeks later for each subject. From a tape recording of each reading, the experimenters counted number of stuttered words, and figured overall reading rate and nonstuttered reading rate. These values were used to obtain adaptation scores, by a standard procedure, for frequency of stuttering, nonstuttered reading rate, and overall reading rate for each subject at each reading. The data shown in the two tables below for mean frequency of stuttered words per reading and mean overall reading rates in syllables per minute for the 23 subjects were estimated from the graphic presentation of results in Cullinan on pp. 73 and 78, respectively. For each set of data, test the null hypothesis that the average scores for the three days are the same.

| | Mean Frequency of Stuttered Words on Days | | |
|---|---|---|---|
| Reading | 1 | 2 | 3 |
| 1 | 44 | 38 | 36 |
| 2 | 31 | 28 | 32 |
| 3 | 30 | 27 | 34 |
| 4 | 26 | 27 | 32 |
| 5 | 27 | 28 | 31 |

| | Mean Reading Rate in Syllables/Minute on Days | | |
|---|---|---|---|
| Reading | 1 | 2 | 3 |
| 1 | 183 | 200 | 203 |
| 2 | 203 | 217 | 221 |
| 3 | 207 | 220 | 222 |
| 4 | 212 | 224 | 225 |
| 5 | 212 | 225 | 228 |

7.20 Aborn and Rubinstein (1952) conducted a study applying the concept of information theory to the problem of learning as reflected by immediate recall. Forty potential subjects were asked to memorize a list of nonsense syllables and their classification into one of four classes (for example, syllables starting with a $z$ constituted one class). Twenty-six of these were later deemed proficient enough to participate in the study. Each participant was given three minutes to memorize a passage compiled using only these syllables, and then asked to write what he could recall. This was done for six different passages of approximately equal length (30–32 syllables) with varying degrees of organization, which is inversely proportional to the average rate of information in the passage. Passage I was a random organization of the syllables, passage II was a low degree of organization, and passages III, IV, V, and VI were organized to increasing degrees. The rank of organization, and average rate of information measured in bits per syllable are as follows:

| Passage | Organization Rank | Information |
|---|---|---|
| I | 1 | 4 |
| II | 2 | 3 |
| III | 3 | 2.5 |
| IV | 4 | 2 |
| V | 5 | 1.5 |
| VI | 6 | 1 |

The passages were given to the subjects in random order; in grading each passage for each subject, the number of syllables recalled correctly was counted and the information scores were computed as number of syllables recalled multiplied by the average rate of information per syllable. The mean values of these variables were reported graphically (p. 263), and the data shown below, although artificial, represent results for a comparable experiment using only eight subjects.

(a) Test the null hypothesis that the average number of syllables correct is the same for each passage (that is, it is not affected by the organization level of the passage), and perform a multiple comparisons analysis at level .10 to determine whether subjects are able to recall more when there is an organization pattern to aid memory.

(b) Since the rate of information per syllable is inversely proportional to the degree of organization, the average information scores were expected to be the same for all passages. Do these data support this conclusion?

| Subject | Number of Syllables Correct for Passage | | | | | |
|---|---|---|---|---|---|---|
| | I | II | III | IV | V | VI |
| 1 | 3 | 4 | 10 | 12 | 11 | 17 |
| 2 | 5 | 6 | 10 | 13 | 12 | 21 |
| 3 | 7 | 9 | 12 | 16 | 15 | 20 |
| 4 | 9 | 8 | 13 | 17 | 17 | 18 |
| 5 | 5 | 7 | 11 | 14 | 14 | 16 |
| 6 | 4 | 5 | 8 | 13 | 14 | 13 |
| 7 | 12 | 13 | 9 | 10 | 12 | 14 |
| 8 | 10 | 12 | 7 | 12 | 11 | 12 |

| Subject | Information Score for Passage | | | | | |
|---|---|---|---|---|---|---|
| | I | II | III | IV | V | VI |
| 1 | 12 | 12 | 25 | 24 | 16.5 | 17 |
| 2 | 20 | 18 | 25 | 26 | 18 | 21 |
| 3 | 28 | 27 | 30 | 32 | 22.5 | 20 |
| 4 | 36 | 24 | 32.5 | 34 | 25.5 | 18 |
| 5 | 20 | 21 | 27.5 | 28 | 21 | 16 |
| 6 | 16 | 15 | 20 | 26 | 21 | 13 |
| 7 | 36 | 39 | 22.5 | 20 | 18 | 14 |
| 8 | 40 | 26 | 17.5 | 24 | 16.5 | 12 |

7.21 Mosteller and Wallace (1964) performed an extensive study of the authorship of the Federalist papers. Of the 77 papers written during the period 1768–1787 in an attempt to support the ratification of the Constitution by the citizens of New York, the authorship is generally agreed on most papers, but 15 papers are still in dispute. Of those agreed upon, 5 are attributed to John Jay, 43 to Alexander Hamilton, and 14 to James Madison. In attempting to settle authorship of the disputed papers, Mosteller tried to discriminate among writings on the basis of the rates with which the three authors used high-frequency words, like a, and, be, in, of, the, to, and so forth. The data below (from Table 8.1-1, pp. 244–245) are the rate of occurrence of these seven words per 1000 words in the writings known to be by Hamilton, Madison, and Jay. Analyze the data to determine whether the mean rates of occurrence for these words are the same for the three authors.

| Word | Hamilton | Madison | Jay |
|------|----------|---------|-----|
| a | 22.85 | 20.22 | 13.57 |
| and | 24.50 | 27.55 | 45.36 |
| be | 20.06 | 16.45 | 19.15 |
| in | 24.37 | 23.05 | 20.82 |
| of | 64.65 | 57.80 | 43.87 |
| the | 91.27 | 93.65 | 67.48 |
| to | 40.71 | 35.25 | 35.69 |

7.22 Although speed in reading is of great importance for an educated and informed person in our fast-moving society, comprehension and retention of information from reading are equally important. In an experiment to investigate the effect of passing time on recall of material read, eight subjects are given the same material to read in a fixed period of time. These subjects are then tested to ascertain the retention of information read after four different intervals of time. The scores shown below are from Ferguson (1966, p. 364); higher scores indicate a higher degree of recall.

| Subject | Time Interval I | II | III | IV |
|---------|-----|-----|-----|-----|
| 1 | 4 | 5 | 9 | 3 |
| 2 | 8 | 9 | 14 | 7 |
| 3 | 7 | 13 | 14 | 6 |
| 4 | 16 | 12 | 14 | 10 |
| 5 | 2 | 4 | 7 | 6 |
| 6 | 1 | 4 | 5 | 3 |
| 7 | 2 | 6 | 7 | 9 |
| 8 | 5 | 7 | 8 | 9 |

(a) Test the null hypothesis that there is no difference in amount of material recalled after different time intervals or, equivalently, that different time intervals have no effect on retention rate.
(b) Estimate the time interval at which retention rate is a maximum.
(c) Use the method of multiple comparisons to determine which pairs of time intervals differ significantly at level .15.

7.23 In a collaborative study of dry milk powders, six different types A to F are tested in each of seven different laboratories, and ranked in order of decreasing quality, that is 1 = best, 6 = poorest. The results shown below are from Bliss (1967, p. 339).
(a) Test the null hypothesis that there is no difference between powders.
(b) Use multiple comparisons at level .15 to see which pairs differ significantly.
(c) Estimate the true order of quality of the milk powders.

| Lab | Rank for Powder | | | | | |
|-----|---|-----|-----|---|---|---|
|     | A | B   | C   | D | E | F |
| 1   | 2 | 3   | 6   | 1 | 5 | 4 |
| 2   | 2 | 1   | 3   | 4 | 5 | 6 |
| 3   | 1 | 2   | 3   | 5 | 4 | 6 |
| 4   | 2 | 3   | 1   | 5 | 6 | 4 |
| 5   | 4 | 1.5 | 1.5 | 6 | 3 | 5 |
| 6   | 1 | 3   | 4   | 5 | 2 | 6 |
| 7   | 2 | 4   | 1   | 5 | 6 | 3 |

7.24 Water suspensions of the insecticide benzene hexachloride containing three different concentrations of its $\gamma$-isomer were sprayed on soybeans during the ripening period. Ripened soybeans from each concentration of isomer and from an unsprayed control were ranked from 1 to 4 for absence of off-flavor in a taste test. The data below are from Ishikura and Ozaki (1952).

## RANKING OF OFF-FLAVOR BY TASTER

| Taster | Content of Isomer (Percent) | | | | Taster | Content of Isomer (Percent) | | | |
|--------|------|----|----|---------|--------|------|----|----|---------|
|        | 50.3 | 65 | 95 | Control |        | 50.3 | 65 | 95 | Control |
| 1  | 4 | 3 | 2 | 1 | 11 | 1 | 2 | 4 | 3 |
| 2  | 4 | 1 | 3 | 2 | 12 | 4 | 3 | 2 | 1 |
| 3  | 4 | 3 | 2 | 1 | 13 | 4 | 1 | 3 | 2 |
| 4  | 3 | 4 | 2 | 1 | 14 | 1 | 4 | 2 | 3 |
| 5  | 2 | 1 | 3 | 4 | 15 | 4 | 2 | 3 | 1 |
| 6  | 4 | 3 | 1 | 2 | 16 | 3 | 4 | 2 | 1 |
| 7  | 2 | 4 | 3 | 1 | 17 | 4 | 1 | 3 | 2 |
| 8  | 4 | 3 | 1 | 2 | 18 | 4 | 2 | 3 | 1 |
| 9  | 4 | 3 | 1 | 2 | 19 | 4 | 3 | 2 | 1 |
| 10 | 3 | 4 | 2 | 1 | 20 | 4 | 2 | 3 | 1 |

(a) Test the null hypothesis that the effect on off-taste is the same for the four treatments.

(b) Do a multiple comparisons test to see which pairs of treatments differ significantly at level .20.

7.25 A consumer report periodical needs information on five U.S. makes of cars, all compact models. Data are obtained on price, handling performance, comfort, included extras, and so on. Then one car of each make is randomly selected for supervised tests designed to estimate operating expense per mile, which includes gas, oil, repairs, and scheduled preventive maintenance services.

Since operating expense for any make is affected by past mileage, four consecutive test runs are made with each car; each run consists of 15,000 miles. The data below are the expenses per mile in cents.

| Test Run | Car Make | | | | |
|---|---|---|---|---|---|
| | 1 | 2 | 3 | 4 | 5 |
| A | 5.9 | 6.7 | 6.1 | 6.2 | 5.8 |
| B | 6.1 | 6.2 | 6.3 | 6.2 | 6.4 |
| C | 6.7 | 6.8 | 6.9 | 6.6 | 6.9 |
| D | 7.5 | 7.6 | 7.7 | 7.4 | 7.6 |

(a) Is there a significant difference among the five makes in operating expense?
(b) If in all other comparisons there is no significant difference between makes and if the range in price is $100, with the first make being the least expensive, the second make next lowest, and so on, which make should a prospective consumer purchase? (*Hint:* Use a multiple comparisons test for a selected subset of all possible comparisons.)
(c) Can you explain the difference between the information yielded by the tests in (a) and (b)?

7.26 Mosteller and Tukey (1968, p. 198) report some data collected by Kappauf and Smith in a test of the performance of six randomly selected observers in reading three sizes of dials. Each observer made 60 readings on each size of dial, and the number of errors were counted for all readings on each size. The data are shown below. Test the null hypothesis that the average number of errors is the same for each of the three sizes of dials.

| Observer | Errors in 60 Trials | | |
|---|---|---|---|
| | Size I | Size II | Size III |
| 1 | 7 | 4 | 4 |
| 2 | 15 | 6 | 3 |
| 3 | 10 | 7 | 5 |
| 4 | 28 | 20 | 16 |
| 5 | 11 | 8 | 9 |
| 6 | 25 | 17 | 15 |

7.27 Six rats are trained to run through a maze by experiencing 17 trials. The time required for these rats to run through the maze during their 18th to 22nd trials are shown below, as measured in units of 100(log seconds) − 70 and reported by Kaplan et al (1951). Test the hypothesis that there is no difference between the rats as regards time required to run the maze.

| Trial | Rat | | | | | |
|-------|-----|----|----|----|-----|-----|
|       | 1   | 2  | 3  | 4  | 5   | 6   |
| 18    | 25  | 60 | 62 | 30 | 107 | 81  |
| 19    | 60  | 30 | 90 | 50 | 95  | 95  |
| 20    | 8   | 60 | 59 | 60 | 73  | 84  |
| 21    | 34  | 84 | 70 | 38 | 60  | 108 |
| 22    | 48  | 96 | 64 | 67 | 70  | 112 |

7.28 Warm et al (1964) investigated whether temporal perception is affected by briefly presented visual stimuli related to personal values. Twenty-four words, 12 frequently seen and 12 infrequently seen, were classified into six stimulus categories of personal value according to the Allport, Vernon, Lindzey Study of Values. The categories and their value ranks are (1) theoretical, (2) economic, (3) political, (4) esthetic, (5) religious, and (6) social. For example, the words scientific and physics are in the frequent theoretical category, while hospitality and service are in the frequent social category. The 24 words were flashed on a screen for exactly 1 second, at 24-second intervals, in a randomized order of presentation. The subjects were asked to record their impression of the flash duration for each word without the use of any aid to timing. Suppose that the average time perception for the four words in each value category areas shown below for 10 subjects. Test the null hypothesis that temporal perception is the same for each of the six value categories. If the null hypothesis is rejected, perform a multiple comparisons analysis.

| Subject | Value Category | | | | | |
|---------|-----|-----|-----|-----|-----|-----|
|         | 1   | 2   | 3   | 4   | 5   | 6   |
| 1       | .88 | .92 | .84 | .78 | .99 | .75 |
| 2       | .76 | .73 | .83 | .79 | .85 | .90 |
| 3       | .76 | .81 | .85 | .89 | .88 | .91 |
| 4       | .79 | .78 | .87 | .88 | .90 | .76 |
| 5       | .82 | .85 | .78 | .72 | .75 | .77 |
| 6       | .90 | .87 | .89 | .98 | .85 | .83 |
| 7       | .85 | .83 | .88 | .90 | .83 | .91 |
| 8       | .90 | .90 | .95 | .91 | .90 | .90 |
| 9       | .85 | .90 | .83 | .96 | .91 | .86 |
| 10      | .79 | .83 | .82 | .82 | .90 | .91 |

7.29 In the same article by Warm et al. (1964, p. 245), the mean temporal judgments for 45 subjects were given separately for frequent and infrequent words in each of the six different value categories. Analyze the association between perception for frequent and infrequent words.

| Value Category | 1 | 2 | 3 | 4 | 5 | 6 |
|---|---|---|---|---|---|---|
| Frequent Words | .93 | .94 | 1.01 | .92 | .89 | .91 |
| Infrequent Words | .84 | .81 | .85 | .78 | .80 | .87 |

7.30 The following data were obtained from the *Uniform Crime Reports for the United States* (1969, p. 8):

### CRIME RATE BY REGIONS, 1969
(rate per 100,000 inhabitants)

| Crime Index Offenses | Northeastern | North Central | Southern | Western |
|---|---|---|---|---|
| Murder | 5.2 | 6.5 | 10.4 | 6.1 |
| Forcible rape | 12.6 | 17.2 | 17.0 | 29.2 |
| Robbery | 188.6 | 148.5 | 112.2 | 151.8 |
| Aggravated assault | 124.1 | 121.5 | 186.5 | 176.7 |
| Burglary | 985.3 | 807.2 | 837.4 | 1437.1 |
| Larceny ($50 and over) | 743.7 | 654.2 | 616.2 | 1161.4 |
| Auto theft | 526.4 | 415.7 | 307.5 | 553.5 |

Test the null hypothesis that there is no significant difference between the regions of the United States with regard to the number of criminal offenses.

7.31 Some classic data collected by Goring and quoted by K. Pearson in 1909 and by M. G. Kendall (1943, p. 356) cross-classify criminals according to type of crime committed and drinking habits. Test the null hypothesis of independence between type of crime and drinking habits.

| Offense | Drinker | Abstainer |
|---|---|---|
| Arson | 50 | 43 |
| Rape | 88 | 62 |
| Violence | 155 | 110 |
| Stealing | 379 | 300 |
| Coining | 18 | 14 |
| Fraud | 63 | 144 |

7.32 As part of a larger study, Anderson (1955) classified pebbles as either faceted or unfaceted, and cross-classified them on the basis of two lithologic types, granite and metamorphic.

| Type | Faceted | Unfaceted |
|---|---|---|
| Granite | 41 | 170 |
| Metamorphic | 14 | 42 |

On the basis of these data, would you conclude that faceting is independent of these lithologic types?

7.33 To assist local students preparing for the CPA examination, the accounting department of a university offers a review course at a reasonable cost. On the editorial page of a recent issue of the college newspaper, several students who failed the exam claimed that the review course was inadequate. The department collected the following data in a random sample from their records of the past years.

|              | Passed | Failed | Total |
|--------------|--------|--------|-------|
| Had Review   | 79     | 21     | 100   |
| Had No Review| 61     | 39     | 100   |
| Total:       | 140    | 60     | 200   |

What criterion is being used here to measure the effectiveness of the review? Will the data above enable them to refute the claim? Give a statistical analysis.

7.34 Use the chi-square test to determine whether there is an association between educational level and the type of magazine a person claims to prefer.

| Magazine Type | College | High School |
|---------------|---------|-------------|
| Sports        | 15      | 30          |
| News          | 5       | 40          |
| Business      | 20      | 10          |

7.35 By making an intensive study of the published literature from 1967 to 1969, Spates and Levin (1972) showed that the underground counterculture in the United States has had almost no effect on the dominant cultural values in recent years. This conclusion is in flat contradiction to the many sociologists who claim that the hippie movement has caused a major change in socio-cultural values, namely away from the "instrumental" values of achievement and toward the "expressive" ones, and therefore a shift toward counter-culture values. The findings were based in part on a chi-square analysis of the following sets of data. The data in the first table (from p. 343) were used to establish that the underground press is committed to expressive value themes in their publications and hence that the contraculture is a strong force, and the second table (from p. 349) was used to show that the value themes in the literature published by the so-called "middle class" magazines have not changed significantly in the 10-year period in which the counterculture was developing. Verify that these data support the conclusions reached.

| Value Theme | Type of Publication | |
|---|---|---|
| | Underground | Middle Class |
| Expressive | 314 | 280 |
| Instrumental | 104 | 259 |
| | 418 | 539 |

| Value Theme | Middle Class Magazines | |
|---|---|---|
| | 1957–59 | 1967–69 |
| Expressive | 144 | 136 |
| Instrumental | 128 | 131 |
| | 272 | 267 |

7.36 In a study of job satisfaction among 72 persons with middle income occupations, Morse and Weiss (1955, p. 198) reported the data shown below. Test the null hypothesis that degree of satisfaction is independent of type of occupation.

| Occupation | Satisfied and Very Satisfied | Indifferent and Dissatisfied |
|---|---|---|
| Professional | 25 | 3 |
| Manager | 14 | 8 |
| Salesman | 18 | 4 |
| | 57 | 15 |

7.37 Most lawyers aspire to become a partner in a legal firm rather than a permanent associate, since partners have more prestige and authority, in addition to higher income. Smigel (1964, p. 133) reported the following classification of 46 lawyers according to position in a legal firm and type of secondary school education, public or private. Regarding these categories as nominal scale variables, calculate the phi coefficient.

| | Partner | Associate | Total |
|---|---|---|---|
| Private School | 15 | 10 | 25 |
| Public School | 9 | 12 | 21 |
| Total: | 24 | 22 | 46 |

7.38 In a medical experiment, 30 patients with high blood pressure (hypertension) were randomly divided into two groups of size 15 each. The members of one group were given a placebo (P) while the other patients were each given a new drug (D) claimed effective against hypertension. Each trial was double-blind, meaning that neither the patient nor the physician knew whether the drug or the placebo was being administered, and the two treatments were given with the same frequency for the same period of time. The physician then gave each patient a series of blood pressure tests and made a judgment concerning whether the prognosis was favorable (1) or unfavorable (0). The data are shown below.

| Patient | 1 | 2 | 3 | 4 | 5 | 6 | 7 | 8 | 9 | 10 | 11 | 12 | 13 | 14 | 15 |
|---------|---|---|---|---|---|---|---|---|---|----|----|----|----|----|----|
| Treatment | P | P | P | P | P | P | P | P | P | P | P | P | P | P | P |
| Prognosis | 0 | 0 | 1 | 0 | 0 | 1 | 0 | 0 | 0 | 0 | 0 | 0 | 1 | 0 | 0 |

| Patient | 16 | 17 | 18 | 19 | 20 | 21 | 22 | 23 | 24 | 25 | 26 | 27 | 28 | 29 | 30 |
|---------|----|----|----|----|----|----|----|----|----|----|----|----|----|----|----|
| Treatment | D | D | D | D | D | D | D | D | D | D | D | D | D | D | D |
| Prognosis | 1 | 1 | 1 | 0 | 1 | 1 | 0 | 1 | 1 | 1 | 1 | 0 | 1 | 0 | 1 |

(a) Treat both sets of observations as ordinal data, considering the placebo as a low dosage and the drug as a higher dosage, and calculate the Kendall Tau coefficient $T$ with correction for ties using (2.19).

(b) Cast the data in a $2 \times 2$ contingency table, calculate $Q$ and the Cramér coefficient $C$ from (4.9), showing that $T = C$.

7.39 Suppose that 16 newly married couples were given a questionnaire in 1960. The questions covered all relevant background and interest factors. On the basis of this information each couple was given a prediction score for marital happiness. These persons were contacted in 1972 and asked whether they were still married to the same person. The answers were recorded as 0 for divorced and 1 for still married. Find the point-biserial correlation coefficient for these data.

| 1960 Score | 1972 Status | 1960 Score | 1972 Status |
|------------|-------------|------------|-------------|
| 75 | 1 | 65 | 1 |
| 72 | 1 | 69 | 1 |
| 51 | 0 | 82 | 1 |
| 95 | 1 | 45 | 0 |
| 90 | 0 | 56 | 1 |
| 45 | 0 | 15 | 0 |
| 21 | 0 | 79 | 1 |
| 38 | 1 | 73 | 1 |

7.40 Schiffman (1972) reported a study of decision-making processes of elderly consumers in considering purchase of a new product. The subjects all resided in a community of geriatrics in New York, and the product was a new salt substitute. A 30-cents-off coupon was distributed to encourage the subjects to try the product, and 100 households were interviewed some time later. The writers formed three hypotheses to investigate risk policies of elderly consumers, and the statistical analysis used to support their findings was the $\gamma$ coefficient for the various pairs of relevant variables observed. These variables were defined as perceived taste risk, perceived health risk, perceived error tolerance as measured in two different ways, trial of the new salt substitute, and past history of new food trial. Perceived taste risk, health risk and past new food trial were both measured as low, medium or high and are thus ordinal variables; the two error tolerance variables were measured on a nominal scale, since an observation was determined according to the answer to a two-choice question. Since the article does not give the actual data, use the following artificial.

(a) To determine the value of the $\phi$ coefficient for the 100 subjects when one or both variables are nominal.

(b) To determine the values of the $\gamma$ coefficient when both variables are ordinal. (Consider trial of the new salt substitute as ordinal.)

| Taste Risk | Health Risk | | | Total |
|---|---|---|---|---|
| | Low | Medium | High | |
| Low | 12 | 15 | 8 | 35 |
| Medium | 10 | 18 | 7 | 35 |
| High | 8 | 17 | 5 | 30 |
| Total: | 30 | 50 | 20 | 100 |

| Perceived Error Tolerance II | Perceived Error Tolerance I | | Total |
|---|---|---|---|
| | 0 | 1 | |
| 0 | 24 | 18 | 42 |
| 1 | 23 | 35 | 58 |
| Total: | 47 | 53 | 100 |

| Trial of the Salt Substitute | Taste Risk | | | Total |
|---|---|---|---|---|
| | Low | Medium | High | |
| 0 | 20 | 23 | 40 | 83 |
| 1 | 9 | 4 | 4 | 17 |
| Total: | 29 | 27 | 44 | 100 |

7.41 The relationship between birth rank and ability to cope with anxiety conditions is a subject of great interest to psychologists. Schachter (1959, p. 75) presents

data on 29 men, all coming from families with four or more children who were fighter pilots during the Korean War. The categories are effectiveness as a fighter pilot and birth rank. Measure the association by computing
(a) The Goodman-Kruskal $\gamma$ coefficient,
(b) The Kendall Tau coefficient with correction for ties, and
(c) The Cramér coefficient.

| Birth Rank | Fighter Pilot Effectiveness | | | Total |
|---|---|---|---|---|
| | Ace | Near-Ace | Nonace | |
| Firstborn | 1 | 1 | 3 | 5 |
| Second born | 1 | 4 | 1 | 6 |
| Third born | 4 | 4 | 2 | 10 |
| Fourth born | 4 | 2 | 2 | 8 |
| Total: | 10 | 11 | 8 | 29 |

# 8

# Tests for Randomness

## 1 Introduction

Suppose that 10 persons are waiting in line to purchase a ticket at a movie theater. If these persons are five males and five females and the order of the sexes in the queue is M F M F M F M F M F, can this be considered a random arrangement according to sex? Intuitively, the answer is no, since the complete alternation of the two types of symbols suggests intentional mixing by sex. This arrangement is an extreme case, as is the configuration M M M M M F F F F F, with complete clustering according to sex. Although such patterns could occur by chance in a random arrangement of symbols, the likelihood is small.

Alternatively, suppose that we have a set of temporal or sequential observations, where each observation occupies a unique position in the sequence. If the observations are produced by a random process, they are independent and identically distributed. If the value of a variable in the sequence is influenced by its position in the sequence, or by the observation which precedes it, the process is not truly random. If the observations are transformed into symbols that reflect some property of their position or relative magnitude, the resulting pattern of symbols in the sequence can be used to study the randomness of the process.

Patterns of symbols in an ordered arrangement can be analyzed using runs. The term "run" is used by gamblers and other players of games of chance, as in a "run of heads" or a "run of bad luck." We are using the term in exactly the same sense. In any ordered sequence of some two types of symbols, a run is defined as a succession of one or more identical symbols, which are followed and preceded by a different symbol or no symbol at all.

Thus, in the theater example above, the arrangement with sexes alternating contains ten runs, each of length one, while the clustered arrangement has two runs, each of length five.

Both the total number of runs and their lengths provide clues to lack of randomness in an ordered sequence of two types of symbols. Too few runs, too many runs, a run of excessive length, and so on, occur rarely in a truly random sequence. Thus any of these criteria could be used as statistical evidence against the null hypothesis of randomness. The number of runs and the lengths of the runs are highly interrelated quantities. If there are too few runs, some must be long; if all runs are short, there would be many runs. The most convenient criterion for statistical analysis is the total number of runs; we consider only tests based on this quantity.

The alternative to randomness is often simply nonrandomness. In a test based on the total number of runs, either too few or too many runs suggests dependence between the observations generated by the process, and therefore lack of randomness. However, the type of nonrandomness can be more clearly identified if the two situations are distinguished, since a tendency to cluster produces few runs, whereas a tendency to mix results in many runs. This information can be useful in developing a theoretical model to simulate some observed pattern. Analyses based on runs are also useful in quality control studies, since recognition of a pattern may help to identify the cause of nonrandomness. A significant departure from randomness in a theoretically random process can only be due to some factor that introduces a regularity into the system. If this source can then be located through a study of the specific pattern of the arrangement, the process can hopefully be restored to randomness. Runs tests are also useful in time series studies when the object is to study the behavior of observations over a period of past time. Although this may be called trend analysis, the conclusions apply to recorded past behavior only and may not provide a good basis for extrapolating future behavior. This is always true of forecasting, but especially so when the criterion is runs. Such analyses can identify a form of nonrandomness, but not an overall or even current trend.

Tests for randomness may also be useful as a preliminary to other statistical analyses. All inference procedures concerning population characteristics, whether parametric or nonparametric, begin with the assumption of a random sample of observations for analysis. And yet, in many practical applications, the observations cannot be collected in a truly random fashion, as in temporal or sequentially generated data. Then the information about the order in which the observations occur can be used to test a hypothesis that the data can be considered as generated by a random process. If the observed sequential order supports randomness, then, for practical purposes, observations from the process may be considered a set of independent and identically distributed random variables. Then statistical inferences that are valid only if the process is random can be performed with some degree of confidence.

Runs tests for randomness are not particularly powerful, mainly because of the generality of the inference situation and the very weak assumptions required for validity. However, they occupy a unique position in statistical inference as the only tests that focus on the order in a sequence. Runs tests are quite versatile, and they have some important practical applications.

We limit our study to two different runs tests for randomness. Both are applicable to analysis of sequences of two types of symbols: they differ according to whether successive symbols in the sequence analyzed have a relationship. The ordinary runs test is appropriate when the symbols are independent under the hypothesis of randomness, whereas the runs up and down test applies to symbols that are dependent even though observations in the original sequence may be independent. This dependency is present because the sequence of two types of symbols is formed by comparing the magnitude of each observation with the one immediately preceding it in the sequence.

## 2   Ordinary Runs Test

*Inference situation*   The null hypothesis is that an ordered sequence of two types of symbols can be considered a random arrangement, or that the process generating the two types of symbols is a random process. By random we mean that all distinguishable arrangements of symbols are equally likely to result, as occurs when the random variables producing the two types of symbols are independent and have exactly the same distribution. The statistical interpretation of nonrandomness is either that the variables are dependent, or that they are not identically distributed, or both. Nonrandomness is reflected in the sequence of symbols by either a tendency to cluster or a tendency to mix. The hypothesis set then is one of the following:

*Two-sided alternative*
   $H:$   Sequence is random
   $A:$   Sequence is not random.

*One-sided alternatives*
   $H:$   Sequence is random
   $A_+:$   Sequence exhibits tendency to mix
or
   $H:$   Sequence is random
   $A_-:$   Sequence exhibits tendency to cluster.

*Sampling situation*   The data for analysis consist of an ordered sequence of two types of symbols, $m$ of the first type and $n$ of the second type. The total number of observations is $m + n = N$. Hence the measurement scale of the data for analysis need be only nominal and dichotomous, but their order must be meaningful.

*Rationale and test statistic*   Under the null hypothesis of randomness, every arrangement of the two types of symbols is equally likely to be observed. The number of runs of one type of element is either equal to, one more than, or one less than the number of runs of the other type of element, since the two types of runs must alternate in the sequence. The total number of runs is a reflection of the randomness of the sequence, and the test statistic is defined as

$$U = \text{total number of runs of either type.}$$

Too few runs would occur if there is a tendency for the like types of symbols to cluster together, whereas too many runs reflect a tendency for like types of symbols to be isolated, that is, intentional mixing of the two types. Since either of these situations is indicative of nonrandomness, the two-sided alternative would be supported if $U$ is either too small or too large.

*Finding the P-value*   The sampling distribution of $U$ is given in Table L for all $m \leq n$ such that $m + n \leq 20$, and other $m$ and $n$ values such that $m \leq n \leq 12$, cumulated for right and left tails where $P \leq .500$. The smaller number of symbols should be labeled type 1 so that this table covers all cases with no more than 20 symbols of both types combined, plus those cases with 11 or 12 of either type symbol. The total number of runs possible ranges from 2 to $2m$ for $m = n$ and from 2 to $2m + 1$ for $m < n$.

  Since too few runs is indicative of intentional clustering, a left-tail probability for $U$ is appropriate for the one-sided alternative $A_-$; on the other hand, a right-tail probability for $U$ should be used with the one-sided alternative of intentional mixing, that is, $A_+$.

  When $m = n$, the P-value for the two-sided alternative is found by doubling the smaller tail probability, which is right tail if $U$ exceeds its median value and left tail if $U$ is less than its median value. Since Table L gives only tail probabilities not exceeding .5, the tail probability to be doubled is the only one given in the table. When $m \neq n$, we adopt the same convention for finding the P-value for the two-sided alternative. However, it should be pointed out that there is some debate among statisticians as to whether this is the best procedure to follow, since the distribution of $U$ is not symmetric if $m \neq n$.

  These methods for finding the appropriate P-values are summarized in the guide below.

| Alternative | P-value (Table L) |
|---|---|
| $A_+$:  Sequence exhibits mixing | Right-tail probability for $U$ |
| $A_-$:  Sequence exhibits clustering | Left-tail probability for $U$ |
| $A$:  Sequence is not random | 2(smaller tail probability for $U$) |

The asymptotic sampling distribution of a standardized $U$ is the normal probability function. The mean of $U$ is $1 + 2mn/N$ and the standard deviation is

$$\sqrt{\frac{2mn(2mn - N)}{N^2(N - 1)}}.$$

For sample sizes outside the range of Table L, $P$-values can be found from Table A. $U$ is standardized, and a continuity correction of 0.5 is introduced since $U$ can take on only integer values. Thus the $z$ statistics are defined as

$$z_L = \frac{U + 0.5 - 1 - 2mn/N}{\sqrt{\dfrac{2mn(2mn - N)}{N^2(N - 1)}}} \qquad z_R = \frac{U - 0.5 - 1 - 2mn/N}{\sqrt{\dfrac{2mn(2mn - N)}{N^2(N - 1)}}}$$

$$z = \begin{cases} -z_L & \text{if } U < 1 + 2mn/N \\ z_R & \text{if } U > 1 + 2mn/N. \end{cases}$$

Using the symmetry of the normal distribution, the large-sample, asymptotic approximate $P$-values can be conveniently found as follows:

| Alternative | P-value (Table A) |
|---|---|
| $A_+$:  Sequence exhibits mixing | Right-tail probability for $z_R$ |
| $A_-$:  Sequence exhibits clustering | Left-tail probability for $z_L$ |
| $A$:  Sequence is not random | 2(right-tail probability for $z$) |

**example   2.1**
     A quality control analyst at a newly built manufacturing plant needs to devise a sampling procedure for parts produced in order to establish an overall quality control program. The sampling procedure is crucial to any quality control program, since sample results are the basis for many important decisions, such as adjusting the production process when too many defectives are being produced or deciding on the number of extra items that should be produced to ensure that orders are filled on time with goods that meet at least minimum quality standards. Since each item is classified as

simply defective or good, many of these decisions are based on the proportion of defectives produced during any given time period. The testing of items is an expensive and time-consuming process, but the cost is based only on the total number of items tested. Thus the cost is the same whether testing is carried out on a few large batches or many small batches of items. The question then is whether frequent small samples or less frequent large samples provide more information about the quality of a day's production.

The answer depends upon whether defective items come from the production line in clusters or are randomly distributed among the good items. If the defectives seem to be clustered, small samples should be taken at frequent intervals over the day in order to obtain a reliable estimate of the proportion of defectives being produced. If the defectives are produced at random, larger samples taken at less frequent intervals over the day should produce adequate estimates.

The analyst took several samples of size 30, at frequent but random times during a randomly chosen day of production. The defectives were coded with a zero, and the good items with a one, and the results reported in the order in which the items were taken from the assembly line. For example, two of the samples were reported as follows:

Sample 2:

$$0\ 0\ 0\ 0\ 1\ 1\ 1\ 1\ 1\ 1\ 1\ 1\ 1\ 1\ 1\ 1\ 1\ 1\ 1\ 1\ 1\ 0\ 0\ 0\ 1\ 1\ 1\ 1\ 1\ 1$$
$$m = 7 \qquad n = 23 \qquad U = 4$$

Sample 5:

$$1\ 1\ 1\ 1\ 1\ 1\ 1\ 1\ 0\ 0\ 0\ 1\ 1\ 1\ 1\ 1\ 1\ 1\ 1\ 1\ 1\ 1\ 1\ 1\ 1\ 0\ 0\ 1\ 1$$
$$m = 5 \qquad n = 25 \qquad U = 5$$

For each sample obtained, the following hypothesis set is to be tested.

$H$:    The process generates defective and good items randomly
$A_-$:    The process generates defectives in clusters.

We use the data of sample 2 to illustrate the calculation of $P$-values. Since the $(m, n)$ combination is outside the range of Table L, the normal approximation must be used. For the stated alternative, the appropriate $P$-value is a right-tail probability for $-z_L$. The calculations are as follows:

$$z_L = \frac{4 + 0.5 - 1 - 2(7)(23)/30}{\sqrt{\dfrac{2(7)(23)[2(7)(23) - 30]}{30^2(29)}}}$$

$$= -\frac{217}{\sqrt{3242.21}} = \frac{-217}{57} = -3.81.$$

The left-tail probability for $z_L = -3.81$ from Table A is .000, and this is the asymptotic approximate $P$-value. The null hypothesis is not supported.

Since each one of the samples had extremely small $P$-values, the analyst concluded that the defective items were produced in clusters. Further, these clusters seemed to have no regularity in length or frequency. Therefore, the plant adopted a sampling procedure of many small samples, taken at frequent but random times during a production day.

*Additional notes*   If the level of measurement is higher than nominal, or the data are not dichotomous as collected, the symbols studied for pattern may be artificially imposed according to some dichotomizing criterion. If the measurement scale is at least ordinal, the dichotomy is frequently effected by comparing the magnitude of each observation with a single focal point, commonly the median of the sample (although the mean can also be used for interval scale data), and noting whether each observation exceeds or is exceeded by this value. Any observations equal to this value may be ignored in the analysis. This technique is called runs above and below the median. Since $m$ and $n$ are then equal or nearly so, the exact sampling distribution of $U$ is symmetric or nearly so; this improves the accuracy of the asymptotic normal approximation. Other dichotomizing criteria may be equally appropriate, depending on the context of the problem and possibly also on the type of nonrandomness that is of interest. In the following example, the data are dichotomized by comparing each observation with the population median.

**example  2.2**

It is claimed that successive last digits of telephone numbers in an alphabetical listing by name constitute a set of random numbers. Opening a telephone directory at random and starting with a randomly selected point, a sequence of 20 last digits is observed. Since the last digit is equally likely to be any integer between 0 and 9 inclusive, the population median is 4.5. A sequence of two types of symbols can be artificially imposed by noting whether each last digit is above (A) or below (B) this value. Suppose the results are as follows:

Last digit   7 0 3 4 8 5 3 1 5 0 2 6 8 0 9 4 0 1 8 1
Symbol       A B B B A A B B A B B A A B A B B B A B.

**SOLUTION**

Since there is no reason to expect any particular type of nonrandomness here, we form the hypothesis set with a two-sided alternative.

$H$:  These 20 numbers appear in a random order

$A$:  These 20 numbers do not appear in a random order.

Since there are eight digits larger than 4.5 and 12 smaller than 4.5, the A's should be called the type 1 elements. Then the sample results are

$$m = 8 \qquad n = 12 \qquad U = 12.$$

Consulting Table L for $m = 8, n = 12$ we find the only tail probability given for $U = 12$ is right tail, and this value is .337. Hence the two-tailed $P$-value is $2(.337) = .674$. The statistical conclusion is that these 20 numbers can be considered a sequence of random digits.

**example   2.3**

A psychological experiment is conducted to see whether an animal can "learn" its way through a maze. The maze has four choice points; at each point the animal must go either right or left, and only one way is correct since the other is a dead end. The investigator counts the number of times the animal chooses the correct path first at the four choice points. Suppose that the animal is put through the maze 64 consecutive times, and that the time-ordered observations on number of correct choices are as follows:

0, 1, 2, 1, 1, 2, 3, 2, 2, 2, 1, 1, 3, 2, 1, 2, 1, 2, 2, 1, 1,
2, 2, 1, 4, 3, 1, 2, 2, 1, 2, 2, 2, 2, 3, 2, 2, 3, 4, 3, 2, 3,
3, 2, 3, 3, 2, 3, 3, 2, 3, 4, 3, 3, 4, 2, 3, 3, 4, 3, 4, 4, 4, 4.

If these results exhibit a tendency for the higher numbers of correct choices to cluster, we might conclude support for the research hypothesis. Perform an appropriate test.

**SOLUTION**
We use the number of runs test for the null hypothesis of randomness against the alternative of clustering. A logical dichotomizing criterion is that 0, 1, or 2 correct choices imply no learning (type 2 elements), while 3 or 4 correct choices are indicative of learning (type 1 elements). Then we have $m = 27, n = 37$, and a total of $U = 20$ runs. Then

$$z_L = \frac{20 + 0.5 - 1 - 2(27)(37)/64}{\sqrt{\dfrac{2(27)(37)(1998 - 64)}{64^2(63)}}} = -3.10.$$

The appropriate $P$-value is a left-tail probability from Table A for $-3.10$, or $P = .0010$. Thus we reject the null hypothesis. Since the majority of runs of type 1 elements occur for later repetitions, we conclude that the data do support a research conclusion that learning occurs with time.

*Other inferences*   The ordinary runs test can also be used as a general two-sample test for the null hypothesis of identical populations as long as the measurement scale is at least ordinal for both samples. Data from two mutually independent samples are arranged in increasing order of magnitude, keeping track of which observation is from which sample. If the observations in this arrangement are replaced by symbols to indicate which sample they belong to, say 1 for the first sample and 2 for the second sample, we have an ordered sequence of two types of symbols. If the two populations are identical, the observations from the two samples should be well mixed, and the 1s and 2s would then constitute a random sequence. If, for example, the observations from the second population tend to be larger on the average than the observations from the first population, most of the 1s should precede most of the 2s and there will be few runs. If, on the other hand, the two populations have the same locations but the second population has a larger scale, most of the 1s should be in the middle of the arrangement with the 2s on either side, and again there will be fewer runs than expected from identical populations. Therefore the number of runs in the sequence of 1s and 2s can be used as a test criterion when the alternative is two-sided, stating a simple inequality between populations. Since small values of $U$ do not support $H$, the appropriate $P$-value is a left-tail probability from Table L even though the alternative is two-sided.

When used for this situation, the ordinary runs test is usually called the *Wald-Wolfowitz runs test* after the two statisticians who proposed it. The test is not particularly powerful, since it is sensitive to all types of differences between the populations. The test criterion cannot distinguish a location difference from a scale difference, or even a direction for such differences. In a rather unique data situation, however, the runs test is the only one appropriate. This is where it is not possible to make even an ordinal measurement on the variable of interest, but the data from the two samples occur in some meaningful kind of order.

# 3    Runs Up and Down

In the preceding section we explained that numerical observations could be transformed into two types of symbols by comparing the magnitude of each observation with some fixed value. In Example 2.2 the dichotomy was effected by noting whether an observation was above or below 4.5, and the resulting sequence of A's and B's was studied for randomness. With such a procedure, certain information is lost that might be useful in identifying a pattern of nonrandomness. Some of this relevant information is retained when numerical observations are analyzed by a procedure based on "runs up and down." With this method, the magnitude of each observation is not compared with a single number; rather, the magnitude of each observation is compared with that of the one immediately preceding it in the sequence.

If the preceding value is smaller, a "run up" is started; if it is larger, a "run down" commences. Thus we can make use of information about when the sequence is increasing, and for how long, or when it is decreasing, and for how long. These characteristics are all reflected in the total number of runs, whether up or down, and this quantity can be used as a statistical criterion to test the null hypothesis of randomness.

For example, if an ordered sequence of six observations is 8, 13, 1, 3, 4, 7, there is a run up of length one, followed by a run down of length one, followed by a run up of length three. If the original sequence of six observations is replaced by five plus or minus signs, which reflect the direction of change between consecutive values, the result is $+, -, +, +, +$. The symbols then indicate the magnitude of each observation relative to the immediately preceding one, and the total number of runs of symbols is easily counted, as in Section 2.

*Inference situation*    The null hypothesis is identical to that of the ordinary runs test of the Section 2; that is, an ordered sequence of variables can be considered a random arrangement, or generated by a random process.

*Two-sided alternative*
> *H:*   Sequence is random
> *A:*   Sequence is not random.

*One-sided alternatives*
> *H:*   Sequence is random
> $A_+$:   Sequence exhibits tendency to mix

or
> $A_-$:   Sequence exhibits tendency to cluster.

*Sampling situation*    The data consist of an ordered sequence of observations measured on at least an ordinal scale, but representing values on a continuum.

*Rationale*    Under the null hypothesis of randomness, every possible arrangement of symbols indicating the sign of the difference between successive observations (each either plus or minus) is equally likely to be observed. The pattern of arrangement in general and the total number of runs in particular show the movement of the series. If the sequence is increasing (or decreasing) most of the time, as for a trend in a constant direction, like signs would tend to cluster; this pattern is reflected by too few runs. On the other hand, frequent fluctuations, cyclical movements, or erratic variations would produce many switches of signs, which would be reflected by too many runs.

*Test statistic*    An arrangement of $N$ different observations is recorded as a sequence of $N - 1$ plus or minus signs according to relative magnitude of successive observations. The test statistic is the total number of runs in this derived sequence, denoted by $V$; that is,

$$V = \text{total number of runs, whether up or down.}$$

*Finding the P-value*   The distribution of the test statistic $V$ is not the same as the distribution of $U$ in the preceding section, since even in a random arrangement of independent observations the symbols (signs) here are not independent, and the probability of a plus (or minus) symbol is not constant. For example, if a number in a random sequence is quite large, the probability that the observation following it will be even larger is smaller than the probability that the next observation will be smaller; this affects the probability of a plus sign.

The null sampling distribution of $V$ is given in Table M for $N \leq 25$ as left-tail probabilities for $P \leq .500$ and right-tail probabilities for all other values, that is, left tail for $V$ less than or equal to its median value and right tail otherwise. The possible values of $V$ range from 1 to $N - 1$. Note that the $N$ in the table is the number of observations, which will always be one *more* than the number of algebraic signs from which the number of runs is counted.

According to the discussion above, the appropriate $P$-values are right tail for $A_+$ and left tail for $A_-$. For a two-sided test, we follow the same convention as in Section 2 and define the $P$-value as two times the probability given in the table. These results are summarized below.

| Alternative | | $P$-value (Table M) |
|---|---|---|
| $A_+$: | Sequence exhibits mixing | Right-tail probability for $V$ |
| $A_-$: | Sequence exhibits clustering | Left-tail probability for $V$ |
| $A$: | Sequence is nonrandom | 2(smaller tail probability for $V$) |

For sample sizes exceeding 25, the asymptotic sampling distribution of a standardized $V$ can be used to find approximate $P$-values. The mean of $V$ is $(2N - 1)/3$, and the standard deviation is $\sqrt{(16N - 29)/90}$. Thus the standardized variables, with a continuity correction of 0.5, are

$$z_L = \frac{V + 0.5 - (2N - 1)/3}{\sqrt{(16N - 29)/90}} \qquad z_R = \frac{V - 0.5 - (2N - 1)/3}{\sqrt{(16N - 29)/90}}$$

$$z = \begin{cases} -z_L & \text{if } V < (2N - 1)/3 \\ z_R & \text{if } V > (2N - 1)/3. \end{cases}$$

Then the appropriate asymptotic approximate $P$-values are found from Table A as shown below.

| | Alternative | $P$-value (Table A) |
|---|---|---|
| $A_+$: | Sequence exhibits mixing | Right-tail probability for $z_R$ |
| $A_-$: | Sequence exhibits clustering | Left-tail probability for $z_L$ |
| $A$: | Sequence is nonrandom | 2(right-tail probability for $z$) |

*Ties*   Under the assumption of a continuous population, the probability is zero that any two observations are equal. Unless equal values appear consecutively in the given ordered sequence, they present no problem for analysis based on runs up and down. If a tie occurs between adjacent observations in the ordered sequence, a zero should be recorded, and a zero is neither plus nor minus. If the number of zeros is small relative to $N$, zeros are ignored and $N$ is reduced correspondingly to be one more than the actual number of signs entering the analysis. If it is not desirable to reduce $N$, the conservative approach would be to report the largest $P$-value of those found when all possible numbers of runs are counted according to what happens to the arrangement when the zeros are replaced by either a plus or a minus sign. If a zero occurs between two opposite algebraic signs, the number of runs is unique and unchanged. But if a zero occurs between two like signs, the number of runs is either unchanged or increased by two, depending on which sign is assigned to the zero difference. The larger number of runs gives the larger left-tail probability, while the smaller number gives the larger right-tail probability.

*Quality control applications*   The test based on the number of runs up and down is particularly useful in quality control studies, where the question of interest is not only whether nonrandomness exists, but also whether any significant nonrandomness can be attributed to some particular difficulty in the production process, called an "assignable cause." Diagnosing the source of the variation in some product is frequently important, since such information may be useful in tracking down and eliminating the factors contributing to the variation. Once an "assignable cause" is identified and corrected, randomness can be restored to the process. If the observations are made on only a nominal scale of measurement, the number of runs test in Section 2 can be used in this same type of application.

**example   3.1**
    The battle tank is an effective fighting machine only when its crew is capable of delivering fire with speed and accuracy. A measure of each crew's

capability is the score it achieves on the "Tank Crew Qualification Firing Exercise," an exercise required annually by all Army tank crews.

The crews scheduled to fire on any given day perform the exercise in sequence; that is, each crew occupies a unique position in the firing sequence for that day. Although the assigned sequence is not determined in a random manner by statistical standards, it can be considered purposeless. For the firing exercises to be an equitable measure of qualification, a crew's performance should be independent of its position in the firing sequence. The scores as arranged according to position in the sequence should be independent and identically distributed in a random sequence. Test the following scores of 24 crews for randomness. The 24 numbers are arranged according to position occupied by each crew in the firing line on a particular day; that is,

205, 300, 285, 325, 320, 415, 345, 240, 330, 495, 375, 265,
410, 440, 395, 465, 445, 535, 160, 305, 465, 235, 405, 465.

**SOLUTION**

The hypothesis set is

$H:$   The sequence of scores is random
$A:$   The sequence of scores is not random.

In order to use the runs up and down test for analysis, each score is compared with the score for that crew occupying the position to its immediate left. The result is recorded as plus for a larger score and minus for a smaller score; the 23 signs are as follows.

$$+, -, +, -, +, -, -, +, +, -, -,$$
$$+, +, -, +, -, +, -, +, +, -, +, +$$
$$N = 24 \qquad V = 17.$$

From Table M, the tail probability given for $V = 17$ when $N = 24$ is .3405. The two-sided $P$-value is then 2(.3405) = .6810. The null hypothesis is supported and the statistical conclusion is that the arrangement of scores is random.

An extension of this conclusion to any arbitrary day's firing would be valid only if this day can be considered typical in the sense that firing on all other days is conducted in the same manner and under the same conditions. If the test results had supported nonrandomness, it might be advisable to obtain data for other days as well so that the sequences could be examined for patterns. The manner in which the firing exercise was conducted should also be scrutinized in an effort to determine the source of nonrandomness so that corrective action could be taken.

**example   3.2**

In recent years the policy of lead production in the United States has been to produce at total capacity rate but with no overtime permitted. Thus the production in any month should be independent of production in the following month. A U.S. Senator has expressed concern over a suspected increase in the national production of lead, but the monthly production figures over the past few years have shown large variation. Before a large-scale investigation of production is launched, the monthly figures shown in Table 3.1 will be studied for patterns in the fluctuations.

**TABLE 3.1**
**MINE PRODUCTION (IN THOUSANDS OF SHORT TONS)**

| Month | 1969 | 1970 | 1971 |
|-------|------|------|------|
| Jan.  | 37.2 | 47.8 | 45.3 |
| Feb.  | 35.1 | 46.9 | 41.9 |
| Mar.  | 39.3 | 52.5 | 52.7 |
| Apr.  | 42.6 | 49.7 | 47.1 |
| May   | 44.2 | 51.3 | 45.6 |
| June  | 45.5 | 47.4 | 45.4 |
| July  | 44.4 | 46.6 |      |
| Aug.  | 45.4 | 48.1 |      |
| Sept. | 43.5 | 48.6 |      |
| Oct.  | 41.7 | 46.5 |      |
| Nov.  | 41.5 | 48.5 |      |
| Dec.  | 41.4 | 45.3 |      |

**SOLUTION**

The mining operation can be considered a process generating a sequence of observations, and the question is whether the monthly production outputs are in fact independent. Since an alternative of trend seems the most reasonable, we use a one-sided alternative with the hypothesis set

$H$:   Monthly production is a random process
$A_-$:   Monthly production figures exhibit a pattern of clustering.

Each monthly production figure is subtracted from that of the immediately preceding month to generate the following sequence of 29 plus and minus signs, and one zero, as follows:

$$-, +, +, +, +, -, +, -, -, -, -,$$
$$+, -, +, -, +, -, -, +, +, -, +, -,$$
$$0, -, +, -, -, -.$$

Ignoring the zero, the effective sample size is 29 and the number of runs is 17, or

$$V = 17.$$

Since $N = 29$ is outside the range of Table M, we use the normal approximation, where

$$z_L = \frac{17 + 0.5 - (57/3)}{\sqrt{\dfrac{16(29) - 29}{90}}} = \frac{-1.5}{2.20} = -.68.$$

From Table A, we find that the left-tail probability for $z_L = -.68$ is .2483, and this is the asymptotic approximate $P$-value. The null hypothesis of randomness in fluctuations is supported by these data.

## 4   Summary

The two procedures covered in this chapter are both designed for testing the null hypothesis that an ordered sequence of observations can be considered random. Even though these tests have limited applications, they are an important addition to the repertory of statistical methods.

The basic difference between the procedures is that analysis of runs up and down is based on the pattern of relative magnitudes of successive observations, while ordinary runs analysis is based on the pattern of a dichotomization of any kinds of events. Thus, with ordinary runs, the null sampling distribution is conditional upon the number of observations of each of the two defined outcomes, and these outcomes occur with constant probability. Neither of these properties is true for runs up and down. Further, runs up and down are defined only for observations that are measured on at least an ordinal scale and represent a variable that is continuous. The only requirement for application of ordinary runs is that we have two types of symbols arranged in some particular order. As a result, this method has a broader possible range of applicability.

When either procedure is applicable, the dichotomy for the ordinary runs test is usually effected by comparing each observation with the median of all observations. This is a somewhat artificial dichotomy that seems to discard a large amount of information about the pattern of relative magnitudes. When this information is relevant, runs up and down should provide a more sensitive method of analysis.

When the alternative to randomness is completely general, these tests have no parametric counterpart. Power cannot be calculated without a specific alternative, and thus efficiency cannot be described in this case.

Both of the runs tests have asymptotic relative efficiency zero against normal regression alternatives, and also against normal location alternatives. (Therefore, the Wald-Wolfowitz runs test should be considered a poor substitute for the Mann-Whitney-Wilcoxon test in the two-sample location model.) Their primary usefulness is in situations where no other test procedure is applicable.

## PROBLEMS

8.1  A row of tobacco plants in a field has the following arrangement of diseased (D) and healthy (H) plants. Test the null hypothesis of randomness against the alternative of clustering.

$$H\ H\ H\ H\ H\ D\ D\ D\ D\ D\ H\ H\ H\ D\ H\ H\ H\ H$$

8.2  The data below represent the numbers of persons passing through the turnstile at a subway entrance on 14 successive Mondays between 8:00 and 8:30 a.m. Use the method of runs above and below the median to see if these data can be considered as generated by a random process.

$$23,\ 20,\ 29,\ 14,\ 20,\ 21,\ 26,\ 16,\ 25,\ 28,\ 19,\ 27,\ 28,\ 30$$

8.3  Analyze the data of Example 3.2, using the ordinary runs test and dichotomizing at the monthly production average for 30 months, which is $1360/30 = 45.33$.

8.4  Rework Example 2.2, but using even and odd digits as the dichotomizing criterion (zero is considered an even digit).

8.5  Flip a balanced coin 50 times and test the results for randomness of occurrence of H and T.

8.6  A time-and-motion study was made in the permanent mold department of a foundry. The question of primary interest was whether the workers are less efficient after lunch than before, where efficiency is determined by the amount of time required to pour the molten metal into the die and form a casting. If the study supports this conclusion, variations in pouring time are effected by assignable causes that might be corrected, eliminated, or taken into account. Extensive data collection has determined that the average pouring time is 12 seconds. For the present investigation, a date is chosen at random and eight observations are taken both before and after lunch. The pouring times measured in seconds are given below, listed in order of observation.

| Before lunch | 9.4, | 9.8, | 11.1, | 11.2, | 14.0, | 11.8, | 12.1, | 12.5 |
| After lunch | 16.4, | 15.8, | 15.4, | 14.0, | 12.2, | 11.8, | 11.2, | 11.4 |

Test the combined ordered sequence for randomness against an alternative of clustering, using the procedures based on

(a) The total number of runs, where 12 seconds is the focal point of comparison.

(b) The total number of runs up and down.

8.7 The winning time for the Olympic Games champion in the 1500-meter run for men has been recorded for each Olympic Games since 1896. For the period 1896–1972, the historic record of these times in minutes and seconds is shown below (*The World Almanac and Book of Facts* 1973). Analyze these time-ordered observations for randomness, against an alternative of insufficient fluctuation, by using the runs up and down test.

4:33.2, 4:06.0, 4:05.2, 4:12.0, 4:03.2, 3:56.8, 4:01.4, 3:53.6, 3:53.2, 3:51.2, 3:47.8, 3:49.8, 3:45.2, 3:41.2, 3:35.6, 3:38.1, 3:34.9, 3:36.3

8.8 Suppose it is hypothesized that anyone can list numbers in such a manner that they can be considered statistically random. Make an arbitrary list of at least 20 digits, selected from the digits 0–9, consciously aiming at a sequence that could be considered random, and use an appropriate runs test for analysis.

8.9 Numbers in a telephone directory were frequently used to select random samples before random number tables were widely available. In order to see whether your local telephone directory could legitimately be used for this purpose, select a page at random and determine the number of runs of even numbers in the last digit. Does the sequence appear to be random?

8.10 In numerous problem situations it is either impossible or very difficult to obtain analytical solutions. In such cases a simulation technique termed the Monte Carlo method is often employed to simulate original data with random number generators. The generator algorithms used in computer solutions to Monte Carlo problems generate random numbers in a deterministic manner. If five-digit random numbers are desired, a five-digit number seed is utilized to initiate the generation process. Thus the random numbers generated are dependent upon the seed number. In most cases, the so-called pseudorandom numbers should be analyzed in order to determine whether the sequence may be considered statistically random. Given the following sequence of five-digit random numbers generated by the Monte Carlo method, perform a runs test and state conclusions.

54321, 67826, 21465, 89211, 06321, 46313, 57644, 12085, 06412, 78787, 92634, 36545, 01766, 31122, 13344, 71235, 21876, 46329

8.11 In Problem 4.13 of Chapter 4, the 12 observations given on pouring times are listed in order of time of day when read down successive columns. If this ordered sequence of observations can be considered random, a quality control engineer would interpret the variation in pouring times as inherent to the process and due to random causes only. If it is not random, there may be an assignable cause and a means of restoring efficiency to the process. Test the null hypothesis or randomness using runs up and down.

8.12 The U.S. Selective Service began the Draft Lottery in 1970. By this procedure, each calendar date in the year is randomly assigned a unique number between 1 and 365 (366 in a leap year), called the draft priority number. Persons who are 19 years old as of December 31 and otherwise eligible are then drafted according to the number associated with their birth day. Since each year the procedure is repeated with a new group of 19-year-olds, any men not called in the previous year because of their high draft number are essentially "safe" from the draft for good, in the sense that all younger eligible men would be called first. The draft priority numbers are published at the beginning of each year. The procedure was criticized in 1970 as nonrandom, since the earlier birthdays (before July 1) appeared to have more of the higher numbers assigned (and hence lower draft priorities) than did the later birthdays. An excerpt of the 1970 draft numbers, from *U.S. News & World Report* (December 15, 1969, p. 34), is shown below. Using 183 as the line of division between "safe" and "unsafe" birthdays, that is, the midpoint of a calendar year, the draft numbers can be designated as "a" for above 183 and "b" for below or equal to 183. For example, for the first 10 days of the year the sequence of a's and b's is abaabaaaaa. Using the numbers for the entire year and taking the calendar days in sequence, there were 86 runs of a's and 86 runs of b's, or a total of 172 runs produced from 183 b's and 183 a's. Is there any justification for the claim of lack of randomness?

## 1970 DRAFT PRIORITY NUMBERS FOR JANUARY

| Birthday | Number | Birthday | Number | Birthday | Number |
|----------|--------|----------|--------|----------|--------|
| 1/1  | 305 | 1/11 | 329 | 1/21 | 186 |
| 1/2  | 159 | 1/12 | 221 | 1/22 | 337 |
| 1/3  | 251 | 1/13 | 318 | 1/23 | 118 |
| 1/4  | 215 | 1/14 | 238 | 1/24 | 59  |
| 1/5  | 101 | 1/15 | 17  | 1/25 | 52  |
| 1/6  | 224 | 1/16 | 121 | 1/26 | 92  |
| 1/7  | 306 | 1/17 | 235 | 1/27 | 355 |
| 1/8  | 199 | 1/18 | 140 | 1/28 | 77  |
| 1/9  | 194 | 1/19 | 58  | 1/29 | 349 |
| 1/10 | 325 | 1/20 | 280 | 1/30 | 164 |
|      |     |      |     | 1/31 | 211 |

8.13 A tapeworm grows from the head down by sections called proglottids. Each sexually mature proglottid has characteristics of both sexes and can produce eggs of new worms independently of other proglottids. An important genetic characteristic of each proglottid is the number of testes and their configuration in relation to the ovary. The testes may be anterior or posterior, with total number varying between 0 and 6 inclusive. Of the 28 possible classifications of configurations of testes, the most common one is two anterior and one posterior. This is called Normal, while any of the other 27 configurations is called Variant.

As part of a study in parasitology, Dr. B. P. Adhikari (1973) of the Indian Statistical Institute investigated the randomness of Normal and Variant configurations in successive proglottids using data collected by Dr. Everett L. Schiller of The Johns Hopkins University for a very large number of tapeworms of different ages. The overall conclusion was that the sequence is random in younger tapeworms (age 16 days or less) and nonrandom in older tapeworms (30 days or more). The implication of this result is that there must be some stabilization of configuration after a certain age. These patterns are now being investigated in an attempt to set up a genetic-statistical model to simulate them.

Data on one tapeworm from each age group are shown below. Use the Normal approximation to the distribution of the ordinary number of runs test statistic to find a one-tailed $P$-value for the null hypothesis of randomness in each age group.

| Tapeworm Number | Age Group | Number of Normals | Number of Variants | Runs of Normals | Runs of Variants |
|---|---|---|---|---|---|
| 16 | 8 days | 284 | 35 | 32 | 31 |
| 30 | 16 days | 676 | 73 | 67 | 66 |
| 23 | 30 days | 502 | 55 | 46 | 45 |
| 83 | 91 days | 530 | 32 | 28 | 27 |

8.14 Quality control is very important in clinical laboratories, and the blind acceptance of a manufacturer's suggested control limits is dangerous. Rather, the laboratory should make its own statistical evaluations of quality control limits and correct for drift. Suppose that samples of sera are split randomly and run each day for 24 days for serum cholesterol determination. The differences, sample 1 minus sample 2, in readings on successive days are shown below. If the laboratory remains in reasonable control over this period, the differences should be randomly distributed about zero. Test the data for randomness using the number of runs test, with plus and minus signs as the dichotomizing criterion.

$$-1.83,\ -0.17,\ -5.17,\ +3.67,\ +3.83,\ +1.33,\ +1.67,\ +2.67,$$
$$-2.33,\ +4.17,\ -2.33,\ +3.00,\ -1.50,\ -2.50,\ +2.17,\ +2.00,$$
$$+2.50,\ +1.67,\ +1.33,\ +0.33,\ +4.50,\ +0.67,\ +0.17,\ +1.36$$

8.15 For the example on life-testing in Problem 4.8 of Chapter 4, suppose the time to failure is not measured for the components observed under normal and accelerated conditions. Rather, we simply note, as the failures occur, which environmental group the component was in, control (C) or exposure (E). Then the data would have been recorded as

E E E C E E E C E C C E C C E E C C.

If the average times to failure are the same for both the control and exposure groups, the E and C symbols in this sequence should be well mixed. If one group, say the E group, failed sooner on the average, there should be a predominance of E's occurring toward the beginning of the sequence. If the

E group has longer life on the average, most of the E's should occur toward the end. Thus too few runs could indicate a difference in the central tendency of times to failure of the two groups, although it cannot indicate the direction of the difference. Analyze the data in this form by the Wald-Wolfowitz runs test. Note that when observations are to be analyzed in this way, the process of collecting data can be terminated once all items in one of the two groups have failed, and hence may result in a saving of time and expense.

8.16   For the example on fatigue testing in Problem 4.14 of Chapter 4, suppose that the failures are simply noted, in order of their occurrence, by which machine was producing the stress. Write the data, and use the Wald-Wolfowitz runs test to test the hypothesis that average times to failure are the same for both machines.

# Appendix

## Review of Sigma ($\Sigma$) Notation

The symbol

$$\sum_{i=1}^{r}$$

is a mathematical shorthand notation to indicate that the items following the symbol are to be summed, through all successive indicated integer values of the index of summation $i$, here $i = 1, 2, \ldots, r$. For example,

$$\sum_{i=1}^{3} x_i^2$$

is to be interpreted as the sum $x_1^2 + x_2^2 + x_3^2$, in which the appropriate numbers for the variables $x_i$ are then substituted for calculation. When there is no possibility of confusion, it is frequently more convenient to drop the subscripts and the indication of the values of the index. For example, $\Sigma x^2$ means the sum is to be taken over all values of $x^2$, where the meaning of "all" would be evident from the context. Three useful algebraic properties of the $\Sigma$ notation are:

1. The $\Sigma$ of a sum (or difference) is the sum (or difference) of the $\Sigma$'s. In symbols, for example,

$$\sum_{i=1}^{r} (x_i \pm y_i) = \sum_{i=1}^{r} x_i \pm \sum_{i=1}^{r} y_i.$$

2. The $\Sigma$ of a constant times a variable is the constant times the $\Sigma$ of the variable. In symbols, if $c$ is any constant and $x$ is a variable,

$$\sum_{i=1}^{r} cx_i = c \sum_{i=1}^{r} x_i.$$

3. The $\Sigma$ of a constant is $r$ times the constant if $r$ is the number of values of the index of summation. In symbols, for any constant $c$, we have

$$\sum_{i=1}^{r} c = rc.$$

## TABLE A
## NORMAL DISTRIBUTION
Table entries are the cumulative probability, right tail from the value of $z$ to plus infinity, and also left tail from minus infinity to the value of $-z$ for all $P \le .50$. $z$ is the standardized normal variable, $z = (x - \mu)/\sigma$. Read down the first column to the correct first decimal value of $z$, and over to the correct column for the second decimal value. The number at the intersection is the value of $P$.

| z | .00 | .01 | .02 | .03 | .04 | .05 | .06 | .07 | .08 | .09 |
|---|---|---|---|---|---|---|---|---|---|---|
| 0.0 | .5000 | .4960 | .4920 | .4880 | .4840 | .4801 | .4761 | .4721 | .4681 | .4641 |
| 0.1 | .4602 | .4562 | .4522 | .4483 | .4443 | .4404 | .4364 | .4325 | .4286 | .4247 |
| 0.2 | .4207 | .4168 | .4129 | .4090 | .4052 | .4013 | .3974 | .3936 | .3897 | .3859 |
| 0.3 | .3821 | .3783 | .3745 | .3707 | .3669 | .3632 | .3594 | .3557 | .3520 | .3483 |
| 0.4 | .3446 | .3409 | .3372 | .3336 | .3300 | .3264 | .3228 | .3192 | .3156 | .3121 |
| 0.5 | .3085 | .3050 | .3015 | .2981 | .2946 | .2912 | .2877 | .2843 | .2810 | .2776 |
| 0.6 | .2743 | .2709 | .2676 | .2643 | .2611 | .2578 | .2546 | .2514 | .2483 | .2451 |
| 0.7 | .2420 | .2389 | .2358 | .2327 | .2296 | .2266 | .2236 | .2206 | .2177 | .2148 |
| 0.8 | .2119 | .2090 | .2061 | .2033 | .2005 | .1977 | .1949 | .1922 | .1894 | .1867 |
| 0.9 | .1841 | .1814 | .1788 | .1762 | .1736 | .1711 | .1685 | .1660 | .1635 | .1611 |
| 1.0 | .1587 | .1562 | .1539 | .1515 | .1492 | .1469 | .1446 | .1423 | .1401 | .1379 |
| 1.1 | .1357 | .1335 | .1314 | .1292 | .1271 | .1251 | .1230 | .1210 | .1190 | .1170 |
| 1.2 | .1151 | .1131 | .1112 | .1093 | .1075 | .1056 | .1038 | .1020 | .1003 | .0985 |
| 1.3 | .0968 | .0951 | .0934 | .0918 | .0901 | .0885 | .0869 | .0853 | .0838 | .0823 |
| 1.4 | .0808 | .0793 | .0778 | .0764 | .0749 | .0735 | .0721 | .0708 | .0694 | .0681 |
| 1.5 | .0668 | .0655 | .0643 | .0630 | .0618 | .0606 | .0594 | .0582 | .0571 | .0559 |
| 1.6 | .0548 | .0537 | .0526 | .0516 | .0505 | .0495 | .0485 | .0475 | .0465 | .0455 |
| 1.7 | .0446 | .0436 | .0427 | .0418 | .0409 | .0401 | .0392 | .0384 | .0375 | .0367 |
| 1.8 | .0359 | .0351 | .0344 | .0336 | .0329 | .0322 | .0314 | .0307 | .0301 | .0294 |
| 1.9 | .0287 | .0281 | .0274 | .0268 | .0262 | .0256 | .0250 | .0244 | .0239 | .0233 |
| 2.0 | .0228 | .0222 | .0217 | .0212 | .0207 | .0202 | .0197 | .0192 | .0188 | .0183 |
| 2.1 | .0179 | .0174 | .0170 | .0166 | .0162 | .0158 | .0154 | .0150 | .0146 | .0143 |
| 2.2 | .0139 | .0136 | .0132 | .0129 | .0125 | .0122 | .0119 | .0116 | .0113 | .0110 |
| 2.3 | .0107 | .0104 | .0102 | .0099 | .0096 | .0094 | .0091 | .0089 | .0087 | .0084 |
| 2.4 | .0082 | .0080 | .0078 | .0075 | .0073 | .0071 | .0069 | .0068 | .0066 | .0064 |
| 2.5 | .0062 | .0060 | .0059 | .0057 | .0055 | .0054 | .0052 | .0051 | .0049 | .0048 |
| 2.6 | .0047 | .0045 | .0044 | .0043 | .0041 | .0040 | .0039 | .0038 | .0037 | .0036 |
| 2.7 | .0035 | .0034 | .0033 | .0032 | .0031 | .0030 | .0029 | .0028 | .0027 | .0026 |
| 2.8 | .0026 | .0025 | .0024 | .0023 | .0023 | .0022 | .0021 | .0021 | .0020 | .0019 |
| 2.9 | .0019 | .0018 | .0018 | .0017 | .0016 | .0016 | .0015 | .0015 | .0014 | .0014 |
| 3.0 | .0013 | .0013 | .0013 | .0012 | .0012 | .0011 | .0011 | .0011 | .0010 | .0010 |
| 3.1 | .0010 | .0009 | .0009 | .0009 | .0008 | .0008 | .0008 | .0008 | .0007 | .0007 |
| 3.2 | .0007 | .0007 | .0006 | .0006 | .0006 | .0006 | .0006 | .0005 | .0005 | .0005 |
| 3.3 | .0005 | .0005 | .0005 | .0004 | .0004 | .0004 | .0004 | .0004 | .0004 | .0003 |
| 3.4 | .0003 | .0003 | .0003 | .0003 | .0003 | .0003 | .0003 | .0003 | .0003 | .0002 |
| 3.5 | .0002 | .0002 | .0002 | .0002 | .0002 | .0002 | .0002 | .0002 | .0002 | .0002 |

Source: Adapted from Table 1 of E. S. Pearson and H. O. Hartley, eds. (1966), *Biometrika Tables for Statisticians*, Vol. 1, 3rd ed., Cambridge University Press, Cambridge, England, with permission of the Biometrika Trustees.

## TABLE B
## CHI-SQUARE DISTRIBUTION

Table entries on all df lines are the values of a chi-square random variable for which the right-tail probability is as given on the top row.

Right-Tail Probability

| df | .99 | .98 | .95 | .90 | .80 | .70 | .50 | .30 | .20 | .10 | .05 | .02 | .01 | .001 |
|---|---|---|---|---|---|---|---|---|---|---|---|---|---|---|
| 1 | .00016 | .00063 | .0039 | .016 | .064 | .15 | .46 | 1.07 | 1.64 | 2.71 | 3.84 | 5.41 | 6.64 | 10.83 |
| 2 | .02 | .04 | .10 | .21 | .45 | .71 | 1.39 | 2.41 | 3.22 | 4.60 | 5.99 | 7.82 | 9.21 | 13.82 |
| 3 | .12 | .18 | .35 | .58 | 1.00 | 1.42 | 2.37 | 3.66 | 4.64 | 6.25 | 7.82 | 9.84 | 11.34 | 16.27 |
| 4 | .30 | .43 | .71 | 1.06 | 1.65 | 2.20 | 3.36 | 4.88 | 5.99 | 7.78 | 9.49 | 11.67 | 13.28 | 18.46 |
| 5 | .55 | .75 | 1.14 | 1.61 | 2.34 | 3.00 | 4.35 | 6.06 | 7.29 | 9.24 | 11.07 | 13.39 | 15.09 | 20.52 |
| 6 | .87 | 1.13 | 1.64 | 2.20 | 3.07 | 3.83 | 5.35 | 7.23 | 8.56 | 10.64 | 12.59 | 15.03 | 16.81 | 22.46 |
| 7 | 1.24 | 1.56 | 2.17 | 2.83 | 3.82 | 4.67 | 6.35 | 8.38 | 9.80 | 12.02 | 14.07 | 16.62 | 18.48 | 24.32 |
| 8 | 1.65 | 2.03 | 2.73 | 3.49 | 4.59 | 5.53 | 7.34 | 9.52 | 11.03 | 13.36 | 15.51 | 18.17 | 20.09 | 26.12 |
| 9 | 2.09 | 2.53 | 3.32 | 4.17 | 5.38 | 6.39 | 8.34 | 10.66 | 12.24 | 14.68 | 16.92 | 19.68 | 21.67 | 27.88 |
| 10 | 2.56 | 3.06 | 3.94 | 4.86 | 6.18 | 7.27 | 9.34 | 11.78 | 13.44 | 15.99 | 18.31 | 21.16 | 23.21 | 29.59 |
| 11 | 3.05 | 3.61 | 4.58 | 5.58 | 6.99 | 8.15 | 10.34 | 12.90 | 14.63 | 17.28 | 19.68 | 22.62 | 24.72 | 31.26 |
| 12 | 3.57 | 4.18 | 5.23 | 6.30 | 7.81 | 9.03 | 11.34 | 14.01 | 15.81 | 18.55 | 21.03 | 24.05 | 26.22 | 32.91 |
| 13 | 4.11 | 4.76 | 5.89 | 7.04 | 8.63 | 9.93 | 12.34 | 15.12 | 16.98 | 19.81 | 22.36 | 25.47 | 27.69 | 34.53 |
| 14 | 4.66 | 5.37 | 6.57 | 7.79 | 9.47 | 10.82 | 13.34 | 16.22 | 18.15 | 21.06 | 23.68 | 26.87 | 29.14 | 36.12 |
| 15 | 5.23 | 5.98 | 7.26 | 8.55 | 10.31 | 11.72 | 14.34 | 17.32 | 19.31 | 22.31 | 25.00 | 28.26 | 30.58 | 37.70 |

## TABLE B (Continued)

### Right-Tail Probability

| df | .99 | .98 | .95 | .90 | .80 | .70 | .50 | .30 | .20 | .10 | .05 | .02 | .01 | .001 |
|----|-----|-----|-----|-----|-----|-----|-----|-----|-----|-----|-----|-----|-----|------|
| 16 | 5.81 | 6.61 | 7.96 | 9.31 | 11.15 | 12.62 | 15.34 | 18.42 | 20.46 | 23.54 | 26.30 | 29.63 | 32.00 | 39.29 |
| 17 | 6.41 | 7.26 | 8.67 | 10.08 | 12.00 | 13.53 | 16.34 | 19.51 | 21.62 | 24.77 | 27.59 | 31.00 | 33.41 | 40.75 |
| 18 | 7.02 | 7.91 | 9.39 | 10.86 | 12.86 | 14.44 | 17.34 | 20.60 | 22.76 | 25.99 | 28.87 | 32.35 | 34.80 | 42.31 |
| 19 | 7.63 | 8.57 | 10.12 | 11.65 | 13.72 | 15.35 | 18.34 | 21.69 | 23.90 | 27.20 | 30.14 | 33.69 | 36.19 | 43.82 |
| 20 | 8.26 | 9.24 | 10.85 | 12.44 | 14.58 | 16.27 | 19.34 | 22.78 | 25.04 | 28.41 | 31.41 | 35.02 | 37.57 | 45.32 |
| 21 | 8.90 | 9.92 | 11.59 | 13.24 | 15.44 | 17.18 | 20.34 | 23.86 | 26.17 | 29.62 | 32.67 | 36.34 | 38.93 | 46.80 |
| 22 | 9.54 | 10.60 | 12.34 | 14.04 | 16.31 | 18.10 | 21.34 | 24.94 | 27.30 | 30.81 | 33.92 | 37.66 | 40.29 | 48.27 |
| 23 | 10.20 | 11.29 | 13.09 | 14.85 | 17.19 | 19.02 | 22.34 | 26.02 | 28.43 | 32.01 | 35.17 | 38.97 | 41.64 | 49.73 |
| 24 | 10.86 | 11.99 | 13.85 | 15.66 | 18.06 | 19.94 | 23.34 | 27.10 | 29.55 | 33.20 | 36.42 | 40.27 | 42.98 | 51.18 |
| 25 | 11.52 | 12.70 | 14.61 | 16.47 | 18.94 | 20.87 | 24.34 | 28.17 | 30.68 | 34.38 | 37.65 | 41.57 | 44.31 | 52.62 |
| 26 | 12.20 | 13.41 | 15.38 | 17.29 | 19.82 | 21.79 | 25.34 | 29.25 | 31.80 | 35.56 | 38.88 | 42.86 | 45.64 | 54.05 |
| 27 | 12.88 | 14.12 | 16.15 | 18.11 | 20.70 | 22.72 | 26.34 | 30.32 | 32.91 | 36.74 | 40.11 | 44.14 | 46.96 | 55.48 |
| 28 | 13.56 | 14.85 | 16.93 | 18.94 | 21.59 | 23.65 | 27.34 | 31.39 | 34.03 | 37.92 | 41.34 | 45.42 | 48.28 | 56.89 |
| 29 | 14.26 | 15.57 | 17.71 | 19.77 | 22.48 | 24.58 | 28.34 | 32.46 | 35.14 | 39.09 | 42.56 | 46.69 | 49.59 | 58.30 |
| 30 | 14.95 | 16.31 | 18.49 | 20.60 | 23.36 | 25.51 | 29.34 | 33.53 | 36.25 | 40.26 | 43.77 | 47.96 | 50.89 | 59.70 |

For df > 30, the probabilities based on the asymptotic distribution are approximated as follows:

Let $Q$ be a chi-square random variable with degrees of freedom df. A right- or left-tail probability for $Q$ is approximated by a right- or left-tail probability, respectively, from Table A for z, where

$$z = \sqrt{2Q} - \sqrt{2(\text{df}) - 1}$$

Source: Adapted from Table IV of R. A. Fisher and F. Yates (1963), *Statistical Tables for Biological, Agricultural and Medical Research*, 6th ed., Hafner Publishing Company, New York, with permission of Professor Yates and of Longman Group Ltd., 6th ed., 1974, previously published by Oliver & Boyd Ltd.

## TABLE C
## KOLMOGOROV-SMIRNOV ONE-SAMPLE STATISTIC

Table entries for any sample size $N$ are the values of a Kolmogorov-Smirnov one-sample random variable for which the right-tail probability for a two-sided test is as given on the top row, and the right-tail probability for a one-sided test is as given on the bottom row.

| | Right-Tail Probability for Two-Sided Test | | | | | | | | | | |
|---|---|---|---|---|---|---|---|---|---|---|---|
| $N$ | .200 | .100 | .050 | .020 | .010 | $N$ | .200 | .100 | .050 | .020 | .010 |
| 1 | .900 | .950 | .975 | .990 | .995 | 21 | .226 | .259 | .287 | .321 | .344 |
| 2 | .684 | .776 | .842 | .900 | .929 | 22 | .221 | .253 | .281 | .314 | .337 |
| 3 | .565 | .636 | .708 | .785 | .829 | 23 | .216 | .247 | .275 | .307 | .330 |
| 4 | .493 | .565 | .624 | .689 | .734 | 24 | .212 | .242 | .269 | .301 | .323 |
| 5 | .447 | .509 | .563 | .627 | .669 | 25 | .208 | .238 | .264 | .295 | .317 |
| 6 | .410 | .468 | .519 | .577 | .617 | 26 | .204 | .233 | .259 | .290 | .311 |
| 7 | .381 | .436 | .483 | .538 | .576 | 27 | .200 | .229 | .254 | .284 | .305 |
| 8 | .358 | .410 | .454 | .507 | .542 | 28 | .197 | .225 | .250 | .279 | .300 |
| 9 | .339 | .387 | .430 | .480 | .513 | 29 | .193 | .221 | .246 | .275 | .295 |
| 10 | .323 | .369 | .409 | .457 | .489 | 30 | .190 | .218 | .242 | .270 | .290 |
| 11 | .308 | .352 | .391 | .437 | .468 | 31 | .187 | .214 | .238 | .266 | .285 |
| 12 | .296 | .338 | .375 | .419 | .449 | 32 | .184 | .211 | .234 | .262 | .281 |
| 13 | .285 | .325 | .361 | .404 | .432 | 33 | .182 | .208 | .231 | .258 | .277 |
| 14 | .275 | .314 | .349 | .390 | .418 | 34 | .179 | .205 | .227 | .254 | .273 |
| 15 | .266 | .304 | .338 | .377 | .404 | 35 | .177 | .202 | .224 | .251 | .269 |
| 16 | .258 | .295 | .327 | .366 | .392 | 36 | .174 | .199 | .221 | .247 | .265 |
| 17 | .250 | .286 | .318 | .355 | .381 | 37 | .172 | .196 | .218 | .244 | .262 |
| 18 | .244 | .279 | .309 | .346 | .371 | 38 | .170 | .194 | .215 | .241 | .258 |
| 19 | .237 | .271 | .301 | .337 | .361 | 39 | .168 | .191 | .213 | .238 | .255 |
| 20 | .232 | .265 | .294 | .329 | .352 | 40 | .165 | .189 | .210 | .235 | .252 |
| | .100 | .050 | .025 | .010 | .005 | | .100 | .050 | .025 | .010 | .005 |

**Right-Tail Probability for One-Sided Test**

For $N > 40$, the table entries based on the asymptotic distribution are approximated by calculating the following for the appropriate value of $N$:

| | Right-Tail Probability for Two-Sided Test | | | |
|---|---|---|---|---|
| .200 | .100 | .050 | .020 | .010 |
| $1.07/\sqrt{N}$ | $1.22/\sqrt{N}$ | $1.36/\sqrt{N}$ | $1.52/\sqrt{N}$ | $1.63/\sqrt{N}$ |
| .100 | .050 | .025 | .010 | .005 |

**Right-Tail Probability for One-Sided Test**

Source: Adapted from L. H. Miller (1956), Table of percentage points of Kolmogorov statistics, *Journal of the American Statistical Association*, 51, 111-121 with permission.

## TABLE D
## KOLMOGOROV-SMIRNOV TWO-SAMPLE STATISTIC

Entries labeled $P$ in the table are the cumulative probability, right tail from the value of $mnD$ to plus infinity, for small $P$, all $2 \le m \le n \le 12$ or $m + n \le 16$, whichever occurs first. The same probability for $mnD_+$ or $mnD_-$ will equal $P/2$ if $P$ is very small. In the second portion of the table, where $9 \le (m = n) \le 20$, the table entries are the smallest values of $mnD$ for which the right-tail probability is less than or equal to the selected values .010, .020, .050, .100, and .200 shown on the top row. These values are approximately correct for $mnD_+$ or $mnD_-$ for the right-tail probabilities shown on the bottom row.

| $m$ | $n$ | $mnD$ | $P$ | $m$ | $n$ | $mnD$ | $P$ | $m$ | $n$ | $mnD$ | $P$ |
|---|---|---|---|---|---|---|---|---|---|---|---|
| 2 | 2 | 4 | .333 | 3 | 6 | 18 | .024 | 4 | 5 | 20 | .016 |
| 2 | 3 | 6 | .200 | | | 15 | .095 | | | 16 | .079 |
| 2 | 4 | 8 | .133 | | | 12 | .333 | | | 15 | .143 |
| 2 | 5 | 10 | .095 | 3 | 7 | 21 | .017 | 4 | 6 | 24 | .010 |
| | | 8 | .286 | | | 18 | .067 | | | 20 | .048 |
| 2 | 6 | 12 | .071 | | | 15 | .167 | | | 18 | .095 |
| | | 10 | .214 | 3 | 8 | 24 | .012 | | | 16 | .181 |
| 2 | 7 | 14 | .056 | | | 21 | .048 | 4 | 7 | 28 | .006 |
| | | 12 | .167 | | | 18 | .121 | | | 24 | .030 |
| 2 | 8 | 16 | .044 | 3 | 9 | 27 | .009 | | | 21 | .067 |
| | | 14 | .133 | | | 24 | .036 | | | 20 | .121 |
| 2 | 9 | 18 | .036 | | | 21 | .091 | 4 | 8 | 32 | .004 |
| | | 16 | .109 | | | 18 | .236 | | | 28 | .020 |
| 2 | 10 | 20 | .030 | 3 | 10 | 30 | .007 | | | 24 | .085 |
| | | 18 | .091 | | | 27 | .028 | | | 20 | .222 |
| | | 16 | .182 | | | 24 | .070 | 4 | 9 | 36 | .003 |
| 2 | 11 | 22 | .026 | | | 21 | .140 | | | 32 | .014 |
| | | 20 | .077 | 3 | 11 | 33 | .005 | | | 28 | .042 |
| | | 18 | .154 | | | 30 | .022 | | | 27 | .062 |
| 2 | 12 | 24 | .022 | | | 27 | .055 | | | 24 | .115 |
| | | 22 | .066 | | | 24 | .110 | 4 | 10 | 40 | .002 |
| | | 20 | .132 | 3 | 12 | 36 | .004 | | | 36 | .010 |
| 3 | 3 | 9 | .100 | | | 33 | .018 | | | 32 | .030 |
| 3 | 4 | 12 | .057 | | | 30 | .044 | | | 30 | .046 |
| | | 9 | .229 | | | 27 | .088 | | | 28 | .084 |
| 3 | 5 | 15 | .036 | | | 24 | .189 | | | 26 | .126 |
| | | 12 | .143 | 4 | 4 | 16 | .029 | | | | |
| | | | | | | 12 | .229 | | | | |

### TABLE D (Continued)

| $m$ | $n$ | $mnD$ | $P$ | $m$ | $n$ | $mnD$ | $P$ | $m$ | $n$ | $mnD$ | $P$ |
|---|---|---|---|---|---|---|---|---|---|---|---|
| 4 | 11 | 44 | .001 | 5 | 10 | 50 | .001 | 6 | 10 | 60 | .000 |
|   |    | 40 | .007 |   |    | 45 | .004 |   |    | 54 | .002 |
|   |    | 36 | .022 |   |    | 40 | .019 |   |    | 50 | .004 |
|   |    | 33 | .035 |   |    | 35 | .061 |   |    | 48 | .009 |
|   |    | 32 | .063 |   |    | 30 | .166 |   |    | 44 | .019 |
|   |    | 29 | .098 | 5 | 11 | 55 | .000 |   |    | 42 | .031 |
|   |    | 28 | .144 |   |    | 50 | .003 |   |    | 40 | .042 |
| 4 | 12 | 48 | .001 |   |    | 45 | .010 |   |    | 38 | .066 |
|   |    | 44 | .005 |   |    | 44 | .014 |   |    | 36 | .092 |
|   |    | 40 | .016 |   |    | 40 | .029 |   |    | 34 | .125 |
|   |    | 36 | .048 |   |    | 39 | .044 | 7 | 7 | 49 | .001 |
|   |    | 32 | .112 |   |    | 35 | .074 |   |    | 42 | .008 |
| 5 | 5 | 25 | .008 |   |    | 34 | .106 |   |    | 35 | .053 |
|   |    | 20 | .079 | 6 | 6 | 36 | .002 |   |    | 28 | .212 |
|   |    | 15 | .357 |   |    | 30 | .026 | 7 | 8 | 56 | .000 |
| 5 | 6 | 30 | .004 |   |    | 24 | .143 |   |    | 49 | .002 |
|   |    | 25 | .026 | 6 | 7 | 42 | .001 |   |    | 48 | .005 |
|   |    | 24 | .048 |   |    | 36 | .008 |   |    | 42 | .013 |
|   |    | 20 | .108 |   |    | 35 | .015 |   |    | 41 | .024 |
| 5 | 7 | 35 | .003 |   |    | 30 | .038 |   |    | 40 | .033 |
|   |    | 30 | .015 |   |    | 29 | .068 |   |    | 35 | .056 |
|   |    | 28 | .030 |   |    | 28 | .091 |   |    | 34 | .087 |
|   |    | 25 | .066 |   |    | 24 | .147 |   |    | 33 | .118 |
|   |    | 23 | .116 | 6 | 8 | 48 | .001 | 7 | 9 | 63 | .000 |
| 5 | 8 | 40 | .002 |   |    | 42 | .005 |   |    | 56 | .001 |
|   |    | 35 | .009 |   |    | 40 | .009 |   |    | 54 | .003 |
|   |    | 32 | .020 |   |    | 36 | .023 |   |    | 49 | .008 |
|   |    | 30 | .042 |   |    | 34 | .043 |   |    | 47 | .015 |
|   |    | 27 | .079 |   |    | 32 | .061 |   |    | 45 | .021 |
|   |    | 25 | .126 |   |    | 30 | .093 |   |    | 42 | .034 |
| 5 | 9 | 45 | .001 |   |    | 28 | .139 |   |    | 40 | .055 |
|   |    | 40 | .006 | 6 | 9 | 54 | .000 |   |    | 38 | .079 |
|   |    | 36 | .014 |   |    | 48 | .003 |   |    | 36 | .098 |
|   |    | 35 | .028 |   |    | 45 | .006 |   |    | 35 | .127 |
|   |    | 31 | .056 |   |    | 42 | .014 | 8 | 8 | 64 | .000 |
|   |    | 30 | .086 |   |    | 39 | .028 |   |    | 56 | .002 |
|   |    | 27 | .119 |   |    | 36 | .061 |   |    | 48 | .019 |
|   |    |    |      |   |    | 33 | .095 |   |    | 40 | .087 |
|   |    |    |      |   |    | 30 | .176 |   |    | 32 | .283 |

TABLE D (Continued)

| $m = n$ | Right-Tail Probability for mnD (Two-Sided Statistic) | | | | |
|---|---|---|---|---|---|
| | .200 | .100 | .050 | .020 | .010 |
| 9 | 45 | 54 | 54 | 63 | 63 |
| 10 | 50 | 60 | 70 | 70 | 80 |
| 11 | 66 | 66 | 77 | 88 | 88 |
| 12 | 72 | 72 | 84 | 96 | 96 |
| 13 | 78 | 91 | 91 | 104 | 117 |
| 14 | 84 | 98 | 112 | 112 | 126 |
| 15 | 90 | 105 | 120 | 135 | 135 |
| 16 | 112 | 112 | 128 | 144 | 160 |
| 17 | 119 | 136 | 136 | 153 | 170 |
| 18 | 126 | 144 | 162 | 180 | 180 |
| 19 | 133 | 152 | 171 | 190 | 190 |
| 20 | 140 | 160 | 180 | 200 | 220 |
| | .100 | .050 | .025 | .010 | .005 |
| | Approximate Right-Tail Probability for mnD$_+$ or mnD$_-$ (One-Sided Statistic) | | | | |

For sample sizes outside the range of this table, the quantile points based on the asymptotic distribution are approximated by calculating the following for the appropriate values of $m$, $n$ and $N = m + n$:

| Right-Tail Probability for D | | | | |
|---|---|---|---|---|
| .200 | .100 | .050 | .020 | .010 |
| $1.07\sqrt{N/mn}$ | $1.22\sqrt{N/mn}$ | $1.36\sqrt{N/mn}$ | $1.52\sqrt{N/mn}$ | $1.63\sqrt{N/mn}$ |
| .100 | .050 | .025 | .010 | .005 |
| Right-Tail Probability for D$_+$ or D$_-$ | | | | |

Source: Adapted from Table I of H. L. Harter and D. B. Owen, eds. (1970), *Selected Tables in Mathematical Statistics*, Vol. 2, Markham Publishing Company, Chicago, with permission of The Institute of Mathematical Statistics.

## TABLE E
## CUMULATIVE BINOMIAL DISTRIBUTION

Table entries for any $n$, $x$, and $\theta$ are the left-tail cumulative probability of $x$ or less "successes" in $n$ Bernoulli trials, where $\theta$ is the probability of success.

| $n$ | $x$ | .05 | .10 | .15 | .20 | $\theta$<br>.25 | .30 | .35 | .40 | .45 |
|---|---|---|---|---|---|---|---|---|---|---|
| 1 | 0 | .9500 | .9000 | .8500 | .8000 | .7500 | .7000 | .6500 | .6000 | .5500 |
|   | 1 | 1.0000 | 1.0000 | 1.0000 | 1.0000 | 1.0000 | 1.0000 | 1.0000 | 1.0000 | 1.0000 |
| 2 | 0 | .9025 | .8100 | .7225 | .6400 | .5625 | .4900 | .4225 | .3600 | .3025 |
|   | 1 | .9975 | .9900 | .9775 | .9600 | .9375 | .9100 | .8775 | .8400 | .7975 |
|   | 2 | 1.0000 | 1.0000 | 1.0000 | 1.0000 | 1.0000 | 1.0000 | 1.0000 | 1.0000 | 1.0000 |
| 3 | 0 | .8574 | .7290 | .6141 | .5120 | .4219 | .3430 | .2746 | .2160 | .1664 |
|   | 1 | .9928 | .9720 | .9392 | .8960 | .8438 | .7840 | .7182 | .6480 | .5748 |
|   | 2 | .9999 | .9990 | .9966 | .9920 | .9844 | .9730 | .9571 | .9360 | .9089 |
|   | 3 | 1.0000 | 1.0000 | 1.0000 | 1.0000 | 1.0000 | 1.0000 | 1.0000 | 1.0000 | 1.0000 |
| 4 | 0 | .8145 | .6561 | .5220 | .4096 | .3164 | .2401 | .1785 | .1296 | .0915 |
|   | 1 | .9860 | .9477 | .8905 | .8192 | .7383 | .6517 | .5630 | .4752 | .3910 |
|   | 2 | .9995 | .9963 | .9880 | .9728 | .9492 | .9163 | .8735 | .8208 | .7585 |
|   | 3 | 1.0000 | .9999 | .9995 | .9984 | .9961 | .9919 | .9850 | .9744 | .9590 |
|   | 4 | 1.0000 | 1.0000 | 1.0000 | 1.0000 | 1.0000 | 1.0000 | 1.0000 | 1.0000 | 1.0000 |
| 5 | 0 | .7738 | .5905 | .4437 | .3277 | .2373 | .1681 | .1160 | .0778 | .0503 |
|   | 1 | .9774 | .9185 | .8352 | .7373 | .6328 | .5282 | .4284 | .3370 | .2562 |
|   | 2 | .9988 | .9914 | .9734 | .9421 | .8965 | .8369 | .7648 | .6826 | .5931 |
|   | 3 | 1.0000 | .9995 | .9978 | .9933 | .9844 | .9692 | .9460 | .9130 | .8688 |
|   | 4 | 1.0000 | 1.0000 | .9999 | .9997 | .9990 | .9976 | .9947 | .9898 | .9815 |
|   | 5 | 1.0000 | 1.0000 | 1.0000 | 1.0000 | 1.0000 | 1.0000 | 1.0000 | 1.0000 | 1.0000 |
| 6 | 0 | .7351 | .5314 | .3771 | .2621 | .1780 | .1176 | .0754 | .0467 | .0277 |
|   | 1 | .9672 | .8857 | .7765 | .6554 | .5339 | .4202 | .3191 | .2333 | .1636 |
|   | 2 | .9978 | .9842 | .9527 | .9011 | .8306 | .7443 | .6471 | .5443 | .4415 |
|   | 3 | .9999 | .9987 | .9941 | .9830 | .9624 | .9295 | .8826 | .8208 | .7447 |
|   | 4 | 1.0000 | .9999 | .9996 | .9984 | .9954 | .9891 | .9777 | .9590 | .9308 |
|   | 5 | 1.0000 | 1.0000 | 1.0000 | .9999 | .9998 | .9993 | .9982 | .9959 | .9917 |
|   | 6 | 1.0000 | 1.0000 | 1.0000 | 1.0000 | 1.0000 | 1.0000 | 1.0000 | 1.0000 | 1.0000 |
| 7 | 0 | .6983 | .4783 | .3206 | .2097 | .1335 | .0824 | .0490 | .0280 | .0152 |
|   | 1 | .9556 | .8503 | .7166 | .5767 | .4449 | .3294 | .2338 | .1586 | .1024 |
|   | 2 | .9962 | .9743 | .9262 | .8520 | .7564 | .6471 | .5323 | .4199 | .3164 |
|   | 3 | .9998 | .9973 | .9879 | .9667 | .9294 | .8740 | .8002 | .7102 | .6083 |
|   | 4 | 1.0000 | .9998 | .9988 | .9953 | .9871 | .9712 | .9444 | .9037 | .8471 |
|   | 5 | 1.0000 | 1.0000 | .9999 | .9996 | .9987 | .9962 | .9910 | .9812 | .9643 |
|   | 6 | 1.0000 | 1.0000 | 1.0000 | 1.0000 | .9999 | .9998 | .9994 | .9984 | .9963 |
|   | 7 | 1.0000 | 1.0000 | 1.0000 | 1.0000 | 1.0000 | 1.0000 | 1.0000 | 1.0000 | 1.0000 |

## TABLE E (Continued)

| n | x | .50 | .55 | .60 | .65 | θ .70 | .75 | .80 | .85 | .90 | .95 |
|---|---|-----|-----|-----|-----|-----|-----|-----|-----|-----|-----|
| 1 | 0 | .5000 | .4500 | .4000 | .3500 | .3000 | .2500 | .2000 | .1500 | .1000 | .0500 |
|   | 1 | 1.0000 | 1.0000 | 1.0000 | 1.0000 | 1.0000 | 1.0000 | 1.0000 | 1.0000 | 1.0000 | 1.0000 |
| 2 | 0 | .2500 | .2025 | .1600 | .1225 | .0900 | .0625 | .0400 | .0225 | .0100 | .0025 |
|   | 1 | .7500 | .6975 | .6400 | .5775 | .5100 | .4375 | .3600 | .2775 | .1900 | .0975 |
|   | 2 | 1.0000 | 1.0000 | 1.0000 | 1.0000 | 1.0000 | 1.0000 | 1.0000 | 1.0000 | 1.0000 | 1.0000 |
| 3 | 0 | .1250 | .0911 | .0640 | .0429 | .0270 | .0156 | .0080 | .0034 | .0010 | .0001 |
|   | 1 | .5000 | .4252 | .3520 | .2818 | .2160 | .1562 | .1040 | .0608 | .0280 | .0072 |
|   | 2 | .8750 | .8336 | .7840 | .7254 | .6570 | .5781 | .4880 | .3859 | .2710 | .1426 |
|   | 3 | 1.0000 | 1.0000 | 1.0000 | 1.0000 | 1.0000 | 1.0000 | 1.0000 | 1.0000 | 1.0000 | 1.0000 |
| 4 | 0 | .0625 | .0410 | .0256 | .0150 | .0081 | .0039 | .0016 | .0005 | .0001 | .0000 |
|   | 1 | .3125 | .2415 | .1792 | .1265 | .0837 | .0508 | .0272 | .0120 | .0037 | .0005 |
|   | 2 | .6875 | .6090 | .5248 | .4370 | .3483 | .2617 | .1808 | .1095 | .0523 | .0140 |
|   | 3 | .9375 | .9085 | .8704 | .8215 | .7599 | .6836 | .5904 | .4780 | .3439 | .1855 |
|   | 4 | 1.0000 | 1.0000 | 1.0000 | 1.0000 | 1.0000 | 1.0000 | 1.0000 | 1.0000 | 1.0000 | 1.0000 |
| 5 | 0 | .0312 | .0185 | .0102 | .0053 | .0024 | .0010 | .0003 | .0001 | .0000 | .0000 |
|   | 1 | .1875 | .1312 | .0870 | .0540 | .0308 | .0156 | .0067 | .0022 | .0005 | .0000 |
|   | 2 | .5000 | .4069 | .3174 | .2352 | .1631 | .1035 | .0579 | .0266 | .0086 | .0012 |
|   | 3 | .8125 | .7438 | .6630 | .5716 | .4718 | .3672 | .2627 | .1648 | .0815 | .0226 |
|   | 4 | .9688 | .9497 | .9222 | .8840 | .8319 | .7627 | .6723 | .5563 | .4095 | .2262 |
|   | 5 | 1.0000 | 1.0000 | 1.0000 | 1.0000 | 1.0000 | 1.0000 | 1.0000 | 1.0000 | 1.0000 | 1.0000 |
| 6 | 0 | .0156 | .0083 | .0041 | .0018 | .0007 | .0002 | .0001 | .0000 | .0000 | .0000 |
|   | 1 | .1094 | .0692 | .0410 | .0223 | .0109 | .0046 | .0016 | .0004 | .0001 | .0000 |
|   | 2 | .3438 | .2553 | .1792 | .1174 | .0705 | .0376 | .0170 | .0059 | .0013 | .0001 |
|   | 3 | .6562 | .5585 | .4557 | .3529 | .2557 | .1694 | .0989 | .0473 | .0158 | .0022 |
|   | 4 | .8906 | .8364 | .7667 | .6809 | .5798 | .4661 | .3446 | .2235 | .1143 | .0328 |
|   | 5 | .9844 | .9723 | .9533 | .9246 | .8824 | .8220 | .7379 | .6229 | .4686 | .2649 |
|   | 6 | 1.0000 | 1.0000 | 1.0000 | 1.0000 | 1.0000 | 1.0000 | 1.0000 | 1.0000 | 1.0000 | 1.0000 |
| 7 | 0 | .0078 | .0037 | .0016 | .0006 | .0002 | .0001 | .0000 | .0000 | .0000 | .0000 |
|   | 1 | .0625 | .0357 | .0188 | .0090 | .0038 | .0013 | .0004 | .0001 | .0000 | .0000 |
|   | 2 | .2266 | .1529 | .0963 | .0556 | .0288 | .0129 | .0047 | .0012 | .0002 | .0000 |
|   | 3 | .5000 | .3917 | .2898 | .1998 | .1260 | .0706 | .0333 | .0121 | .0027 | .0002 |
|   | 4 | .7734 | .6836 | .5801 | .4677 | .3529 | .2436 | .1480 | .0738 | .0257 | .0038 |
|   | 5 | .9375 | .8976 | .8414 | .7662 | .6706 | .5551 | .4233 | .2834 | .1497 | .0444 |
|   | 6 | .9922 | .9848 | .9720 | .9510 | .9176 | .8665 | .7903 | .6794 | .5217 | .3017 |
|   | 7 | 1.0000 | 1.0000 | 1.0000 | 1.0000 | 1.0000 | 1.0000 | 1.0000 | 1.0000 | 1.0000 | 1.0000 |

TABLE E (Continued)

| n | x | .05 | .10 | .15 | .20 | θ<br>.25 | .30 | .35 | .40 | .45 |
|---|---|-----|-----|-----|-----|-----|-----|-----|-----|-----|
| 8 | 0 | .6634 | .4305 | .2725 | .1678 | .1001 | .0576 | .0319 | .0168 | .0084 |
|   | 1 | .9428 | .8131 | .6572 | .5033 | .3671 | .2553 | .1691 | .1064 | .0632 |
|   | 2 | .9942 | .9619 | .8948 | .7969 | .6785 | .5518 | .4278 | .3154 | .2201 |
|   | 3 | .9996 | .9950 | .9786 | .9437 | .8862 | .8059 | .7064 | .5941 | .4770 |
|   | 4 | 1.0000 | .9996 | .9971 | .9896 | .9727 | .9420 | .8939 | .8263 | .7396 |
|   | 5 | 1.0000 | 1.0000 | .9998 | .9988 | .9958 | .9887 | .9747 | .9502 | .9115 |
|   | 6 | 1.0000 | 1.0000 | 1.0000 | .9999 | .9996 | .9987 | .9964 | .9915 | .9819 |
|   | 7 | 1.0000 | 1.0000 | 1.0000 | 1.0000 | 1.0000 | .9999 | .9998 | .9993 | .9983 |
|   | 8 | 1.0000 | 1.0000 | 1.0000 | 1.0000 | 1.0000 | 1.0000 | 1.0000 | 1.0000 | 1.0000 |
| 9 | 0 | .6302 | .3874 | .2316 | .1342 | .0751 | .0404 | .0207 | .0101 | .0046 |
|   | 1 | .9288 | .7748 | .5995 | .4362 | .3003 | .1960 | .1211 | .0705 | .0385 |
|   | 2 | .9916 | .9470 | .8591 | .7382 | .6007 | .4628 | .3373 | .2318 | .1495 |
|   | 3 | .9994 | .9917 | .9661 | .9144 | .8343 | .7297 | .6089 | .4826 | .3614 |
|   | 4 | 1.0000 | .9991 | .9944 | .9804 | .9511 | .9012 | .8283 | .7334 | .6214 |
|   | 5 | 1.0000 | .9999 | .9994 | .9969 | .9900 | .9747 | .9464 | .9006 | .8342 |
|   | 6 | 1.0000 | 1.0000 | 1.0000 | .9997 | .9987 | .9957 | .9888 | .9750 | .9502 |
|   | 7 | 1.0000 | 1.0000 | 1.0000 | 1.0000 | .9999 | .9996 | .9986 | .9962 | .9909 |
|   | 8 | 1.0000 | 1.0000 | 1.0000 | 1.0000 | 1.0000 | 1.0000 | .9999 | .9997 | .9992 |
|   | 9 | 1.0000 | 1.0000 | 1.0000 | 1.0000 | 1.0000 | 1.0000 | 1.0000 | 1.0000 | 1.0000 |
| 10 | 0 | .5987 | .3487 | .1969 | .1074 | .0563 | .0282 | .0135 | .0060 | .0025 |
|   | 1 | .9139 | .7361 | .5443 | .3758 | .2440 | .1493 | .0860 | .0464 | .0233 |
|   | 2 | .9885 | .9298 | .8202 | .6778 | .5256 | .3828 | .2616 | .1673 | .0996 |
|   | 3 | .9990 | .9872 | .9500 | .8791 | .7759 | .6496 | .5138 | .3823 | .2660 |
|   | 4 | .9999 | .9984 | .9901 | .9672 | .9219 | .8497 | .7515 | .6331 | .5044 |
|   | 5 | 1.0000 | .9999 | .9986 | .9936 | .9803 | .9527 | .9051 | .8338 | .7384 |
|   | 6 | 1.0000 | 1.0000 | .9999 | .9991 | .9965 | .9894 | .9740 | .9452 | .8980 |
|   | 7 | 1.0000 | 1.0000 | 1.0000 | .9999 | .9996 | .9984 | .9952 | .9877 | .9726 |
|   | 8 | 1.0000 | 1.0000 | 1.0000 | 1.0000 | 1.0000 | .9999 | .9995 | .9983 | .9955 |
|   | 9 | 1.0000 | 1.0000 | 1.0000 | 1.0000 | 1.0000 | 1.0000 | 1.0000 | .9999 | .9997 |
|   | 10 | 1.0000 | 1.0000 | 1.0000 | 1.0000 | 1.0000 | 1.0000 | 1.0000 | 1.0000 | 1.0000 |
| 11 | 0 | .5688 | .3138 | .1673 | .0859 | .0422 | .0198 | .0088 | .0036 | .0014 |
|   | 1 | .8981 | .6974 | .4922 | .3221 | .1971 | .1130 | .0606 | .0302 | .0139 |
|   | 2 | .9848 | .9104 | .7788 | .6174 | .4552 | .3127 | .2001 | .1189 | .0652 |
|   | 3 | .9984 | .9815 | .9306 | .8389 | .7133 | .5696 | .4256 | .2963 | .1911 |
|   | 4 | .9999 | .9972 | .9841 | .9496 | .8854 | .7897 | .6683 | .5328 | .3971 |
|   | 5 | 1.0000 | .9997 | .9973 | .9883 | .9657 | .9218 | .8513 | .7535 | .6331 |
|   | 6 | 1.0000 | 1.0000 | .9997 | .9980 | .9924 | .9784 | .9499 | .9006 | .8262 |
|   | 7 | 1.0000 | 1.0000 | 1.0000 | .9998 | .9988 | .9957 | .9878 | .9707 | .9390 |
|   | 8 | 1.0000 | 1.0000 | 1.0000 | 1.0000 | .9999 | .9994 | .9980 | .9941 | .9852 |
|   | 9 | 1.0000 | 1.0000 | 1.0000 | 1.0000 | 1.0000 | 1.0000 | .9998 | .9993 | .9978 |
|   | 10 | 1.0000 | 1.0000 | 1.0000 | 1.0000 | 1.0000 | 1.0000 | 1.0000 | 1.0000 | .9998 |
|   | 11 | 1.0000 | 1.0000 | 1.0000 | 1.0000 | 1.0000 | 1.0000 | 1.0000 | 1.0000 | 1.0000 |

TABLE E (Continued)

| n | x | $\theta$ .50 | .55 | .60 | .65 | .70 | .75 | .80 | .85 | .90 | .95 |
|---|---|------|------|------|------|------|------|------|------|------|------|
| 8 | 0 | .0039 | .0017 | .0007 | .0002 | .0001 | .0000 | .0000 | .0000 | .0000 | .0000 |
|   | 1 | .0352 | .0181 | .0085 | .0036 | .0013 | .0004 | .0001 | .0000 | .0000 | .0000 |
|   | 2 | .1445 | .0885 | .0498 | .0253 | .0113 | .0042 | .0012 | .0002 | .0000 | .0000 |
|   | 3 | .3633 | .2604 | .1737 | .1061 | .0580 | .0273 | .0104 | .0029 | .0004 | .0000 |
|   | 4 | .6367 | .5230 | .4059 | .2936 | .1941 | .1138 | .0563 | .0214 | .0050 | .0004 |
|   | 5 | .8555 | .7799 | .6846 | .5722 | .4482 | .3215 | .2031 | .1052 | .0381 | .0058 |
|   | 6 | .9648 | .9368 | .8936 | .8309 | .7447 | .6329 | .4967 | .3428 | .1869 | .0572 |
|   | 7 | .9961 | .9916 | .9832 | .9681 | .9424 | .8999 | .8322 | .7275 | .5695 | .3366 |
|   | 8 | 1.0000 | 1.0000 | 1.0000 | 1.0000 | 1.0000 | 1.0000 | 1.0000 | 1.0000 | 1.0000 | 1.0000 |
| 9 | 0 | .0020 | .0008 | .0003 | .0001 | .0000 | .0000 | .0000 | .0000 | .0000 | .0000 |
|   | 1 | .0195 | .0091 | .0038 | .0014 | .0004 | .0001 | .0000 | .0000 | .0000 | .0000 |
|   | 2 | .0898 | .0498 | .0250 | .0112 | .0043 | .0013 | .0003 | .0000 | .0000 | .0000 |
|   | 3 | .2539 | .1658 | .0994 | .0536 | .0253 | .0100 | .0031 | .0006 | .0001 | .0000 |
|   | 4 | .5000 | .3786 | .2666 | .1717 | .0988 | .0489 | .0196 | .0056 | .0009 | .0000 |
|   | 5 | .7461 | .6386 | .5174 | .3911 | .2703 | .1657 | .0856 | .0339 | .0083 | .0006 |
|   | 6 | .9102 | .8505 | .7682 | .6627 | .5372 | .3993 | .2618 | .1409 | .0530 | .0084 |
|   | 7 | .9805 | .9615 | .9295 | .8789 | .8040 | .6997 | .5638 | .4005 | .2252 | .0712 |
|   | 8 | .9980 | .9954 | .9899 | .9793 | .9596 | .9249 | .8658 | .7684 | .6126 | .3698 |
|   | 9 | 1.0000 | 1.0000 | 1.0000 | 1.0000 | 1.0000 | 1.0000 | 1.0000 | 1.0000 | 1.0000 | 1.0000 |
| 10 | 0 | .0010 | .0003 | .0001 | .0000 | .0000 | .0000 | .0000 | .00C0 | .0000 | .0000 |
|   | 1 | .0107 | .0045 | .0017 | .0005 | .0001 | .0000 | .0000 | .0000 | .0000 | .0000 |
|   | 2 | .0547 | .0274 | .0123 | .0048 | .0016 | .0004 | .0001 | .0000 | .0000 | .0000 |
|   | 3 | .1719 | .1020 | .0548 | .0260 | .0106 | .0035 | .0009 | .0001 | .0000 | .0000 |
|   | 4 | .3770 | .2616 | .1662 | .0949 | .0473 | .0197 | .0064 | .0014 | .0001 | .0000 |
|   | 5 | .6230 | .4956 | .3669 | .2485 | .1503 | .0781 | .0328 | .0099 | .0016 | .0001 |
|   | 6 | .8281 | .7340 | .6177 | .4862 | .3504 | .2241 | .1209 | .0500 | .0128 | .0010 |
|   | 7 | .9453 | .9004 | .8327 | .7384 | .6172 | .4744 | .3222 | .1798 | .0702 | .0115 |
|   | 8 | .9893 | .9767 | .9536 | .9140 | .8507 | .7560 | .6242 | .4557 | .2639 | .0861 |
|   | 9 | .9990 | .9975 | .9940 | .9865 | .9718 | .9437 | .8926 | .8031 | .6513 | .4013 |
|   | 10 | 1.0000 | 1.0000 | 1.0000 | 1.0000 | 1.0000 | 1.0000 | 1.0000 | 1.0000 | 1.0000 | 1.0000 |
| 11 | 0 | .0005 | .0002 | .0000 | .0000 | .0000 | .0000 | .0000 | .0000 | .0000 | .0000 |
|   | 1 | .0059 | .0022 | .0007 | .0002 | .0000 | .0000 | .0000 | .0000 | .0000 | .0000 |
|   | 2 | .0327 | .0148 | .0059 | .0020 | .0006 | .0001 | .0000 | .0000 | .0000 | .0000 |
|   | 3 | .1133 | .0610 | .0293 | .0122 | .0043 | .0012 | .0002 | .0000 | .0000 | .0000 |
|   | 4 | .2744 | .1738 | .0994 | .0501 | .0216 | .0076 | .0020 | .0003 | .0000 | .0000 |
|   | 5 | .5000 | .3669 | .2465 | .1487 | .0782 | .0343 | .0117 | .0027 | .0003 | .0000 |
|   | 6 | .7256 | .6029 | .4672 | .3317 | .2103 | .1146 | .0504 | .0159 | .0028 | .0001 |
|   | 7 | .8867 | .8089 | .7037 | .5744 | .4304 | .2867 | .1611 | .0694 | .0185 | .0016 |
|   | 8 | .9673 | .9348 | .8811 | .7999 | .6873 | .5448 | .3826 | .2212 | .0896 | .0152 |
|   | 9 | .9941 | .9861 | .9698 | .9394 | .8870 | .8029 | .6779 | .5078 | .3026 | .1019 |
|   | 10 | .9995 | .9986 | .9964 | .9912 | .9802 | .9578 | .9141 | .8327 | .6862 | .4312 |
|   | 11 | 1.0000 | 1.0000 | 1.0000 | 1.0000 | 1.0000 | 1.0000 | 1.0000 | 1.0000 | 1.0000 | 1.0000 |

**TABLE E (Continued)**

| | | θ | | | | | | | | |
|---|---|---|---|---|---|---|---|---|---|---|
| *n* | *x* | .05 | .10 | .15 | .20 | .25 | .30 | .35 | .40 | .45 |
| 12 | 0 | .5404 | .2824 | .1422 | .0687 | .0317 | .0138 | .0057 | .0022 | .0008 |
| | 1 | .8816 | .6590 | .4435 | .2749 | .1584 | .0850 | .0424 | .0424 | .0083 |
| | 2 | .9804 | .8891 | .7358 | .5583 | .3907 | .2528 | .1513 | .0834 | .0421 |
| | 3 | .9978 | .9744 | .9078 | .7946 | .6488 | .4925 | .3467 | .2253 | .1345 |
| | 4 | .9998 | .9957 | .9761 | .9274 | .8424 | .7237 | .5833 | .4382 | .3044 |
| | 5 | 1.0000 | .9995 | .9954 | .9806 | .9456 | .8822 | .7873 | .6652 | .5269 |
| | 6 | 1.0000 | .9999 | .9993 | .9961 | .9857 | .9614 | .9154 | .8418 | .7393 |
| | 7 | 1.0000 | 1.0000 | .9999 | .9994 | .9972 | .9905 | .9745 | .9427 | .8883 |
| | 8 | 1.0000 | 1.0000 | 1.0000 | .9999 | .9996 | .9983 | .9944 | .9847 | .9644 |
| | 9 | 1.0000 | 1.0000 | 1.0000 | 1.0000 | 1.0000 | .9998 | .9992 | .9972 | .9921 |
| | 10 | 1.0000 | 1.0000 | 1.0000 | 1.0000 | 1.0000 | 1.0000 | .9999 | .9997 | .9989 |
| | 11 | 1.0000 | 1.0000 | 1.0000 | 1.0000 | 1.0000 | 1.0000 | 1.0000 | 1.0000 | .9999 |
| | 12 | 1.0000 | 1.0000 | 1.0000 | 1.0000 | 1.0000 | 1.0000 | 1.0000 | 1.0000 | 1.0000 |
| 13 | 0 | .5133 | .2542 | .1209 | .0550 | .0238 | .0097 | .0037 | .0013 | .0004 |
| | 1 | .8646 | .6213 | .3983 | .2336 | .1267 | .0637 | .0296 | .0126 | .0049 |
| | 2 | .9755 | .8661 | .7296 | .5017 | .3326 | .2025 | .1132 | .0579 | .0269 |
| | 3 | .9969 | .9658 | .9033 | .7473 | .5843 | .4206 | .2783 | .1686 | .0929 |
| | 4 | .9997 | .9935 | .9740 | .9009 | .7940 | .6543 | .5005 | .3530 | .2279 |
| | 5 | 1.0000 | .9991 | .9947 | .9700 | .9198 | .8346 | .7159 | .5744 | .4268 |
| | 6 | 1.0000 | .9999 | .9987 | .9930 | .9757 | .9376 | .8705 | .7712 | .6437 |
| | 7 | 1.0000 | 1.0000 | .9998 | .9988 | .9944 | .9818 | .9538 | .9023 | .8212 |
| | 8 | 1.0000 | 1.0000 | 1.0000 | .9998 | .9990 | .9960 | .9874 | .9679 | .9302 |
| | 9 | 1.0000 | 1.0000 | 1.0000 | 1.0000 | .9999 | .9993 | .9975 | .9922 | .9797 |
| | 10 | 1.0000 | 1.0000 | 1.0000 | 1.0000 | 1.0000 | .9999 | .9997 | .9987 | .9959 |
| | 11 | 1.0000 | 1.0000 | 1.0000 | 1.0000 | 1.0000 | 1.0000 | 1.0000 | .9999 | .9995 |
| | 12 | 1.0000 | 1.0000 | 1.0000 | 1.0000 | 1.0000 | 1.0000 | 1.0000 | 1.0000 | 1.0000 |
| | 13 | 1.0000 | 1.0000 | 1.0000 | 1.0000 | 1.0000 | 1.0000 | 1.0000 | 1.0000 | 1.0000 |
| 14 | 0 | .4877 | .2288 | .1028 | .0440 | .0178 | .0068 | .0024 | .0008 | .0002 |
| | 1 | .8470 | .5846 | .3567 | .1979 | .1010 | .0475 | .0205 | .0081 | .0029 |
| | 2 | .9699 | .8416 | .6479 | .4481 | .2811 | .1608 | .0839 | .0398 | .0170 |
| | 3 | .9958 | .9559 | .8535 | .6982 | .5213 | .3552 | .2205 | .1243 | .0632 |
| | 4 | .9996 | .9908 | .9533 | .8702 | .7415 | .5842 | .4227 | .2793 | .1672 |
| | 5 | 1.0000 | .9985 | .9885 | .9561 | .8883 | .7805 | .6405 | .4859 | .3373 |
| | 6 | 1.0000 | .9998 | .9978 | .9884 | .9617 | .9067 | .8164 | .6925 | .5461 |
| | 7 | 1.0000 | 1.0000 | .9997 | .9976 | .9897 | .9685 | .9247 | .8499 | .7414 |
| | 8 | 1.0000 | 1.0000 | 1.0000 | .9996 | .9978 | .9917 | .9757 | .9417 | .8811 |
| | 9 | 1.0000 | 1.0000 | 1.0000 | 1.0000 | .9997 | .9983 | .9940 | .9825 | .9574 |
| | 10 | 1.0000 | 1.0000 | 1.0000 | 1.0000 | 1.0000 | .9998 | .9989 | .9961 | .9886 |
| | 11 | 1.0000 | 1.0000 | 1.0000 | 1.0000 | 1.0000 | 1.0000 | .9999 | .9994 | .9978 |
| | 12 | 1.0000 | 1.0000 | 1.0000 | 1.0000 | 1.0000 | 1.0000 | 1.0000 | .9999 | .9997 |
| | 13 | 1.0000 | 1.0000 | 1.0000 | 1.0000 | 1.0000 | 1.0000 | 1.0000 | 1.0000 | 1.0000 |
| | 14 | 1.0000 | 1.0000 | 1.0000 | 1.0000 | 1.0000 | 1.0000 | 1.0000 | 1.0000 | 1.0000 |

TABLE E (Continued)

| n | x | .50 | .55 | .60 | .65 | θ .70 | .75 | .80 | .85 | .90 | .95 |
|---|---|-----|-----|-----|-----|-----|-----|-----|-----|-----|-----|
| 12 | 0 | .0002 | .0001 | .0000 | .0000 | .0000 | .0000 | .0000 | .0000 | .0000 | .0000 |
| | 1 | .0032 | .0011 | .0003 | .0001 | .0000 | .0000 | .0000 | .0000 | .0000 | .0000 |
| | 2 | .0193 | .0079 | .0028 | .0008 | .0002 | .0000 | .0000 | .0000 | .0000 | .0000 |
| | 3 | .0730 | .0356 | .0153 | .0056 | .0017 | .0004 | .0001 | .0000 | .0000 | .0000 |
| | 4 | .1938 | .1117 | .0573 | .0255 | .0095 | .0028 | .0006 | .0001 | .0000 | .0000 |
| | 5 | .3872 | .2607 | .1582 | .0846 | .0386 | .0143 | .0039 | .0007 | .0001 | .0000 |
| | 6 | .6128 | .4731 | .3348 | .2127 | .1178 | .0544 | .0194 | .0046 | .0005 | .0000 |
| | 7 | .8062 | .6956 | .5618 | .4167 | .2763 | .1576 | .0726 | .0239 | .0043 | .0002 |
| | 8 | .9270 | .8655 | .7747 | .6533 | .5075 | .3512 | .2054 | .0922 | .0256 | .0022 |
| | 9 | .9807 | .9579 | .9166 | .8487 | .7472 | .6093 | .4417 | .2642 | .1109 | .0196 |
| | 10 | .9968 | .9917 | .9804 | .9576 | .9150 | .8416 | .7251 | .5565 | .3410 | .1184 |
| | 11 | .9998 | .9992 | .9978 | .9943 | .9862 | .9683 | .9313 | .8578 | .7176 | .4596 |
| | 12 | 1.0000 | 1.0000 | 1.0000 | 1.0000 | 1.0000 | 1.0000 | 1.0000 | 1.0000 | 1.0000 | 1.0000 |
| 13 | 0 | .0001 | .0000 | .0000 | .0000 | .0000 | .0000 | .0000 | .0000 | .0000 | .0000 |
| | 1 | .0017 | .0005 | .0001 | .0000 | .0000 | .0000 | .0000 | .0000 | .0000 | .0000 |
| | 2 | .0112 | .0041 | .0013 | .0003 | .0001 | .0000 | .0000 | .0000 | .0000 | .0000 |
| | 3 | .0461 | .0203 | .0078 | .0025 | .0007 | .0001 | .0000 | .0000 | .0000 | .0000 |
| | 4 | .1334 | .0698 | .0321 | .0126 | .0040 | .0010 | .0002 | .0000 | .0000 | .0000 |
| | 5 | .2905 | .1788 | .0977 | .0462 | .0182 | .0056 | .0012 | .0002 | .0000 | .0000 |
| | 6 | .5000 | .3563 | .2288 | .1295 | .0624 | .0243 | .0070 | .0013 | .0001 | .0000 |
| | 7 | .7095 | .5732 | .4256 | .2841 | .1654 | .0802 | .0300 | .0053 | .0009 | .0000 |
| | 8 | .8666 | .7721 | .6470 | .4995 | .3457 | .2060 | .0991 | .0260 | .0065 | .0003 |
| | 9 | .9539 | .9071 | .8314 | .7217 | .5794 | .4157 | .2527 | .0967 | .0342 | .0031 |
| | 10 | .9888 | .9731 | .9421 | .8868 | .7975 | .6674 | .4983 | .2704 | .1339 | .0245 |
| | 11 | .9983 | .9951 | .9874 | .9704 | .9363 | .8733 | .7664 | .6017 | .3787 | .1354 |
| | 12 | .9999 | .9996 | .9987 | .9963 | .9903 | .9762 | .9450 | .8791 | .7458 | .4867 |
| | 13 | 1.0000 | 1.0000 | 1.0000 | 1.0000 | 1.0000 | 1.0000 | 1.0000 | 1.0000 | 1.0000 | 1.0000 |
| 14 | 0 | .0000 | .0000 | .0000 | .0000 | .0000 | .0000 | .0000 | .0000 | .0000 | .0000 |
| | 1 | .0009 | .0003 | .0001 | .0000 | .0000 | .0000 | .0000 | .0000 | .0000 | .0000 |
| | 2 | .0065 | .0022 | .0006 | .0001 | .0000 | .0000 | .0000 | .0000 | .0000 | .0000 |
| | 3 | .0287 | .0114 | .0039 | .0011 | .0002 | .0000 | .0000 | .0000 | .0000 | .0000 |
| | 4 | .0898 | .0462 | .0175 | .0060 | .0017 | .0003 | .0000 | .0000 | .0000 | .0000 |
| | 5 | .2120 | .1189 | .0583 | .0243 | .0083 | .0022 | .0004 | .0000 | .0000 | .0000 |
| | 6 | .3953 | .2586 | .1501 | .0753 | .0315 | .0103 | .0024 | .0003 | .0000 | .0000 |
| | 7 | .6047 | .4539 | .3075 | .1836 | .0933 | .0383 | .0116 | .0022 | .0002 | .0000 |
| | 8 | .7880 | .6627 | .5141 | .3595 | .2195 | .1117 | .0439 | .0115 | .0015 | .0000 |
| | 9 | .9102 | .8328 | .7207 | .5773 | .4158 | .2585 | .1298 | .0467 | .0092 | .0004 |
| | 10 | .9713 | .9368 | .8757 | .7795 | .6448 | .4787 | .3018 | .1465 | .0441 | .0042 |
| | 11 | .9935 | .9830 | .9602 | .9161 | .8392 | .7189 | .5519 | .3521 | .1584 | .0301 |
| | 12 | .9991 | .9971 | .9919 | .9795 | .9525 | .8990 | .8021 | .6433 | .4154 | .1530 |
| | 13 | .9999 | .9998 | .9992 | .9976 | .9932 | .9822 | .9560 | .8972 | .7712 | .5123 |
| | 14 | 1.0000 | 1.0000 | 1.0000 | 1.0000 | 1.0000 | 1.0000 | 1.0000 | 1.0000 | 1.0000 | 1.0000 |

TABLE E (Continued)

| n | x | .05 | .10 | .15 | .20 | .25 | .30 | .35 | .40 | .45 |
|---|---|-----|-----|-----|-----|-----|-----|-----|-----|-----|
| 15 | 0 | .4633 | .2059 | .0874 | .0352 | .0134 | .0047 | .0016 | .0005 | .0001 |
| | 1 | .8290 | .5490 | .3186 | .1671 | .0802 | .0353 | .0142 | .0052 | .0017 |
| | 2 | .9638 | .8159 | .6042 | .3980 | .2361 | .1268 | .0617 | .0271 | .0107 |
| | 3 | .9945 | .9444 | .8227 | .6482 | .4613 | .2969 | .1727 | .0905 | .0424 |
| | 4 | .9994 | .9873 | .9383 | .8358 | .6865 | .5155 | .3519 | .2173 | .1204 |
| | 5 | .9999 | .9978 | .9832 | .9389 | .8516 | .7216 | .5643 | .4032 | .2608 |
| | 6 | 1.0000 | .9997 | .9964 | .9819 | .9434 | .8689 | .7548 | .6098 | .4522 |
| | 7 | 1.0000 | 1.0000 | .9994 | .9958 | .9827 | .9500 | .8868 | .7869 | .6535 |
| | 8 | 1.0000 | 1.0000 | .9999 | .9992 | .9958 | .9848 | .9578 | .9050 | .8182 |
| | 9 | 1.0000 | 1.0000 | 1.0000 | .9999 | .9992 | .9963 | .9876 | .9662 | .9231 |
| | 10 | 1.0000 | 1.0000 | 1.0000 | 1.0000 | .9999 | .9993 | .9972 | .9907 | .9745 |
| | 11 | 1.0000 | 1.0000 | 1.0000 | 1.0000 | 1.0000 | .9999 | .9995 | .9981 | .9937 |
| | 12 | 1.0000 | 1.0000 | 1.0000 | 1.0000 | 1.0000 | 1.0000 | .9999 | .9997 | .9989 |
| | 13 | 1.0000 | 1.0000 | 1.0000 | 1.0000 | 1.0000 | 1.0000 | 1.0000 | 1.0000 | .9999 |
| | 14 | 1.0000 | 1.0000 | 1.0000 | 1.0000 | 1.0000 | 1.0000 | 1.0000 | 1.0000 | 1.0000 |
| | 15 | 1.0000 | 1.0000 | 1.0000 | 1.0000 | 1.0000 | 1.0000 | 1.0000 | 1.0000 | 1.0000 |
| 16 | 0 | .4401 | .1853 | .0743 | .0281 | .0100 | .0033 | .0010 | .0003 | .0001 |
| | 1 | .8108 | .5147 | .2839 | .1407 | .0635 | .0261 | .0098 | .0033 | .0010 |
| | 2 | .9571 | .7892 | .5614 | .3518 | .1971 | .0994 | .0451 | .0183 | .0066 |
| | 3 | .9930 | .9316 | .7899 | .5981 | .4050 | .2459 | .1339 | .0651 | .0281 |
| | 4 | .9991 | .9830 | .9209 | .7982 | .6302 | .4499 | .2892 | .1666 | .0853 |
| | 5 | .9999 | .9967 | .9765 | .9183 | .8103 | .6598 | .4900 | .3288 | .1976 |
| | 6 | 1.0000 | .9995 | .9944 | .9733 | .9204 | .8247 | .6881 | .5272 | .3660 |
| | 7 | 1.0000 | .9999 | .9989 | .9930 | .9729 | .9256 | .8406 | .7161 | .5629 |
| | 8 | 1.0000 | 1.0000 | .9998 | .9985 | .9925 | .9743 | .9329 | .8577 | .7441 |
| | 9 | 1.0000 | 1.0000 | 1.0000 | .9998 | .9984 | .9929 | .9771 | .9417 | .8759 |
| | 10 | 1.0000 | 1.0000 | 1.0000 | 1.0000 | .9997 | .9984 | .9938 | .9809 | .9514 |
| | 11 | 1.0000 | 1.0000 | 1.0000 | 1.0000 | 1.0000 | .9997 | .9987 | .9951 | .9851 |
| | 12 | 1.0000 | 1.0000 | 1.0000 | 1.0000 | 1.0000 | 1.0000 | .9998 | .9991 | .9965 |
| | 13 | 1.0000 | 1.0000 | 1.0000 | 1.0000 | 1.0000 | 1.0000 | 1.0000 | .9999 | .9994 |
| | 14 | 1.0000 | 1.0000 | 1.0000 | 1.0000 | 1.0000 | 1.0000 | 1.0000 | 1.0000 | .9999 |
| | 15 | 1.0000 | 1.0000 | 1.0000 | 1.0000 | 1.0000 | 1.0000 | 1.0000 | 1.0000 | 1.0000 |
| | 16 | 1.0000 | 1.0000 | 1.0000 | 1.0000 | 1.0000 | 1.0000 | 1.0000 | 1.0000 | 1.0000 |

**TABLE E (Continued)**

| n | x | .50 | .55 | .60 | .65 | θ .70 | .75 | .80 | .85 | .90 | .95 |
|---|---|-----|-----|-----|-----|-----|-----|-----|-----|-----|-----|
| 15 | 0 | .0000 | .0000 | .0000 | .0000 | .0000 | .0000 | .0000 | .0000 | .0000 | .0000 |
| | 1 | .0005 | .0001 | .0000 | .0000 | .0000 | .0000 | .0000 | .0000 | .0000 | .0000 |
| | 2 | .0037 | .0011 | .0003 | .0001 | .0000 | .0000 | .0000 | .0000 | .0000 | .0000 |
| | 3 | .0176 | .0063 | .0019 | .0005 | .0001 | .0000 | .0000 | .0000 | .0000 | .0000 |
| | 4 | .0592 | .0255 | .0093 | .0028 | .0007 | .0001 | .0000 | .0000 | .0000 | .0000 |
| | 5 | .1509 | .0769 | .0338 | .0124 | .0037 | .0008 | .0001 | .0000 | .0000 | .0000 |
| | 6 | .3036 | .1818 | .0950 | .0422 | .0152 | .0042 | .0008 | .0001 | .0000 | .0000 |
| | 7 | .5000 | .3465 | .2131 | .1132 | .0500 | .0173 | .0042 | .0006 | .0000 | .0000 |
| | 8 | .6964 | .5478 | .3902 | .2452 | .1311 | .0566 | .0181 | .0036 | .0003 | .0000 |
| | 9 | .8491 | .7392 | .5968 | .4357 | .2784 | .1484 | .0611 | .0168 | .0022 | .0001 |
| | 10 | .9408 | .8796 | .7827 | .6481 | .4845 | .3135 | .1642 | .0617 | .0127 | .0006 |
| | 11 | .9824 | .9576 | .9095 | .8273 | .7031 | .5387 | .3518 | .1773 | .0556 | .0055 |
| | 12 | .9963 | .9893 | .9729 | .9383 | .8732 | .7639 | .6020 | .3958 | .1841 | .0362 |
| | 13 | .9995 | .9983 | .9948 | .9858 | .9647 | .9198 | .8329 | .6814 | .4510 | .1710 |
| | 14 | 1.0000 | .9999 | .9995 | .9984 | .9953 | .9866 | .9648 | .9126 | .7941 | .5367 |
| | 15 | 1.0000 | 1.0000 | 1.0000 | 1.0000 | 1.0000 | 1.0000 | 1.0000 | 1.0000 | 1.0000 | 1.0000 |
| 16 | 0 | .0000 | .0000 | .0000 | .0000 | .0000 | .0000 | .0000 | .0000 | .0000 | .0000 |
| | 1 | .0003 | .0001 | .0000 | .0000 | .0000 | .0000 | .0000 | .0000 | .0000 | .0000 |
| | 2 | .0021 | .0006 | .0001 | .0000 | .0000 | .0000 | .0000 | .0000 | .0000 | .0000 |
| | 3 | .0106 | .0035 | .0009 | .0002 | .0000 | .0000 | .0000 | .0000 | .0000 | .0000 |
| | 4 | .0384 | .0149 | .0049 | .0013 | .0003 | .0000 | .0000 | .0000 | .0000 | .0000 |
| | 5 | .1051 | .0486 | .0191 | .0062 | .0016 | .0003 | .0000 | .0000 | .0000 | .0000 |
| | 6 | .2272 | .1241 | .0583 | .0229 | .0071 | .0016 | .0002 | .0000 | .0000 | .0000 |
| | 7 | .4018 | .2559 | .1423 | .0671 | .0257 | .0075 | .0015 | .0002 | .0000 | .0000 |
| | 8 | .5982 | .4371 | .2839 | .1594 | .0744 | .0271 | .0070 | .0011 | .0001 | .0000 |
| | 9 | .7228 | .6340 | .4728 | .3119 | .1753 | .0796 | .0267 | .0056 | .0005 | .0000 |
| | 10 | .8949 | .8024 | .6712 | .5100 | .3402 | .1897 | .0817 | .0235 | .0033 | .0001 |
| | 11 | .9616 | .9147 | .8334 | .7108 | .5501 | .3698 | .2018 | .0791 | .0170 | .0009 |
| | 12 | .9894 | .9719 | .9349 | .8661 | .7541 | .5950 | .4019 | .2101 | .0684 | .0070 |
| | 13 | .9979 | .9934 | .9817 | .9549 | .9006 | .8729 | .6482 | .4386 | .2108 | .0429 |
| | 14 | .9997 | .9990 | .9967 | .9902 | .9739 | .9365 | .8593 | .7161 | .4853 | .1892 |
| | 15 | 1.0000 | .9999 | .9997 | .9990 | .9967 | .9900 | .9719 | .9257 | .8147 | .5599 |
| | 16 | 1.0000 | 1.0000 | 1.0000 | 1.0000 | 1.0000 | 1.0000 | 1.0000 | 1.0000 | 1.0000 | 1.0000 |

TABLE E (Continued)

| n | x | .05 | .10 | .15 | .20 | θ .25 | .30 | .35 | .40 | .45 |
|---|---|---|---|---|---|---|---|---|---|---|
| 17 | 0 | .4181 | .1668 | .0631 | .0225 | .0075 | .0023 | .0007 | .0002 | .0000 |
| | 1 | .7922 | .4818 | .2525 | .1182 | .0501 | .0193 | .0067 | .0021 | .0006 |
| | 2 | .9497 | .7618 | .5198 | .3096 | .1637 | .0774 | .0327 | .0123 | .0041 |
| | 3 | .9912 | .9174 | .7556 | .5489 | .3530 | .2019 | .1028 | .0464 | .0184 |
| | 4 | .9988 | .9779 | .9013 | .7582 | .5739 | .3887 | .2348 | .1260 | .0596 |
| | 5 | .9999 | .9953 | .9681 | .8943 | .7653 | .5968 | .4197 | .2639 | .1471 |
| | 6 | 1.0000 | .9992 | .9917 | .9623 | .8929 | .7752 | .6188 | .4478 | .2902 |
| | 7 | 1.0000 | .9999 | .9983 | .9891 | .9598 | .8954 | .7872 | .6405 | .4743 |
| | 8 | 1.0000 | 1.0000 | .9997 | .9974 | .9876 | .9597 | .9006 | .8011 | .6626 |
| | 9 | 1.0000 | 1.0000 | 1.0000 | .9995 | .9969 | .9873 | .9617 | .9081 | .8166 |
| | 10 | 1.0000 | 1.0000 | 1.0000 | .9999 | .9994 | .9968 | .9880 | .9652 | .9174 |
| | 11 | 1.0000 | 1.0000 | 1.0000 | 1.0000 | .9999 | .9993 | .9970 | .9894 | .9699 |
| | 12 | 1.0000 | 1.0000 | 1.0000 | 1.0000 | 1.0000 | .9999 | .9994 | .9975 | .9914 |
| | 13 | 1.0000 | 1.0000 | 1.0000 | 1.0000 | 1.0000 | 1.0000 | .9999 | .9995 | .9981 |
| | 14 | 1.0000 | 1.0000 | 1.0000 | 1.0000 | 1.0000 | 1.0000 | 1.0000 | .9999 | .9997 |
| | 15 | 1.0000 | 1.0000 | 1.0000 | 1.0000 | 1.0000 | 1.0000 | 1.0000 | 1.0000 | 1.0000 |
| | 16 | 1.0000 | 1.0000 | 1.0000 | 1.0000 | 1.0000 | 1.0000 | 1.0000 | 1.0000 | 1.0000 |
| | 17 | 1.0000 | 1.0000 | 1.0000 | 1.0000 | 1.0000 | 1.0000 | 1.0000 | 1.0000 | 1.0000 |
| 18 | 0 | .3972 | .1501 | .0536 | .0180 | .0056 | .0016 | .0004 | .0001 | .0000 |
| | 1 | .7735 | .4503 | .2241 | .0991 | .0395 | .0142 | .0046 | .0013 | .0003 |
| | 2 | .9419 | .7338 | .4797 | .2713 | .1353 | .0600 | .0236 | .0082 | .0025 |
| | 3 | .9891 | .9018 | .7202 | .5010 | .3057 | .1646 | .0783 | .0328 | .0120 |
| | 4 | .9985 | .9718 | .8794 | .7164 | .5187 | .3327 | .1886 | .0942 | .0411 |
| | 5 | .9998 | .9936 | .9581 | .8671 | .7175 | .5344 | .3550 | .2088 | .1077 |
| | 6 | 1.0000 | .9988 | .9882 | .9487 | .8610 | .7217 | .5491 | .3743 | .2258 |
| | 7 | 1.0000 | .9998 | .9973 | .9837 | .9431 | .8593 | .7283 | .5634 | .3915 |
| | 8 | 1.0000 | 1.0000 | .9995 | .9957 | .9807 | .9404 | .8609 | .7368 | .5778 |
| | 9 | 1.0000 | 1.0000 | .9999 | .9991 | .9946 | .9790 | .9403 | .8653 | .7473 |
| | 10 | 1.0000 | 1.0000 | 1.0000 | .9998 | .9988 | .9939 | .9788 | .9424 | .8720 |
| | 11 | 1.0000 | 1.0000 | 1.0000 | 1.0000 | .9998 | .9986 | .9938 | .9797 | .9463 |
| | 12 | 1.0000 | 1.0000 | 1.0000 | 1.0000 | 1.0000 | .9997 | .9986 | .9942 | .9817 |
| | 13 | 1.0000 | 1.0000 | 1.0000 | 1.0000 | 1.0000 | 1.0000 | .9997 | .9987 | .9951 |
| | 14 | 1.0000 | 1.0000 | 1.0000 | 1.0000 | 1.0000 | 1.0000 | 1.0000 | .9998 | .9990 |
| | 15 | 1.0000 | 1.0000 | 1.0000 | 1.0000 | 1.0000 | 1.0000 | 1.0000 | 1.0000 | .9999 |
| | 16 | 1.0000 | 1.0000 | 1.0000 | 1.0000 | 1.0000 | 1.0000 | 1.0000 | 1.0000 | 1.0000 |
| | 17 | 1.0000 | 1.0000 | 1.0000 | 1.0000 | 1.0000 | 1.0000 | 1.0000 | 1.0000 | 1.0000 |
| | 18 | 1.0000 | 1.0000 | 1.0000 | 1.0000 | 1.0000 | 1.0000 | 1.0000 | 1.0000 | 1.0000 |

## TABLE E (Continued)

| n | x | .50 | .55 | .60 | .65 | θ .70 | .75 | .80 | .85 | .90 | .95 |
|---|---|-----|-----|-----|-----|-----|-----|-----|-----|-----|-----|
| 17 | 0 | .0000 | .0000 | .0000 | .0000 | .0000 | .0000 | .0000 | .0000 | .0000 | .0000 |
|  | 1 | .0001 | .0000 | .0000 | .0000 | .0000 | .0000 | .0000 | .0000 | .0000 | .0000 |
|  | 2 | .0012 | .0003 | .0001 | .0000 | .0000 | .0000 | .0000 | .0000 | .0000 | .0000 |
|  | 3 | .0064 | .0019 | .0005 | .0001 | .0000 | .0000 | .0000 | .0000 | .0000 | .0000 |
|  | 4 | .0245 | .0086 | .0025 | .0006 | .0001 | .0000 | .0000 | .0000 | .0000 | .0000 |
|  | 5 | .0717 | .0301 | .0106 | .0030 | .0007 | .0001 | .0000 | .0000 | .0000 | .0000 |
|  | 6 | .1662 | .0826 | .0348 | .0120 | .0032 | .0006 | .0001 | .0000 | .0000 | .0000 |
|  | 7 | .3145 | .1834 | .0919 | .0383 | .0127 | .0031 | .0005 | .0000 | .0000 | .0000 |
|  | 8 | .5000 | .3374 | .1989 | .0994 | .0403 | .0124 | .0026 | .0003 | .0000 | .0000 |
|  | 9 | .6855 | .5257 | .3595 | .2128 | .1046 | .0402 | .0109 | .0017 | .0001 | .0000 |
|  | 10 | .8338 | .7098 | .5522 | .3812 | .2248 | .1071 | .0377 | .0083 | .0008 | .0000 |
|  | 11 | .9283 | .8529 | .7361 | .5803 | .4032 | .2347 | .1057 | .0319 | .0047 | .0001 |
|  | 12 | .9755 | .9404 | .8740 | .7652 | .6113 | .4261 | .2418 | .0987 | .0221 | .0012 |
|  | 13 | .9936 | .9816 | .9536 | .8972 | .7981 | .6470 | .4511 | .2444 | .0826 | .0088 |
|  | 14 | .9988 | .9959 | .9877 | .9673 | .9226 | .8363 | .6904 | .4802 | .2382 | .0503 |
|  | 15 | .9999 | .9994 | .9979 | .9933 | .9807 | .9499 | .8818 | .7475 | .5182 | .2078 |
|  | 16 | 1.0000 | 1.0000 | .9998 | .9993 | .9977 | .9925 | .9775 | .9369 | .8332 | .5819 |
|  | 17 | 1.0000 | 1.0000 | 1.0000 | 1.0000 | 1.0000 | 1.0000 | 1.0000 | 1.0000 | 1.0000 | 1.0000 |
| 18 | 0 | .0000 | .0000 | .0000 | .0000 | .0000 | .0000 | .0000 | .0000 | .0000 | .0000 |
|  | 1 | .0001 | .0000 | .0000 | .0000 | .0000 | .0000 | .0000 | .0000 | .0000 | .0000 |
|  | 2 | .0007 | .0001 | .0000 | .0000 | .0000 | .0000 | .0000 | .0000 | .0000 | .0000 |
|  | 3 | .0038 | .0010 | .0002 | .0000 | .0000 | .0000 | .0000 | .0000 | .0000 | .0000 |
|  | 4 | .0154 | .0049 | .0013 | .0003 | .0000 | .0000 | .0000 | .0000 | .0000 | .0000 |
|  | 5 | .0481 | .0183 | .0058 | .0014 | .0003 | .0000 | .0000 | .0000 | .0000 | .0000 |
|  | 6 | .1189 | .0537 | .0203 | .0062 | .0014 | .0002 | .0000 | .0000 | .0000 | .0000 |
|  | 7 | .2403 | .1280 | .0576 | .0212 | .0061 | .0012 | .0002 | .0000 | .0000 | .0000 |
|  | 8 | .4073 | .2527 | .1347 | .0597 | .0210 | .0054 | .0009 | .0001 | .0000 | .0000 |
|  | 9 | .5927 | .4222 | .2632 | .1391 | .0596 | .0193 | .0043 | .0005 | .0000 | .0000 |
|  | 10 | .7597 | .6085 | .4366 | .2717 | .1407 | .0569 | .0163 | .0027 | .0002 | .0000 |
|  | 11 | .8811 | .7742 | .6257 | .4509 | .2783 | .1390 | .0513 | .0118 | .0012 | .0000 |
|  | 12 | .9519 | .8923 | .7912 | .6450 | .4656 | .2825 | .1329 | .0419 | .0064 | .0002 |
|  | 13 | .9846 | .9589 | .9058 | .8114 | .6673 | .4813 | .2836 | .1206 | .0282 | .0015 |
|  | 14 | .9962 | .9880 | .9672 | .9217 | .8354 | .6943 | .4990 | .2798 | .0982 | .0109 |
|  | 15 | .9993 | .9975 | .9918 | .9764 | .9400 | .8647 | .7287 | .5203 | .2662 | .0581 |
|  | 16 | .9999 | .9997 | .9987 | .9954 | .9858 | .9605 | .9009 | .7759 | .5497 | .2265 |
|  | 17 | 1.0000 | 1.0000 | .9999 | .9996 | .9984 | .9944 | .9820 | .9464 | .8499 | .6028 |
|  | 18 | 1.0000 | 1.0000 | 1.0000 | 1.0000 | 1.0000 | 1.0000 | 1.0000 | 1.0000 | 1.0000 | 1.0000 |

**TABLE E (Continued)**

| n | x | .05 | .10 | .15 | .20 | θ .25 | .30 | .35 | .40 | .45 |
|---|---|-----|-----|-----|-----|-----|-----|-----|-----|-----|
| 19 | 0 | .3774 | .1351 | .0456 | .0144 | .0042 | .0011 | .0003 | .0001 | .0000 |
| | 1 | .7547 | .4203 | .1985 | .0829 | .0310 | .0104 | .0031 | .0008 | .0002 |
| | 2 | .9335 | .7054 | .4413 | .2369 | .1113 | .0462 | .0170 | .0055 | .0015 |
| | 3 | .9868 | .8850 | .6841 | .4551 | .2631 | .1332 | .0591 | .0230 | .0077 |
| | 4 | .9980 | .9648 | .8556 | .6733 | .4654 | .2822 | .1500 | .0696 | .0280 |
| | 5 | .9998 | .9914 | .9463 | .8369 | .6678 | .4739 | .2968 | .1629 | .0777 |
| | 6 | 1.0000 | .9983 | .9837 | .9324 | .8251 | .6655 | .4812 | .3081 | .1727 |
| | 7 | 1.0000 | .9997 | .9959 | .9767 | .9225 | .8180 | .6656 | .4878 | .3169 |
| | 8 | 1.0000 | 1.0000 | .9992 | .9933 | .9713 | .9161 | .8145 | .6675 | .4940 |
| | 9 | 1.0000 | 1.0000 | .9999 | .9984 | .9911 | .9674 | .9125 | .8139 | .6710 |
| | 10 | 1.0000 | 1.0000 | 1.0000 | .9997 | .9977 | .9895 | .9653 | .9115 | .8159 |
| | 11 | 1.0000 | 1.0000 | 1.0000 | 1.0000 | .9995 | .9972 | .9886 | .9648 | .9129 |
| | 12 | 1.0000 | 1.0000 | 1.0000 | 1.0000 | .9999 | .9994 | .9969 | .9884 | .9658 |
| | 13 | 1.0000 | 1.0000 | 1.0000 | 1.0000 | 1.0000 | .9999 | .9993 | .9969 | .9891 |
| | 14 | 1.0000 | 1.0000 | 1.0000 | 1.0000 | 1.0000 | 1.0000 | .9999 | .9994 | .9972 |
| | 15 | 1.0000 | 1.0000 | 1.0000 | 1.0000 | 1.0000 | 1.0000 | 1.0000 | .9999 | .9995 |
| | 16 | 1.0000 | 1.0000 | 1.0000 | 1.0000 | 1.0000 | 1.0000 | 1.0000 | 1.0000 | .9999 |
| | 17 | 1.0000 | 1.0000 | 1.0000 | 1.0000 | 1.0000 | 1.0000 | 1.0000 | 1.0000 | 1.0000 |
| | 18 | 1.0000 | 1.0000 | 1.0000 | 1.0000 | 1.0000 | 1.0000 | 1.0000 | 1.0000 | 1.0000 |
| | 19 | 1.0000 | 1.0000 | 1.0000 | 1.0000 | 1.0000 | 1.0000 | 1.0000 | 1.0000 | 1.0000 |
| 20 | 0 | .3585 | .1216 | .0388 | .0115 | .0032 | .0008 | .0002 | .0000 | .0000 |
| | 1 | .7358 | .3917 | .1756 | .0692 | .0243 | .0076 | .0021 | .0005 | .0001 |
| | 2 | .9245 | .6769 | .4049 | .2061 | .0913 | .0355 | .0121 | .0036 | .0009 |
| | 3 | .9841 | .8670 | .6477 | .4114 | .2252 | .1071 | .0444 | .0160 | .0049 |
| | 4 | .9974 | .9568 | .8298 | .6296 | .4148 | .2375 | .1182 | .0510 | .0189 |
| | 5 | .9997 | .9887 | .9327 | .8042 | .6172 | .4164 | .2454 | .1256 | .0553 |
| | 6 | 1.0000 | .9976 | .9781 | .9133 | .7858 | .6080 | .4166 | .2500 | .1299 |
| | 7 | 1.0000 | .9996 | .9941 | .9679 | .8982 | .7723 | .6010 | .4159 | .2520 |
| | 8 | 1.0000 | .9999 | .9987 | .9900 | .9591 | .8867 | .7624 | .5956 | .4143 |
| | 9 | 1.0000 | 1.0000 | .9998 | .9974 | .9861 | .9520 | .8782 | .7553 | .5914 |
| | 10 | 1.0000 | 1.0000 | 1.0000 | .9994 | .9961 | .9829 | .9468 | .8725 | .7507 |
| | 11 | 1.0000 | 1.0000 | 1.0000 | .9999 | .9991 | .9949 | .9804 | .9435 | .8692 |
| | 12 | 1.0000 | 1.0000 | 1.0000 | 1.0000 | .9998 | .9987 | .9940 | .9790 | .9420 |
| | 13 | 1.0000 | 1.0000 | 1.0000 | 1.0000 | 1.0000 | .9997 | .9985 | .9935 | .9786 |
| | 14 | 1.0000 | 1.0000 | 1.0000 | 1.0000 | 1.0000 | 1.0000 | .9997 | .9984 | .9936 |
| | 15 | 1.0000 | 1.0000 | 1.0000 | 1.0000 | 1.0000 | 1.0000 | 1.0000 | .9997 | .9985 |
| | 16 | 1.0000 | 1.0000 | 1.0000 | 1.0000 | 1.0000 | 1.0000 | 1.0000 | 1.0000 | .9997 |
| | 17 | 1.0000 | 1.0000 | 1.0000 | 1.0000 | 1.0000 | 1.0000 | 1.0000 | 1.0000 | 1.0000 |
| | 18 | 1.0000 | 1.0000 | 1.0000 | 1.0000 | 1.0000 | 1.0000 | 1.0000 | 1.0000 | 1.0000 |
| | 19 | 1.0000 | 1.0000 | 1.0000 | 1.0000 | 1.0000 | 1.0000 | 1.0000 | 1.0000 | 1.0000 |
| | 20 | 1.0000 | 1.0000 | 1.0000 | 1.0000 | 1.0000 | 1.0000 | 1.0000 | 1.0000 | 1.0000 |

TABLE E (Continued)

| n | x | .50 | .55 | .60 | .65 | θ .70 | .75 | .80 | .85 | .90 | .95 |
|---|---|-----|-----|-----|-----|-------|-----|-----|-----|-----|-----|
| 19 | 0 | .0000 | .0000 | .0000 | .0000 | .0000 | .0000 | .0000 | .0000 | .0000 | .0000 |
| | 1 | .0000 | .0000 | .0000 | .0000 | .0000 | .0000 | .0000 | .0000 | .0000 | .0000 |
| | 2 | .0004 | .0001 | .0000 | .0000 | .0000 | .0000 | .0000 | .0000 | .0000 | .0000 |
| | 3 | .0022 | .0005 | .0001 | .0000 | .0000 | .0000 | .0000 | .0000 | .0000 | .0000 |
| | 4 | .0096 | .0028 | .0006 | .0001 | .0000 | .0000 | .0000 | .0000 | .0000 | .0000 |
| | 5 | .0318 | .0109 | .0031 | .0007 | .0001 | .0000 | .0000 | .0000 | .0000 | .0000 |
| | 6 | .0835 | .0342 | .0116 | .0031 | .0006 | .0001 | .0000 | .0000 | .0000 | .0000 |
| | 7 | .1796 | .0871 | .0352 | .0114 | .0028 | .0005 | .0000 | .0000 | .0000 | .0000 |
| | 8 | .3238 | .1841 | .0885 | .0347 | .0105 | .0023 | .0003 | .0000 | .0000 | .0000 |
| | 9 | .5000 | .3290 | .1861 | .0875 | .0326 | .0089 | .0016 | .0001 | .0000 | .0000 |
| | 10 | .6762 | .5060 | .3325 | .1855 | .0839 | .0287 | .0067 | .0008 | .0000 | .0000 |
| | 11 | .8204 | .6831 | .5122 | .3344 | .1820 | .0775 | .0233 | .0041 | .0003 | .0000 |
| | 12 | .9165 | .8273 | .6919 | .5188 | .3345 | .1749 | .0676 | .0163 | .0017 | .0000 |
| | 13 | .9682 | .9223 | .8371 | .7032 | .5261 | .3322 | .1631 | .0537 | .0086 | .0002 |
| | 14 | .9904 | .9720 | .9304 | .8500 | .7178 | .5346 | .3267 | .1444 | .0352 | .0020 |
| | 15 | .9978 | .9923 | .9770 | .9409 | .8668 | .7369 | .5449 | .3159 | .1150 | .0132 |
| | 16 | .9996 | .9985 | .9945 | .9830 | .9538 | .8887 | .7631 | .5587 | .2946 | .0665 |
| | 17 | 1.0000 | .9998 | .9992 | .9969 | .9896 | .9690 | .9171 | .8015 | .5797 | .2453 |
| | 18 | 1.0000 | 1.0000 | .9999 | .9997 | .9989 | .9958 | .9856 | .9544 | .8649 | .6226 |
| | 19 | 1.0000 | 1.0000 | 1.0000 | 1.0000 | 1.0000 | 1.0000 | 1.0000 | 1.0000 | 1.0000 | 1.0000 |
| 20 | 0 | .0000 | .0000 | .0000 | .0000 | .0000 | .0000 | .0000 | .0000 | .0000 | .0000 |
| | 1 | .0000 | .0000 | .0000 | .0000 | .0000 | .0000 | .0000 | .0000 | .0000 | .0000 |
| | 2 | .0002 | .0000 | .0000 | .0000 | .0000 | .0000 | .0000 | .0000 | .0000 | .0000 |
| | 3 | .0013 | .0003 | .0000 | .0000 | .0000 | .0000 | .0000 | .0000 | .0000 | .0000 |
| | 4 | .0059 | .0015 | .0003 | .0000 | .0000 | .0000 | .0000 | .0000 | .0000 | .0000 |
| | 5 | .0207 | .0064 | .0016 | .0003 | .0000 | .0000 | .0000 | .0000 | .0000 | .0000 |
| | 6 | .0577 | .0214 | .0065 | .0015 | .0003 | .0000 | .0000 | .0000 | .0000 | .0000 |
| | 7 | .1316 | .0580 | .0210 | .0060 | .0013 | .0002 | .0000 | .0000 | .0000 | .0000 |
| | 8 | .2517 | .1308 | .0565 | .0196 | .0051 | .0009 | .0001 | .0000 | .0000 | .0000 |
| | 9 | .4119 | .2493 | .1275 | .0532 | .0171 | .0039 | .0006 | .0000 | .0000 | .0000 |
| | 10 | .5881 | .4086 | .2447 | .1218 | .0480 | .0139 | .0026 | .0002 | .0000 | .0000 |
| | 11 | .7483 | .5857 | .4044 | .2376 | .1133 | .0409 | .0100 | .0013 | .0001 | .0000 |
| | 12 | .8684 | .7480 | .5841 | .3990 | .2277 | .1018 | .0321 | .0059 | .0004 | .0000 |
| | 13 | .9423 | .8701 | .7500 | .5834 | .3920 | .2142 | .0867 | .0219 | .0024 | .0000 |
| | 14 | .9793 | .9447 | .8744 | .7546 | .5836 | .3828 | .1958 | .0673 | .0113 | .0003 |
| | 15 | .9941 | .9811 | .9490 | .8818 | .7625 | .5852 | .3704 | .1702 | .0432 | .0026 |
| | 16 | .9987 | .9951 | .9840 | .9556 | .8929 | .7748 | .5886 | .3523 | .1330 | .0159 |
| | 17 | .9998 | .9991 | .9964 | .9879 | .9645 | .9087 | .7939 | .5951 | .3231 | .0755 |
| | 18 | 1.0000 | .9999 | .9995 | .9979 | .9924 | .9757 | .9308 | .8244 | .6083 | .2642 |
| | 19 | 1.0000 | 1.0000 | 1.0000 | .9998 | .9992 | .9968 | .9885 | .9612 | .8784 | .6415 |
| | 20 | 1.0000 | 1.0000 | 1.0000 | 1.0000 | 1.0000 | 1.0000 | 1.0000 | 1.0000 | 1.0000 | 1.0000 |

Source: Adapted from Table 2 of *Tables of the Binomial Distribution*, (January 1950 with *Corrigenda* 1952 and 1958), National Bureau of Standards, U.S. Government Printing Office, Washington, D.C., with permission.

## TABLE F
## CUMULATIVE BINOMIAL DISTRIBUTION FOR $\theta = .5$

Entries labeled $P$ in the table are the cumulative probability from each extreme to the value of $S$, for the given $n$ when $\theta = .5$. Left-tail probabilities are given for $S \leq .5n$, and right-tail for $S \geq .5n$. $S$ is interpreted as either $S_+$ or $S_-$.

| n | Left S | P | Right S | n | Left S | P | Right S | n | Left S | P | Right S |
|---|---|---|---|---|---|---|---|---|---|---|---|
| 1 | 0 | .5000 | 1 | 12 | 0 | .0002 | 12 | 17 | 0 | .0000 | 17 |
| 2 | 0 | .2500 | 2 |  | 1 | .0032 | 11 |  | 1 | .0001 | 16 |
|  | 1 | .7500 | 1 |  | 2 | .0193 | 10 |  | 2 | .0012 | 15 |
| 3 | 0 | .1250 | 3 |  | 3 | .0730 | 9 |  | 3 | .0064 | 14 |
|  | 1 | .5000 | 2 |  | 4 | .1938 | 8 |  | 4 | .0245 | 13 |
| 4 | 0 | .0625 | 4 |  | 5 | .3872 | 7 |  | 5 | .0717 | 12 |
|  | 1 | .3125 | 3 |  | 6 | .6128 | 6 |  | 6 | .1662 | 11 |
|  | 2 | .6875 | 2 | 13 | 0 | .0001 | 13 |  | 7 | .3145 | 10 |
| 5 | 0 | .0312 | 5 |  | 1 | .0017 | 12 |  | 8 | .5000 | 9 |
|  | 1 | .1875 | 4 |  | 2 | .0112 | 11 | 18 | 0 | .0000 | 18 |
|  | 2 | .5000 | 3 |  | 3 | .0461 | 10 |  | 1 | .0001 | 17 |
| 6 | 0 | .0156 | 6 |  | 4 | .1334 | 9 |  | 2 | .0007 | 16 |
|  | 1 | .1094 | 5 |  | 5 | .2905 | 8 |  | 3 | .0038 | 15 |
|  | 2 | .3438 | 4 |  | 6 | .5000 | 7 |  | 4 | .0154 | 14 |
|  | 3 | .6562 | 3 | 14 | 0 | .0000 | 14 |  | 5 | .0481 | 13 |
| 7 | 0 | .0078 | 7 |  | 1 | .0009 | 13 |  | 6 | .1189 | 12 |
|  | 1 | .0625 | 6 |  | 2 | .0065 | 12 |  | 7 | .2403 | 11 |
|  | 2 | .2266 | 5 |  | 3 | .0287 | 11 |  | 8 | .4073 | 10 |
|  | 3 | .5000 | 4 |  | 4 | .0898 | 10 |  | 9 | .5927 | 9 |
| 8 | 0 | .0039 | 8 |  | 5 | .2120 | 9 | 19 | 0 | .0000 | 19 |
|  | 1 | .0352 | 7 |  | 6 | .3953 | 8 |  | 1 | .0000 | 18 |
|  | 2 | .1445 | 6 |  | 7 | .6047 | 7 |  | 2 | .0004 | 17 |
|  | 3 | .3633 | 5 | 15 | 0 | .0000 | 15 |  | 3 | .0022 | 16 |
|  | 4 | .6367 | 4 |  | 1 | .0005 | 14 |  | 4 | .0096 | 15 |
| 9 | 0 | .0020 | 9 |  | 2 | .0037 | 13 |  | 5 | .0318 | 14 |
|  | 1 | .0195 | 8 |  | 3 | .0176 | 12 |  | 6 | .0835 | 13 |
|  | 2 | .0898 | 7 |  | 4 | .0592 | 11 |  | 7 | .1796 | 12 |
|  | 3 | .2539 | 6 |  | 5 | .1509 | 10 |  | 8 | .3238 | 11 |
|  | 4 | .5000 | 5 |  | 6 | .3036 | 9 |  | 9 | .5000 | 10 |
| 10 | 0 | .0010 | 10 |  | 7 | .5000 | 8 | 20 | 0 | .0000 | 20 |
|  | 1 | .0107 | 9 | 16 | 0 | .0000 | 16 |  | 1 | .0000 | 19 |
|  | 2 | .0547 | 8 |  | 1 | .0003 | 15 |  | 2 | .0002 | 18 |
|  | 3 | .1719 | 7 |  | 2 | .0021 | 14 |  | 3 | .0013 | 17 |
|  | 4 | .3770 | 6 |  | 3 | .0106 | 13 |  | 4 | .0059 | 16 |
|  | 5 | .6230 | 5 |  | 4 | .0384 | 12 |  | 5 | .0207 | 15 |
| 11 | 0 | .0005 | 11 |  | 5 | .1051 | 11 |  | 6 | .0577 | 14 |
|  | 1 | .0059 | 10 |  | 6 | .2272 | 10 |  | 7 | .1316 | 13 |
|  | 2 | .0327 | 9 |  | 7 | .4018 | 9 |  | 8 | .2517 | 12 |
|  | 3 | .1133 | 8 |  | 8 | .5982 | 8 |  | 9 | .4119 | 11 |
|  | 4 | .2744 | 7 |  |  |  |  |  | 10 | .5881 | 10 |
|  | 5 | .5000 | 6 |  |  |  |  |  |  |  |  |

## TABLE F  (Continued)

For $n > 20$, the probabilities are found from Table A as follows:

$$z_{+,R} = \frac{S_+ - 0.5 - .5n}{.5\sqrt{n}} \qquad z_{-,R} = \frac{S_- - 0.5 - .5n}{.5\sqrt{n}}$$

| Desired | Approximated by |
|---------|-----------------|
| Right-tail probability for $S_+$ | Right-tail probability for $z_{+,R}$ |
| Right-tail probability for $S_-$ | Right-tail probability for $z_{-,R}$ |

Source: Adapted from Table 2 of *Tables of the Binomial Distribution*, (January 1950 with *Corrigenda* 1952 and 1958), National Bureau of Standards, U.S. Government Printing Office, Washington, D.C., with permission.

## TABLE G
## CUMULATIVE PROBABILITIES FOR WILCOXON SIGNED RANK STATISTIC

Entries labeled $P$ in the table are the cumulative probability from each extreme to the value of the signed rank statistic for the given $n$. Left-tail probabilities are given for $T \leq n(n+1)/4$, and right-tail for $T \geq n(n+1)/4$. $T$ is interpreted as either $T_+$ or $T_-$.

| | Left | | Right | | Left | | Right | | Left | | Right |
|---|---|---|---|---|---|---|---|---|---|---|---|
| n | T | P | T | n | T | P | T | n | T | P | T |
| 2 | 0 | .250 | 3 | 7 | 0 | .008 | 28 | 9 | 0 | .002 | 45 |
| | 1 | .500 | 2 | | 1 | .016 | 27 | | 1 | .004 | 44 |
| 3 | 0 | .125 | 6 | | 2 | .023 | 26 | | 2 | .006 | 43 |
| | 1 | .250 | 5 | | 3 | .039 | 25 | | 3 | .010 | 42 |
| | 2 | .375 | 4 | | 4 | .055 | 24 | | 4 | .014 | 41 |
| | 3 | .625 | 3 | | 5 | .078 | 23 | | 5 | .020 | 40 |
| 4 | 0 | .062 | 10 | | 6 | .109 | 22 | | 6 | .027 | 39 |
| | 1 | .125 | 9 | | 7 | .148 | 21 | | 7 | .037 | 38 |
| | 2 | .188 | 8 | | 8 | .188 | 20 | | 8 | .049 | 37 |
| | 3 | .312 | 7 | | 9 | .234 | 19 | | 9 | .064 | 36 |
| | 4 | .438 | 6 | | 10 | .289 | 18 | | 10 | .082 | 35 |
| | 5 | .562 | 5 | | 11 | .344 | 17 | | 11 | .102 | 34 |
| 5 | 0 | .031 | 15 | | 12 | .406 | 16 | | 12 | .125 | 33 |
| | 1 | .062 | 14 | | 13 | .469 | 15 | | 13 | .150 | 32 |
| | 2 | .094 | 13 | | 14 | .531 | 14 | | 14 | .180 | 31 |
| | 3 | .156 | 12 | 8 | 0 | .004 | 36 | | 15 | .213 | 30 |
| | 4 | .219 | 11 | | 1 | .008 | 35 | | 16 | .248 | 29 |
| | 5 | .312 | 10 | | 2 | .012 | 34 | | 17 | .285 | 28 |
| | 6 | .406 | 9 | | 3 | .020 | 33 | | 18 | .326 | 27 |
| | 7 | .500 | 8 | | 4 | .027 | 32 | | 19 | .367 | 26 |
| 6 | 0 | .016 | 21 | | 5 | .039 | 31 | | 20 | .410 | 25 |
| | 1 | .031 | 20 | | 6 | .055 | 30 | | 21 | .455 | 24 |
| | 2 | .047 | 19 | | 7 | .074 | 29 | | 22 | .500 | 23 |
| | 3 | .078 | 18 | | 8 | .098 | 28 | 10 | 0 | .001 | 55 |
| | 4 | .109 | 17 | | 9 | .125 | 27 | | 1 | .002 | 54 |
| | 5 | .156 | 16 | | 10 | .156 | 26 | | 2 | .003 | 53 |
| | 6 | .219 | 15 | | 11 | .191 | 25 | | 3 | .005 | 52 |
| | 7 | .281 | 14 | | 12 | .230 | 24 | | 4 | .007 | 51 |
| | 8 | .344 | 13 | | 13 | .273 | 23 | | 5 | .010 | 50 |
| | 9 | .422 | 12 | | 14 | .320 | 22 | | 6 | .014 | 49 |
| | 10 | .500 | 11 | | 15 | .371 | 21 | | 7 | .019 | 48 |
| | | | | | 16 | .422 | 20 | | 8 | .024 | 47 |
| | | | | | 17 | .473 | 19 | | 9 | .032 | 46 |
| | | | | | 18 | .527 | 18 | | 10 | .042 | 45 |

| n | Left T | P | Right T | n | Left T | P | Right T | n | Left T | P | Right T |
|---|---|---|---|---|---|---|---|---|---|---|---|
| 10 | 11 | .053 | 44 | 12 | 0 | .000 | 78 | 13 | 11 | .007 | 80 |
|  | 12 | .065 | 43 |  | 1 | .000 | 77 |  | 12 | .009 | 79 |
|  | 13 | .080 | 42 |  | 2 | .001 | 76 |  | 13 | .011 | 78 |
|  | 14 | .097 | 41 |  | 3 | .001 | 75 |  | 14 | .013 | 77 |
|  | 15 | .116 | 40 |  | 4 | .002 | 74 |  | 15 | .016 | 76 |
|  | 16 | .138 | 39 |  | 5 | .002 | 73 |  | 16 | .020 | 75 |
|  | 17 | .161 | 38 |  | 6 | .003 | 72 |  | 17 | .024 | 74 |
|  | 18 | .188 | 37 |  | 7 | .005 | 71 |  | 18 | .029 | 73 |
|  | 19 | .216 | 36 |  | 8 | .006 | 70 |  | 19 | .034 | 72 |
|  | 20 | .246 | 35 |  | 9 | .008 | 69 |  | 20 | .040 | 71 |
|  | 21 | .278 | 34 |  | 10 | .010 | 68 |  | 21 | .047 | 70 |
|  | 22 | .312 | 33 |  | 11 | .013 | 67 |  | 22 | .055 | 69 |
|  | 23 | .348 | 32 |  | 12 | .017 | 66 |  | 23 | .064 | 68 |
|  | 24 | .385 | 31 |  | 13 | .021 | 65 |  | 24 | .073 | 67 |
|  | 25 | .423 | 30 |  | 14 | .026 | 64 |  | 25 | .084 | 66 |
|  | 26 | .461 | 29 |  | 15 | .032 | 63 |  | 26 | .095 | 65 |
|  | 27 | .500 | 28 |  | 16 | .039 | 62 |  | 27 | .108 | 64 |
| 11 | 0 | .000 | 66 |  | 17 | .046 | 61 |  | 28 | .122 | 63 |
|  | 1 | .001 | 65 |  | 18 | .055 | 60 |  | 29 | .137 | 62 |
|  | 2 | .001 | 64 |  | 19 | .065 | 59 |  | 30 | .153 | 61 |
|  | 3 | .002 | 63 |  | 20 | .076 | 58 |  | 31 | .170 | 60 |
|  | 4 | .003 | 62 |  | 21 | .088 | 57 |  | 32 | .188 | 59 |
|  | 5 | .005 | 61 |  | 22 | .102 | 56 |  | 33 | .207 | 58 |
|  | 6 | .007 | 60 |  | 23 | .117 | 55 |  | 34 | .227 | 57 |
|  | 7 | .009 | 59 |  | 24 | .133 | 54 |  | 35 | .249 | 56 |
|  | 8 | .012 | 58 |  | 25 | .151 | 53 |  | 36 | .271 | 55 |
|  | 9 | .016 | 57 |  | 26 | .170 | 52 |  | 37 | .294 | 54 |
|  | 10 | .021 | 56 |  | 27 | .190 | 51 |  | 38 | .318 | 53 |
|  | 11 | .027 | 55 |  | 28 | .212 | 50 |  | 39 | .342 | 52 |
|  | 12 | .034 | 54 |  | 29 | .235 | 49 |  | 40 | .368 | 51 |
|  | 13 | .042 | 53 |  | 30 | .259 | 48 |  | 41 | .393 | 50 |
|  | 14 | .051 | 52 |  | 31 | .285 | 47 |  | 42 | .420 | 49 |
|  | 15 | .062 | 51 |  | 32 | .311 | 46 |  | 43 | .446 | 48 |
|  | 16 | .074 | 50 |  | 33 | .339 | 45 |  | 44 | .473 | 47 |
|  | 17 | .087 | 49 |  | 34 | .367 | 44 |  | 45 | .500 | 46 |
|  | 18 | .103 | 48 |  | 35 | .396 | 43 | 14 | 0 | .000 | 105 |
|  | 19 | .120 | 47 |  | 36 | .425 | 42 |  | 1 | .000 | 104 |
|  | 20 | .139 | 46 |  | 37 | .455 | 41 |  | 2 | .000 | 103 |
|  | 21 | .160 | 45 |  | 38 | .485 | 40 |  | 3 | .000 | 102 |
|  | 22 | .183 | 44 |  | 39 | .515 | 39 |  | 4 | .000 | 101 |
|  | 23 | .207 | 43 | 13 | 0 | .000 | 91 |  | 5 | .001 | 100 |
|  | 24 | .232 | 42 |  | 1 | .000 | 90 |  | 6 | .001 | 99 |
|  | 25 | .260 | 41 |  | 2 | .000 | 89 |  | 7 | .001 | 98 |
|  | 26 | .289 | 40 |  | 3 | .001 | 88 |  | 8 | .002 | 97 |
|  | 27 | .319 | 39 |  | 4 | .001 | 87 |  | 9 | .002 | 96 |
|  | 28 | .350 | 38 |  | 5 | .001 | 86 |  | 10 | .003 | 95 |
|  | 29 | .382 | 37 |  | 6 | .002 | 85 |  | 11 | .003 | 94 |
|  | 30 | .416 | 36 |  | 7 | .002 | 84 |  | 12 | .004 | 93 |
|  | 31 | .449 | 35 |  | 8 | .003 | 83 |  | 13 | .005 | 92 |
|  | 32 | .483 | 34 |  | 9 | .004 | 82 |  | 14 | .007 | 91 |
|  | 33 | .517 | 33 |  | 10 | .005 | 81 |  | 15 | .008 | 90 |

| | Left | | Right | | Left | | Right | | Left | | Right |
|---|---|---|---|---|---|---|---|---|---|---|---|
| n | T | P | T | n | T | P | T | n | T | P | T |
| 14 | 16 | .010 | 89 | 15 | 0 | .000 | 120 | 15 | 31 | .053 | 89 |
| | 17 | .012 | 88 | | 1 | .000 | 119 | | 32 | .060 | 88 |
| | 18 | .015 | 87 | | 2 | .000 | 118 | | 33 | .068 | 87 |
| | 19 | .018 | 86 | | 3 | .000 | 117 | | 34 | .076 | 86 |
| | 20 | .021 | 85 | | 4 | .000 | 116 | | 35 | .084 | 85 |
| | 21 | .025 | 84 | | 5 | .000 | 115 | | 36 | .094 | 84 |
| | 22 | .029 | 83 | | 6 | .000 | 114 | | 37 | .104 | 83 |
| | 23 | .034 | 82 | | 7 | .001 | 113 | | 38 | .115 | 82 |
| | 24 | .039 | 81 | | 8 | .001 | 112 | | 39 | .126 | 81 |
| | 25 | .045 | 80 | | 9 | .001 | 111 | | 40 | .138 | 80 |
| | 26 | .052 | 79 | | 10 | .001 | 110 | | 41 | .151 | 79 |
| | 27 | .059 | 78 | | 11 | .002 | 109 | | 42 | .165 | 78 |
| | 28 | .068 | 77 | | 12 | .002 | 108 | | 43 | .180 | 77 |
| | 29 | .077 | 76 | | 13 | .003 | 107 | | 44 | .195 | 76 |
| | 30 | .086 | 75 | | 14 | .003 | 106 | | 45 | .211 | 75 |
| | 31 | .097 | 74 | | 15 | .004 | 105 | | 46 | .227 | 74 |
| | 32 | .108 | 73 | | 16 | .005 | 104 | | 47 | .244 | 73 |
| | 33 | .121 | 72 | | 17 | .006 | 103 | | 48 | .262 | 72 |
| | 34 | .134 | 71 | | 18 | .008 | 102 | | 49 | .281 | 71 |
| | 35 | .148 | 70 | | 19 | .009 | 101 | | 50 | .300 | 70 |
| | 36 | .163 | 69 | | 20 | .011 | 100 | | 51 | .319 | 69 |
| | 37 | .179 | 68 | | 21 | .013 | 99 | | 52 | .339 | 68 |
| | 38 | .196 | 67 | | 22 | .015 | 98 | | 53 | .360 | 67 |
| | 39 | .213 | 66 | | 23 | .018 | 97 | | 54 | .381 | 66 |
| | 40 | .232 | 65 | | 24 | .021 | 96 | | 55 | .402 | 65 |
| | 41 | .251 | 64 | | 25 | .024 | 95 | | 56 | .423 | 64 |
| | 42 | .271 | 63 | | 26 | .028 | 94 | | 57 | .445 | 63 |
| | 43 | .292 | 62 | | 27 | .032 | 93 | | 58 | .467 | 62 |
| | 44 | .313 | 61 | | 28 | .036 | 92 | | 59 | .489 | 61 |
| | 45 | .335 | 60 | | 29 | .042 | 91 | | 60 | .511 | 60 |
| | 46 | .357 | 59 | | 30 | .047 | 90 | | | | |
| | 47 | .380 | 58 | | | | | | | | |
| | 48 | .404 | 57 | | | | | | | | |
| | 49 | .428 | 56 | | | | | | | | |
| | 50 | .452 | 55 | | | | | | | | |
| | 51 | .476 | 54 | | | | | | | | |
| | 52 | .500 | 53 | | | | | | | | |

For $n > 15$, the probabilities are found from Table A as follows.

$$z_{+,R} = \frac{T_+ - 0.5 - n(n+1)/4}{\sqrt{n(n+1)(2n+1)/24}} \qquad z_{-,R} = \frac{T_- - 0.5 - n(n+1)/4}{\sqrt{n(n+1)(2n+1)/24}}$$

| Desired | Approximated by |
|---|---|
| Right-tail probability for $T_+$ | Right-tail probability for $z_{+,R}$ |
| Right-tail probability for $T_-$ | Right-tail probability for $z_{-,R}$ |

Source: Adapted from Table II of H. L. Harter and D. B. Owen, eds. (1972), *Selected Tables in Mathematical Statistics*, Vol. 1, Markham Publishing Co., Chicago, with permission of The Institute of Mathematical Statistics.

## TABLE H
## MANN-WHITNEY-WILCOXON DISTRIBUTION

Entries labeled $P$ in the table are the cumulative probability from each extreme to the value of $T_x$ for the given sample sizes $m \leq n$ ($m$ is the size of the $X$ sample). Left-tail probabilities are given for $T_x \leq m(N + 1)/2$ and right-tail probabilities for $T_x \geq m(N + 1)/2$, where $N = m + n$.

| $n$ | Left $T_x$ | $P$ | Right $T_x$ | $n$ | Left $T_x$ | $P$ | Right $T_x$ | $n$ | Left $T_x$ | $P$ | Right $T_x$ |
|---|---|---|---|---|---|---|---|---|---|---|---|
| | $m = 1$ | | | | $m = 2$ | | | | $m = 2$ | | |
| 1 | 1 | .500 | 2 | 2 | 3 | .167 | 7 | 8 | 3 | .022 | 19 |
| 2 | 1 | .333 | 3 | | 4 | .333 | 6 | | 4 | .044 | 18 |
| | 2 | .667 | 2 | | 5 | .667 | 5 | | 5 | .089 | 17 |
| 3 | 1 | .250 | 4 | 3 | 3 | .100 | 9 | | 6 | .133 | 16 |
| | 2 | .500 | 3 | | 4 | .200 | 8 | | 7 | .200 | 15 |
| 4 | 1 | .200 | 5 | | 5 | .400 | 7 | | 8 | .267 | 14 |
| | 2 | .400 | 4 | | 6 | .600 | 6 | | 9 | .356 | 13 |
| | 3 | .600 | 3 | 4 | 3 | .067 | 11 | | 10 | .444 | 12 |
| 5 | 1 | .167 | 6 | | 4 | .133 | 10 | | 11 | .556 | 11 |
| | 2 | .333 | 5 | | 5 | .267 | 9 | 9 | 3 | .018 | 21 |
| | 3 | .500 | 4 | | 6 | .400 | 8 | | 4 | .036 | 20 |
| 6 | 1 | .143 | 7 | | 7 | .600 | 7 | | 5 | .073 | 19 |
| | 2 | .286 | 6 | 5 | 3 | .048 | 13 | | 6 | .109 | 18 |
| | 3 | .429 | 5 | | 4 | .095 | 12 | | 7 | .164 | 17 |
| | 4 | .571 | 4 | | 5 | .190 | 11 | | 8 | .218 | 16 |
| 7 | 1 | .125 | 8 | | 6 | .286 | 10 | | 9 | .291 | 15 |
| | 2 | .250 | 7 | | 7 | .429 | 9 | | 10 | .364 | 14 |
| | 3 | .375 | 6 | | 8 | .571 | 8 | | 11 | .455 | 13 |
| | 4 | .500 | 5 | 6 | 3 | .036 | 15 | | 12 | .545 | 12 |
| 8 | 1 | .111 | 9 | | 4 | .071 | 14 | 10 | 3 | .015 | 23 |
| | 2 | .222 | 8 | | 5 | .143 | 13 | | 4 | .030 | 22 |
| | 3 | .333 | 7 | | 6 | .214 | 12 | | 5 | .061 | 21 |
| | 4 | .444 | 6 | | 7 | .321 | 11 | | 6 | .091 | 20 |
| | 5 | .556 | 5 | | 8 | .429 | 10 | | 7 | .136 | 19 |
| 9 | 1 | .100 | 10 | | 9 | .571 | 9 | | 8 | .182 | 18 |
| | 2 | .200 | 9 | 7 | 3 | .028 | 17 | | 9 | .242 | 17 |
| | 3 | .300 | 8 | | 4 | .056 | 16 | | 10 | .303 | 16 |
| | 4 | .400 | 7 | | 5 | .111 | 15 | | 11 | .379 | 15 |
| | 5 | .500 | 6 | | 6 | .167 | 14 | | 12 | .455 | 14 |
| 10 | 1 | .091 | 11 | | 7 | .250 | 13 | | 13 | .545 | 13 |
| | 2 | .182 | 10 | | 8 | .333 | 12 | | | | |
| | 3 | .273 | 9 | | 9 | .444 | 11 | | | | |
| | 4 | .364 | 8 | | 10 | .556 | 10 | | | | |
| | 5 | .455 | 7 | | | | | | | | |
| | 6 | .545 | 6 | | | | | | | | |

**TABLE H (Continued)**

| $n$ | Left $T_x$ | $P$ | Right $T_x$ | $n$ | Left $T_x$ | $P$ | Right $T_x$ | $n$ | Left $T_x$ | $P$ | Right $T_x$ |
|---|---|---|---|---|---|---|---|---|---|---|---|
| | | $m = 3$ | | | | $m = 3$ | | | | $m = 4$ | |
| 3 | 6 | .050 | 15 | 8 | 6 | .006 | 30 | 4 | 10 | .014 | 26 |
| | 7 | .100 | 14 | | 7 | .012 | 29 | | 11 | .029 | 25 |
| | 8 | .200 | 13 | | 8 | .024 | 28 | | 12 | .057 | 24 |
| | 9 | .350 | 12 | | 9 | .042 | 27 | | 13 | .100 | 23 |
| | 10 | .500 | 11 | | 10 | .067 | 26 | | 14 | .171 | 22 |
| 4 | 6 | .029 | 18 | | 11 | .097 | 25 | | 15 | .243 | 21 |
| | 7 | .057 | 17 | | 12 | .139 | 24 | | 16 | .343 | 20 |
| | 8 | .114 | 16 | | 13 | .188 | 23 | | 17 | .443 | 19 |
| | 9 | .200 | 15 | | 14 | .248 | 22 | | 18 | .557 | 18 |
| | 10 | .314 | 14 | | 15 | .315 | 21 | 5 | 10 | .008 | 30 |
| | 11 | .429 | 13 | | 16 | .388 | 20 | | 11 | .016 | 29 |
| | 12 | .571 | 12 | | 17 | .461 | 19 | | 12 | .032 | 28 |
| 5 | 6 | .018 | 21 | | 18 | .539 | 18 | | 13 | .056 | 27 |
| | 7 | .036 | 20 | 9 | 6 | .005 | 33 | | 14 | .095 | 26 |
| | 8 | .071 | 19 | | 7 | .009 | 32 | | 15 | .143 | 25 |
| | 9 | .125 | 18 | | 8 | .018 | 31 | | 16 | .206 | 24 |
| | 10 | .196 | 17 | | 9 | .032 | 30 | | 17 | .278 | 23 |
| | 11 | .286 | 16 | | 10 | .050 | 29 | | 18 | .365 | 22 |
| | 12 | .393 | 15 | | 11 | .073 | 28 | | 19 | .452 | 21 |
| | 13 | .500 | 14 | | 12 | .105 | 27 | | 20 | .548 | 20 |
| 6 | 6 | .012 | 24 | | 13 | .141 | 26 | 6 | 10 | .005 | 34 |
| | 7 | .024 | 23 | | 14 | .186 | 25 | | 11 | .010 | 33 |
| | 8 | .048 | 22 | | 15 | .241 | 24 | | 12 | .019 | 32 |
| | 9 | .083 | 21 | | 16 | .300 | 23 | | 13 | .033 | 31 |
| | 10 | .131 | 20 | | 17 | .364 | 22 | | 14 | .057 | 30 |
| | 11 | .190 | 19 | | 18 | .432 | 21 | | 15 | .086 | 29 |
| | 12 | .274 | 18 | | 19 | .500 | 20 | | 16 | .129 | 28 |
| | 13 | .357 | 17 | 10 | 6 | .003 | 36 | | 17 | .176 | 27 |
| | 14 | .452 | 16 | | 7 | .007 | 35 | | 18 | .238 | 26 |
| | 15 | .548 | 15 | | 8 | .014 | 34 | | 19 | .305 | 25 |
| 7 | 6 | .008 | 27 | | 9 | .024 | 33 | | 20 | .381 | 24 |
| | 7 | .017 | 26 | | 10 | .038 | 32 | | 21 | .457 | 23 |
| | 8 | .033 | 25 | | 11 | .056 | 31 | | 22 | .543 | 22 |
| | 9 | .058 | 24 | | 12 | .080 | 30 | | | | |
| | 10 | .092 | 23 | | 13 | .108 | 29 | | | | |
| | 11 | .133 | 22 | | 14 | .143 | 28 | | | | |
| | 12 | .192 | 21 | | 15 | .185 | 27 | | | | |
| | 13 | .258 | 20 | | 16 | .234 | 26 | | | | |
| | 14 | .333 | 19 | | 17 | .287 | 25 | | | | |
| | 15 | .417 | 18 | | 18 | .346 | 24 | | | | |
| | 16 | .500 | 17 | | 19 | .406 | 23 | | | | |
| | | | | | 20 | .469 | 22 | | | | |
| | | | | | 21 | .531 | 21 | | | | |

## TABLE H  (Continued)

| n | Left $T_x$ | P | Right $T_x$ | n | Left $T_x$ | P | Right $T_x$ | n | Left $T_x$ | P | Right $T_x$ |
|---|---|---|---|---|---|---|---|---|---|---|---|
| | | $m = 4$ | | | | $m = 4$ | | | | $m = 5$ | |
| 7 | 10 | .003 | 38 | 9 | 10 | .001 | 46 | 5 | 15 | .004 | 40 |
| | 11 | .006 | 37 | | 11 | .003 | 45 | | 16 | .008 | 39 |
| | 12 | .012 | 36 | | 12 | .006 | 44 | | 17 | .016 | 38 |
| | 13 | .021 | 35 | | 13 | .010 | 43 | | 18 | .028 | 37 |
| | 14 | .036 | 34 | | 14 | .017 | 42 | | 19 | .048 | 36 |
| | 15 | .055 | 33 | | 15 | .025 | 41 | | 20 | .075 | 35 |
| | 16 | .082 | 32 | | 16 | .038 | 40 | | 21 | .111 | 34 |
| | 17 | .115 | 31 | | 17 | .053 | 39 | | 22 | .155 | 33 |
| | 18 | .158 | 30 | | 18 | .074 | 38 | | 23 | .210 | 32 |
| | 19 | .206 | 29 | | 19 | .099 | 37 | | 24 | .274 | 31 |
| | 20 | .264 | 28 | | 20 | .130 | 36 | | 25 | .345 | 30 |
| | 21 | .324 | 27 | | 21 | .165 | 35 | | 26 | .421 | 29 |
| | 22 | .394 | 26 | | 22 | .207 | 34 | | 27 | .500 | 28 |
| | 23 | .464 | 25 | | 23 | .252 | 33 | 6 | 15 | .002 | 45 |
| | 24 | .536 | 24 | | 24 | .302 | 32 | | 16 | .004 | 44 |
| 8 | 10 | .002 | 42 | | 25 | .355 | 31 | | 17 | .009 | 43 |
| | 11 | .004 | 41 | | 26 | .413 | 30 | | 18 | .015 | 42 |
| | 12 | .008 | 40 | | 27 | .470 | 29 | | 19 | .026 | 41 |
| | 13 | .014 | 39 | | 28 | .530 | 28 | | 20 | .041 | 40 |
| | 14 | .024 | 38 | 10 | 10 | .001 | 50 | | 21 | .063 | 39 |
| | 15 | .036 | 37 | | 11 | .002 | 49 | | 22 | .089 | 38 |
| | 16 | .055 | 36 | | 12 | .004 | 48 | | 23 | .123 | 37 |
| | 17 | .077 | 35 | | 13 | .007 | 47 | | 24 | .165 | 36 |
| | 18 | .107 | 34 | | 14 | .012 | 46 | | 25 | .214 | 35 |
| | 19 | .141 | 33 | | 15 | .018 | 45 | | 26 | .268 | 34 |
| | 20 | .184 | 32 | | 16 | .027 | 44 | | 27 | .331 | 33 |
| | 21 | .230 | 31 | | 17 | .038 | 43 | | 28 | .396 | 32 |
| | 22 | .285 | 30 | | 18 | .053 | 42 | | 29 | .465 | 31 |
| | 23 | .341 | 29 | | 19 | .071 | 41 | | 30 | .535 | 30 |
| | 24 | .404 | 28 | | 20 | .094 | 40 | | | | |
| | 25 | .467 | 27 | | 21 | .120 | 39 | | | | |
| | 26 | .533 | 26 | | 22 | .152 | 38 | | | | |
| | | | | | 23 | .187 | 37 | | | | |
| | | | | | 24 | .227 | 36 | | | | |
| | | | | | 25 | .270 | 35 | | | | |
| | | | | | 26 | .318 | 34 | | | | |
| | | | | | 27 | .367 | 33 | | | | |
| | | | | | 28 | .420 | 32 | | | | |
| | | | | | 29 | .473 | 31 | | | | |
| | | | | | 30 | .527 | 30 | | | | |

| n | Left $T_x$ | P | Right $T_x$ | n | Left $T_x$ | P | Right $T_x$ | n | Left $T_x$ | P | Right $T_x$ |
|---|---|---|---|---|---|---|---|---|---|---|---|
| | | $m = 5$ | | | | $m = 5$ | | | | $m = 6$ | |
| 7 | 15 | .001 | 50 | 9 | 15 | .000 | 60 | 6 | 21 | .001 | 57 |
| | 16 | .003 | 49 | | 16 | .001 | 59 | | 22 | .002 | 56 |
| | 17 | .005 | 48 | | 17 | .002 | 58 | | 23 | .004 | 55 |
| | 18 | .009 | 47 | | 18 | .003 | 57 | | 24 | .008 | 54 |
| | 19 | .015 | 46 | | 19 | .006 | 56 | | 25 | .013 | 53 |
| | 20 | .024 | 45 | | 20 | .009 | 55 | | 26 | .021 | 52 |
| | 21 | .037 | 44 | | 21 | .014 | 54 | | 27 | .032 | 51 |
| | 22 | .053 | 43 | | 22 | .021 | 53 | | 28 | .047 | 50 |
| | 23 | .074 | 42 | | 23 | .030 | 52 | | 29 | .066 | 49 |
| | 24 | .101 | 41 | | 24 | .041 | 51 | | 30 | .090 | 48 |
| | 25 | .134 | 40 | | 25 | .056 | 50 | | 31 | .120 | 47 |
| | 26 | .172 | 39 | | 26 | .073 | 49 | | 32 | .155 | 46 |
| | 27 | .216 | 38 | | 27 | .095 | 48 | | 33 | .197 | 45 |
| | 28 | .265 | 37 | | 28 | .120 | 47 | | 34 | .242 | 44 |
| | 29 | .319 | 36 | | 29 | .149 | 46 | | 35 | .294 | 43 |
| | 30 | .378 | 35 | | 30 | .182 | 45 | | 36 | .350 | 42 |
| | 31 | .438 | 34 | | 31 | .219 | 44 | | 37 | .409 | 41 |
| | 32 | .500 | 33 | | 32 | .259 | 43 | | 38 | .469 | 40 |
| 8 | 15 | .001 | 55 | | 33 | .303 | 42 | | 39 | .531 | 39 |
| | 16 | .002 | 54 | | 34 | .350 | 41 | 7 | 21 | .001 | 63 |
| | 17 | .003 | 53 | | 35 | .399 | 40 | | 22 | .001 | 62 |
| | 18 | .005 | 52 | | 36 | .449 | 39 | | 23 | .002 | 61 |
| | 19 | .009 | 51 | | 37 | .500 | 38 | | 24 | .004 | 60 |
| | 20 | .015 | 50 | 10 | 15 | .000 | 65 | | 25 | .007 | 59 |
| | 21 | .023 | 49 | | 16 | .001 | 64 | | 26 | .011 | 58 |
| | 22 | .033 | 48 | | 17 | .001 | 63 | | 27 | .017 | 57 |
| | 23 | .047 | 47 | | 18 | .002 | 62 | | 28 | .026 | 56 |
| | 24 | .064 | 46 | | 19 | .004 | 61 | | 29 | .037 | 55 |
| | 25 | .085 | 45 | | 20 | .006 | 60 | | 30 | .051 | 54 |
| | 26 | .111 | 44 | | 21 | .010 | 59 | | 31 | .069 | 53 |
| | 27 | .142 | 43 | | 22 | .014 | 58 | | 32 | .090 | 52 |
| | 28 | .177 | 42 | | 23 | .020 | 57 | | 33 | .117 | 51 |
| | 29 | .218 | 41 | | 24 | .028 | 56 | | 34 | .147 | 50 |
| | 30 | .262 | 40 | | 25 | .038 | 55 | | 35 | .183 | 49 |
| | 31 | .311 | 39 | | 26 | .050 | 54 | | 36 | .223 | 48 |
| | 32 | .362 | 38 | | 27 | .065 | 53 | | 37 | .267 | 47 |
| | 33 | .416 | 37 | | 28 | .082 | 52 | | 38 | .314 | 46 |
| | 34 | .472 | 36 | | 29 | .103 | 51 | | 39 | .365 | 45 |
| | 35 | .528 | 35 | | 30 | .127 | 50 | | 40 | .418 | 44 |
| | | | | | 31 | .155 | 49 | | 41 | .473 | 43 |
| | | | | | 32 | .185 | 48 | | 42 | .527 | 42 |
| | | | | | 33 | .220 | 47 | | | | |
| | | | | | 34 | .257 | 46 | | | | |
| | | | | | 35 | .297 | 45 | | | | |
| | | | | | 36 | .339 | 44 | | | | |
| | | | | | 37 | .384 | 43 | | | | |
| | | | | | 38 | .430 | 42 | | | | |
| | | | | | 39 | .477 | 41 | | | | |
| | | | | | 40 | .523 | 40 | | | | |

TABLE H (Continued)

| n | Left $T_x$ | P | Right $T_x$ | n | Left $T_x$ | P | Right $T_x$ | n | Left $T_x$ | P | Right $T_x$ |
|---|---|---|---|---|---|---|---|---|---|---|---|
| | | $m = 6$ | | | | $m = 6$ | | | | $m = 7$ | |
| 8 | 21 | .000 | 69 | 9 | 41 | .228 | 55 | 7 | 28 | .000 | 77 |
| | 22 | .001 | 68 | | 42 | .264 | 54 | | 29 | .001 | 76 |
| | 23 | .001 | 67 | | 43 | .303 | 53 | | 30 | .001 | 75 |
| | 24 | .002 | 66 | | 44 | .344 | 52 | | 31 | .002 | 74 |
| | 25 | .004 | 65 | | 45 | .388 | 51 | | 32 | .003 | 73 |
| | 26 | .006 | 64 | | 46 | .432 | 50 | | 33 | .006 | 72 |
| | 27 | .010 | 63 | | 47 | .477 | 49 | | 34 | .009 | 71 |
| | 28 | .015 | 62 | | 48 | .523 | 48 | | 35 | .013 | 70 |
| | 29 | .021 | 61 | 10 | 21 | .000 | 81 | | 36 | .019 | 69 |
| | 30 | .030 | 60 | | 22 | .000 | 80 | | 37 | .027 | 68 |
| | 31 | .041 | 59 | | 23 | .000 | 79 | | 38 | .036 | 67 |
| | 32 | .054 | 58 | | 24 | .001 | 78 | | 39 | .049 | 66 |
| | 33 | .071 | 57 | | 25 | .001 | 77 | | 40 | .064 | 65 |
| | 34 | .091 | 56 | | 26 | .002 | 76 | | 41 | .082 | 64 |
| | 35 | .114 | 55 | | 27 | .004 | 75 | | 42 | .104 | 63 |
| | 36 | .141 | 54 | | 28 | .005 | 74 | | 43 | .130 | 62 |
| | 37 | .172 | 53 | | 29 | .008 | 73 | | 44 | .159 | 61 |
| | 38 | .207 | 52 | | 30 | .011 | 72 | | 45 | .191 | 60 |
| | 39 | .245 | 51 | | 31 | .016 | 71 | | 46 | .228 | 59 |
| | 40 | .286 | 50 | | 32 | .021 | 70 | | 47 | .267 | 58 |
| | 41 | .331 | 49 | | 33 | .028 | 69 | | 48 | .310 | 57 |
| | 42 | .377 | 48 | | 34 | .036 | 68 | | 49 | .355 | 56 |
| | 43 | .426 | 47 | | 35 | .047 | 67 | | 50 | .402 | 55 |
| | 44 | .475 | 46 | | 36 | .059 | 66 | | 51 | .451 | 54 |
| | 45 | .525 | 45 | | 37 | .074 | 65 | | 52 | .500 | 53 |
| 9 | 21 | .000 | 75 | | 38 | .090 | 64 | 8 | 28 | .000 | 84 |
| | 22 | .000 | 74 | | 39 | .110 | 63 | | 29 | .000 | 83 |
| | 23 | .001· | 73 | | 40 | .132 | 62 | | 30 | .001 | 82 |
| | 24 | .001 | 72 | | 41 | .157 | 61 | | 31 | .001 | 81 |
| | 25 | .002 | 71 | | 42 | .184 | 60 | | 32 | .002 | 80 |
| | 26 | .004 | 70 | | 43 | .214 | 59 | | 33 | .003 | 79 |
| | 27 | .006 | 69 | | 44 | .246 | 58 | | 34 | .005 | 78 |
| | 28 | .009 | 68 | | 45 | .281 | 57 | | 35 | .007 | 77 |
| | 29 | .013 | 67 | | 46 | .318 | 56 | | 36 | .010 | 76 |
| | 30 | .018 | 66 | | 47 | .356 | 55 | | 37 | .014 | 75 |
| | 31 | .025 | 65 | | 48 | .396 | 54 | | 38 | .020 | 74 |
| | 32 | .033 | 64 | | 49 | .437 | 53 | | 39 | .027 | 73 |
| | 33 | .044 | 63 | | 50 | .479 | 52 | | 40 | .036 | 72 |
| | 34 | .057 | 62 | | 51 | .521 | 51 | | 41 | .047 | 71 |
| | 35 | .072 | 61 | | | | | | 42 | .060 | 70 |
| | 36 | .091 | 60 | | | | | | 43 | .076 | 69 |
| | 37 | .112 | 59 | | | | | | 44 | .095 | 68 |
| | 38 | .136 | 58 | | | | | | 45 | .116 | 67 |
| | 39 | .164 | 57 | | | | | | 46 | .140 | 66 |
| | 40 | .194 | 56 | | | | | | 47 | .168 | 65 |

## TABLE H (Continued)

| n | Left $T_x$ | P | Right $T_x$ | n | Left $T_x$ | P | Right $T_x$ | n | Left $T_x$ | P | Right $T_x$ |
|---|---|---|---|---|---|---|---|---|---|---|---|
| | | $m = 7$ | | | | $m = 7$ | | | | $m = 8$ | |
| 8 | 48 | .198 | 64 | 10 | 28 | .000 | 98 | 8 | 36 | .000 | 100 |
| | 49 | .232 | 63 | | 29 | .000 | 97 | | 37 | .000 | 99 |
| | 50 | .268 | 62 | | 30 | .000 | 96 | | 38 | .000 | 98 |
| | 51 | .306 | 61 | | 31 | .000 | 95 | | 39 | .001 | 97 |
| | 52 | .347 | 60 | | 32 | .001 | 94 | | 40 | .001 | 96 |
| | 53 | .389 | 59 | | 33 | .001 | 93 | | 41 | .001 | 95 |
| | 54 | .433 | 58 | | 34 | .002 | 92 | | 42 | .002 | 94 |
| | 55 | .478 | 57 | | 35 | .002 | 91 | | 43 | .003 | 93 |
| | 56 | .522 | 56 | | 36 | .003 | 90 | | 44 | .005 | 92 |
| 9 | 28 | .000 | 91 | | 37 | .005 | 89 | | 45 | .007 | 91 |
| | 29 | .000 | 90 | | 38 | .007 | 88 | | 46 | .010 | 90 |
| | 30 | .000 | 89 | | 39 | .009 | 87 | | 47 | .014 | 89 |
| | 31 | .001 | 88 | | 40 | .012 | 86 | | 48 | .019 | 88 |
| | 32 | .001 | 87 | | 41 | .017 | 85 | | 49 | .025 | 87 |
| | 33 | .002 | 86 | | 42 | .022 | 84 | | 50 | .032 | 86 |
| | 34 | .003 | 85 | | 43 | .028 | 83 | | 51 | .041 | 85 |
| | 35 | .004 | 84 | | 44 | .035 | 82 | | 52 | .052 | 84 |
| | 36 | .006 | 83 | | 45 | .044 | 81 | | 53 | .065 | 83 |
| | 37 | .008 | 82 | | 46 | .054 | 80 | | 54 | .080 | 82 |
| | 38 | .011 | 81 | | 47 | .067 | 79 | | 55 | .097 | 81 |
| | 39 | .016 | 80 | | 48 | .081 | 78 | | 56 | .117 | 80 |
| | 40 | .021 | 79 | | 49 | .097 | 77 | | 57 | .139 | 79 |
| | 41 | .027 | 78 | | 50 | .115 | 76 | | 58 | .164 | 78 |
| | 42 | .036 | 77 | | 51 | .135 | 75 | | 59 | .191 | 77 |
| | 43 | .045 | 76 | | 52 | .157 | 74 | | 60 | .221 | 76 |
| | 44 | .057 | 75 | | 53 | .182 | 73 | | 61 | .253 | 75 |
| | 45 | .071 | 74 | | 54 | .209 | 72 | | 62 | .287 | 74 |
| | 46 | .087 | 73 | | 55 | .237 | 71 | | 63 | .323 | 73 |
| | 47 | .105 | 72 | | 56 | .268 | 70 | | 64 | .360 | 72 |
| | 48 | .126 | 71 | | 57 | .300 | 69 | | 65 | .399 | 71 |
| | 49 | .150 | 70 | | 58 | .335 | 68 | | 66 | .439 | 70 |
| | 50 | .176 | 69 | | 59 | .370 | 67 | | 67 | .480 | 69 |
| | 51 | .204 | 68 | | 60 | .406 | 66 | | 68 | .520 | 68 |
| | 52 | .235 | 67 | | 61 | .443 | 65 | 9 | 36 | .000 | 108 |
| | 53 | .268 | 66 | | 62 | .481 | 64 | | 37 | .000 | 107 |
| | 54 | .303 | 65 | | 63 | .519 | 63 | | 38 | .000 | 106 |
| | 55 | .340 | 64 | | | | | | 39 | .000 | 105 |
| | 56 | .379 | 63 | | | | | | 40 | .000 | 104 |
| | 57 | .419 | 62 | | | | | | 41 | .001 | 103 |
| | 58 | .459 | 61 | | | | | | 42 | .001 | 102 |
| | 59 | .500 | 60 | | | | | | 43 | .002 | 101 |

TABLE  H  (Continued)

| n | Left $T_x$ | P | Right $T_x$ | n | Left $T_x$ | P | Right $T_x$ | n | Left $T_x$ | P | Right $T_x$ |
|---|---|---|---|---|---|---|---|---|---|---|---|
| | | $m = 8$ | | | | $m = 8$ | | | | $m = 9$ | |
| 9 | 44 | .003 | 100 | 10 | 36 | .000 | 116 | 9 | 45 | .000 | 126 |
| | 45 | .004 | 99 | | 37 | .000 | 115 | | 46 | .000 | 125 |
| | 46 | .006 | 98 | | 38 | .000 | 114 | | 47 | .000 | 124 |
| | 47 | .008 | 97 | | 39 | .000 | 113 | | 48 | .000 | 123 |
| | 48 | .010 | 96 | | 40 | .000 | 112 | | 49 | .000 | 122 |
| | 49 | .014 | 95 | | 41 | .000 | 111 | | 50 | .000 | 121 |
| | 50 | .018 | 94 | | 42 | .001 | 110 | | 51 | .001 | 120 |
| | 51 | .023 | 93 | | 43 | .001 | 109 | | 52 | .001 | 119 |
| | 52 | .030 | 92 | | 44 | .002 | 108 | | 53 | .001 | 118 |
| | 53 | .037 | 91 | | 45 | .002 | 107 | | 54 | .002 | 117 |
| | 54 | .046 | 90 | | 46 | .003 | 106 | | 55 | .003 | 116 |
| | 55 | .057 | 89 | | 47 | .004 | 105 | | 56 | .004 | 115 |
| | 56 | .069 | 88 | | 48 | .006 | 104 | | 57 | .005 | 114 |
| | 57 | .084 | 87 | | 49 | .008 | 103 | | 58 | .007 | 113 |
| | 58 | .100 | 86 | | 50 | .010 | 102 | | 59 | .009 | 112 |
| | 59 | .118 | 85 | | 51 | .013 | 101 | | 60 | .012 | 111 |
| | 60 | .138 | 84 | | 52 | .017 | 100 | | 61 | .016 | 110 |
| | 61 | .161 | 83 | | 53 | .022 | 99 | | 62 | .020 | 109 |
| | 62 | .185 | 82 | | 54 | .027 | 98 | | 63 | .025 | 108 |
| | 63 | .212 | 81 | | 55 | .034 | 97 | | 64 | .031 | 107 |
| | 64 | .240 | 80 | | 56 | .042 | 96 | | 65 | .039 | 106 |
| | 65 | .271 | 79 | | 57 | .051 | 95 | | 66 | .047 | 105 |
| | 66 | .303 | 78 | | 58 | .061 | 94 | | 67 | .057 | 104 |
| | 67 | .336 | 77 | | 59 | .073 | 93 | | 68 | .068 | 103 |
| | 68 | .371 | 76 | | 60 | .086 | 92 | | 69 | .081 | 102 |
| | 69 | .407 | 75 | | 61 | .102 | 91 | | 70 | .095 | 101 |
| | 70 | .444 | 74 | | 62 | .118 | 90 | | 71 | .111 | 100 |
| | 71 | .481 | 73 | | 63 | .137 | 89 | | 72 | .129 | 99 |
| | 72 | .519 | 72 | | 64 | .158 | 88 | | 73 | .149 | 98 |
| | | | | | 65 | .180 | 87 | | 74 | .170 | 97 |
| | | | | | 66 | .204 | 86 | | 75 | .193 | 96 |
| | | | | | 67 | .230 | 85 | | 76 | .218 | 95 |
| | | | | | 68 | .257 | 84 | | 77 | .245 | 94 |
| | | | | | 69 | .286 | 83 | | 78 | .273 | 93 |
| | | | | | 70 | .317 | 82 | | 79 | .302 | 92 |
| | | | | | 71 | .348 | 81 | | 80 | .333 | 91 |
| | | | | | 72 | .381 | 80 | | 81 | .365 | 90 |
| | | | | | 73 | .414 | 79 | | 82 | .398 | 89 |
| | | | | | 74 | .448 | 78 | | 83 | .432 | 88 |
| | | | | | 75 | .483 | 77 | | 84 | .466 | 87 |
| | | | | | 76 | .517 | 76 | | 85 | .500 | 86 |

| $n$ | Left $T_x$ | $P$ | Right $T_x$ | $n$ | Left $T_x$ | $P$ | Right $T_x$ | $n$ | Left $T_x$ | $P$ | Right $T_x$ |
|---|---|---|---|---|---|---|---|---|---|---|---|
| | $m = 9$ | | | | $m = 9$ | | | | $m = 10$ | | |
| 10 | 45 | .000 | 135 | 10 | 78 | .178 | 102 | 10 | 73 | .007 | 137 |
| | 46 | .000 | 134 | | 79 | .200 | 101 | | 74 | .009 | 136 |
| | 47 | .000 | 133 | | 80 | .223 | 100 | | 75 | .012 | 135 |
| | 48 | .000 | 132 | | 81 | .248 | 99 | | 76 | .014 | 134 |
| | 49 | .000 | 131 | | 82 | .274 | 98 | | 77 | .018 | 133 |
| | 50 | .000 | 130 | | 83 | .302 | 97 | | 78 | .022 | 132 |
| | 51 | .000 | 129 | | 84 | .330 | 96 | | 79 | .026 | 131 |
| | 52 | .000 | 128 | | 85 | .360 | 95 | | 80 | .032 | 130 |
| | 53 | .001 | 127 | | 86 | .390 | 94 | | 81 | .038 | 129 |
| | 54 | .001 | 126 | | 87 | .421 | 93 | | 82 | .045 | 128 |
| | 55 | .001 | 125 | | 88 | .452 | 92 | | 83 | .053 | 127 |
| | 56 | .002 | 124 | | 89 | .484 | 91 | | 84 | .062 | 126 |
| | 57 | .003 | 123 | | 90 | .516 | 90 | | 85 | .072 | 125 |
| | 58 | .004 | 122 | | | | | | 86 | .083 | 124 |
| | 59 | .005 | 121 | | | | | | 87 | .095 | 123 |
| | 60 | .007 | 120 | | | $m = 10$ | | | 88 | .109 | 122 |
| | 61 | .009 | 119 | | | | | | 89 | .124 | 121 |
| | 62 | .011 | 118 | 10 | 55 | .000 | 155 | | 90 | .140 | 120 |
| | 63 | .014 | 117 | | 56 | .000 | 154 | | 91 | .157 | 119 |
| | 64 | .017 | 116 | | 57 | .000 | 153 | | 92 | .176 | 118 |
| | 65 | .022 | 115 | | 58 | .000 | 152 | | 93 | .197 | 117 |
| | 66 | .027 | 114 | | 59 | .000 | 151 | | 94 | .218 | 116 |
| | 67 | .033 | 113 | | 60 | .000 | 150 | | 95 | .241 | 115 |
| | 68 | .039 | 112 | | 61 | .000 | 149 | | 96 | .264 | 114 |
| | 69 | .047 | 111 | | 62 | .000 | 148 | | 97 | .289 | 113 |
| | 70 | .056 | 110 | | 63 | .000 | 147 | | 98 | .315 | 112 |
| | 71 | .067 | 109 | | 64 | .001 | 146 | | 99 | .342 | 111 |
| | 72 | .078 | 108 | | 65 | .001 | 145 | | 100 | .370 | 110 |
| | 73 | .091 | 107 | | 66 | .001 | 144 | | 101 | .398 | 109 |
| | 74 | .106 | 106 | | 67 | .001 | 143 | | 102 | .427 | 108 |
| | 75 | .121 | 105 | | 68 | .002 | 142 | | 103 | .456 | 107 |
| | 76 | .139 | 104 | | 69 | .003 | 141 | | 104 | .485 | 106 |
| | 77 | .158 | 103 | | 70 | .003 | 140 | | 105 | .515 | 105 |
| | | | | | 71 | .004 | 139 | | | | |
| | | | | | 72 | .006 | 138 | | | | |

For $m$ or $n$ larger than 10, the probabilities are found from Table A as follows:

$$z_{x,L} = \frac{T_x + 0.5 - m(N + 1)/2}{\sqrt{mn(N + 1)/12}} \qquad z_{x,R} = \frac{T_x - 0.5 - m(N + 1)/2}{\sqrt{mn(N + 1)/12}}$$

| Desired | Approximated by |
|---|---|
| Left-tail probability for $T_x$ | Left-tail probability for $z_{x,L}$ |
| Right-tail probability for $T_x$ | Right-tail probability for $z_{x,R}$ |

Source: Adapted from Table B of C. H. Kraft and C. Van Eeden (1969), *A Nonparametric Introduction to Statistics*, Macmillan Publishing Co., New York, with permission.

## TABLE I
## SPEARMAN RANK CORRELATION STATISTIC

Entries labeled $P$ in the table are the cumulative probability, right-tail from the value of $R$ to its maximum value one, for all $R \geq 0, n \leq 10$. The same probability is a cumulative left-tail probability, from the minimum value minus one to the value $-R$. For $10 < n \leq 30$, the table gives the smallest value of $R$ (largest value of $-R$) for which the right-tail (left-tail) probability for a one-sided test is less than or equal to selected values, .100, .050, .025, .010, .005, .001, shown on the top row. These same values apply to $|R|$ for a two-sided test with tail probability .200, .100, .050, .020, .010, .002, shown on the bottom row.

| $n$ | $R$ | $P$ | $n$ | $R$ | $P$ | $n$ | $R$ | $P$ | $n$ | $R$ | $P$ |
|---|---|---|---|---|---|---|---|---|---|---|---|
| 3 | 1.000 | .167 | 7 | 1.000 | .000 | 8 | .810 | .011 | 9 | 1.000 | .000 |
|   | .500 | .500 |   | .964 | .001 |   | .786 | .014 |   | .983 | .000 |
| 4 | 1.000 | .042 |   | .929 | .003 |   | .762 | .018 |   | .967 | .000 |
|   | .800 | .167 |   | .893 | .006 |   | .738 | .023 |   | .950 | .000 |
|   | .600 | .208 |   | .857 | .012 |   | .714 | .029 |   | .933 | .000 |
|   | .400 | .375 |   | .821 | .017 |   | .690 | .035 |   | .917 | .001 |
|   | .200 | .458 |   | .786 | .024 |   | .667 | .042 |   | .900 | .001 |
|   | .000 | .542 |   | .750 | .033 |   | .643 | .048 |   | .883 | .002 |
| 5 | 1.000 | .008 |   | .714 | .044 |   | .619 | .057 |   | .867 | .002 |
|   | .900 | .042 |   | .679 | .055 |   | .595 | .066 |   | .850 | .003 |
|   | .800 | .067 |   | .643 | .069 |   | .571 | .076 |   | .833 | .004 |
|   | .700 | .117 |   | .607 | .083 |   | .548 | .085 |   | .817 | .005 |
|   | .600 | .175 |   | .571 | .100 |   | .524 | .098 |   | .800 | .007 |
|   | .500 | .225 |   | .536 | .118 |   | .500 | .108 |   | .783 | .009 |
|   | .400 | .258 |   | .500 | .133 |   | .476 | .122 |   | .767 | .011 |
|   | .300 | .342 |   | .464 | .151 |   | .452 | .134 |   | .750 | .013 |
|   | .200 | .392 |   | .429 | .177 |   | .429 | .150 |   | .733 | .016 |
|   | .100 | .475 |   | .393 | .198 |   | .405 | .163 |   | .717 | .018 |
|   | .000 | .525 |   | .357 | .222 |   | .381 | .180 |   | .700 | .022 |
| 6 | 1.000 | .001 |   | .321 | .249 |   | .357 | .195 |   | .683 | .025 |
|   | .943 | .008 |   | .286 | .278 |   | .333 | .214 |   | .667 | .029 |
|   | .886 | .017 |   | .250 | .297 |   | .310 | .231 |   | .650 | .033 |
|   | .829 | .029 |   | .214 | .331 |   | .286 | .250 |   | .633 | .038 |
|   | .771 | .051 |   | .179 | .357 |   | .262 | .268 |   | .617 | .043 |
|   | .714 | .068 |   | .143 | .391 |   | .238 | .291 |   | .600 | .048 |
|   | .657 | .088 |   | .107 | .420 |   | .214 | .310 |   | .583 | .054 |
|   | .600 | .121 |   | .071 | .453 |   | .190 | .332 |   | .567 | .060 |
|   | .543 | .149 |   | .036 | .482 |   | .167 | .352 |   | .550 | .066 |
|   | .486 | .178 |   | .000 | .518 |   | .143 | .376 |   | .533 | .074 |
|   | .429 | .210 | 8 | 1.000 | .000 |   | .119 | .397 |   | .517 | .081 |
|   | .371 | .249 |   | .976 | .000 |   | .095 | .420 |   | .500 | .089 |
|   | .314 | .282 |   | .952 | .001 |   | .071 | .441 |   | .483 | .097 |
|   | .257 | .329 |   | .929 | .001 |   | .048 | .467 |   | .467 | .106 |
|   | .200 | .357 |   | .905 | .002 |   | .024 | .488 |   | .450 | .115 |
|   | .143 | .401 |   | .881 | .004 |   | .000 | .512 |   | .433 | .125 |
|   | .086 | .460 |   | .857 | .005 |   |   |   |   | .417 | .135 |
|   | .029 | .500 |   | .833 | .008 |   |   |   |   | .400 | .146 |

TABLE 1 (Continued)

| n | R | P | n | R | P | n | R | P | n | R | P |
|---|---|---|---|---|---|---|---|---|---|---|---|
| 9 | .383 | .156 | 10 | .964 | .000 | 10 | .636 | .027 | 10 | .309 | .193 |
|   | .367 | .168 |    | .952 | .000 |    | .624 | .030 |    | .297 | .203 |
|   | .350 | .179 |    | .939 | .000 |    | .612 | .033 |    | .285 | .214 |
|   | .333 | .193 |    | .927 | .000 |    | .600 | .037 |    | .273 | .224 |
|   | .317 | .205 |    | .915 | .000 |    | .588 | .040 |    | .261 | .235 |
|   | .300 | .218 |    | .903 | .000 |    | .576 | .044 |    | .248 | .246 |
|   | .283 | .231 |    | .891 | .001 |    | .564 | .048 |    | .236 | .257 |
|   | .267 | .247 |    | .879 | .001 |    | .552 | .052 |    | .224 | .268 |
|   | .250 | .260 |    | .867 | .001 |    | .539 | .057 |    | .212 | .280 |
|   | .233 | .276 |    | .855 | .001 |    | .527 | .062 |    | .200 | .292 |
|   | .217 | .290 |    | .842 | .002 |    | .515 | .067 |    | .188 | .304 |
|   | .200 | .307 |    | .830 | .002 |    | .503 | .072 |    | .176 | .316 |
|   | .183 | .322 |    | .818 | .003 |    | .491 | .077 |    | .164 | .328 |
|   | .167 | .339 |    | .806 | .004 |    | .479 | .083 |    | .152 | .341 |
|   | .150 | .354 |    | .794 | .004 |    | .467 | .089 |    | .139 | .354 |
|   | .133 | .372 |    | .782 | .005 |    | .455 | .096 |    | .127 | .367 |
|   | .117 | .388 |    | .770 | .007 |    | .442 | .102 |    | .115 | .379 |
|   | .100 | .405 |    | .758 | .008 |    | .430 | .109 |    | .103 | .393 |
|   | .083 | .422 |    | .745 | .009 |    | .418 | .116 |    | .091 | .406 |
|   | .067 | .440 |    | .733 | .010 |    | .406 | .124 |    | .079 | .419 |
|   | .050 | .456 |    | .721 | .012 |    | .394 | .132 |    | .067 | .433 |
|   | .033 | .474 |    | .709 | .013 |    | .382 | .139 |    | .055 | .446 |
|   | .017 | .491 |    | .697 | .015 |    | .370 | .148 |    | .042 | .459 |
|   | .000 | .509 |    | .685 | .017 |    | .358 | .156 |    | .030 | .473 |
| 10 | 1.000 | .000 |   | .673 | .019 |    | .345 | .165 |    | .018 | .486 |
|   | .988 | .000 |    | .661 | .022 |    | .333 | .174 |    | .006 | .500 |
|   | .976 | .000 |    | .648 | .025 |    | .321 | .184 |    |      |      |

TABLE I (Continued)

| $n$ | Right-Tail (Left-Tail) Probability on $R$ $(-R)$ for One-Sided Test | | | | | |
|---|---|---|---|---|---|---|
| | .100 | .050 | .025 | .010 | .005 | .001 |
| 11 | .427 | .536 | .618 | .709 | .764 | .855 |
| 12 | .406 | .503 | .587 | .678 | .734 | .825 |
| 13 | .385 | .484 | .560 | .648 | .703 | .797 |
| 14 | .367 | .464 | .538 | .626 | .679 | .771 |
| 15 | .354 | .446 | .521 | .604 | .657 | .750 |
| 16 | .341 | .429 | .503 | .585 | .635 | .729 |
| 17 | .329 | .414 | .488 | .566 | .618 | .711 |
| 18 | .317 | .401 | .474 | .550 | .600 | .692 |
| 19 | .309 | .391 | .460 | .535 | .584 | .675 |
| 20 | .299 | .380 | .447 | .522 | .570 | .660 |
| 21 | .292 | .370 | .436 | .509 | .556 | .647 |
| 22 | .284 | .361 | .425 | .497 | .544 | .633 |
| 23 | .278 | .353 | .416 | .486 | .532 | .620 |
| 24 | .275 | .344 | .407 | .476 | .521 | .608 |
| 25 | .265 | .337 | .398 | .466 | .511 | .597 |
| 26 | .260 | .331 | .390 | .457 | .501 | .586 |
| 27 | .255 | .324 | .383 | .449 | .492 | .576 |
| 28 | .250 | .318 | .376 | .441 | .483 | .567 |
| 29 | .245 | .312 | .369 | .433 | .475 | .557 |
| 30 | .241 | .307 | .363 | .426 | .467 | .548 |
| | .200 | .100 | .050 | .020 | .010 | .002 |
| | Tail Probability on $|R|$ for Two-Sided Test | | | | | |

For $n > 30$, the probabilities are found from Table A by calculating $z = R\sqrt{n-1}$. The left- or right-tail probability for $R$ can be approximated by the left or right tail probability for $z$ respectively.

Source: Part I ($n \le 10$) is adapted from Table 44 of E. S. Pearson and H. O. Hartley, eds. (1966), *Biometrika Tables for Statisticians*, Vol. 1, 3rd ed., Cambridge University Press, Cambridge, England, with permission of the Biometrika Trustees, and from Table 2 of M. G. Kendall (1948, 4th ed. 1970), *Rank Correlation Methods*, Charles Griffin & Co., Ltd., London & High Wycombe, with permission of the author and publisher. Part II ($10 < n \le 30$) is adapted from G. J. Glasser and R. F. Winter (1961), Critical values of the rank correlation coefficient for testing the hypothesis of independence, *Biometrika*, 48, 444-448, with permission of the Editor of Biometrika Auxiliary Publications and the authors.

## TABLE J
## KENDALL TAU STATISTIC

Entries labeled $P$ in the table are the cumulative probability, right-tail from the value of $T$ to its maximum value one, for all $T \geq 0$, $n \leq 10$. The same probability is a cumulative left-tail probability, from the minimum value minus one to the value $-T$. For $10 < n \leq 30$, the table gives the smallest value of $T$ (largest value of $-T$) for which the right-tail (left-tail) probability for a one-sided test is less than or equal to selected values, .100, .050, .025, .010, .005, shown on the top row. These same values apply to $|T|$ for a two-sided test with tail probability .200, .100, .050, .020, .010, shown on the bottom row.

| n | T | P | n | T | P | n | T | P | n | T | P |
|---|---|---|---|---|---|---|---|---|---|---|---|
| 3 | 1.000 | .167 | 7 | 1.000 | .000 | 9 | 1.000 | .000 | 10 | 1.000 | .000 |
|   | .333 | .500 |   | .905 | .001 |   | .944 | .000 |   | .956 | .000 |
| 4 | 1.000 | .042 |   | .810 | .005 |   | .889 | .000 |   | .911 | .000 |
|   | .667 | .167 |   | .714 | .015 |   | .833 | .000 |   | .867 | .000 |
|   | .333 | .375 |   | .619 | .035 |   | .778 | .001 |   | .822 | .000 |
|   | .000 | .625 |   | .524 | .068 |   | .722 | .003 |   | .778 | .000 |
| 5 | 1.000 | .008 |   | .429 | .119 |   | .667 | .006 |   | .733 | .001 |
|   | .800 | .042 |   | .333 | .191 |   | .611 | .012 |   | .689 | .002 |
|   | .600 | .117 |   | .238 | .281 |   | .556 | .022 |   | .644 | .005 |
|   | .400 | .242 |   | .143 | .386 |   | .500 | .038 |   | .600 | .008 |
|   | .200 | .408 |   | .048 | .500 |   | .444 | .060 |   | .556 | .014 |
|   | .000 | .592 | 8 | 1.000 | .000 |   | .389 | .090 |   | .511 | .023 |
| 6 | 1.000 | .001 |   | .929 | .000 |   | .333 | .130 |   | .467 | .036 |
|   | .867 | .008 |   | .857 | .001 |   | .278 | .179 |   | .422 | .054 |
|   | .733 | .028 |   | .786 | .003 |   | .222 | .238 |   | .378 | .078 |
|   | .600 | .068 |   | .714 | .007 |   | .167 | .306 |   | .333 | .108 |
|   | .467 | .136 |   | .643 | .016 |   | .111 | .381 |   | .289 | .146 |
|   | .333 | .235 |   | .571 | .031 |   | .056 | .460 |   | .244 | .190 |
|   | .200 | .360 |   | .500 | .054 |   | .000 | .540 |   | .200 | .242 |
|   | .067 | .500 |   | .429 | .089 |   |   |   |   | .156 | .300 |
|   |   |   |   | .357 | .138 |   |   |   |   | .111 | .364 |
|   |   |   |   | .286 | .199 |   |   |   |   | .067 | .431 |
|   |   |   |   | .214 | .274 |   |   |   |   | .022 | .500 |
|   |   |   |   | .143 | .360 |   |   |   |   |   |   |
|   |   |   |   | .071 | .452 |   |   |   |   |   |   |
|   |   |   |   | .000 | .548 |   |   |   |   |   |   |

**TABLE  J**  (Continued)

| $n$ | Right-Tail (Left-Tail) Probability on $T\,(-T)$ for One-Sided Test | | | | |
|---|---|---|---|---|---|
|  | .100 | .050 | .025 | .010 | .005 |
| 11 | .345 | .418 | .491 | .564 | .600 |
| 12 | .303 | .394 | .455 | .545 | .576 |
| 13 | .308 | .359 | .436 | .513 | .564 |
| 14 | .275 | .363 | .407 | .473 | .516 |
| 15 | .276 | .333 | .390 | .467 | .505 |
| 16 | .250 | .317 | .383 | .433 | .483 |
| 17 | .250 | .309 | .368 | .426 | .471 |
| 18 | .242 | .294 | .346 | .412 | .451 |
| 19 | .228 | .287 | .333 | .392 | .439 |
| 20 | .221 | .274 | .326 | .379 | .421 |
| 21 | .210 | .267 | .314 | .371 | .410 |
| 22 | .195 | .253 | .295 | .344 | .378 |
| 23 | .202 | .257 | .296 | .352 | .391 |
| 24 | .196 | .246 | .290 | .341 | .377 |
| 25 | .193 | .240 | .287 | .333 | .367 |
| 26 | .188 | .237 | .280 | .329 | .360 |
| 27 | .179 | .231 | .271 | .322 | .356 |
| 28 | .180 | .228 | .265 | .312 | .344 |
| 29 | .172 | .222 | .261 | .310 | .340 |
| 30 | .172 | .218 | .255 | .301 | .333 |
|  | .200 | .100 | .050 | .020 | .010 |
|  | Tail Probability on $|T|$ for Two-Sided Test | | | | |

For $n > 30$, the probabilities are found from Table A by calculating $z = 3T\sqrt{n(n-1)}/\sqrt{2(2n+5)}$. The left- or right-tail probability for $T$ can be approximated by the left- or right-tail probability for $z$, respectively.

Source: Part I ($n \leq 10$) is adapted from Table 45 of E. S. Pearson and H. O. Hartley, eds. (1966), *Biometrika Tables for Statisticians*, Vol. 1, 3rd ed., Cambridge University Press, Cambridge, England, with permission of the Biometrika Trustees, and from Table 1 of M. G. Kendall (1948, 4th ed. 1970), *Rank Correlation Methods*, Charles Griffin & Co., Ltd., London & High Wycombe, with permission of the author and publisher. Part II ($10 < n \leq 30$) is adapted from L. Kaarsemaker and A. van Wijngaarden (1953), Tables for use in rank correlation, *Statistica Neerlandica*, 7, 41–54, with permission.

## TABLE K
## KENDALL COEFFICIENT OF CONCORDANCE AND
## FRIEDMAN ANALYSIS OF VARIANCE STATISTICS

Entries labeled $P$ in the table are the cumulative probability, right-tail from the value of $S$ to its maximum value, for all $P \leq .50$, $k \leq 8$ for $n = 3$, $k \leq 4$ for $n = 4$.

| n | k | S | P | n | k | S | P | n | k | S | P | n | k | S | P |
|---|---|---|---|---|---|---|---|---|---|---|---|---|---|---|---|
| 3 | 2 | 8 | .167 | 3 | 7 | 98 | .000 | 4 | 2 | 20 | .042 | 4 | 4 | 80 | .000 |
|   |   | 6 | .500 |   |   | 96 | .000 |   |   | 18 | .167 |   |   | 78 | .001 |
|   | 3 | 18 | .028 |   |   | 86 | .000 |   |   | 16 | .208 |   |   | 76 | .001 |
|   |   | 14 | .194 |   |   | 78 | .001 |   |   | 14 | .375 |   |   | 74 | .001 |
|   |   | 8 | .361 |   |   | 74 | .003 |   |   | 12 | .458 |   |   | 72 | .002 |
|   | 4 | 32 | .005 |   |   | 72 | .004 |   | 3 | 45 | .002 |   |   | 70 | .003 |
|   |   | 26 | .042 |   |   | 62 | .008 |   |   | 43 | .002 |   |   | 68 | .003 |
|   |   | 24 | .069 |   |   | 56 | .016 |   |   | 41 | .017 |   |   | 66 | .006 |
|   |   | 18 | .125 |   |   | 54 | .021 |   |   | 37 | .033 |   |   | 64 | .007 |
|   |   | 14 | .273 |   |   | 50 | .027 |   |   | 35 | .054 |   |   | 62 | .012 |
|   |   | 8 | .431 |   |   | 42 | .051 |   |   | 33 | .075 |   |   | 58 | .014 |
|   | 5 | 50 | .001 |   |   | 38 | .085 |   |   | 29 | .148 |   |   | 56 | .019 |
|   |   | 42 | .008 |   |   | 32 | .112 |   |   | 27 | .175 |   |   | 54 | .033 |
|   |   | 38 | .024 |   |   | 26 | .192 |   |   | 25 | .207 |   |   | 52 | .036 |
|   |   | 32 | .039 |   |   | 24 | .237 |   |   | 21 | .300 |   |   | 50 | .052 |
|   |   | 26 | .093 |   |   | 18 | .305 |   |   | 19 | .342 |   |   | 48 | .054 |
|   |   | 24 | .124 |   |   | 14 | .486 |   |   | 17 | .446 |   |   | 46 | .068 |
|   |   | 18 | .182 |   | 8 | 128 | .000 |   |   |   |   |   |   | 44 | .077 |
|   |   | 14 | .367 |   |   | 126 | .000 |   |   |   |   |   |   | 42 | .094 |
|   | 6 | 72 | .000 |   |   | 122 | .000 |   |   |   |   |   |   | 40 | .105 |
|   |   | 62 | .002 |   |   | 114 | .000 |   |   |   |   |   |   | 38 | .141 |
|   |   | 56 | .006 |   |   | 104 | .000 |   |   |   |   |   |   | 36 | .158 |
|   |   | 54 | .008 |   |   | 98 | .001 |   |   |   |   |   |   | 34 | .190 |
|   |   | 50 | .012 |   |   | 96 | .001 |   |   |   |   |   |   | 32 | .200 |
|   |   | 42 | .029 |   |   | 86 | .002 |   |   |   |   |   |   | 30 | .242 |
|   |   | 38 | .052 |   |   | 78 | .005 |   |   |   |   |   |   | 26 | .324 |
|   |   | 32 | .072 |   |   | 74 | .008 |   |   |   |   |   |   | 24 | .355 |
|   |   | 26 | .142 |   |   | 72 | .010 |   |   |   |   |   |   | 22 | .389 |
|   |   | 24 | .184 |   |   | 62 | .018 |   |   |   |   |   |   | 20 | .432 |
|   |   | 18 | .252 |   |   | 56 | .030 |   |   |   |   |   |   |   |   |
|   |   | 14 | .430 |   |   | 54 | .038 |   |   |   |   |   |   |   |   |
|   |   |   |   |   |   | 50 | .047 |   |   |   |   |   |   |   |   |
|   |   |   |   |   |   | 42 | .079 |   |   |   |   |   |   |   |   |
|   |   |   |   |   |   | 38 | .120 |   |   |   |   |   |   |   |   |
|   |   |   |   |   |   | 32 | .149 |   |   |   |   |   |   |   |   |
|   |   |   |   |   |   | 26 | .236 |   |   |   |   |   |   |   |   |
|   |   |   |   |   |   | 24 | .285 |   |   |   |   |   |   |   |   |
|   |   |   |   |   |   | 18 | .355 |   |   |   |   |   |   |   |   |

**TABLE K  (Continued)**

For $n$ and $k$ outside the range of this table, right-tail probabilities are found from Table B as follows:

$$Q = \frac{12S}{kn(n+1)}$$

| Desired | Approximated by |
| --- | --- |
| Right-tail probability for $S$ or $W$ | Right-tail probability for $Q$ with $n - 1$ degrees of freedom |

Source: Adapted from Table 46 of E. S. Pearson and H. O. Hartley, eds. (1966), *Biometrika Tables for Statisticians*, Vol. 1, 3rd ed., Cambridge University Press, Cambridge, England, with permission of the Biometrika Trustees, and from Table 5 of M. G. Kendall (1948, 4th ed., 1970), *Rank Correlation Methods*, Charles Griffin & Co., Ltd., London & High Wycombe, with permission of the author and publisher.

## TABLE L
## NUMBER OF RUNS DISTRIBUTION

Entries labeled P in the table are the cumulative probability from each extreme to the value of U for the given $m \le n$. Tail probabilities are given only for $P \le .50$.

**Left-Tail Probabilities**

| m | n | U | P | m | n | U | P | m | n | U | P | m | n | U | P |
|---|---|---|---|---|---|---|---|---|---|---|---|---|---|---|---|
| 2 | 2 | 2 | .333 | 2 | 18 | 2 | .011 | 3 | 14 | 2 | .003 | 4 | 10 | 2 | .002 |
| 2 | 3 | 2 | .200 |   |    | 3 | .105 |   |    | 3 | .025 |   |    | 3 | .014 |
|   |   | 3 | .500 |   |    | 4 | .284 |   |    | 4 | .101 |   |    | 4 | .068 |
| 2 | 4 | 2 | .133 | 3 | 3  | 2 | .100 |   |    | 5 | .350 |   |    | 5 | .203 |
|   |   | 3 | .400 |   |    | 3 | .300 | 3 | 15 | 2 | .002 |   |    | 6 | .419 |
| 2 | 5 | 2 | .095 | 3 | 4  | 2 | .057 |   |    | 3 | .022 | 4 | 11 | 2 | .001 |
|   |   | 3 | .333 |   |    | 3 | .200 |   |    | 4 | .091 |   |    | 3 | .011 |
| 2 | 6 | 2 | .071 | 3 | 5  | 2 | .036 |   |    | 5 | .331 |   |    | 4 | .055 |
|   |   | 3 | .286 |   |    | 3 | .143 | 3 | 16 | 2 | .002 |   |    | 5 | .176 |
| 2 | 7 | 2 | .056 |   |    | 4 | .429 |   |    | 3 | .020 |   |    | 6 | .374 |
|   |   | 3 | .250 | 3 | 6  | 2 | .024 |   |    | 4 | .082 | 4 | 12 | 2 | .001 |
| 2 | 8 | 2 | .044 |   |    | 3 | .107 |   |    | 5 | .314 |   |    | 3 | .009 |
|   |   | 3 | .222 |   |    | 4 | .345 | 3 | 17 | 2 | .002 |   |    | 4 | .045 |
| 2 | 9 | 2 | .036 | 3 | 7  | 2 | .017 |   |    | 3 | .018 |   |    | 5 | .154 |
|   |   | 3 | .200 |   |    | 3 | .083 |   |    | 4 | .074 |   |    | 6 | .335 |
|   |   | 4 | .491 |   |    | 4 | .283 |   |    | 5 | .298 | 4 | 13 | 2 | .001 |
| 2 | 10| 2 | .030 | 3 | 8  | 2 | .012 | 4 | 4  | 2 | .029 |   |    | 3 | .007 |
|   |   | 3 | .182 |   |    | 3 | .067 |   |    | 3 | .114 |   |    | 4 | .037 |
|   |   | 4 | .455 |   |    | 4 | .236 |   |    | 4 | .371 |   |    | 5 | .136 |
| 2 | 11| 2 | .026 | 3 | 9  | 2 | .009 | 4 | 5  | 2 | .016 |   |    | 6 | .302 |
|   |   | 3 | .167 |   |    | 3 | .055 |   |    | 3 | .071 | 4 | 14 | 2 | .001 |
|   |   | 4 | .423 |   |    | 4 | .200 |   |    | 4 | .262 |   |    | 3 | .006 |
| 2 | 12| 2 | .022 |   |    | 5 | .491 |   |    | 5 | .500 |   |    | 4 | .031 |
|   |   | 3 | .154 | 3 | 10 | 2 | .007 | 4 | 6  | 2 | .010 |   |    | 5 | .121 |
|   |   | 4 | .396 |   |    | 3 | .045 |   |    | 3 | .048 |   |    | 6 | .274 |
| 2 | 13| 2 | .019 |   |    | 4 | .171 |   |    | 4 | .190 | 4 | 15 | 2 | .001 |
|   |   | 3 | .143 |   |    | 5 | .455 |   |    | 5 | .405 |   |    | 3 | .005 |
|   |   | 4 | .371 | 3 | 11 | 2 | .005 | 4 | 7  | 2 | .006 |   |    | 4 | .027 |
| 2 | 14| 2 | .017 |   |    | 3 | .038 |   |    | 3 | .033 |   |    | 5 | .108 |
|   |   | 3 | .133 |   |    | 4 | .148 |   |    | 4 | .142 |   |    | 6 | .249 |
|   |   | 4 | .350 |   |    | 5 | .423 |   |    | 5 | .333 | 4 | 16 | 2 | .000 |
| 2 | 15| 2 | .015 | 3 | 12 | 2 | .004 | 4 | 8  | 2 | .004 |   |    | 3 | .004 |
|   |   | 3 | .125 |   |    | 3 | .033 |   |    | 3 | .024 |   |    | 4 | .023 |
|   |   | 4 | .331 |   |    | 4 | .130 |   |    | 4 | .109 |   |    | 5 | .097 |
| 2 | 16| 2 | .013 |   |    | 5 | .396 |   |    | 5 | .279 |   |    | 6 | .227 |
|   |   | 3 | .118 | 3 | 13 | 2 | .004 | 4 | 9  | 2 | .003 | 5 | 5  | 2 | .008 |
|   |   | 4 | .314 |   |    | 3 | .029 |   |    | 3 | .018 |   |    | 3 | .040 |
| 2 | 17| 2 | .012 |   |    | 4 | .114 |   |    | 4 | .085 |   |    | 4 | .167 |
|   |   | 3 | .111 |   |    | 5 | .371 |   |    | 5 | .236 |   |    | 5 | .357 |
|   |   | 4 | .298 |   |    |   |      |   |    | 6 | .471 |   |    |   |      |

TABLE L (Continued)

## Left-Tail Probabilities

| m | n | U | P | m | n | U | P | m | n | U | P | m | n | U | P |
|---|---|---|---|---|---|---|---|---|---|---|---|---|---|---|---|
| 5 | 6 | 2 | .004 | 5 | 14 | 2 | .000 | 6 | 11 | 2 | .000 | 7 | 9 | 2 | .000 |
|   |   | 3 | .024 |   |    | 3 | .002 |   |    | 3 | .001 |   |   | 3 | .001 |
|   |   | 4 | .110 |   |    | 4 | .011 |   |    | 4 | .009 |   |   | 4 | .010 |
|   |   | 5 | .262 |   |    | 5 | .044 |   |    | 5 | .036 |   |   | 5 | .035 |
| 5 | 7 | 2 | .003 |   |    | 6 | .125 |   |    | 6 | .108 |   |   | 6 | .108 |
|   |   | 3 | .015 |   |    | 7 | .299 |   |    | 7 | .242 |   |   | 7 | .231 |
|   |   | 4 | .076 |   |    | 8 | .496 |   |    | 8 | .436 |   |   | 8 | .427 |
|   |   | 5 | .197 | 5 | 15 | 2 | .000 | 6 | 12 | 2 | .000 | 7 | 10 | 2 | .000 |
|   |   | 6 | .424 |   |    | 3 | .001 |   |    | 3 | .001 |   |    | 3 | .001 |
| 5 | 8 | 2 | .002 |   |    | 4 | .009 |   |    | 4 | .007 |   |    | 4 | .006 |
|   |   | 3 | .010 |   |    | 5 | .037 |   |    | 5 | .028 |   |    | 5 | .024 |
|   |   | 4 | .054 |   |    | 6 | .108 |   |    | 6 | .087 |   |    | 6 | .080 |
|   |   | 5 | .152 |   |    | 7 | .272 |   |    | 7 | .205 |   |    | 7 | .182 |
|   |   | 6 | .347 |   |    | 8 | .460 |   |    | 8 | .383 |   |    | 8 | .355 |
| 5 | 9 | 2 | .001 | 6 | 6  | 2 | .002 | 6 | 13 | 2 | .000 | 7 | 11 | 2 | .000 |
|   |   | 3 | .007 |   |    | 3 | .013 |   |    | 3 | .001 |   |    | 3 | .001 |
|   |   | 4 | .039 |   |    | 4 | .067 |   |    | 4 | .005 |   |    | 4 | .004 |
|   |   | 5 | .119 |   |    | 5 | .175 |   |    | 5 | .022 |   |    | 5 | .018 |
|   |   | 6 | .287 |   |    | 6 | .392 |   |    | 6 | .070 |   |    | 6 | .060 |
| 5 | 10 | 2 | .001 | 6 | 7 | 2 | .001 |   |    | 7 | .176 |   |    | 7 | .145 |
|   |    | 3 | .005 |   |   | 3 | .008 |   |    | 8 | .338 |   |    | 8 | .296 |
|   |    | 4 | .029 |   |   | 4 | .043 | 6 | 14 | 2 | .000 |   |    | 9 | .484 |
|   |    | 5 | .095 |   |   | 5 | .121 |   |    | 3 | .001 | 7 | 12 | 2 | .000 |
|   |    | 6 | .239 |   |   | 6 | .296 |   |    | 4 | .004 |   |    | 3 | .000 |
|   |    | 7 | .455 |   |   | 7 | .500 |   |    | 5 | .017 |   |    | 4 | .003 |
| 5 | 11 | 2 | .000 | 6 | 8 | 2 | .001 |   |    | 6 | .058 |   |    | 5 | .013 |
|   |    | 3 | .004 |   |   | 3 | .005 |   |    | 7 | .151 |   |    | 6 | .046 |
|   |    | 4 | .022 |   |   | 4 | .028 |   |    | 8 | .299 |   |    | 7 | .117 |
|   |    | 5 | .077 |   |   | 5 | .086 | 7 | 7  | 2 | .001 |   |    | 8 | .247 |
|   |    | 6 | .201 |   |   | 6 | .226 |   |    | 3 | .004 |   |    | 9 | .428 |
|   |    | 7 | .407 |   |   | 7 | .413 |   |    | 4 | .025 | 7 | 13 | 2 | .000 |
| 5 | 12 | 2 | .000 | 6 | 9 | 2 | .000 |   |    | 5 | .078 |   |    | 3 | .000 |
|   |    | 3 | .003 |   |   | 3 | .003 |   |    | 6 | .209 |   |    | 4 | .002 |
|   |    | 4 | .017 |   |   | 4 | .019 |   |    | 7 | .383 |   |    | 5 | .010 |
|   |    | 5 | .063 |   |   | 5 | .063 | 7 | 8  | 2 | .000 |   |    | 6 | .035 |
|   |    | 6 | .170 |   |   | 6 | .175 |   |    | 3 | .002 |   |    | 7 | .095 |
|   |    | 7 | .365 |   |   | 7 | .343 |   |    | 4 | .015 |   |    | 8 | .208 |
| 5 | 13 | 2 | .000 | 6 | 10 | 2 | .000 |   |   | 5 | .051 |   |    | 9 | .378 |
|   |    | 3 | .002 |   |    | 3 | .002 |   |   | 6 | .149 | 8 | 8  | 2 | .000 |
|   |    | 4 | .013 |   |    | 4 | .013 |   |   | 7 | .296 |   |    | 3 | .001 |
|   |    | 5 | .053 |   |    | 5 | .047 |   |   |   |      |   |    | 4 | .009 |
|   |    | 6 | .145 |   |    | 6 | .137 |   |   |   |      |   |    | 5 | .032 |
|   |    | 7 | .330 |   |    | 7 | .287 |   |   |   |      |   |    | 6 | .100 |
|   |    |   |      |   |    | 8 | .497 |   |   |   |      |   |    | 7 | .214 |
|   |    |   |      |   |    |   |      |   |   |   |      |   |    | 8 | .405 |

**TABLE L (Continued)**

**Left-Tail Probabilities**

| m | n | U | P | m | n | U | P | m | n | U | P | m | n | U | P |
|---|---|---|---|---|---|---|---|---|---|---|---|---|---|---|---|
| 8 | 9 | 2 | .000 | 9 | 9 | 2 | .000 | 10 | 10 | 2 | .000 | 11 | 11 | 2 | .000 |
|   |   | 3 | .001 |   |   | 3 | .000 |   |   | 3 | .000 |   |   | 3 | .000 |
|   |   | 4 | .005 |   |   | 4 | .003 |   |   | 4 | .001 |   |   | 4 | .000 |
|   |   | 5 | .020 |   |   | 5 | .012 |   |   | 5 | .004 |   |   | 5 | .002 |
|   |   | 6 | .069 |   |   | 6 | .044 |   |   | 6 | .019 |   |   | 6 | .007 |
|   |   | 7 | .157 |   |   | 7 | .109 |   |   | 7 | .051 |   |   | 7 | .023 |
|   |   | 8 | .319 |   |   | 8 | .238 |   |   | 8 | .128 |   |   | 8 | .063 |
|   |   | 9 | .500 |   |   | 9 | .399 |   |   | 9 | .242 |   |   | 9 | .135 |
| 8 | 10 | 2 | .000 | 9 | 10 | 2 | .000 |   |   | 10 | .414 |   |   | 10 | .260 |
|   |   | 3 | .000 |   |   | 3 | .000 | 10 | 11 | 2 | .000 |   |   | 11 | .410 |
|   |   | 4 | .003 |   |   | 4 | .002 |   |   | 3 | .000 | 11 | 12 | 2 | .000 |
|   |   | 5 | .013 |   |   | 5 | .008 |   |   | 4 | .001 |   |   | 3 | .000 |
|   |   | 6 | .048 |   |   | 6 | .029 |   |   | 5 | .003 |   |   | 4 | .000 |
|   |   | 7 | .117 |   |   | 7 | .077 |   |   | 6 | .012 |   |   | 5 | .001 |
|   |   | 8 | .251 |   |   | 8 | .179 |   |   | 7 | .035 |   |   | 6 | .005 |
|   |   | 9 | .419 |   |   | 9 | .319 |   |   | 8 | .092 |   |   | 7 | .015 |
| 8 | 11 | 2 | .000 | 9 | 11 | 2 | .000 |   |   | 9 | .185 |   |   | 8 | .044 |
|   |   | 3 | .000 |   |   | 3 | .000 |   |   | 10 | .335 |   |   | 9 | .099 |
|   |   | 4 | .002 |   |   | 4 | .001 |   |   | 11 | .500 |   |   | 10 | .202 |
|   |   | 5 | .009 |   |   | 5 | .005 | 10 | 12 | 2 | .000 |   |   | 11 | .335 |
|   |   | 6 | .034 |   |   | 6 | .020 |   |   | 3 | .000 | 12 | 12 | 2 | .000 |
|   |   | 7 | .088 |   |   | 7 | .055 |   |   | 4 | .000 |   |   | 3 | .000 |
|   |   | 8 | .199 |   |   | 8 | .135 |   |   | 5 | .002 |   |   | 4 | .000 |
|   |   | 9 | .352 |   |   | 9 | .255 |   |   | 6 | .008 |   |   | 5 | .001 |
| 8 | 12 | 2 | .000 |   |   | 10 | .430 |   |   | 7 | .024 |   |   | 6 | .003 |
|   |   | 3 | .000 | 9 | 12 | 2 | .000 |   |   | 8 | .067 |   |   | 7 | .009 |
|   |   | 4 | .001 |   |   | 3 | .000 |   |   | 9 | .142 |   |   | 8 | .030 |
|   |   | 5 | .006 |   |   | 4 | .001 |   |   | 10 | .271 |   |   | 9 | .070 |
|   |   | 6 | .025 |   |   | 5 | .003 |   |   | 11 | .425 |   |   | 10 | .150 |
|   |   | 7 | .067 |   |   | 6 | .014 |   |   |   |   |   |   | 11 | .263 |
|   |   | 8 | .159 |   |   | 7 | .040 |   |   |   |   |   |   | 12 | .421 |
|   |   | 9 | .297 |   |   | 8 | .103 |   |   |   |   |   |   |   |   |
|   |   | 10 | .480 |   |   | 9 | .205 |   |   |   |   |   |   |   |   |
|   |   |   |   |   |   | 10 | .362 |   |   |   |   |   |   |   |   |

TABLE L (Continued)

**Right-Tail Probabilities**

| m | n | U | P | m | n | U | P | m | n | U | P | m | n | U | P |
|---|---|---|---|---|---|---|---|---|---|---|---|---|---|---|---|
| 2 | 2 | 4 | .333 | 4 | 10 | 9 | .126 | 5 | 14 | 11 | .111 | 7 | 7 | 14 | .001 |
| 2 | 3 | 5 | .100 |   |   | 8 | .294 |   |   | 10 | .234 |   |   | 13 | .004 |
|   |   | 4 | .500 | 4 | 11 | 9 | .154 | 5 | 15 | 11 | .129 |   |   | 12 | .025 |
| 2 | 4 | 5 | .200 |   |   | 8 | .330 |   |   | 10 | .258 |   |   | 11 | .078 |
| 2 | 5 | 5 | .286 | 4 | 12 | 9 | .181 | 6 | 6 | 12 | .002 |   |   | 10 | .209 |
| 2 | 6 | 5 | .357 |   |   | 8 | .363 |   |   | 11 | .013 |   |   | 9 | .383 |
| 2 | 7 | 5 | .417 | 4 | 13 | 9 | .208 |   |   | 10 | .067 | 7 | 8 | 15 | .000 |
| 2 | 8 | 5 | .467 |   |   | 8 | .393 |   |   | 9 | .175 |   |   | 14 | .002 |
| 3 | 3 | 6 | .100 | 4 | 14 | 9 | .234 |   |   | 8 | .392 |   |   | 13 | .012 |
|   |   | 5 | .300 |   |   | 8 | .421 | 6 | 7 | 13 | .001 |   |   | 12 | .051 |
| 3 | 4 | 7 | .029 | 4 | 15 | 9 | .258 |   |   | 12 | .008 |   |   | 11 | .133 |
|   |   | 6 | .200 |   |   | 8 | .446 |   |   | 11 | .034 |   |   | 10 | .296 |
|   |   | 5 | .457 | 4 | 16 | 9 | .282 |   |   | 10 | .121 |   |   | 9 | .486 |
| 3 | 5 | 7 | .071 |   |   | 8 | .470 |   |   | 9 | .267 | 7 | 9 | 15 | .001 |
|   |   | 6 | .286 | 5 | 5 | 10 | .008 |   |   | 8 | .500 |   |   | 14 | .006 |
| 3 | 6 | 7 | .119 |   |   | 9 | .040 | 6 | 8 | 13 | .002 |   |   | 13 | .025 |
|   |   | 6 | .357 |   |   | 8 | .167 |   |   | 12 | .016 |   |   | 12 | .084 |
| 3 | 7 | 7 | .167 |   |   | 7 | .357 |   |   | 11 | .063 |   |   | 11 | .194 |
|   |   | 6 | .417 | 5 | 6 | 11 | .002 |   |   | 10 | .179 |   |   | 10 | .378 |
| 3 | 8 | 7 | .212 |   |   | 10 | .024 |   |   | 9 | .354 | 7 | 10 | 15 | .002 |
|   |   | 6 | .467 |   |   | 9 | .089 | 6 | 9 | 13 | .006 |   |   | 14 | .010 |
| 3 | 9 | 7 | .255 |   |   | 8 | .262 |   |   | 12 | .028 |   |   | 13 | .043 |
| 3 | 10 | 7 | .294 |   |   | 7 | .478 |   |   | 11 | .098 |   |   | 12 | .121 |
| 3 | 11 | 7 | .330 | 5 | 7 | 11 | .008 |   |   | 10 | .238 |   |   | 11 | .257 |
| 3 | 12 | 7 | .363 |   |   | 10 | .045 |   |   | 9 | .434 |   |   | 10 | .451 |
| 3 | 13 | 7 | .393 |   |   | 9 | .146 | 6 | 10 | 13 | .010 | 7 | 11 | 15 | .004 |
| 3 | 14 | 7 | .421 |   |   | 8 | .348 |   |   | 12 | .042 |   |   | 14 | .017 |
| 3 | 15 | 7 | .446 | 5 | 8 | 11 | .016 |   |   | 11 | .136 |   |   | 13 | .064 |
| 3 | 16 | 7 | .470 |   |   | 10 | .071 |   |   | 10 | .294 |   |   | 12 | .160 |
| 3 | 17 | 7 | .491 |   |   | 9 | .207 | 6 | 11 | 13 | .017 |   |   | 11 | .318 |
| 4 | 4 | 8 | .029 |   |   | 8 | .424 |   |   | 12 | .058 | 7 | 12 | 15 | .007 |
|   |   | 7 | .114 | 5 | 9 | 11 | .028 |   |   | 11 | .176 |   |   | 14 | .025 |
|   |   | 6 | .371 |   |   | 10 | .098 |   |   | 10 | .346 |   |   | 13 | .089 |
| 4 | 5 | 9 | .008 |   |   | 9 | .266 | 6 | 12 | 13 | .025 |   |   | 12 | .199 |
|   |   | 8 | .071 |   |   | 8 | .490 |   |   | 12 | .075 |   |   | 11 | .376 |
|   |   | 7 | .214 | 5 | 10 | 11 | .042 |   |   | 11 | .217 | 7 | 13 | 15 | .010 |
|   |   | 6 | .500 |   |   | 10 | .126 |   |   | 10 | .395 |   |   | 14 | .034 |
| 4 | 6 | 9 | .024 |   |   | 9 | .322 | 6 | 13 | 13 | .034 |   |   | 13 | .116 |
|   |   | 8 | .119 | 5 | 11 | 11 | .058 |   |   | 12 | .092 |   |   | 12 | .238 |
|   |   | 7 | .310 |   |   | 10 | .154 |   |   | 11 | .257 |   |   | 11 | .430 |
| 4 | 7 | 9 | .045 |   |   | 9 | .374 |   |   | 10 | .439 | 8 | 8 | 16 | .000 |
|   |   | 8 | .167 | 5 | 12 | 11 | .075 | 6 | 14 | 13 | .044 |   |   | 15 | .001 |
|   |   | 7 | .394 |   |   | 10 | .181 |   |   | 12 | .111 |   |   | 14 | .009 |
| 4 | 8 | 9 | .071 |   |   | 9 | .421 |   |   | 11 | .295 |   |   | 13 | .032 |
|   |   | 8 | .212 | 5 | 13 | 11 | .092 |   |   | 10 | .480 |   |   | 12 | .100 |
|   |   | 7 | .467 |   |   | 10 | .208 |   |   |   |   |   |   | 11 | .214 |
| 4 | 9 | 9 | .098 |   |   | 9 | .465 |   |   |   |   |   |   | 10 | .405 |
|   |   | 8 | .255 |   |   |   |   |   |   |   |   |   |   |   |   |

TABLE L  (Continued)

**Right-Tail Probabilities**

| m | n | U | P | m | n | U | P | m | n | U | P | m | n | U | P |
|---|---|---|---|---|---|---|---|---|---|---|---|---|---|---|---|
| 8 | 9 | 17 | .000 | 9 | 9 | 18 | .000 | 10 | 10 | 20 | .000 | 11 | 11 | 22 | .000 |
|   |   | 16 | .001 |   |   | 17 | .000 |   |   | 19 | .000 |   |   | 21 | .000 |
|   |   | 15 | .004 |   |   | 16 | .003 |   |   | 18 | .000 |   |   | 20 | .000 |
|   |   | 14 | .020 |   |   | 15 | .012 |   |   | 17 | .001 |   |   | 19 | .002 |
|   |   | 13 | .061 |   |   | 14 | .044 |   |   | 16 | .004 |   |   | 18 | .007 |
|   |   | 12 | .157 |   |   | 13 | .109 |   |   | 15 | .019 |   |   | 17 | .023 |
|   |   | 11 | .298 |   |   | 12 | .238 |   |   | 14 | .051 |   |   | 16 | .063 |
|   |   | 10 | .500 |   |   | 11 | .399 |   |   | 13 | .128 |   |   | 15 | .135 |
| 8 | 10 | 17 | .000 | 9 | 10 | 19 | .000 |   |   | 12 | .242 |   |   | 14 | .260 |
|   |   | 16 | .002 |   |   | 18 | .000 |   |   | 11 | .414 |   |   | 13 | .410 |
|   |   | 15 | .010 |   |   | 17 | .001 | 10 | 11 | 21 | .000 | 11 | 12 | 23 | .000 |
|   |   | 14 | .036 |   |   | 16 | .008 |   |   | 20 | .000 |   |   | 22 | .000 |
|   |   | 13 | .097 |   |   | 15 | .026 |   |   | 19 | .000 |   |   | 21 | .000 |
|   |   | 12 | .218 |   |   | 14 | .077 |   |   | 18 | .003 |   |   | 20 | .001 |
|   |   | 11 | .379 |   |   | 13 | .166 |   |   | 17 | .010 |   |   | 19 | .004 |
| 8 | 11 | 17 | .001 |   |   | 12 | .319 |   |   | 16 | .035 |   |   | 18 | .015 |
|   |   | 16 | .004 |   |   | 11 | .490 |   |   | 15 | .085 |   |   | 17 | .041 |
|   |   | 15 | .018 | 9 | 11 | 19 | .000 |   |   | 14 | .185 |   |   | 16 | .099 |
|   |   | 14 | .057 |   |   | 18 | .001 |   |   | 13 | .320 |   |   | 15 | .191 |
|   |   | 13 | .138 |   |   | 17 | .003 |   |   | 12 | .500 |   |   | 14 | .335 |
|   |   | 12 | .278 |   |   | 16 | .015 | 10 | 12 | 21 | .000 |   |   | 13 | .493 |
|   |   | 11 | .453 |   |   | 15 | .045 |   |   | 20 | .000 | 12 | 12 | 24 | .000 |
| 8 | 12 | 17 | .001 |   |   | 14 | .115 |   |   | 19 | .001 |   |   | 23 | .000 |
|   |   | 16 | .007 |   |   | 13 | .227 |   |   | 18 | .006 |   |   | 22 | .000 |
|   |   | 15 | .029 |   |   | 12 | .395 |   |   | 17 | .020 |   |   | 21 | .001 |
|   |   | 14 | .080 |   |   |   |   |   |   | 16 | .056 |   |   | 20 | .003 |
|   |   | 13 | .183 |   |   |   |   |   |   | 15 | .125 |   |   | 19 | .009 |
|   |   | 12 | .337 |   |   |   |   |   |   | 14 | .245 |   |   | 18 | .030 |
|   |   |   |   |   |   |   |   |   |   | 13 | .395 |   |   | 17 | .070 |
|   |   |   |   |   |   |   |   |   |   |   |   |   |   | 16 | .150 |
|   |   |   |   |   |   |   |   |   |   |   |   |   |   | 15 | .263 |
|   |   |   |   |   |   |   |   |   |   |   |   |   |   | 14 | .421 |

For $m + n = N > 20$ and $m > 12$, $n > 12$, the probabilities are found from Table A as follows:

$$z_L = \frac{U + 0.5 - 1 - 2mn/N}{\sqrt{\dfrac{2mn(2mn - N)}{N^2(N - 1)}}} \qquad z_R = \frac{U - 0.5 - 1 - 2mn/N}{\sqrt{\dfrac{2mn(2mn - N)}{N^2(N - 1)}}}$$

| **Desired** | **Approximated by** |
|---|---|
| Left-tail probability for $U$ | Left-tail probability for $z_L$ |
| Right-tail probability for $U$ | Right-tail probability for $z_R$ |

Source: Adapted from Frieda S. Swed and C. Eisenhart (1943), Tables for testing randomness of grouping in a sequence of alternatives, *Annals of Mathematical Statistics*, 14, 66–87, with permission.

## TABLE M
## NUMBER OF RUNS UP AND DOWN DISTRIBUTION

Entries in the table labeled $P$ are the cumulative probability from each extreme to the value of $V$. Left-tail probabilities are given only for $V$ less than or equal to its median; right-tail probabilities are given for all other values of $V$. $N$ is the number of observations for which successive differences are nonzero, and hence one more than the number of symbols of signs of differences.

| N | V | Left-tail P | V | Right-tail P | N | V | Left-tail P | V | Right-tail P |
|---|---|---|---|---|---|---|---|---|---|
| 3 | 1 | .3333 | 2 | .6667 | 13 | 1 | .0000 | | |
| 4 | | | 3 | .4167 | | 2 | .0000 | | |
| | 1 | .0833 | 2 | .9167 | | 3 | .0001 | 12 | .0072 |
| 5 | 1 | .0167 | 4 | .2667 | | 4 | .0026 | 11 | .0568 |
| | 2 | .2500 | 3 | .7500 | | 5 | .0213 | 10 | .2058 |
| 6 | 1 | .0028 | | | | 6 | .0964 | 9 | .4587 |
| | 2 | .0861 | 5 | .1694 | | 7 | .2749 | 8 | .7251 |
| | 3 | .4139 | 4 | .5861 | 14 | 1 | .0000 | | |
| 7 | 1 | .0004 | 6 | .1079 | | 2 | .0000 | | |
| | 2 | .0250 | 5 | .4417 | | 3 | .0000 | | |
| | 3 | .1909 | 4 | .8091 | | 4 | .0007 | 13 | .0046 |
| 8 | 1 | .0000 | | | | 5 | .0079 | 12 | .0391 |
| | 2 | .0063 | 7 | .0687 | | 6 | .0441 | 11 | .1536 |
| | 3 | .0749 | 6 | .3250 | | 7 | .1534 | 10 | .3722 |
| | 4 | .3124 | 5 | .6876 | | 8 | .3633 | 9 | .6367 |
| 9 | 1 | .0000 | | | 15 | 1 | .0000 | | |
| | 2 | .0014 | | | | 2 | .0000 | | |
| | 3 | .0257 | 8 | .0437 | | 3 | .0000 | | |
| | 4 | .1500 | 7 | .2347 | | 4 | .0002 | | |
| | 5 | .4347 | 6 | .5653 | | 5 | .0027 | 14 | .0029 |
| 10 | 1 | .0000 | | | | 6 | .0186 | 13 | .0267 |
| | 2 | .0003 | 9 | .0278 | | 7 | .0782 | 12 | .1134 |
| | 3 | .0079 | 8 | .1671 | | 8 | .2216 | 11 | .2970 |
| | 4 | .0633 | 7 | .4524 | | 9 | .4520 | 10 | .5480 |
| | 5 | .2427 | 6 | .7573 | 16 | 1 | .0000 | | |
| 11 | 1 | .0000 | | | | 2 | .0000 | | |
| | 2 | .0001 | | | | 3 | .0000 | | |
| | 3 | .0022 | 10 | .0177 | | 4 | .0001 | 15 | .0019 |
| | 4 | .0239 | 9 | .1177 | | 5 | .0009 | 14 | .0182 |
| | 5 | .1196 | 8 | .3540 | | 6 | .0072 | 13 | .0828 |
| | 6 | .3438 | 7 | .6562 | | 7 | .0367 | 12 | .2335 |
| 12 | 1 | .0000 | | | | 8 | .1238 | 11 | .4631 |
| | 2 | .0000 | | | | 9 | .2975 | 10 | .7025 |
| | 3 | .0005 | | | | | | | |
| | 4 | .0082 | 11 | .0113 | | | | | |
| | 5 | .0529 | 10 | .0821 | | | | | |
| | 6 | .1918 | 9 | .2720 | | | | | |
| | 7 | .4453 | 8 | .5547 | | | | | |

**TABLE M (Continued)**

| N | V | Left-tail P | V | Right-tail P | N | V | Left-tail P | V | Right-tail P |
|---|---|---|---|---|---|---|---|---|---|
| 17 | 1 | .0000 | | | 21 | 1 | .0000 | | |
| | 2 | .0000 | | | | 2 | .0000 | | |
| | 3 | .0000 | | | | 3 | .0000 | | |
| | 4 | .0000 | | | | 4 | .0000 | | |
| | 5 | .0003 | 16 | .0012 | | 5 | .0000 | | |
| | 6 | .0026 | 15 | .0123 | | 6 | .0000 | | |
| | 7 | .0160 | 14 | .0600 | | 7 | .0003 | 20 | .0002 |
| | 8 | .0638 | 13 | .1812 | | 8 | .0023 | 19 | .0025 |
| | 9 | .1799 | 12 | .3850 | | 9 | .0117 | 18 | .0154 |
| | 10 | .3770 | 11 | .6230 | | 10 | .0431 | 17 | .0591 |
| 18 | 1 | .0000 | | | | 11 | .1202 | 16 | .1602 |
| | 2 | .0000 | | | | 12 | .2622 | 15 | .3293 |
| | 3 | .0000 | | | | 13 | .4603 | 14 | .5397 |
| | 4 | .0000 | | | 22 | 1 | .0000 | | |
| | 5 | .0001 | | | | 2 | .0000 | | |
| | 6 | .0009 | 17 | .0008 | | 3 | .0000 | | |
| | 7 | .0065 | 16 | .0083 | | 4 | .0000 | | |
| | 8 | .0306 | 15 | .0431 | | 5 | .0000 | | |
| | 9 | .1006 | 14 | .1389 | | 6 | .0000 | 21 | .0001 |
| | 10 | .2443 | 13 | .3152 | | 7 | .0001 | 20 | .0017 |
| | 11 | .4568 | 12 | .5432 | | 8 | .0009 | 19 | .0108 |
| 19 | 1 | .0000 | | | | 9 | .0050 | 18 | .0437 |
| | 2 | .0000 | | | | 10 | .0213 | 17 | .1251 |
| | 3 | .0000 | | | | 11 | .0674 | 16 | .2714 |
| | 4 | .0000 | | | | 12 | .1661 | 15 | .4688 |
| | 5 | .0000 | 18 | .0005 | | 13 | .3276 | 14 | .6724 |
| | 6 | .0003 | 17 | .0056 | 23 | 1 | .0000 | | |
| | 7 | .0025 | 16 | .0308 | | 2 | .0000 | | |
| | 8 | .0137 | 15 | .1055 | | 3 | .0000 | | |
| | 9 | .0523 | 14 | .2546 | | 4 | .0000 | | |
| | 10 | .1467 | 13 | .4663 | | 5 | .0000 | | |
| | 11 | .3144 | 12 | .6856 | | 6 | .0000 | | |
| 20 | 1 | .0000 | | | | 7 | .0000 | 22 | .0001 |
| | 2 | .0000 | | | | 8 | .0003 | 21 | .0011 |
| | 3 | .0000 | | | | 9 | .0021 | 20 | .0076 |
| | 4 | .0000 | | | | 10 | .0099 | 19 | .0321 |
| | 5 | .0000 | | | | 11 | .0356 | 18 | .0968 |
| | 6 | .0001 | 19 | .0003 | | 12 | .0988 | 17 | .2211 |
| | 7 | .0009 | 18 | .0038 | | 13 | .2188 | 16 | .4020 |
| | 8 | .0058 | 17 | .0218 | | 14 | .3953 | 15 | .6047 |
| | 9 | .0255 | 16 | .0793 | | | | | |
| | 10 | .0821 | 15 | .2031 | | | | | |
| | 11 | .2012 | 14 | .3945 | | | | | |
| | 12 | .3873 | 13 | .6127 | | | | | |

TABLE M (Continued)

| N | V | Left-tail P | V | Right-tail P | N | V | Left-tail P | V | Right-tail P |
|---|---|---|---|---|---|---|---|---|---|
| 24 | 1 | .0000 | | | 25 | 1 | .0000 | | |
| | 2 | .0000 | | | | 2 | .0000 | | |
| | 3 | .0000 | | | | 3 | .0000 | | |
| | 4 | .0000 | | | | 4 | .0000 | | |
| | 5 | .0000 | | | | 5 | .0000 | | |
| | 6 | .0000 | | | | 6 | .0000 | | |
| | 7 | .0000 | | | | 7 | .0000 | 24 | .0000 |
| | 8 | .0001 | 23 | .0000 | | 8 | .0000 | 23 | .0005 |
| | 9 | .0008 | 22 | .0007 | | 9 | .0003 | 22 | .0037 |
| | 10 | .0044 | 21 | .0053 | | 10 | .0018 | 21 | .0170 |
| | 11 | .0177 | 20 | .0235 | | 11 | .0084 | 20 | .0564 |
| | 12 | .0554 | 19 | .0742 | | 12 | .0294 | 19 | .1423 |
| | 13 | .1374 | 18 | .1783 | | 13 | .0815 | 18 | .2852 |
| | 14 | .2768 | 17 | .3405 | | 14 | .1827 | 17 | .4708 |
| | 15 | .4631 | 16 | .5369 | | 15 | .3384 | 16 | .6616 |

For $N > 25$, the probabilities are found from Table A as follows:

$$z_L = \frac{V + 0.5 - (2N - 1)/3}{\sqrt{(16N - 29)/90}} \qquad z_R = \frac{V - 0.5 - (2N - 1)/3}{\sqrt{(16N - 29)/90}}$$

| Desired | Approximated by |
|---|---|
| Left-tail probability for $V$ | Left-tail probability for $z_L$ |
| Right-tail probability for $V$ | Right-tail probability for $z_R$ |

Source: Adapted from E. S. Edgington (1961), Probability table for number of runs of signs of first differences, *Journal of the American Statistical Association*, 56, 156–159 with permission.

---

**TABLE N**
**CRITICAL $z$ VALUES FOR $p$ MULTIPLE COMPARISONS**
    An entry in the table for a given $p$ and level of significance $\alpha$ is the quantile point of the standard normal distribution such that the right-tail probability is equal to $\alpha/2p$. For values of $p$ outside the range of this table, $z$ can be found from Table A by linear interpolation.

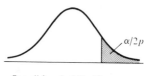

Overall Level of Significance $\alpha$

|  |  |  | $\alpha$ |  |  |  |
| --- | --- | --- | --- | --- | --- | --- |
| $p$ | .30 | .25 | .20 | .15 | .10 | .05 |
| 1 | 1.036 | 1.150 | 1.282 | 1.440 | 1.645 | 1.960 |
| 2 | 1.440 | 1.534 | 1.645 | 1.780 | 1.960 | 2.241 |
| 3 | 1.645 | 1.732 | 1.834 | 1.960 | 2.128 | 2.394 |
| 4 | 1.780 | 1.863 | 1.960 | 2.080 | 2.241 | 2.498 |
| 5 | 1.881 | 1.960 | 2.054 | 2.170 | 2.326 | 2.576 |
| 6 | 1.960 | 2.037 | 2.128 | 2.241 | 2.394 | 2.638 |
| 7 | 2.026 | 2.100 | 2.189 | 2.300 | 2.450 | 2.690 |
| 8 | 2.080 | 2.154 | 2.241 | 2.350 | 2.498 | 2.734 |
| 9 | 2.128 | 2.200 | 2.287 | 2.394 | 2.539 | 2.773 |
| 10 | 2.170 | 2.241 | 2.326 | 2.432 | 2.576 | 2.807 |
| 11 | 2.208 | 2.278 | 2.362 | 2.467 | 2.608 | 2.838 |
| 12 | 2.241 | 2.301 | 2.394 | 2.498 | 2.638 | 2.866 |
| 15 | 2.326 | 2.394 | 2.475 | 2.576 | 2.713 | 2.935 |
| 21 | 2.450 | 2.515 | 2.593 | 2.690 | 2.823 | 3.038 |
| 28 | 2.552 | 2.615 | 2.690 | 2.785 | 2.913 | 3.125 |

    Source: Adapted from Table IX of R. A. Fisher and F. A. Yates (1963), *Statistical Tables for Biological, Agricultural and Medical Research*, 6th ed., Hafner Publishing Company, New York, with permission of Professor Yates and of Longman Group Ltd., 6th ed., 1974, previously published by Oliver & Boyd Ltd.

# TABLE O
## RANDOM NUMBERS

This table may be entered at any point and read in any direction and for any number of digits at a time.

| 10480 | 15011 | 01536 | 02011 | 81647 | 91646 | 69179 | 14194 | 62590 | 36207 | 20969 | 99570 | 91291 | 90700 |
| 22368 | 46573 | 25595 | 85393 | 30995 | 89198 | 27982 | 53402 | 93965 | 34095 | 52666 | 19174 | 39615 | 99505 |
| 24130 | 48360 | 22527 | 97265 | 76393 | 64809 | 15179 | 24830 | 49340 | 32081 | 30680 | 19655 | 63348 | 58629 |
| 42167 | 93093 | 06243 | 61680 | 07856 | 16376 | 39440 | 53537 | 71341 | 57004 | 00849 | 74917 | 97758 | 16379 |
| 37570 | 39975 | 81837 | 16656 | 06121 | 91782 | 60468 | 81305 | 49684 | 60672 | 14110 | 06927 | 01263 | 54613 |
| | | | | | | | | | | | | | |
| 77921 | 06907 | 11008 | 42751 | 27756 | 53498 | 18602 | 70659 | 90655 | 15053 | 21916 | 81825 | 44394 | 42880 |
| 99562 | 72905 | 56420 | 69994 | 98872 | 31016 | 71194 | 18738 | 44013 | 48840 | 63213 | 21069 | 10634 | 12952 |
| 96301 | 91977 | 05463 | 07972 | 18876 | 20922 | 94595 | 56869 | 69014 | 60045 | 18425 | 84903 | 42508 | 32307 |
| 89579 | 14342 | 63661 | 10281 | 17453 | 18103 | 57740 | 84378 | 25331 | 12566 | 58678 | 44947 | 05585 | 56941 |
| 85475 | 36857 | 53342 | 53988 | 53060 | 59533 | 38867 | 62300 | 08158 | 17983 | 16439 | 11458 | 18593 | 64952 |
| | | | | | | | | | | | | | |
| 28918 | 69578 | 88231 | 33276 | 70997 | 79936 | 56865 | 05859 | 90106 | 31595 | 01547 | 85590 | 91610 | 78188 |
| 63553 | 40961 | 48235 | 03427 | 49626 | 69445 | 18663 | 72695 | 52180 | 20847 | 12243 | 90511 | 33703 | 90322 |
| 09429 | 93969 | 52636 | 92737 | 88974 | 33488 | 36320 | 17617 | 30015 | 08272 | 84115 | 27156 | 30613 | 74952 |
| 10365 | 61129 | 87529 | 85689 | 48237 | 52267 | 67689 | 93394 | 01511 | 26358 | 85104 | 20285 | 29975 | 89868 |
| 07119 | 97336 | 71048 | 08178 | 77233 | 13916 | 47564 | 81056 | 97735 | 85977 | 29372 | 74461 | 28551 | 90707 |
| | | | | | | | | | | | | | |
| 51085 | 12765 | 51821 | 51259 | 77452 | 16308 | 60756 | 92144 | 49442 | 53900 | 70960 | 63990 | 75601 | 40719 |
| 02368 | 21382 | 52404 | 60268 | 89368 | 19885 | 55322 | 44819 | 01188 | 65255 | 64835 | 44919 | 05944 | 55157 |
| 01011 | 54092 | 33362 | 94904 | 31273 | 04146 | 18594 | 29852 | 71585 | 85030 | 51132 | 01915 | 92747 | 64951 |
| 52162 | 53916 | 46369 | 58586 | 23216 | 14513 | 83149 | 98736 | 23495 | 64350 | 94738 | 17752 | 35156 | 35749 |
| 07056 | 97628 | 33787 | 09998 | 42698 | 06691 | 76988 | 13602 | 51851 | 46104 | 88916 | 19509 | 25625 | 58104 |

## TABLE O  (Continued)

| | | | | | | | | | | | | | |
|---|---|---|---|---|---|---|---|---|---|---|---|---|---|
| 48663 | 91245 | 85828 | 14346 | 09172 | 30168 | 90229 | 04734 | 59193 | 22178 | 30421 | 61666 | 99904 | 32812 |
| 54164 | 58492 | 22421 | 74103 | 47070 | 25306 | 76468 | 26384 | 58151 | 06646 | 21524 | 15227 | 96909 | 44592 |
| 32639 | 32363 | 05597 | 24200 | 13363 | 38005 | 94342 | 28728 | 35806 | 06912 | 17012 | 64161 | 18296 | 22851 |
| 29334 | 27001 | 87637 | 87308 | 58731 | 00256 | 45834 | 15398 | 46557 | 41135 | 10367 | 07684 | 36188 | 18510 |
| 02488 | 33062 | 28834 | 07351 | 19731 | 92420 | 60952 | 61280 | 50001 | 67658 | 32586 | 86679 | 50720 | 94953 |
| 81525 | 72295 | 04839 | 96423 | 24878 | 82651 | 66566 | 14778 | 76797 | 14780 | 13300 | 87074 | 79666 | 95725 |
| 29676 | 20591 | 68086 | 26432 | 46901 | 20849 | 89768 | 81536 | 86645 | 12659 | 92259 | 57102 | 80428 | 25280 |
| 00742 | 57392 | 39064 | 66432 | 84673 | 40027 | 32832 | 61362 | 98947 | 96067 | 64760 | 64584 | 96096 | 98253 |
| 05366 | 04213 | 25669 | 26422 | 44407 | 44048 | 37937 | 63904 | 45766 | 66134 | 75470 | 66520 | 34693 | 90449 |
| 91921 | 26418 | 64117 | 94305 | 26766 | 25940 | 39972 | 22209 | 71500 | 64568 | 91402 | 42416 | 07844 | 69618 |
| 00582 | 04711 | 87917 | 77341 | 42206 | 35126 | 74087 | 99547 | 81817 | 42607 | 43808 | 76655 | 62028 | 76630 |
| 00725 | 69884 | 62797 | 56170 | 86324 | 88072 | 76222 | 36086 | 84637 | 93161 | 76038 | 65855 | 77919 | 88006 |
| 69011 | 65795 | 95876 | 55293 | 18988 | 27354 | 26575 | 08615 | 40801 | 59920 | 29841 | 80150 | 12777 | 48501 |
| 25976 | 57948 | 29888 | 88604 | 67917 | 48708 | 18912 | 82271 | 65424 | 69774 | 33611 | 54262 | 85963 | 03547 |
| 09763 | 83473 | 73577 | 12908 | 30883 | 18317 | 28290 | 35797 | 05998 | 41688 | 34952 | 37888 | 38917 | 88050 |
| 91567 | 42595 | 27958 | 30134 | 04024 | 86385 | 29880 | 99730 | 55536 | 84855 | 29080 | 09250 | 79656 | 73211 |
| 17955 | 56349 | 90999 | 49127 | 20044 | 59931 | 06115 | 20542 | 18059 | 02008 | 73708 | 83517 | 36103 | 42791 |
| 46503 | 18584 | 18845 | 49618 | 02304 | 51038 | 20655 | 58727 | 28168 | 15475 | 56942 | 53389 | 20562 | 87338 |
| 92157 | 89634 | 94824 | 78171 | 84610 | 82834 | 09922 | 25417 | 44137 | 84813 | 25555 | 21246 | 35509 | 20468 |
| 14577 | 62765 | 35605 | 81263 | 39667 | 47358 | 56873 | 56307 | 61607 | 49518 | 89656 | 20103 | 77490 | 18062 |

## TABLE O (Continued)

| 98427 | 07523 | 33362 | 64270 | 01638 | 92477 | 66969 | 98420 | 04880 | 45585 | 46565 | 04102 | 46880 | 45709 |
| 34914 | 63976 | 88720 | 82765 | 34476 | 17032 | 87589 | 40836 | 32427 | 70002 | 70663 | 88863 | 77775 | 69348 |
| 70060 | 28277 | 39475 | 46473 | 23219 | 53416 | 94970 | 25832 | 69975 | 94484 | 19661 | 72828 | 00102 | 66794 |
| 53976 | 54914 | 06990 | 67245 | 68350 | 82948 | 11398 | 42878 | 80287 | 88267 | 47363 | 46634 | 06541 | 97809 |
| 76072 | 29515 | 40980 | 07391 | 58745 | 25774 | 22987 | 80059 | 39911 | 96189 | 41151 | 14222 | 60697 | 59583 |
| 90725 | 52210 | 83974 | 29992 | 65831 | 38857 | 50490 | 83765 | 55657 | 14361 | 31720 | 57375 | 56228 | 41546 |
| 64364 | 67412 | 33339 | 31926 | 14883 | 24413 | 59744 | 92351 | 97473 | 89286 | 35931 | 04110 | 23726 | 51900 |
| 08962 | 00358 | 31662 | 25388 | 61642 | 34072 | 81249 | 35648 | 56891 | 69352 | 48373 | 45578 | 78547 | 81788 |
| 95012 | 68379 | 93526 | 70765 | 10592 | 04542 | 76463 | 54328 | 02349 | 17247 | 28865 | 14777 | 62730 | 92277 |
| 15664 | 10493 | 20492 | 38391 | 91132 | 21999 | 59516 | 81652 | 27195 | 48223 | 46751 | 22923 | 32261 | 85653 |
| 16408 | 81899 | 04153 | 53381 | 79401 | 21438 | 83035 | 92350 | 36693 | 31238 | 59649 | 91754 | 72772 | 02338 |
| 18629 | 81953 | 05520 | 91962 | 04739 | 13092 | 97662 | 24822 | 94730 | 06496 | 35090 | 04822 | 86774 | 98289 |
| 73115 | 35101 | 47498 | 87637 | 99016 | 71060 | 88824 | 71013 | 18735 | 20286 | 23153 | 72924 | 35165 | 43040 |
| 57491 | 16703 | 23167 | 49323 | 45021 | 33132 | 12544 | 41035 | 80780 | 45393 | 44812 | 12515 | 98931 | 91202 |
| 30405 | 83946 | 23792 | 14422 | 15059 | 45799 | 22716 | 19792 | 09983 | 74353 | 68668 | 30429 | 70735 | 25499 |
| 16631 | 35006 | 85900 | 98275 | 32388 | 52390 | 16815 | 69298 | 82732 | 38480 | 73817 | 32523 | 41961 | 44437 |
| 96773 | 20206 | 42559 | 78985 | 05300 | 22164 | 24369 | 54224 | 35083 | 19687 | 11052 | 91491 | 60383 | 19746 |
| 38935 | 64202 | 14349 | 82674 | 66523 | 44133 | 00697 | 35552 | 35970 | 19124 | 63318 | 29686 | 03387 | 59846 |
| 31624 | 76384 | 17403 | 53363 | 44167 | 64486 | 64758 | 75366 | 76554 | 31601 | 12614 | 33072 | 60332 | 92325 |
| 78919 | 19474 | 23632 | 27889 | 47914 | 02584 | 37680 | 20801 | 72152 | 39339 | 34806 | 08930 | 85001 | 87820 |

Source: Adapted from the 30 page table of 105,000 random digits prepared by the Bureau of Economics of the Interstate Commerce Commission, Washington, D.C., with permission.

# TABLE P — SQUARES AND SQUARE ROOTS

For any number in the first column, the corresponding number in the second column is its square, the corresponding number in the third column is its square root, and the corresponding number in the fourth column is the square root of 10 times the number. In order to obtain the square or square root of any number with three significant digits, the decimal must be moved appropriately. For example, since the square root of 1.23 is 1.10905, the square root of 123 11.0905 and the square root of .0123 is .110905. Similarly, since the square root of 12.3 is 3.50714, the square root of 1230 is 35.0714, and the square root of .123 is .350714.

| $n$ | $n^2$ | $\sqrt{n}$ | $\sqrt{10n}$ | $n$ | $n^2$ | $\sqrt{n}$ | $\sqrt{10n}$ |
|---|---|---|---|---|---|---|---|
| 1.00 | 1.0000 | 1.00000 | 3.16228 | 1.50 | 2.2500 | 1.22474 | 3.87298 |
| 1.01 | 1.0201 | 1.00499 | 3.17805 | 1.51 | 2.2801 | 1.22882 | 3.88587 |
| 1.02 | 1.0404 | 1.00995 | 3.19374 | 1.52 | 2.3104 | 1.23288 | 3.89872 |
| 1.03 | 1.0609 | 1.01489 | 3.20936 | 1.53 | 2.3409 | 1.23693 | 3.91152 |
| 1.04 | 1.0816 | 1.01980 | 3.22490 | 1.54 | 2.3716 | 1.24097 | 3.92428 |
| 1.05 | 1.1025 | 1.02470 | 3.24037 | 1.55 | 2.4025 | 1.24499 | 3.93700 |
| 1.06 | 1.1236 | 1.02956 | 3.25576 | 1.56 | 2.4336 | 1.24900 | 3.94968 |
| 1.07 | 1.1449 | 1.03441 | 3.27109 | 1.57 | 2.4649 | 1.25300 | 3.96232 |
| 1.08 | 1.1664 | 1.03923 | 3.28634 | 1.58 | 2.4964 | 1.25698 | 3.97492 |
| 1.09 | 1.1881 | 1.04403 | 3.30151 | 1.59 | 2.5281 | 1.26095 | 3.98748 |
| 1.10 | 1.2100 | 1.04881 | 3.31662 | 1.60 | 2.5600 | 1.26491 | 4.00000 |
| 1.11 | 1.2321 | 1.05357 | 3.33167 | 1.61 | 2.5921 | 1.26886 | 4.01248 |
| 1.12 | 1.2544 | 1.05830 | 3.34664 | 1.62 | 2.6244 | 1.27279 | 4.02492 |
| 1.13 | 1.2769 | 1.06301 | 3.36155 | 1.63 | 2.6569 | 1.27671 | 4.03733 |
| 1.14 | 1.2996 | 1.06771 | 3.37639 | 1.64 | 2.6896 | 1.28062 | 4.04969 |
| 1.15 | 1.3225 | 1.07238 | 3.39116 | 1.65 | 2.7225 | 1.28452 | 4.06202 |
| 1.16 | 1.3456 | 1.07703 | 3.40588 | 1.66 | 2.7556 | 1.28841 | 4.07431 |
| 1.17 | 1.3689 | 1.08167 | 3.42053 | 1.67 | 2.7889 | 1.29228 | 4.08656 |
| 1.18 | 1.3924 | 1.08628 | 3.43511 | 1.68 | 2.8224 | 1.29615 | 4.09878 |
| 1.19 | 1.4161 | 1.09087 | 3.44964 | 1.69 | 2.8561 | 1.30000 | 4.11096 |
| 1.20 | 1.4400 | 1.09545 | 3.46410 | 1.70 | 2.8900 | 1.30384 | 4.12311 |
| 1.21 | 1.4641 | 1.10000 | 3.47851 | 1.71 | 2.9241 | 1.30767 | 4.13521 |
| 1.22 | 1.4884 | 1.10454 | 3.49285 | 1.72 | 2.9584 | 1.31149 | 4.14729 |
| 1.23 | 1.5129 | 1.10905 | 3.50714 | 1.73 | 2.9929 | 1.31529 | 4.15933 |
| 1.24 | 1.5376 | 1.11355 | 3.52136 | 1.74 | 3.0276 | 1.31909 | 4.17133 |
| 1.25 | 1.5625 | 1.11803 | 3.53553 | 1.75 | 3.0625 | 1.32288 | 4.18330 |
| 1.26 | 1.5876 | 1.12250 | 3.54965 | 1.76 | 3.0976 | 1.32665 | 4.19524 |
| 1.27 | 1.6129 | 1.12694 | 3.56371 | 1.77 | 3.1329 | 1.33041 | 4.20714 |
| 1.28 | 1.6384 | 1.13137 | 3.57771 | 1.78 | 3.1684 | 1.33417 | 4.21900 |
| 1.29 | 1.6641 | 1.13578 | 3.59166 | 1.79 | 3.2041 | 1.33791 | 4.23084 |
| 1.30 | 1.6900 | 1.14018 | 3.60555 | 1.80 | 3.2400 | 1.34164 | 4.24264 |
| 1.31 | 1.7161 | 1.14455 | 3.61939 | 1.81 | 3.2761 | 1.34536 | 4.25441 |
| 1.32 | 1.7424 | 1.14891 | 3.63318 | 1.82 | 3.3124 | 1.34907 | 4.26615 |
| 1.33 | 1.7689 | 1.15326 | 3.64692 | 1.83 | 3.3489 | 1.35277 | 4.27785 |
| 1.34 | 1.7956 | 1.15758 | 3.66060 | 1.84 | 3.3856 | 1.35647 | 4.28952 |
| 1.35 | 1.8225 | 1.16190 | 3.67423 | 1.85 | 3.4225 | 1.36015 | 4.30116 |
| 1.36 | 1.8496 | 1.16619 | 3.68782 | 1.86 | 3.4596 | 1.36382 | 4.31277 |
| 1.37 | 1.8769 | 1.17047 | 3.70135 | 1.87 | 3.4969 | 1.36748 | 4.32435 |
| 1.38 | 1.9044 | 1.17473 | 3.71484 | 1.88 | 3.5344 | 1.37113 | 4.33590 |
| 1.39 | 1.9321 | 1.17898 | 3.72827 | 1.89 | 3.5721 | 1.37477 | 4.34741 |
| 1.40 | 1.9600 | 1.18322 | 3.74166 | 1.90 | 3.6100 | 1.37840 | 4.35890 |
| 1.41 | 1.9881 | 1.18743 | 3.75500 | 1.91 | 3.6481 | 1.38203 | 4.37035 |
| 1.42 | 2.0164 | 1.19164 | 3.76829 | 1.92 | 3.6864 | 1.38564 | 4.38178 |
| 1.43 | 2.0449 | 1.19583 | 3.78153 | 1.93 | 3.7249 | 1.38924 | 4.39318 |
| 1.44 | 2.0736 | 1.20000 | 3.79473 | 1.94 | 3.7636 | 1.39284 | 4.40454 |
| 1.45 | 2.1025 | 1.20416 | 3.80789 | 1.95 | 3.8025 | 1.39642 | 4.41588 |
| 1.46 | 2.1316 | 1.20830 | 3.82099 | 1.96 | 3.8416 | 1.40000 | 4.42719 |
| 1.47 | 2.1609 | 1.21244 | 3.83406 | 1.97 | 3.8809 | 1.40357 | 4.43847 |
| 1.48 | 2.1904 | 1.21655 | 3.84708 | 1.98 | 3.9204 | 1.40712 | 4.44972 |
| 1.49 | 2.2201 | 1.22066 | 3.86005 | 1.99 | 3.9601 | 1.41067 | 4.46094 |

| $n$ | $n^2$ | $\sqrt{n}$ | $\sqrt{10n}$ |
|------|--------|---------|----------|
| 2.00 | 4.0000 | 1.41421 | 4.47214 |
| 2.01 | 4.0401 | 1.41774 | 4.48330 |
| 2.02 | 4.0804 | 1.42127 | 4.49444 |
| 2.03 | 4.1209 | 1.42478 | 4.50555 |
| 2.04 | 4.1616 | 1.42829 | 4.51664 |
| 2.05 | 4.2025 | 1.43178 | 4.52769 |
| 2.06 | 4.2436 | 1.43527 | 4.53872 |
| 2.07 | 4.2849 | 1.43875 | 4.54973 |
| 2.08 | 4.3264 | 1.44222 | 4.56070 |
| 2.09 | 4.3681 | 1.44568 | 4.57165 |
| 2.10 | 4.4100 | 1.44914 | 4.58258 |
| 2.11 | 4.4521 | 1.45258 | 4.59347 |
| 2.12 | 4.4944 | 1.45602 | 4.60435 |
| 2.13 | 4.5369 | 1.45945 | 4.61519 |
| 2.14 | 4.5796 | 1.46287 | 4.62601 |
| 2.15 | 4.6225 | 1.46629 | 4.63681 |
| 2.16 | 4.6656 | 1.46969 | 4.64758 |
| 2.17 | 4.7089 | 1.47309 | 4.65833 |
| 2.18 | 4.7524 | 1.47648 | 4.66905 |
| 2.19 | 4.7961 | 1.47986 | 4.67974 |
| 2.20 | 4.8400 | 1.48324 | 4.69042 |
| 2.21 | 4.8841 | 1.48661 | 4.70106 |
| 2.22 | 4.9284 | 1.48997 | 4.71169 |
| 2.23 | 4.9729 | 1.49332 | 4.72229 |
| 2.24 | 5.0176 | 1.49666 | 4.73286 |
| 2.25 | 5.0625 | 1.50000 | 4.74342 |
| 2.26 | 5.1076 | 1.50333 | 4.75395 |
| 2.27 | 5.1529 | 1.50665 | 4.76445 |
| 2.28 | 5.1984 | 1.50997 | 4.77493 |
| 2.29 | 5.2441 | 1.51327 | 4.78539 |
| 2.30 | 5.2900 | 1.51658 | 4.79583 |
| 2.31 | 5.3361 | 1.51987 | 4.80625 |
| 2.32 | 5.3824 | 1.52315 | 4.81664 |
| 2.33 | 5.4289 | 1.52643 | 4.82701 |
| 2.34 | 5.4756 | 1.52971 | 4.83735 |
| 2.35 | 5.5225 | 1.53297 | 4.84768 |
| 2.36 | 5.5696 | 1.53623 | 4.85798 |
| 2.37 | 5.6169 | 1.53948 | 4.86826 |
| 2.38 | 5.6644 | 1.54272 | 4.87852 |
| 2.39 | 5.7121 | 1.54596 | 4.88876 |
| 2.40 | 5.7600 | 1.54919 | 4.89898 |
| 2.41 | 5.8081 | 1.55242 | 4.90918 |
| 2.42 | 5.8564 | 1.55563 | 4.91935 |
| 2.43 | 5.9049 | 1.55885 | 4.92950 |
| 2.44 | 5.9536 | 1.56205 | 4.93964 |
| 2.45 | 6.0025 | 1.56525 | 4.94975 |
| 2.46 | 6.0516 | 1.56844 | 4.95984 |
| 2.47 | 6.1009 | 1.57162 | 4.96991 |
| 2.48 | 6.1504 | 1.57480 | 4.97996 |
| 2.49 | 6.2001 | 1.57797 | 4.98999 |

| $n$ | $n^2$ | $\sqrt{n}$ | $\sqrt{10n}$ |
|------|--------|---------|----------|
| 2.50 | 6.2500 | 1.58114 | 5.00000 |
| 2.51 | 6.3001 | 1.58430 | 5.00999 |
| 2.52 | 6.3504 | 1.58745 | 5.01996 |
| 2.53 | 6.4009 | 1.59060 | 5.02991 |
| 2.54 | 6.4516 | 1.59374 | 5.03984 |
| 2.55 | 6.5025 | 1.59687 | 5.04975 |
| 2.56 | 6.5536 | 1.60000 | 5.05964 |
| 2.57 | 6.6049 | 1.60312 | 5.06952 |
| 2.58 | 6.6564 | 1.60624 | 5.07937 |
| 2.59 | 6.7081 | 1.60935 | 5.08920 |
| 2.60 | 6.7600 | 1.61245 | 5.09902 |
| 2.61 | 6.8121 | 1.61555 | 5.10882 |
| 2.62 | 6.8644 | 1.61864 | 5.11859 |
| 2.63 | 6.9169 | 1.62173 | 5.12835 |
| 2.64 | 6.9696 | 1.62481 | 5.13809 |
| 2.65 | 7.0225 | 1.62788 | 5.14782 |
| 2.66 | 7.0756 | 1.63095 | 5.15752 |
| 2.67 | 7.1289 | 1.63401 | 5.16720 |
| 2.68 | 7.1824 | 1.63707 | 5.17687 |
| 2.69 | 7.2361 | 1.64012 | 5.18652 |
| 2.70 | 7.2900 | 1.64317 | 5.19615 |
| 2.71 | 7.3441 | 1.64621 | 5.20577 |
| 2.72 | 7.3984 | 1.64924 | 5.21536 |
| 2.73 | 7.4529 | 1.65227 | 5.22494 |
| 2.74 | 7.5076 | 1.65529 | 5.23450 |
| 2.75 | 7.5625 | 1.65831 | 5.24404 |
| 2.76 | 7.6176 | 1.66132 | 5.25357 |
| 2.77 | 7.6729 | 1.66433 | 5.26308 |
| 2.78 | 7.7284 | 1.66733 | 5.27257 |
| 2.79 | 7.7841 | 1.67033 | 5.28205 |
| 2.80 | 7.8400 | 1.67332 | 5.29150 |
| 2.81 | 7.8961 | 1.67631 | 5.30094 |
| 2.82 | 7.9524 | 1.67929 | 5.31037 |
| 2.83 | 8.0089 | 1.68226 | 5.31977 |
| 2.84 | 8.0656 | 1.68523 | 5.32917 |
| 2.85 | 8.1225 | 1.68819 | 5.33854 |
| 2.86 | 8.1796 | 1.69115 | 5.34790 |
| 2.87 | 8.2369 | 1.69411 | 5.35724 |
| 2.88 | 8.2944 | 1.69706 | 5.36656 |
| 2.89 | 8.3521 | 1.70000 | 5.37587 |
| 2.90 | 8.4100 | 1.70294 | 5.38516 |
| 2.91 | 8.4681 | 1.70587 | 5.39444 |
| 2.92 | 8.5264 | 1.70880 | 5.40370 |
| 2.93 | 8.5849 | 1.71172 | 5.41295 |
| 2.94 | 8.6436 | 1.71464 | 5.42218 |
| 2.95 | 8.7025 | 1.71756 | 5.43139 |
| 2.96 | 8.7616 | 1.72047 | 5.44059 |
| 2.97 | 8.8209 | 1.72337 | 5.44977 |
| 2.98 | 8.8804 | 1.72627 | 5.45894 |
| 2.99 | 8.9401 | 1.72916 | 5.46809 |

| $n$ | $n^2$ | $\sqrt{n}$ | $\sqrt{10n}$ |
|------|---------|---------|---------|
| **3.00** | 9.0000 | 1.73205 | 5.47723 |
| 3.01 | 9.0601 | 1.73494 | 5.48635 |
| 3.02 | 9.1204 | 1.73781 | 5.49545 |
| 3.03 | 9.1809 | 1.74069 | 5.50454 |
| 3.04 | 9.2416 | 1.74356 | 5.51362 |
| 3.05 | 9.3025 | 1.74642 | 5.52268 |
| 3.06 | 9.3636 | 1.74929 | 5.53173 |
| 3.07 | 9.4249 | 1.75214 | 5.54076 |
| 3.08 | 9.4864 | 1.75499 | 5.54977 |
| 3.09 | 9.5481 | 1.75784 | 5.55878 |
| **3.10** | 9.6100 | 1.76068 | 5.56776 |
| 3.11 | 9.6721 | 1.76352 | 5.57674 |
| 3.12 | 9.7344 | 1.76635 | 5.58570 |
| 3.13 | 9.7969 | 1.76918 | 5.59464 |
| 3.14 | 9.8596 | 1.77200 | 5.60357 |
| 3.15 | 9.9225 | 1.77482 | 5.61249 |
| 3.16 | 9.9856 | 1.77764 | 5.62139 |
| 3.17 | 10.0489 | 1.78045 | 5.63028 |
| 3.18 | 10.1124 | 1.78326 | 5.63915 |
| 3.19 | 10.1761 | 1.78606 | 5.64801 |
| **3.20** | 10.2400 | 1.78885 | 5.65685 |
| 3.21 | 10.3041 | 1.79165 | 5.66569 |
| 3.22 | 10.3684 | 1.79444 | 5.67450 |
| 3.23 | 10.4329 | 1.79722 | 5.68331 |
| 3.24 | 10.4976 | 1.80000 | 5.69210 |
| 3.25 | 10.5625 | 1.80278 | 5.70088 |
| 3.26 | 10.6276 | 1.80555 | 5.70964 |
| 3.27 | 10.6929 | 1.80831 | 5.71839 |
| 3.28 | 10.7584 | 1.81108 | 5.72713 |
| 3.29 | 10.8241 | 1.81384 | 5.73585 |
| **3.30** | 10.8900 | 1.81659 | 5.74456 |
| 3.31 | 10.9561 | 1.81934 | 5.75326 |
| 3.32 | 11.0224 | 1.82209 | 5.76194 |
| 3.33 | 11.0889 | 1.82483 | 5.77062 |
| 3.34 | 11.1556 | 1.82757 | 5.77927 |
| 3.35 | 11.2225 | 1.83030 | 5.78792 |
| 3.36 | 11.2896 | 1.83303 | 5.79655 |
| 3.37 | 11.3569 | 1.83576 | 5.80517 |
| 3.38 | 11.4244 | 1.83848 | 5.81378 |
| 3.39 | 11.4921 | 1.84120 | 5.82237 |
| **3.40** | 11.5600 | 1.84391 | 5.83095 |
| 3.41 | 11.6281 | 1.84662 | 5.83952 |
| 3.42 | 11.6964 | 1.84932 | 5.84808 |
| 3.43 | 11.7649 | 1.85203 | 5.85662 |
| 3.44 | 11.8336 | 1.85472 | 5.86515 |
| 3.45 | 11.9025 | 1.85742 | 5.87367 |
| 3.46 | 11.9716 | 1.86011 | 5.88218 |
| 3.47 | 12.0409 | 1.86279 | 5.89067 |
| 3.48 | 12.1104 | 1.86548 | 5.89915 |
| 3.49 | 12.1801 | 1.86815 | 5.90762 |

| $n$ | $n^2$ | $\sqrt{n}$ | $\sqrt{10n}$ |
|------|---------|---------|---------|
| **3.50** | 12.2500 | 1.87083 | 5.91608 |
| 3.51 | 12.3201 | 1.87350 | 5.92453 |
| 3.52 | 12.3904 | 1.87617 | 5.93296 |
| 3.53 | 12.4609 | 1.87883 | 5.94138 |
| 3.54 | 12.5316 | 1.88149 | 5.94979 |
| 3.55 | 12.6025 | 1.88414 | 5.95819 |
| 3.56 | 12.6736 | 1.88680 | 5.96657 |
| 3.57 | 12.7449 | 1.88944 | 5.97495 |
| 3.58 | 12.8164 | 1.89209 | 5.98331 |
| 3.59 | 12.8881 | 1.89473 | 5.99166 |
| **3.60** | 12.9600 | 1.89737 | 6.00000 |
| 3.61 | 13.0321 | 1.90000 | 6.00833 |
| 3.62 | 13.1044 | 1.90263 | 6.01664 |
| 3.63 | 13.1769 | 1.90526 | 6.02495 |
| 3.64 | 13.2496 | 1.90788 | 6.03324 |
| 3.65 | 13.3225 | 1.91050 | 6.04152 |
| 3.66 | 13.3956 | 1.91311 | 6.04979 |
| 3.67 | 13.4689 | 1.91572 | 6.05805 |
| 3.68 | 13.5424 | 1.91833 | 6.06630 |
| 3.69 | 13.6161 | 1.92094 | 6.07454 |
| **3.70** | 13.6900 | 1.92354 | 6.08276 |
| 3.71 | 13.7641 | 1.92614 | 6.09098 |
| 3.72 | 13.8384 | 1.92873 | 6.09918 |
| 3.73 | 13.9129 | 1.93132 | 6.10737 |
| 3.74 | 13.9876 | 1.93391 | 6.11555 |
| 3.75 | 14.0625 | 1.93649 | 6.12372 |
| 3.76 | 14.1376 | 1.93907 | 6.13188 |
| 3.77 | 14.2129 | 1.94165 | 6.14003 |
| 3.78 | 14.2884 | 1.94422 | 6.14817 |
| 3.79 | 14.3641 | 1.94679 | 6.15630 |
| **3.80** | 14.4400 | 1.94936 | 6.16441 |
| 3.81 | 14.5161 | 1.95192 | 6.17252 |
| 3.82 | 14.5924 | 1.95448 | 6.18061 |
| 3.83 | 14.6689 | 1.95704 | 6.18870 |
| 3.84 | 14.7456 | 1.95959 | 6.19677 |
| 3.85 | 14.8225 | 1.96214 | 6.20484 |
| 3.86 | 14.8996 | 1.96469 | 6.21289 |
| 3.87 | 14.9769 | 1.96723 | 6.22093 |
| 3.88 | 15.0544 | 1.96977 | 6.22896 |
| 3.89 | 15.1321 | 1.97231 | 6.23699 |
| **3.90** | 15.2100 | 1.97484 | 6.24500 |
| 3.91 | 15.2881 | 1.97737 | 6.25300 |
| 3.92 | 15.3664 | 1.97990 | 6.26099 |
| 3.93 | 15.4449 | 1.98242 | 6.26897 |
| 3.94 | 15.5236 | 1.98494 | 6.27694 |
| 3.95 | 15.6025 | 1.98746 | 6.28490 |
| 3.96 | 15.6816 | 1.98997 | 6.29285 |
| 3.97 | 15.7609 | 1.99249 | 6.30079 |
| 3.98 | 15.8404 | 1.99499 | 6.30872 |
| 3.99 | 15.9201 | 1.99750 | 6.31664 |

| $n$ | $n^2$ | $\sqrt{n}$ | $\sqrt{10n}$ |
|------|---------|-----------|-------------|
| **4.00** | 16.0000 | 2.00000 | 6.32456 |
| 4.01 | 16.0801 | 2.00250 | 6.33246 |
| 4.02 | 16.1604 | 2.00499 | 6.34035 |
| 4.03 | 16.2409 | 2.00749 | 6.34823 |
| 4.04 | 16.3216 | 2.00998 | 6.35610 |
| 4.05 | 16.4025 | 2.01246 | 6.36396 |
| 4.06 | 16.4836 | 2.01494 | 6.37181 |
| 4.07 | 16.5649 | 2.01742 | 6.37966 |
| 4.08 | 16.6464 | 2.01990 | 6.38749 |
| 4.09 | 16.7281 | 2.02237 | 6.39531 |
| **4.10** | 16.8100 | 2.02485 | 6.40312 |
| 4.11 | 16.8921 | 2.02731 | 6.41093 |
| 4.12 | 16.9744 | 2.02978 | 6.41872 |
| 4.13 | 17.0569 | 2.03224 | 6.42651 |
| 4.14 | 17.1396 | 2.03470 | 6.43428 |
| 4.15 | 17.2225 | 2.03715 | 6.44205 |
| 4.16 | 17.3056 | 2.03961 | 6.44981 |
| 4.17 | 17.3889 | 2.04206 | 6.45755 |
| 4.18 | 17.4724 | 2.04450 | 6.46529 |
| 4.19 | 17.5561 | 2.04695 | 6.47302 |
| **4.20** | 17.6400 | 2.04939 | 6.48074 |
| 4.21 | 17.7241 | 2.05183 | 6.48845 |
| 4.22 | 17.8084 | 2.05426 | 6.49615 |
| 4.23 | 17.8929 | 2.05670 | 6.50384 |
| 4.24 | 17.9776 | 2.05913 | 6.51153 |
| 4.25 | 18.0625 | 2.06155 | 6.51920 |
| 4.26 | 18.1476 | 2.06398 | 6.52687 |
| 4.27 | 18.2329 | 2.06640 | 6.53452 |
| 4.28 | 18.3184 | 2.06882 | 6.54217 |
| 4.29 | 18.4041 | 2.07123 | 6.54981 |
| **4.30** | 18.4900 | 2.07364 | 6.55744 |
| 4.31 | 18.5761 | 2.07605 | 6.56506 |
| 4.32 | 18.6624 | 2.07846 | 6.57267 |
| 4.33 | 18.7489 | 2.08087 | 6.58027 |
| 4.34 | 18.8356 | 2.08327 | 6.58787 |
| 4.35 | 18.9225 | 2.08567 | 6.59545 |
| 4.36 | 19.0096 | 2.08806 | 6.60303 |
| 4.37 | 19.0969 | 2.09045 | 6.61060 |
| 4.38 | 19.1844 | 2.09284 | 6.61816 |
| 4.39 | 19.2721 | 2.09523 | 6.62571 |
| **4.40** | 19.3600 | 2.09762 | 6.63325 |
| 4.41 | 19.4481 | 2.10000 | 6.64078 |
| 4.42 | 19.5364 | 2.10238 | 6.64831 |
| 4.43 | 19.6249 | 2.10476 | 6.65582 |
| 4.44 | 19.7136 | 2.10713 | 6.66333 |
| 4.45 | 19.8025 | 2.10950 | 6.67083 |
| 4.46 | 19.8916 | 2.11187 | 6.67832 |
| 4.47 | 19.9809 | 2.11424 | 6.68581 |
| 4.48 | 20.0704 | 2.11660 | 6.69328 |
| 4.49 | 20.1601 | 2.11896 | 6.70075 |

| $n$ | $n^2$ | $\sqrt{n}$ | $\sqrt{10n}$ |
|------|---------|-----------|-------------|
| **4.50** | 20.2500 | 2.12132 | 6.70820 |
| 4.51 | 20.3401 | 2.12368 | 6.71565 |
| 4.52 | 20.4304 | 2.12603 | 6.72309 |
| 4.53 | 20.5209 | 2.12838 | 6.73053 |
| 4.54 | 20.6116 | 2.13073 | 6.73795 |
| 4.55 | 20.7025 | 2.13307 | 6.74537 |
| 4.56 | 20.7936 | 2.13542 | 6.75278 |
| 4.57 | 20.8849 | 2.13776 | 6.76018 |
| 4.58 | 20.9764 | 2.14009 | 6.76757 |
| 4.59 | 21.0681 | 2.14243 | 6.77495 |
| **4.60** | 21.1600 | 2.14476 | 6.78233 |
| 4.61 | 21.2521 | 2.14709 | 6.78970 |
| 4.62 | 21.3444 | 2.14942 | 6.79706 |
| 4.63 | 21.4369 | 2.15174 | 6.80441 |
| 4.64 | 21.5296 | 2.15407 | 6.81175 |
| 4.65 | 21.6225 | 2.15639 | 6.81909 |
| 4.66 | 21.7156 | 2.15870 | 6.82642 |
| 4.67 | 21.8089 | 2.16102 | 6.83374 |
| 4.68 | 21.9024 | 2.16333 | 6.84105 |
| 4.69 | 21.9961 | 2.16564 | 6.84836 |
| **4.70** | 22.0900 | 2.16795 | 6.85565 |
| 4.71 | 22.1841 | 2.17025 | 6.86294 |
| 4.72 | 22.2784 | 2.17256 | 6.87023 |
| 4.73 | 22.3729 | 2.17486 | 6.87750 |
| 4.74 | 22.4676 | 2.17715 | 6.88477 |
| 4.75 | 22.5625 | 2.17945 | 6.89202 |
| 4.76 | 22.6576 | 2.18174 | 6.89928 |
| 4.77 | 22.7529 | 2.18403 | 6.90652 |
| 4.78 | 22.8484 | 2.18632 | 6.91375 |
| 4.79 | 22.9441 | 2.18861 | 6.92098 |
| **4.80** | 23.0400 | 2.19089 | 6.92820 |
| 4.81 | 23.1361 | 2.19317 | 6.93542 |
| 4.82 | 23.2324 | 2.19545 | 6.94262 |
| 4.83 | 23.3289 | 2.19773 | 6.94982 |
| 4.84 | 23.4256 | 2.20000 | 6.95701 |
| 4.85 | 23.5225 | 2.20227 | 6.96419 |
| 4.86 | 23.6196 | 2.20454 | 6.97137 |
| 4.87 | 23.7169 | 2.20681 | 6.97854 |
| 4.88 | 23.8144 | 2.20907 | 6.98570 |
| 4.89 | 23.9121 | 2.21133 | 6.99285 |
| **4.90** | 24.0100 | 2.21359 | 7.00000 |
| 4.91 | 24.1081 | 2.21585 | 7.00714 |
| 4.92 | 24.2064 | 2.21811 | 7.01427 |
| 4.93 | 24.3049 | 2.22036 | 7.02140 |
| 4.94 | 24.4036 | 2.22261 | 7.02851 |
| 4.95 | 24.5025 | 2.22486 | 7.03562 |
| 4.96 | 24.6016 | 2.22711 | 7.04273 |
| 4.97 | 24.7009 | 2.22935 | 7.04982 |
| 4.98 | 24.8004 | 2.23159 | 7.05691 |
| 4.99 | 24.9001 | 2.23383 | 7.06399 |

| $n$ | $n^2$ | $\sqrt{n}$ | $\sqrt{10n}$ | $n$ | $n^2$ | $\sqrt{n}$ | $\sqrt{10n}$ |
|---|---|---|---|---|---|---|---|
| **5.00** | 25.0000 | 2.23607 | 7.07107 | **5.50** | 30.2500 | 2.34521 | 7.41620 |
| 5.01 | 25.1001 | 2.23830 | 7.07814 | 5.51 | 30.3601 | 2.34734 | 7.42294 |
| 5.02 | 25.2004 | 2.24054 | 7.08520 | 5.52 | 30.4704 | 2.34947 | 7.42967 |
| 5.03 | 25.3009 | 2.24277 | 7.09225 | 5.53 | 30.5809 | 2.35160 | 7.43640 |
| 5.04 | 25.4016 | 2.24499 | 7.09930 | 5.54 | 30.6916 | 2.35372 | 7.44312 |
| 5.05 | 25.5025 | 2.24722 | 7.10634 | 5.55 | 30.8025 | 2.35584 | 7.44983 |
| 5.06 | 25.6036 | 2.24944 | 7.11337 | 5.56 | 30.9136 | 2.35797 | 7.45654 |
| 5.07 | 25.7049 | 2.25167 | 7.12039 | 5.57 | 31.0249 | 2.36008 | 7.46324 |
| 5.08 | 25.8064 | 2.25389 | 7.12741 | 5.58 | 31.1364 | 2.36220 | 7.46994 |
| 5.09 | 25.9081 | 2.25610 | 7.13442 | 5.59 | 31.2481 | 2.36432 | 7.47663 |
| **5.10** | 26.0100 | 2.25832 | 7.14143 | **5.60** | 31.3600 | 2.36643 | 7.48331 |
| 5.11 | 26.1121 | 2.26053 | 7.14843 | 5.61 | 31.4721 | 2.36854 | 7.48999 |
| 5.12 | 26.2144 | 2.26274 | 7.15542 | 5.62 | 31.5844 | 2.37065 | 7.49667 |
| 5.13 | 26.3169 | 2.26495 | 7.16240 | 5.63 | 31.6969 | 2.37276 | 7.50333 |
| 5.14 | 26.4196 | 2.26716 | 7.16938 | 5.64 | 31.8096 | 2.37487 | 7.50999 |
| 5.15 | 26.5225 | 2.26936 | 7.17635 | 5.65 | 31.9225 | 2.37697 | 7.51665 |
| 5.16 | 26.6256 | 2.27156 | 7.18331 | 5.66 | 32.0356 | 2.37908 | 7.52330 |
| 5.17 | 26.7289 | 2.27376 | 7.19027 | 5.67 | 32.1489 | 2.38118 | 7.52994 |
| 5.18 | 26.8324 | 2.27596 | 7.19722 | 5.68 | 32.2624 | 2.38238 | 7.53658 |
| 5.19 | 26.9361 | 2.27816 | 7.20417 | 5.69 | 32.3761 | 2.38537 | 7.54321 |
| **5.20** | 27.0400 | 2.28035 | 7.21110 | **5.70** | 32.4900 | 2.38747 | 7.54983 |
| 5.21 | 27.1441 | 2.28254 | 7.21803 | 5.71 | 32.6041 | 2.38956 | 7.55645 |
| 5.22 | 27.2484 | 2.28473 | 7.22496 | 5.72 | 32.7184 | 2.39165 | 7.56307 |
| 5.23 | 27.3529 | 2.28692 | 7.23187 | 5.73 | 32.8329 | 2.39374 | 7.56968 |
| 5.24 | 27.4576 | 2.28910 | 7.23878 | 5.74 | 32.9476 | 2.39583 | 7.57628 |
| 5.25 | 27.5625 | 2.29129 | 7.24569 | 5.75 | 33.0625 | 2.39792 | 7.58288 |
| 5.26 | 27.6676 | 2.29347 | 7.25259 | 5.76 | 33.1776 | 2.40000 | 7.58947 |
| 5.27 | 27.7729 | 2.29565 | 7.25948 | 5.77 | 33.2929 | 2.40208 | 7.59605 |
| 5,28 | 27.8784 | 2.29783 | 7.26636 | 5.78 | 33.4084 | 2.40416 | 7.60263 |
| 5.29 | 27.9841 | 2.30000 | 7.27324 | 5.79 | 33.5241 | 2.40624 | 7.60920 |
| **5.30** | 28.0900 | 2.30217 | 7.28011 | **5.80** | 33.6400 | 2.40832 | 7.61577 |
| 5.31 | 28.1961 | 2.30434 | 7.28697 | 5.81 | 33.7561 | 2.41039 | 7.62234 |
| 5.32 | 28.3024 | 2.30651 | 7.29383 | 5.82 | 33.8724 | 2.41247 | 7.62889 |
| 5.33 | 28.4089 | 2.30868 | 7.30068 | 5.83 | 33.9889 | 2.41454 | 7.63544 |
| 5.34 | 28.5156 | 2.31084 | 7.30753 | 5.84 | 34.1056 | 2.41661 | 7.64199 |
| 5.35 | 28.6225 | 2.31301 | 7.31437 | 5.85 | 34.2225 | 2.41868 | 7.64853 |
| 5.36 | 28.7296 | 2.31517 | 7.32120 | 5.86 | 34.3396 | 2.42074 | 7.65506 |
| 5.37 | 28.8369 | 2.31733 | 7.32803 | 5.87 | 34.4569 | 2.42281 | 7.66159 |
| 5.38 | 28.9444 | 2.31948 | 7.33485 | 5.88 | 34.5744 | 2.42487 | 7.66812 |
| 5.39 | 29.0521 | 2.32164 | 7.34166 | 5.89 | 34.6921 | 2.42693 | 7.67463 |
| **5.40** | 29.1600 | 2.32379 | 7.34847 | **5.90** | 34.8100 | 2.42899 | 7.68115 |
| 5.41 | 29.2681 | 2.32594 | 7.35527 | 5.91 | 34.9281 | 2.43105 | 7.68765 |
| 5.42 | 29.3764 | 2.32809 | 7.36206 | 5.92 | 35.0464 | 2.43311 | 7.69415 |
| 5.43 | 29.4849 | 2.33024 | 7.36885 | 5.93 | 35.1649 | 2.43516 | 7.70065 |
| 5.44 | 29.5936 | 2.33238 | 7.37564 | 5.94 | 35.2836 | 2.43721 | 7.70714 |
| 5.45 | 29.7025 | 2.33452 | 7.38241 | 5.95 | 35.4025 | 2.43926 | 7.71362 |
| 5.46 | 29.8116 | 2.33666 | 7.38918 | 5.96 | 35.5216 | 2.44131 | 7.72010 |
| 5.47 | 29.9209 | 2.33880 | 7.39594 | 5.97 | 35.6409 | 2.44336 | 7.72658 |
| 5.48 | 30.0304 | 2.34094 | 7.40270 | 5.98 | 35.7604 | 2.44540 | 7.73305 |
| 5.49 | 30.1401 | 2.34307 | 7.40945 | 5.99 | 35.8801 | 2.44745 | 7.73951 |

| $n$ | $n^2$ | $\sqrt{n}$ | $\sqrt{10n}$ |
|---|---|---|---|
| 6.00 | 36.0000 | 2.44949 | 7.74597 |
| 6.01 | 36.1201 | 2.45153 | 7.75242 |
| 6.02 | 36.2404 | 2.45357 | 7.75887 |
| 6.03 | 36.3609 | 2.45561 | 7.76531 |
| 6.04 | 36.4816 | 2.45764 | 7.77174 |
| 6.05 | 36.6025 | 2.45967 | 7.77817 |
| 6.06 | 36.7236 | 2.46171 | 7.78460 |
| 6.07 | 36.8449 | 2.46374 | 7.79102 |
| 6.08 | 36.9664 | 2.46577 | 7.79744 |
| 6.09 | 37.0881 | 2.46779 | 7.80385 |
| 6.10 | 37.2100 | 2.46982 | 7.81025 |
| 6.11 | 37.3321 | 2.47184 | 7.81665 |
| 6.12 | 37.4544 | 2.47386 | 7.82304 |
| 6.13 | 37.5769 | 2.47588 | 7.82943 |
| 6.14 | 37.6996 | 2.47790 | 7.83582 |
| 6.15 | 37.8225 | 2.47992 | 7.84219 |
| 6.16 | 37.9456 | 2.48193 | 7.84857 |
| 6.17 | 38.0689 | 2.48395 | 7.85493 |
| 6.18 | 38.1924 | 2.48596 | 7.86130 |
| 6.19 | 38.3161 | 2.48797 | 7.86766 |
| 6.20 | 38.4400 | 2.48998 | 7.87401 |
| 6.21 | 38.5641 | 2.49199 | 7.88036 |
| 6.22 | 38.6884 | 2.49399 | 7.88670 |
| 6.23 | 38.8129 | 2.49600 | 7.89303 |
| 6.24 | 38.9376 | 2.49800 | 7.89937 |
| 6.25 | 39.0625 | 2.50000 | 7.90569 |
| 6.26 | 39.1876 | 2.50200 | 7.91202 |
| 6.27 | 39.3129 | 2.50400 | 7.91833 |
| 6.28 | 39.4384 | 2.50599 | 7.92465 |
| 6.29 | 39.5641 | 2.50799 | 7.93095 |
| 6.30 | 39.6900 | 2.50998 | 7.93725 |
| 6.31 | 39.8161 | 2.51197 | 7.94355 |
| 6.32 | 39.9424 | 2.51396 | 7.94984 |
| 6.33 | 40.0689 | 2.51595 | 7.95613 |
| 6.34 | 40.1956 | 2.51794 | 7.96241 |
| 6.35 | 40.3225 | 2.51992 | 7.96869 |
| 6.36 | 40.4496 | 2.52190 | 7.97496 |
| 6.37 | 40.5769 | 2.52389 | 7.98123 |
| 6.38 | 40.7044 | 2.52587 | 7.98749 |
| 6.39 | 40.8321 | 2.52784 | 7.99375 |
| 6.40 | 40.9600 | 2.52982 | 8.00000 |
| 6.41 | 41.0881 | 2.53180 | 8.00625 |
| 6.42 | 41.2164 | 2.53377 | 8.01249 |
| 6.43 | 41.3449 | 2.53574 | 8.01873 |
| 6.44 | 41.4736 | 2.53772 | 8.02496 |
| 6.45 | 41.6025 | 2.53969 | 8.03119 |
| 6.46 | 41.7316 | 2.54165 | 8.03741 |
| 6.47 | 41.8609 | 2.54362 | 8.04363 |
| 6.48 | 41.9904 | 2.54558 | 8.04984 |
| 6.49 | 42.1201 | 2.54755 | 8.05605 |

| $n$ | $n^2$ | $\sqrt{n}$ | $\sqrt{10n}$ |
|---|---|---|---|
| 6.50 | 42.2500 | 2.54951 | 8.06226 |
| 6.51 | 42.3801 | 2.55147 | 8.06846 |
| 6.52 | 42.5104 | 2.55343 | 8.07465 |
| 6.53 | 42.6409 | 2.55539 | 8.08084 |
| 6.54 | 42.7716 | 2.55734 | 8.08703 |
| 6.55 | 42.9025 | 2.55930 | 8.09321 |
| 6.56 | 43.0336 | 2.56125 | 8.09938 |
| 6.57 | 43.1649 | 2.56320 | 8.10555 |
| 6.58 | 43.2964 | 2.56515 | 8.11172 |
| 6.59 | 43.4281 | 2.56710 | 8.11788 |
| 6.60 | 43.5600 | 2.56905 | 8.12404 |
| 6.61 | 43.6921 | 2.57099 | 8.13019 |
| 6.62 | 43.8244 | 2.57294 | 8.13634 |
| 6.63 | 43.9569 | 2.57488 | 8.14248 |
| 6.64 | 44.0896 | 2.57682 | 8.14862 |
| 6.65 | 44.2225 | 2.57876 | 8.15475 |
| 6.66 | 44.3556 | 2.58070 | 8.16088 |
| 6.67 | 44.4889 | 2.58263 | 8.16701 |
| 6.68 | 44.6224 | 2.58457 | 8.17313 |
| 6.69 | 44.7561 | 2.58650 | 8.17924 |
| 6.70 | 44.8900 | 2.58844 | 8.18535 |
| 6.71 | 45.0241 | 2.59037 | 8.19146 |
| 6.72 | 45.1584 | 2.59230 | 8.19756 |
| 6.73 | 45.2929 | 2.59422 | 8.20366 |
| 6.74 | 45.4276 | 2.59615 | 8.20975 |
| 6.75 | 45.5625 | 2.59808 | 8.21584 |
| 6.76 | 45.6976 | 2.60000 | 8.22192 |
| 6.77 | 45.8329 | 2.60192 | 8.22800 |
| 6.78 | 45.9684 | 2.60384 | 8.23408 |
| 6.79 | 46.1041 | 2.60576 | 8.24015 |
| 6.80 | 46.2400 | 2.60768 | 8.24621 |
| 6.81 | 46.3761 | 2.60960 | 8.25227 |
| 6.82 | 46.5124 | 2.61151 | 8.25833 |
| 6.83 | 46.6489 | 2.61343 | 8.26438 |
| 6.84 | 46.7856 | 2.61534 | 8.27043 |
| 6.85 | 46.9225 | 2.61725 | 8.27647 |
| 6.86 | 47.0596 | 2.61916 | 8.28251 |
| 6.87 | 47.1969 | 2.62107 | 8.28855 |
| 6.88 | 47.3344 | 2.62298 | 8.29458 |
| 6.89 | 47.4721 | 2.62488 | 8.30060 |
| 6.90 | 47.6100 | 2.62679 | 8.30662 |
| 6.91 | 47.7481 | 2.62869 | 8.31264 |
| 6.92 | 47.8864 | 2.63059 | 8.31865 |
| 6.93 | 48.0249 | 2.63249 | 8.32466 |
| 6.94 | 48.1636 | 2.63439 | 8.33067 |
| 6.95 | 48.3025 | 2.63629 | 8.33667 |
| 6.96 | 48.4416 | 2.63818 | 8.34266 |
| 6.97 | 48.5809 | 2.64008 | 8.34865 |
| 6.98 | 48.7204 | 2.64197 | 8.35464 |
| 6.99 | 48.8601 | 2.64386 | 8.36062 |

| $n$ | $n^2$ | $\sqrt{n}$ | $\sqrt{10n}$ |
|---|---|---|---|
| **7.00** | 49.0000 | 2.64575 | 8.36660 |
| 7.01 | 49.1401 | 2.64764 | 8.37257 |
| 7.02 | 49.2804 | 2.64953 | 8.37854 |
| 7.03 | 49.4209 | 2.65141 | 8.38451 |
| 7.04 | 49.5616 | 2.65330 | 8.39047 |
| 7.05 | 49.7025 | 2.65518 | 8.39643 |
| 7.06 | 49.8436 | 2.65707 | 8.40238 |
| 7.07 | 49.9849 | 2.65895 | 8.40833 |
| 7.08 | 50.1264 | 2.66083 | 8.41427 |
| 7.09 | 50.2681 | 2.66271 | 8.42021 |
| **7.10** | 50.4100 | 2.66458 | 8.42615 |
| 7.11 | 50.5521 | 2.66646 | 8.43208 |
| 7.12 | 50.6944 | 2.66833 | 8.43801 |
| 7.13 | 50.8369 | 2.67021 | 8.44393 |
| 7.14 | 50.9796 | 2.67208 | 8.44985 |
| 7.15 | 51.1225 | 2.67395 | 8.45577 |
| 7.16 | 51.2656 | 2.67582 | 8.46168 |
| 7.17 | 51.4089 | 2.67769 | 8.46759 |
| 7.18 | 51.5524 | 2.67955 | 8.47349 |
| 7.19 | 51.6961 | 2.68142 | 8.47939 |
| **7.20** | 51.8400 | 2.68328 | 8.48528 |
| 7.21 | 51.9841 | 2.68514 | 8.49117 |
| 7.22 | 52.1284 | 2.68701 | 8.49706 |
| 7.23 | 52.2729 | 2.68887 | 8.50294 |
| 7.24 | 52.4176 | 2.69072 | 8.50882 |
| 7.25 | 52.5625 | 2.69258 | 8.51469 |
| 7.26 | 52.7076 | 2.69444 | 8.52056 |
| 7.27 | 52.8529 | 2.69629 | 8.52643 |
| 7.28 | 52.9984 | 2.69815 | 8.53229 |
| 7.29 | 53.1441 | 2.70000 | 8.53815 |
| **7.30** | 53.2900 | 2.70185 | 8.54400 |
| 7.31 | 53.4361 | 2.70370 | 8.54985 |
| 7.32 | 53.5824 | 2.70555 | 8.55570 |
| 7.33 | 53.7289 | 2.70740 | 8.56154 |
| 7.34 | 53.8756 | 2.70924 | 8.56738 |
| 7.35 | 54.0225 | 2.71109 | 8.57321 |
| 7.36 | 54.1696 | 2.71293 | 8.57904 |
| 7.37 | 54.3169 | 2.71477 | 8.58487 |
| 7.38 | 54.4644 | 2.71662 | 8.59069 |
| 7.39 | 54.6121 | 2.71846 | 8.59651 |
| **7.40** | 54.7600 | 2.72029 | 8.60233 |
| 7.41 | 54.9081 | 2.72213 | 8.60814 |
| 7.42 | 55.0564 | 2.72397 | 8.61394 |
| 7.43 | 55.2049 | 2.72580 | 8.61974 |
| 7.44 | 55.3536 | 2.72764 | 8.62554 |
| 7.45 | 55.5025 | 2.72947 | 8.63134 |
| 7.46 | 55.6516 | 2.73130 | 8.63713 |
| 7.47 | 55.8009 | 2.73313 | 8.64292 |
| 7.48 | 55.9504 | 2.73496 | 8.64870 |
| 7.49 | 56.1001 | 2.73679 | 8.65448 |

| $n$ | $n^2$ | $\sqrt{n}$ | $\sqrt{10n}$ |
|---|---|---|---|
| **7.50** | 56.2500 | 2.73861 | 8.66025 |
| 7.51 | 56.4001 | 2.74044 | 8.66603 |
| 7.52 | 56.5504 | 2.74226 | 8.67179 |
| 7.53 | 56.7009 | 2.74408 | 8.67756 |
| 7.54 | 56.8516 | 2.74591 | 8.68332 |
| 7.55 | 57.0025 | 2.74773 | 8.68907 |
| 7.56 | 57.1536 | 2.74955 | 8.69483 |
| 7.57 | 57.3049 | 2.75136 | 8.70057 |
| 7.58 | 57.4564 | 2.75318 | 8.70632 |
| 7.59 | 57.6081 | 2.75500 | 8.71206 |
| **7.60** | 57.7600 | 2.75681 | 8.71780 |
| 7.61 | 57.9121 | 2.75862 | 8.72353 |
| 7.62 | 58.0644 | 2.76043 | 8.72926 |
| 7.63 | 58.2169 | 2.76225 | 8.73499 |
| 7.64 | 58.3696 | 2.76405 | 8.74071 |
| 7.65 | 58.5225 | 2.76586 | 8.74643 |
| 7.66 | 58.6756 | 2.76767 | 8.75214 |
| 7.67 | 58.8289 | 2.76948 | 8.75785 |
| 7.68 | 58.9824 | 2.77128 | 8.76356 |
| 7.69 | 59.1361 | 2.77308 | 8.76926 |
| **7.70** | 59.2900 | 2.77489 | 8.77496 |
| 7.71 | 59.4441 | 2.77669 | 8.78066 |
| 7.72 | 59.5984 | 2.77849 | 8.78635 |
| 7.73 | 59.7529 | 2.78029 | 8.79204 |
| 7.74 | 59.9076 | 2.78209 | 8.79773 |
| 7.75 | 60.0625 | 2.78388 | 8.80341 |
| 7.76 | 60.2176 | 2.78568 | 8.80909 |
| 7.77 | 60.3729 | 2.78747 | 8.81476 |
| 7.78 | 60.5284 | 2.78927 | 8.82043 |
| 7.79 | 60.6841 | 2.79106 | 8.82610 |
| **7.80** | 60.8400 | 2.79285 | 8.83176 |
| 7.81 | 60.9961 | 2.79464 | 8.83742 |
| 7.82 | 61.1524 | 2.79643 | 8.84308 |
| 7.83 | 61.3089 | 2.79821 | 8.84873 |
| 7.84 | 61.4656 | 2.80000 | 8.85438 |
| 7.85 | 61.6225 | 2.80179 | 8.86002 |
| 7.86 | 61.7796 | 2.80357 | 8.86566 |
| 7.87 | 61.9369 | 2.80535 | 8.87130 |
| 7.88 | 62.0944 | 2.80713 | 8.87694 |
| 7.89 | 62.2521 | 2.80891 | 8.88257 |
| **7.90** | 62.4100 | 2.81069 | 8.88819 |
| 7.91 | 62.5681 | 2.81247 | 8.89382 |
| 7.92 | 62.7264 | 2.81425 | 8.89944 |
| 7.93 | 62.8849 | 2.81603 | 8.90505 |
| 7.94 | 63.0436 | 2.81780 | 8.91067 |
| 7.95 | 63.2025 | 2.81957 | 8.91628 |
| 7.96 | 63.3616 | 2.82135 | 8.92188 |
| 7.97 | 63.5209 | 2.82312 | 8.92749 |
| 7.98 | 63.6804 | 2.82489 | 8.93308 |
| 7.99 | 63.8401 | 2.82666 | 8.93868 |

| $n$ | $n^2$ | $\sqrt{n}$ | $\sqrt{10n}$ |
|---|---|---|---|
| 8.00 | 64.0000 | 2.82843 | 8.94427 |
| 8.01 | 64.1601 | 2.83019 | 8.94986 |
| 8.02 | 64.3204 | 2.83196 | 8.95545 |
| 8.03 | 64.4809 | 2.83373 | 8.96103 |
| 8.04 | 64.6416 | 2.83549 | 8.96660 |
| 8.05 | 64.8025 | 2.83725 | 8.97218 |
| 8.06 | 64.9636 | 2.83901 | 8.97775 |
| 8.07 | 65.1249 | 2.84077 | 8.98332 |
| 8.08 | 65.2864 | 2.84253 | 8.98888 |
| 8.09 | 65.4481 | 2.84429 | 8.99444 |
| 8.10 | 65.6100 | 2.84605 | 9.00000 |
| 8.11 | 65.7721 | 2.84781 | 9.00555 |
| 8.12 | 65.9344 | 2.84956 | 9.01110 |
| 8.13 | 66.0969 | 2.85132 | 9.01665 |
| 8.14 | 66.2596 | 2.85307 | 9.02219 |
| 8.15 | 66.4225 | 2.85482 | 9.02774 |
| 8.16 | 66.5856 | 2.85657 | 9.03327 |
| 8.17 | 66.7489 | 2.85832 | 9.03881 |
| 8.18 | 66.9124 | 2.86007 | 9.04434 |
| 8.19 | 67.0761 | 2.86182 | 9.04986 |
| 8.20 | 67.2400 | 2.86356 | 9.05539 |
| 8.21 | 67.4041 | 2.86531 | 9.06091 |
| 8.22 | 67.5684 | 2.86705 | 9.06642 |
| 8.23 | 67.7329 | 2.86880 | 9.07193 |
| 8.24 | 67.8976 | 2.87054 | 9.07744 |
| 8.25 | 68.0625 | 2.87228 | 9.08295 |
| 8.26 | 68.2276 | 2.87402 | 9.08845 |
| 8.27 | 68.3929 | 2.87576 | 9.09395 |
| 8.28 | 68.5584 | 2.87750 | 9.09945 |
| 8.29 | 68.7241 | 2.87924 | 9.10494 |
| 8.30 | 68.8900 | 2.88097 | 9.11043 |
| 8.31 | 69.0561 | 2.88271 | 9.11592 |
| 8.32 | 69.2224 | 2.88444 | 9.12140 |
| 8.33 | 69.3889 | 2.88617 | 9.12688 |
| 8.34 | 69.5556 | 2.88791 | 9.13236 |
| 8.35 | 69.7225 | 2.88964 | 9.13783 |
| 8.36 | 69.8896 | 2.89137 | 9.14330 |
| 8.37 | 70.0569 | 2.89310 | 9.14877 |
| 8.38 | 70.2244 | 2.89482 | 9.15423 |
| 8.39 | 70.3921 | 2.89655 | 9.15969 |
| 8.40 | 70.5600 | 2.89828 | 9.16515 |
| 8.41 | 70.7281 | 2.90000 | 9.17061 |
| 8.42 | 70.8964 | 2.90172 | 9.17606 |
| 8.43 | 71.0649 | 2.90345 | 9.18150 |
| 8.44 | 71.2336 | 2.90517 | 9.18695 |
| 8.45 | 71.4025 | 2.90689 | 9.19239 |
| 8.46 | 71.5716 | 2.90861 | 9.19783 |
| 8.47 | 71.7409 | 2.91033 | 9.20326 |
| 8.48 | 71.9104 | 2.91204 | 9.20869 |
| 8.49 | 72.0801 | 2.91376 | 9.21412 |

| $n$ | $n^2$ | $\sqrt{n}$ | $\sqrt{10n}$ |
|---|---|---|---|
| 8.50 | 72.2500 | 2.91548 | 9.21954 |
| 8.51 | 72.4201 | 2.91719 | 9.22497 |
| 8.52 | 72.5904 | 2.91890 | 9.23038 |
| 8.53 | 72.7609 | 2.92062 | 9.23580 |
| 8.54 | 72.9316 | 2.92233 | 9.24121 |
| 8.55 | 73.1025 | 2.92404 | 9.24662 |
| 8.56 | 73.2736 | 2.92575 | 9.25203 |
| 8.57 | 73.4449 | 2.92746 | 9.25743 |
| 8.58 | 73.6164 | 2.92916 | 9.26283 |
| 8.59 | 73.7881 | 2.93087 | 9.26823 |
| 8.60 | 73.9600 | 2.93258 | 9.27362 |
| 8.61 | 74.1321 | 2.93428 | 9.27901 |
| 8.62 | 74.3044 | 2.93598 | 9.28440 |
| 8.63 | 74.4769 | 2.93769 | 9.28978 |
| 8.64 | 74.6496 | 2.93939 | 9.29516 |
| 8.65 | 74.8225 | 2.94109 | 9.30054 |
| 8.66 | 74.9956 | 2.94279 | 9.30591 |
| 8.67 | 75.1689 | 2.94449 | 9.31128 |
| 8.68 | 75.3424 | 2.94618 | 9.31665 |
| 8.69 | 75.5161 | 2.94788 | 9.32202 |
| 8.70 | 75.6900 | 2.94958 | 9.32738 |
| 8.71 | 75.8641 | 2.95127 | 9.33274 |
| 8.72 | 76.0384 | 2.95296 | 9.33809 |
| 8.73 | 76.2129 | 2.95466 | 9.34345 |
| 8.74 | 76.3876 | 2.95635 | 9.34880 |
| 8.75 | 76.5625 | 2.95804 | 9.35414 |
| 8.76 | 76.7376 | 2.95973 | 9.35949 |
| 8.77 | 76.9129 | 2.96142 | 9.36483 |
| 8.78 | 77.0884 | 2.96311 | 9.37017 |
| 8.79 | 77.2641 | 2.96479 | 9.37550 |
| 8.80 | 77.4400 | 2.96648 | 9.38083 |
| 8.81 | 77.6161 | 2.96816 | 9.38616 |
| 8.82 | 77.7924 | 2.96985 | 9.39149 |
| 8.83 | 77.9689 | 2.97153 | 9.39681 |
| 8.84 | 78.1456 | 2.97321 | 9.40213 |
| 8.85 | 78.3225 | 2.97489 | 9.40744 |
| 8.86 | 78.4996 | 2.97658 | 9.41276 |
| 8.87 | 78.6769 | 2.97825 | 9.41807 |
| 8.88 | 78.8544 | 2.97993 | 9.42338 |
| 8.89 | 79.0321 | 2.98161 | 9.42868 |
| 8.90 | 79.2100 | 2.98329 | 9.43398 |
| 8.91 | 79.3881 | 2.98496 | 9.43928 |
| 8.92 | 79.5664 | 2.98664 | 9.44458 |
| 8.93 | 79.7449 | 2.98831 | 9.44987 |
| 8.94 | 79.9236 | 2.98998 | 9.45516 |
| 8.95 | 80.1025 | 2.99166 | 9.46044 |
| 8.96 | 80.2816 | 2.99333 | 9.46573 |
| 8.97 | 80.4609 | 2.99500 | 9.47101 |
| 8.98 | 80.6404 | 2.99666 | 9.47629 |
| 8.99 | 80.8201 | 2.99833 | 9.48156 |

| $n$ | $n^2$ | $\sqrt{n}$ | $\sqrt{10n}$ |
|---|---|---|---|
| 9.00 | 81.0000 | 3.00000 | 9.48683 |
| 9.01 | 81.1801 | 3.00167 | 9.49210 |
| 9.02 | 81.3604 | 3.00333 | 9.49737 |
| 9.03 | 81.5409 | 3.00500 | 9.50263 |
| 9.04 | 81.7216 | 3.00666 | 9.50789 |
| 9.05 | 81.9025 | 3.00832 | 9.51315 |
| 9.06 | 82.0836 | 3.00998 | 9.51840 |
| 9.07 | 82.2649 | 3.01164 | 9.52365 |
| 9.08 | 82.4464 | 3.01330 | 9.52890 |
| 9.09 | 82.6281 | 3.01496 | 9.53415 |
| 9.10 | 82.8100 | 3.01662 | 9.53939 |
| 9.11 | 82.9921 | 3.01828 | 9.54463 |
| 9.12 | 83.1744 | 3.01993 | 9.54987 |
| 9.13 | 83.3569 | 3.02159 | 9.55510 |
| 9.14 | 83.5396 | 3.02324 | 9.56033 |
| 9.15 | 83.7225 | 3.02490 | 9.56556 |
| 9.16 | 83.9056 | 3.02655 | 9.57079 |
| 9.17 | 84.0889 | 3.02820 | 9.57601 |
| 9.18 | 84.2724 | 3.02985 | 9.58123 |
| 9.19 | 84.4561 | 3.03150 | 9.58645 |
| 9.20 | 84.6400 | 3.03315 | 9.59166 |
| 9.21 | 84.8241 | 3.03480 | 9.59687 |
| 9.22 | 85.0084 | 3.03645 | 9.60208 |
| 9.23 | 85.1929 | 3.03809 | 9.60729 |
| 9.24 | 85.3776 | 3.03974 | 9.61249 |
| 9.25 | 85.5625 | 3.04138 | 9.61769 |
| 9.26 | 85.7476 | 3.04302 | 9.62289 |
| 9.27 | 85.9329 | 3.04467 | 9.62808 |
| 9.28 | 86.1184 | 3.04631 | 9.63328 |
| 9.29 | 86.3041 | 3.04795 | 9.63846 |
| 9.30 | 86.4900 | 3.04959 | 9.64365 |
| 9.31 | 86.6761 | 3.05123 | 9.64883 |
| 9.32 | 86.8624 | 3.05287 | 9.65401 |
| 9.33 | 87.0489 | 3.05450 | 9.65919 |
| 9.34 | 87.2356 | 3.05614 | 9.66437 |
| 9.35 | 87.4225 | 3.05778 | 9.66954 |
| 9.36 | 87.6096 | 3.05941 | 9.67471 |
| 9.37 | 87.7969 | 3.06105 | 9.67988 |
| 9.38 | 87.9844 | 3.06268 | 9.68504 |
| 9.39 | 88.1721 | 3.06431 | 9.69020 |
| 9.40 | 88.3600 | 3.06594 | 9.69536 |
| 9.41 | 88.5481 | 3.06757 | 9.70052 |
| 9.42 | 88.7364 | 3.06920 | 9.70567 |
| 9.43 | 88.9249 | 3.07083 | 9.71082 |
| 9.44 | 89.1136 | 3.07246 | 9.71597 |
| 9.45 | 89.3025 | 3.07409 | 9.72111 |
| 9.46 | 89.4916 | 3.07571 | 9.72625 |
| 9.47 | 89.6809 | 3.07734 | 9.73139 |
| 9.48 | 89.8704 | 3.07896 | 9.73653 |
| 9.49 | 90.0601 | 3.08058 | 9.74166 |

| $n$ | $n^2$ | $\sqrt{n}$ | $\sqrt{10n}$ |
|---|---|---|---|
| 9.50 | 90.2500 | 3.08221 | 9.74679 |
| 9.51 | 90.4401 | 3.08383 | 9.75192 |
| 9.52 | 90.6304 | 3.08545 | 9.75705 |
| 9.53 | 90.8209 | 3.08707 | 9.76217 |
| 9.54 | 91.0116 | 3.08869 | 9.76729 |
| 9.55 | 91.2025 | 3.09031 | 9.77241 |
| 9.56 | 91.3936 | 3.09192 | 9.77753 |
| 9.57 | 91.5849 | 3.09354 | 9.78264 |
| 9.58 | 91.7764 | 3.09516 | 9.78775 |
| 9.59 | 91.9681 | 3.09677 | 9.79285 |
| 9.60 | 92.1600 | 3.09839 | 9.79796 |
| 9.61 | 92.3521 | 3.10000 | 9.80306 |
| 9.62 | 92.5444 | 3.10161 | 9.80816 |
| 9.63 | 92.7369 | 3.10322 | 9.81326 |
| 9.64 | 92.9296 | 3.10483 | 9.81835 |
| 9.65 | 93.1225 | 3.10644 | 9.82344 |
| 9.66 | 93.3156 | 3.10805 | 9.82853 |
| 9.67 | 93.5089 | 3.10966 | 9.83362 |
| 9.68 | 93.7024 | 3.11127 | 9.83870 |
| 9.69 | 93.8961 | 3.11288 | 9.84378 |
| 9.70 | 94.0900 | 3.11448 | 9.84886 |
| 9.71 | 94.2841 | 3.11609 | 9.85393 |
| 9.72 | 94.4784 | 3.11769 | 9.85901 |
| 9.73 | 94.6729 | 3.11929 | 9.86408 |
| 9.74 | 94.8676 | 3.12090 | 9.86914 |
| 9.75 | 95.0625 | 3.12250 | 9.87421 |
| 9.76 | 95.2576 | 3.12410 | 9.87927 |
| 9.77 | 95.4529 | 3.12570 | 9.88433 |
| 9.78 | 95.6484 | 3.12730 | 9.88939 |
| 9.79 | 95.8441 | 3.12890 | 9.89444 |
| 9.80 | 96.0400 | 3.13050 | 9.89949 |
| 9.81 | 96.2361 | 3.13209 | 9.90454 |
| 9.82 | 96.4324 | 3.13369 | 9.90959 |
| 9.83 | 96.6289 | 3.13528 | 9.91464 |
| 9.84 | 96.8256 | 3.13688 | 9.91968 |
| 9.85 | 97.0225 | 3.13847 | 9.92472 |
| 9.86 | 97.2196 | 3.14006 | 9.92975 |
| 9.87 | 97.4169 | 3.14166 | 9.93479 |
| 9.88 | 97.6144 | 3.14325 | 9.93982 |
| 9.89 | 97.8121 | 3.14484 | 9.94485 |
| 9.90 | 98.0100 | 3.14643 | 9.94987 |
| 9.91 | 98.2081 | 3.14802 | 9.95490 |
| 9.92 | 98.4064 | 3.14960 | 9.95992 |
| 9.93 | 98.6049 | 3.15119 | 9.96494 |
| 9.94 | 98.8036 | 3.15278 | 9.96995 |
| 9.95 | 99.0025 | 3.15436 | 9.97497 |
| 9.96 | 99.2016 | 3.15595 | 9.97998 |
| 9.97 | 99.4009 | 3.15753 | 9.98499 |
| 9.98 | 99.6004 | 3.15911 | 9.98999 |
| 9.99 | 99.8001 | 3.16070 | 9.99500 |

Source: Adapted from Table 4 of H. D. Larsen, Compiler, *Rinehart Mathematical Tables, Formulas and Curves.* (1948), Rinehart and Company, New York.

Table reprinted with permission of The Macmillan Company from Macmillan Selected Mathematics Tables by E. R. Hedrick. Copyright © 1936 by The Macmillan Company, renewed by Dorothy H. McWilliams, Clyde L. Hedrick, and Elisabeth B. Miller.

# List of References

ANDERSON, N. H. (1961). Scales and statistics: Parametric and nonparametric. *Psychological Bulletin*, **58**, 305–316.

ANDERSON, T. W. and S. L. SCLOVE (1974). *Introductory Statistical Analysis.* Houghton Mifflin Company, Boston.

ASTM TECHNICAL REPORT (1963). *A Guide for Fatigue Testing and the Statistical Analysis of Fatigue Data.* American Society for Testing and Materials, Philadelphia, Pa.

BIRNBAUM, Z. W. and R. A. HALL (1960). Small sample distributions for multi-sample statistics of the Smirnov type. *Annals of Mathematical Statistics*, **22**, 592–596.

BLALOCK, H. M. (1972). *Social Statistics* (2nd ed.). McGraw-Hill Book Company, New York.

BRADLEY, JAMES V. (1968). *Distribution-Free Statistical Tests.* Prentice-Hall, Inc., Englewood Cliffs, N.J.

CHAPIN, F. S. (1960). *Experimental Designs in Social Research*, Harper & Row, New York.

CHATTERJEE, S. K. and P. K. SEN (1973). On Kolmogorov-Smirnov-type tests for symmetry. *Annals of the Institute of Statistical Mathematics*, **25**, 287–299.

CHOU, YA-LUN (1969). *Statistical Analysis.* Holt, Rinehart and Winston, New York.

COCHRAN, W. G. (1963). *Sampling Techniques* (2nd ed.). John Wiley & Sons, New York.

COCHRAN, W. G. and G. M. COX (1957). *Experimental Designs* (2nd ed.). John Wiley & Sons, New York.

CONOVER, W. J. (1971). *Practical Nonparametric Statistics.* John Wiley & Sons, New York.

DANIELS, H. E. (1950). Rank correlation and population models. *Journal of the Royal Statistical Society, B*, **12**, 171–181.

DAVIES, O. L., ed. (1956). *Design and Analysis of Industrial Experiments* (2nd ed.). Hafner Publishing Company, New York.

DEMING, W. E. (1960). *Sample Design in Business Research*. John Wiley & Sons, New York.

DIXON, W. J. and F. J. MASSEY (1969). *Introduction to Statistical Analysis* (3rd ed.). McGraw-Hill Book Company, New York.

DUNN, O. J. (1964). Multiple comparisons using rank sums. *Technometrics*, **6**, 241–252.

DUNNETT, CHARLES W. (1972). Drug screening: The never-ending search for new and better drugs. In Tanur, J., ed., *Statistics: A Guide to the Unknown*, pp. 23–33. Holden-Day, Inc., San Francisco.

DURBIN, J. (1951). Incomplete blocks in ranking experiments. *British Journal of Psychology (Statistics Section)*, **4**, 85–90.

EDGINGTON, E. S. (1969). *Statistical Inference — The Distribution-Free Approach*. McGraw-Hill Book Company, New York.

EDWARDS, A. L. (1968). *Experimental Design in Psychological Research* (3rd ed.). Holt, Rinehart and Winston, New York.

FREUND, JOHN E. (1973). *Modern Elementary Statistics* (4th ed.). Prentice-Hall, Inc., Englewood Cliffs, N.J.

FREUND, J. and F. WILLIAMS (1966). *Dictionary/Outline of Basic Statistics*. McGraw-Hill Book Company, New York.

FRIEDMAN, M. (1937). The use of ranks to avoid the assumption of normality implicit in the analysis of variance. *Journal of the American Statistical Association*, **32**, 675–701.

GAITO, JOHN (1960). Scale classification and statistics. *Psychological Review*, **67**, 277–278.

GIBBONS, JEAN D. (1971). *Nonparametric Statistical Inference*. McGraw-Hill Book Company, New York.

GOODMAN, L. A. and W. H. KRUSKAL (1954). Measures of association for cross classifications. *Journal of the American Statistical Association*, **49**, 732–764.

GUILFORD, J. P. (1954). *Psychometric Methods* (2nd ed.). McGraw-Hill Book Company, New York.

GUPTA, M. K. (1967). An asymptotically nonparametric test of symmetry. *Annals of Mathematical Statistics*, **38**, 849–866.

HAMBURG, MORRIS (1974). *Basic Statistics: A Modern Approach*. Harcourt Brace Jovanovich, Inc., New York.

HANSEN, M. H., W. N. HURWITZ and W. G. MADOW (1953). *Sample Survey Methods and Theory* (2 volumes). John Wiley & Sons, New York.

HARDYCK, C. D. and L. F. PETRINOVICH (1969). *Introduction to Statistics for the Behavioral Sciences*. W. B. Saunders Company, Philadelphia.

HAYS, W. L. (1973). *Statistics for the Social Sciences* (2nd ed.). Holt, Rinehart and Winston, New York.

HODGES, J. L., Jr. and E. L. LEHMANN (1956). The efficiency of some non-parametric competitors of the t-test. *Annals of Mathematical Statistics*, **27**, 324–335.

HOLLANDER, M. and D. A. WOLFE (1973). *Nonparametric Statistical Methods.* John Wiley & Sons, New York.

KENDALL, M. G. (1962). *Rank Correlation Methods* (3rd ed.). Hafner Publishing Company, New York.

KERLINGER, FRED N. (1973). *Foundations of Behavioral Research* (2nd ed.). Holt, Rinehart and Winston, New York.

KIRK, R. (1968). *Experimental Design in Psychological Research.* Brooks/Cole, Belmont, Calif.

KISH, L. (1965). *Survey Sampling.* John Wiley & Sons, New York.

KRAFT, CHARLES and C. VAN EEDEN (1968). *A Nonparametric Introduction to Statistics.* The Macmillan Company, New York.

KRUSKAL, W. H. (1952). A nonparametric test for the several sample problem. *Annals of Mathematical Statistics,* **23,** 525–540.

KRUSKAL, W. H. and W. A. WALLIS (1952). Use of ranks in one-criterion analysis of variance. *Journal of the American Statistical Association,* **47,** 583–621; errata, *ibid.* (1953), **48,** 905–911.

LEVY, SHELDON G. (1968). *Inferential Statistics in the Behavioral Sciences.* Holt, Rinehart and Winston, New York.

LINDQUIST, E. F. (1953). *Design and Analysis of Experiments in Psychology and Education.* Houghton Mifflin Company, Boston.

LINDZEY, GARDNER and E. ARONSON (1968), *The Handbook of Social Psychology,* Vol. II, Research Methods. Addison-Wesley Publishing Company, Reading, Mass.

MANN, H. B. (1945). Nonparametric tests against trend. *Econometrica,* **13,** 245–259.

MANN, H. B. and D. R. WHITNEY (1947). On a test whether one of two random variables is stochastically larger than the other. *Annals of Mathematical Statistics,* **18,** 50–60.

MENDENHALL, W., L. OTT and R. L. SCHEAFFER (1971). *Elementary Survey Sampling.* Wadsworth Publishing Company, Belmont, Calif.

MOORE, G. H. and W. A. WALLIS (1943). Time series significance tests based on signs of differences. *Journal of the American Statistical Association,* **38,** 153–164.

MOSES, L. E. (1963). Rank tests for dispersion. *Annals of Mathematical Statistics,* **34,** 973–983.

MOSES, L. E. (1965). Answer to Query 10: Confidence limits from rank tests. *Techonometrics,* **7,** 257–260.

MOSTELLER, FREDERICK and J. W. TUKEY (1968). Data analysis, including statistics. In Lindzey, Gardner and E. Aronson, eds., *The Handbook of Social Psychology,* Vol. II, pp. 80–203. Addison-Wesley Publishing Company, Reading, Mass.

NOETHER, G. (1971). *Introduction to Statistics: A Fresh Approach.* Houghton Mifflin Company, Boston.

OLMSTEAD, P. S. (1946). Distribution of sample arrangements for runs up and down. *Annals of Mathematical Statistics,* **17,** 24–33.

OSTLE, B. (1963). *Statistics in Research* (2nd ed.). Iowa State University Press, Ames, Iowa.

PIERCE, ALBERT (1970). *Fundamentals of Nonparametric Statistics.* Dickenson Publishing Company, Inc., Belmont, Calif.

PRATT, J. W. and J. D. GIBBONS (1975). *Concepts of Nonparametric Theory.* Manuscript to be published.

RAJ, DES (1972). *The Design of Sample Surveys.* McGraw-Hill Book Company, New York.

SEN, P. K. (1963). On the estimation of relative potency in dilution (-direct) assays by distribution-free methods. *Biometrics,* **19,** 532–552.

SHORACK, G. R. (1969). Testing and estimating ratios of scale parameters. *Journal of the American Statistical Association,* **64,** 999–1013.

SIEGEL, S. (1956). *Nonparametric Statistics for the Behavioral Sciences.* McGraw-Hill book Company, New York.

SIEGEL, S. and J. W. TUKEY (1960). A nonparametric sum of ranks procedure for relative spread in unpaired samples. *Journal of the American Statistical Association,* **55,** 429–445; correction, *ibid.* (1961), **56,** 1005.

STEVENS, S. S. (1946). On the theory of scales of measurement. *Science,* **103,** 677–680.

STEVENS, S. S. (1951). Mathematics, measurement and psychophysics. In Stevens, S. S., ed., *Handbook of Experimental Psychology,* pp. 1–49. John Wiley & Sons, New York.

STUART, A. (1956). The efficiencies of tests of randomness against normal regression. *Journal of the American Statistical Association,* **51,** 285–287.

SUKHATME, B. V. (1957). On certain two sample nonparametric tests for variances. *Annals of Mathematical Statistics,* **28,** 188–194.

TANUR, JUDITH, ed. (1972). *Statistics: A Guide to the Unknown.* Holden-Day, Inc., San Francisco.

TATE, M. W. and R. C. CLELLAND (1957). *Nonparametric and Shortcut Statistics.* The Interstate Publishers & Printers, Danville, Ill.

TUFTE, EDWARD R. (1972). Registration and voting. In Tanur, J., ed., *Statistics: A Guide to the Unknown,* pp. 153–161. Holden-Day, Inc., San Francisco.

WALD, A. and J. WOLFOWITZ (1940). On a test whether two samples are from the same population. *Annals of Mathematical Statistics,* **11,** 147–162.

WALKER, HELEN M. and J. LEV (1953). *Statistical Inference.* Holt, Rinehart and Winston, New York.

WALKER, HELEN M. and J. LEV (1969). *Elementary Statistical Methods* (3rd ed.). Holt, Rinehart and Winston, New York.

WEISS, ROBERT S. (1968). *Statistics in Social Research.* John Wiley & Sons, New York.

WILCOXON, F. (1945). Individual comparisons by ranking methods. *Biometrics,* **1,** 80–83.

WILKS, S. S. (1948). Order statistics. *Bulletin of the American Mathematical Society,* **54,** 6–50.

YAMANE, TARO (1967). *Elementary Sampling Theory.* Prentice-Hall, Inc., Englewood Cliffs, N.J.

# List of Experiments

ABORN, MURRAY, and H. RUBINSTEIN (1952). Information theory and immediate recall. *Journal of Experimental Psychology*, **44**, 260–266.

ABRAMSON, E. E. and R. A. WUNDERLICH (1972). Anxiety, fear and eating: A test of the psychosomatic concept of obesity. *Journal of Abnormal Psychology*, **79**, 317–321.

ADHIKARI, B. P. (1973). Personal communication.

ANDERSON, R. C. (1955). Pebble lithology of the Marseilles Till Sheet in northeastern Illinois. *Journal of Geology*, **63**, 214–227.

ANDERSON, RICHARD C. (1962). Failure imagery in the fantasy of eighth graders as a function of three conditions of induced arousal. *Journal of Educational Psychology*, **53**, 293–298.

ARONSON, E. and J. MILLS (1959). The effect of severity of initiation on liking for a group. *Journal of Abnormal and Social Psychology*, **59**, 177–181.

ASTM TECHNICAL REPORT (1963). *A Guide for Fatigue Testing and the Statistical Analysis of Fatigue Data*. American Society for Testing and Materials, Philadelphia, Pa.

BANNAYAN, G. A. and S. I. HAJDU (1972). Gynecomastia: Clinicopathologic study of 351 cases. *American Journal of Clinical Pathology*, **57**, 431–437.

BARNETT, R. N. and W. J. YOUDEN (1970). A revised scheme for the comparison of quantitative methods. *American Journal of Clinical Pathology*, **54**, 454–462.

BARTLETT, M. S. (1957). Measles periodicity and community size. *Journal of the Royal Statistical Society*, A, **120**, 48–70.

BATEN, W. D., P. I. TACK and H. A. BAEDER (1958). Testing for differences between methods for preparing fish by use of a discriminant function. *Industrial Quality Control*, **14** (January), 6–10.

BAUER, THEODORE J., et al (1954). Do persons lost to long term observation have the same experience as persons observed? *Journal of the American Statistical Association*, **49**, 36–50.

BENDER, IRVING E. (1958). Changes in religious interest: A retest after 15 years. *Journal of Abnormal and Social Psychology*, **57**, 41–46.

BLANK, DAVID M. (1954). Relationship between an index of house prices and building costs. *Journal of the American Statistical Association*, **49**, 67–78.

BLISS, C. I. (1967). *Statistics in Biology*. McGraw-Hill Book Company, New York.

BODERMAN, ALVIN, D. W. FREED and M. T. KINNUCAN (1972). "Touch me, like me": Testing an encounter group assumption. *Journal of Applied Behavioral Science*, **8**, 527–533.

COOPER, L. M. et al (1967). A further attempt to modify hypnotic susceptibility through repeated individualized experience. *International Journal of Clinical and Experimental Hypnosis*, **15**, 118–124.

CULLINAN, WALTER L. (1963). Stability of adaptation in the oral performance of stutterers. *Journal of Speech and Hearing Research*, **6**, 70–83.

CUSHNY, A. R. and A. R. PEEBLES (1905). The action of optical isomers, II: Hyosines. *Journal of Physiology*, **32**, 501–510.

De LINT, JAN (1966). A note on Smart's study of birth rank and affiliation in male University students. *The Journal of Psychology*, **62**, 177–178.

De LINT, JAN (1967). Note on birth order and intelligence test performance. *The Journal of Psychology*, **66**, 15–17.

DUNLOP, A. A. and R. H. HAYMAN (1958). Differences among merino strains in resistance to fleece rot. *Australian Journal of Agricultural Research*, **9**, 260–266.

DUNN, O. J. (1964). Multiple comparisons using rank sums. *Technometrics*, **6**, 241–252.

*Economic Report: The State of South Carolina 1972*. Transmitted to the South Carolina General Assembly by State Budget and Control Board, September 1972.

EDWARDS, A. W. and M. FRACCARO (1960). Distribution and sequences of sexes in a selected sample of Swedish families. *Annals of Human Genetics*, **24**, 245–252.

FERGUSON, G. A. (1966). *Statistical Analysis in Psychology and Education*. McGraw-Hill Book Company, New York.

FINNEY, D. J. (1952). *Statistical Method in Biological Assay*. Charles Griffin and Company, Ltd., London.

FIX, PHILIP F. (1949). Regularity of Old Faithful Geyser, Yellowstone National Park, Wyoming. *American Journal of Science*, **247**, 246–256.

FOSTER, JULIE A. (1962). Kolmogorov-Smirnov test for goodness of fit. *Industrial Quality Control*, **18**, No. 7 (January), 4–8.

FOWLER, M. J., M. J. SULLIVAN and B. R. EKSTRAND (1973). Sleep and memory. *Science*, **179** (January 19), 302–304.

GLENN, G. C. and B. D. KEHN (1972). The biochemical sequelae of RhoGAM inoculation in Rh-positive and Rh-negative post partum women. *American Journal of Clinical Pathology*, **58**, 326–328.

GOODMAN, LEO A. (1954). Some practical techniques in serial number analysis. *Journal of the American Statistical Association*, **49**, 97–112.

GOODMAN, L. A. and W. H. KRUSKAL (1954). Measures of association for cross classifications. *Journal of the American Statistical Association*, **49**, 732–764.

GRIDGEMAN, NORMAN T. (1956). A tasting experiment. *Applied Statistics*, **5**, 106–112.

HILL, H. B. et al (1970). Technical factors affecting the prothrombin time determination. *American Journal of Clinical Pathology*, **54**, 348–360.

*Inpatient Health Facilities as Reported from the 1969 MFI Survey* (1972). DHEW Publication No. (HSM) 73–1801, Rockville, Md.

ISHIKURA, H. and K. OZAKI (1952). Flavour and taste contamination by spraying suspensions of BHC of different purifications. *Botyu-Kagaku*, **17**, 75–82.

JUNG, D. H. and A. C. PAREKH (1970). A semi-micro method for the determination of serum iron-binding capacity without deproteinization. *American Journal of Clinical Pathology*, **54**, 813–817.

KAPLAN, S. J., C. D. TAIT, P. D. WALL and R. B. PAYNE (1951). *Behavioral changes following radiation. 1. Study of retention of a partially learned maze habit.* U.S. Air Force, School of Aviation Medicine, Rep. 1.

KENDALL, M. G. (1943). *The Advanced Theory of Statistics*, Vol. 1. J. B. Lippincott Company, Philadelphia, Pa.

KEPHART, WILLIAM M. (1955). Occupational level and marital disruption. *American Sociological Review*, **20**, 456–465.

KINNEY, JOHN (1973). Poisson updated. *The American Statistician*, **27**, 195.

KOLB, DAVID A. (1965). Achievement motivation training for underachieving high-school boys. *Journal of Personality and Social Psychology*, **2**, 783–792.

KRUMBEIN, W. C. (1954). Applications of statistical methods to sedimentary rocks. *Journal of the American Statistical Association*, **49**, 51–66.

KRUMBEIN, W. C. and F. A. GRAYBILL (1965). *Introduction to Statistical Models in Geology*, McGraw-Hill Book Company, New York.

KRUMBEIN, W. C. and F. J. PETTIJOHN (1938). *Manual of Sedimentary Petrography*. Appleton-Century-Crofts, New York.

MANIS, MELVIN (1955). Social interaction and the self concept. *Journal of Abnormal and Social Psychology*, **51**, 362–370.

McCONVILLE, C. B. and J. K. HEMPHILL (1966). Some effects of communication restraints on problem solving behavior. *Journal of Social Psychology*, **69**, 265–276.

MORSE, N. and R. S. WEISS (1955). Function and meaning of work and the job. *American Sociological Review*, **20**, 191–198.

MOSTELLER, FREDERICK and DAVID L. WALLACE (1964). *Inference and Disputed Authorship: The Federalist.* Addison-Wesley Publishing Company, Reading, Mass.

NATIONAL SCIENCE FOUNDATION (1971). *Research and Development in Industry, 1969.* U.S. Government Printing Office, Washington, D.C.

NESBITT, PAUL D. (1972). Chronic smoking and emotionality. *Journal of Applied Social Psychology*, **2**, 187–196.

OSBORNE, R. T. (1970). Population pollution. *The Journal of Psychology*, **76**, 187–191.

PERROTTA, A. L. and C. A. FINCH (1972). The polychromatophilic erythrocyte. *American Journal of Clinical Pathology*, **57**, 471–477.

ROBERTS, J. A. F. and R. GRIFFITHS (1937). Studies on a child population. II. Retests on the advanced Otis and Stanford-Binet scales with notes on the use of a shortened Binet scale. *Annals of Eugenics*, **8**, 15–45.

ROSENE, H. A. (1950). Ageing and the influx of water into radish roothair cells. *Journal of General Physiology*, **34**, 65–73.

SCHACHTER, G., B. Cohen and H. Goldstein (1972). The demand and supply of college teachers of economics, 1969–1970. *The American Economist*, **26**, No. 1, 126–136.

SCHACHTER, STANLEY (1959). *The Psychology of Affiliation*. Stanford University Press, Stanford, Calif.

SCHIFFMAN, LEON G. (1972). Perceived risk in new product trial by elderly consumers. *Journal of Marketing Research*, **9**, 106–108.

SMIGEL, ERWIN (1964). *The Wall Street Lawyer*. The Free Press, New York.

SPATES, JAMES L. and JACK LEVIN (1972). Beats, hippies, the hip generation, and the American middle class: An analysis of values. *International Social Science Journal*, **24**, 326–353.

*Statistical Abstract of the United States, 1970* (91st annual ed.). U.S. Department of Commerce, Bureau of Census, Washington, D.C.

SURMAN, OWEN S. et al (1973). Hypnosis in a treatment of warts. *Archives of General Psychiatry*, **28**, 439–441.

*The Lancet* (December 16, 1972).

*The 1973 World Almanac and Book of Facts* (1973). Doubleday and Company, New York.

*The World Almanac and Book of Facts* (1973). Newspaper Enterprise Association, New York.

TORRANCE, E. PAUL (1970). Group size and question performance of pre-primary children. *The Journal of Psychology*, **74**, 71–75.

*Uniform Crime Reports for the United States, 1969*. Issued by J. Edgar Hoover, Director of the FBI, Washington, D.C.

*U.S. News & World Report* (December 15, 1969).

VROOM, VICTOR H. and BERND PAHL (1971). Relationship between age and risk taking among managers. *Journal of Applied Psychology*, **55**, 399–405.

WALLIS, W. ALLEN (1936). The Poisson distribution and the Supreme Court. *Journal of the American Statistical Association*, **31**, 376–380.

WARM, Joel S., L. F. GREENBERG and C. S. DUBE (1964). Stimulus and motivational determinants in temporal perception. *The Journal of Psychology*, **58**, 243–248.

WHELPTON, P. K. and C. V. KISER (1950). *Social and Psychological Factors Affecting Fertility*, Vol. 2. Milbank Memorial Fund, New York.

WILCOXON, Frank and R. A. WILCOX (1964). *Some Rapid Approximate Statistical Procedures*. Lederle Laboratories, Pearl River, N.Y.

WILLIAMS, A. O. and M. C. PATH (1970). Ultra-structural changes in murine leukemia cells exposed to L-asparaginase in vitro. *American Journal of Clinical Pathology*, **54**, 658–666.

# Answers to Even-numbered Problems

## Chapter 2

2.4 $Q = 7.2$, df $= 4$, $.10 < P < .20$

2.6 $Q = 380.08$, df $= 9$, $P < .001$

2.8 Sweetness: $Q = 1.84$, df $= 4$, $.70 < P < .80$
   or $Q = 2.13$, df $= 3$, $.50 < P < .70$
   Taste: $Q = 0.34$, df $= 4$, $.98 < P < .99$
   or $Q = 0.24$, df $= 3$, $.95 < P < .98$

2.10 $Q = 1.72$, df $= 5$, $.80 < P < .90$

2.12 $Q = 12.20$, df $= 2$, $.001 < P < .01$

2.14 $Q = 13.29$, df $= 7$, $.05 < P < .10$

2.16 $Q = 14.64$, df $= 5$, $.01 < P < .02$

2.18 (a) $Q = 5.00$, df $= 4$, $.20 < P < .30$
   (b) $Q = 3.29$, df $= 3$, $.30 < P < .50$

2.20 (a) $D = .3264$, $P < .01$
   (b) $D = .2470$, $P < .01$

## Chapter 3

3.2 $z_{+,R} = 3.29$, $P = .0005$

3.4 (a) $z_{+,R} = 7.3$, $P < .0001$
   (b) $P = .0207$

3.6 $P = .0013$

3.8 $P = .031$

3.10 (a) $P = .1719$
   (b) $-4 \leq M_D \leq 1$

3.12 Signed rank: $P = .014$
3.14 Sign test:   Religious   $P = .0032$
              Aesthetic   $P = .0032$
     Signed rank test:   Religious   $P = .000$
                     Aesthetic   $P = .002$
3.16 Stability:       $-4.1 \leq M_D \leq -2.7$
     Temperature:   $-0.9 \leq M_D \leq 0.6$
3.18 $.047 < P < .078$
3.20 $P = .014$
3.22 $P = .002$
3.24 (a) $.492 < P < .556$
     (b) $P = .3438$
3.26 $94 \leq M_D \leq 170.5$
3.28 $P = .0536$
3.30 (a) $z_{+,R} = 2.17, P = .0150$
     (b) $Q = 5.11$, df $= 5, .30 < P < .50$
3.32 $P = .2375$

# Chapter 4

4.2 $T_x = 85.0, P = .082, 0 \leq M_x - M_y \leq 40$
4.4 $T_x = 54, P = .433$
4.6 $T_x = 60, P = .086$
4.8 $T_x = 95, P = .051$
4.10 (a) $z_{x,L} = -2.04, P = .0207$
     (b) $z_{x,L} = -2.20, P = .0139$
4.12 $T_x = 198, z_{x,R} = 2.74, P = .0062$
4.14 $T_x = 123, z_{x,L} = -1.07, .14 < P < .15, -25 \leq M_A - M_B \leq 5$
4.16 $H = 3.53, .10 < P < .20$
4.18 $H = 0.88, .50 < P < .70$
4.20 $H = 2.42, .30 < P < .50$
4.22 $H = 5.15, .05 < P < .10$

# Chapter 5

5.2 $T_x = 40, P = .021$
5.4 $T_x = 88.5, .109 < P < .124$
5.6 $0.978 \leq \sigma_A/\sigma_B \leq 1.020$ at exact level $.892$
5.8 $T_x = 82.17, P = .16$
5.10 $T_x = 53, P = .022$
5.12 $.6129 \leq \sigma_{\text{pre}}/\sigma_{\text{post}} \leq .7398$
5.14 $T_x = 30, P = .090$

# Chapter 6

6.2 $D = .090, P > .20$

6.4 $D = .030, P > .20$

6.6 $Q = 15.28$, df $= 5, P < .01$

6.8 $Q = 6381.21$, df $= 11, P < .001$

6.10 $D = .250, mnD = 16, P > .283$

6.12 $Q = 4.08$, df $= 11, .95 < P < .98$

6.14 $D = .097, P > .200$

6.16 $D = .275, .10 < P < .20$

6.18 $Q = 19.23$, df $= 3, P < .001$

6.20 $Q = 11.33$, df $= 2, P < .01$

6.22 $D = .500, mnD = 32, P = .283$

# Chapter 7

7.2 $T = -.822, P = .000$

7.4 $T = -.69, R = -.813$

7.6 $R = -.76$, one-sided $P < .001$
    $R = -.59$, one-sided $P < .005$

7.10 $R = .907$, one-sided $P < .001$
     $T = .752$, one-sided $P < .005$

7.12 (a) $T_{x_1 x_2} = .484$, two-sided $.020 < P < .050$
         $T_{x_1 x_3} = .303$, two-sided $P = .200$
     (b) $T_{x_1 x_2 \cdot x_3} = .397$
         $T_{x_1 x_3 \cdot x_2} = .029$

7.14 $W = .138, Q = 2.898$, df $= 7, .80 < P < .90, R = -.738\ P = .023$

7.16 (a) $Q = 28.3$, df $= 9, P < .001$
     (b) Best to worst: 9, 8, 5, 2, 1, 4, 3, $= 7, 6, 10$
     (c) Not too much

7.18 $Q = 17.49$, df $= 3, P < .001, |R_i - R_j| \le 13.82$ at level .10, $D$ more efficient than $A$ or $B$, $C$ more efficient than $A$ or $B$.

7.20 (a) $Q = 26.6$, df $= 5, P < .001, |R_i - R_j| \le 20.29$; VI better than I, II, and III; V better than I; IV better than I
     (b) $Q = 17.26$, df $= 5, .001 < P < .01$

7.22 (a) $Q = 9.45$, df $= 3, .02 < P < .05$
     (b) III
     (c) $|R_i - R_j| \le 11.57$; III better than I and IV

7.24 (a) $Q = 17.88$, df $= 3, P < .001$
     (b) $|R_i - R_j| \le 17.38$; group 50.3 different from control and from 95; group 65 different from control

7.26 $S = 58.5, .002 < P < .006$

7.28 $Q = 3.00$, df $= 5, P = .70$

7.30 $Q = 5.228$, df $= 3, .10 < P < .20$

7.32 $Q = 0.84$, df $= 1$, $.30 < P < .50$
7.34 $Q = 25.0$, df $= 2$, $P < .001$
7.36 $Q = 5.05$, df $= 2$, $.05 < P < .10$
7.38 (a) $T^2 = 2/7$
    (b) $Q = 60/7$, $C^2 = 2/7$
7.40 (a) $\phi = .17$
    (b) Health risk and Taste risk $\gamma = .0025$, Taste risk and Trial $\gamma = -.485$

# Chapter 8

8.2 $U = 8$, $P = 1.00$
8.4 $U = 11$, $P = 1.00$
8.6 (a) $U = 5$, $P = .032$
    (b) $V = 5$, $P = .0009$
8.10 $V = 13$, $P = .6304$
8.12 $z_L = -1.21$, $P = .2262$
8.14 $z_L = -1.22$, $P = .2224$
8.16 $z_L = 0.07$, $P = .528$

# Index